Advanced Microeconomics

Harald Wiese

Advanced Microeconomics

Springer Gabler

Harald Wiese
Fakultät Wirtschaftswissenschaften
Universität Leipzig
Leipzig, Germany

ISBN 978-3-658-34958-5 ISBN 978-3-658-34959-2 (eBook)
https://doi.org/10.1007/978-3-658-34959-2

This Springer Gabler imprint is published by the registered company Springer Fachmedien Wiesbaden GmbH part of Springer Nature.
The registered company address is: Abraham-Lincoln-Str. 46, 65189 Wiesbaden, Germany

Für Corinna, Ben, Jasper und Samuel

Preface

What is this book about?

This is a course on advanced microeconomics. It covers a lot of ground, from decision theory to game theory, from bargaining to auction theory, from household theory to oligopoly theory, and from the theory of general equilibrium to regulation theory. It has been used for several years at the university of Leipzig in the Master program "Economics". A lecturer may find advisable to make some more difficult material non-obligatory, as I do in Leipzig.

What about mathematics ... ?

A course in *advanced* microeconomics can use more advanced mathematics, such as open sets or Brouwer's fix-point theorem. I decided not to relegate these concepts to an appendix but to deal with them where they are needed.

Exercises and solutions

The main text is interspersed with questions and problems. Solutions or hints are given at the end of each chapter. On top, we add a few exercises without solutions.

Thank you!

Frank Hüttner translated some of the material from German to English. He and Andreas Tutic suggested both interesting and boring exercises. Franziska Beltz improved the quality of the many figures in the textbook. Some generations of students also provided feedback that helped to improve the manuscript. Michael Diemer, Pavel Brendler, Mathias Klein, Hendrik Kohrs, Max Lillack, Katharina Lotzen, and Katharina Zalewski deserve special mention. Hendrik Kohrs and Katharina Lotzen checked the manuscript and the corresponding slides in detail. The latter also produced the index. Vital last-minute help with technical matters was provided by Alexander Singer. He and Johanna Freistühler checked the manuscript for remaining mistakes.

Leipzig, June 2021 Harald Wiese

Preface

What is this book about?

Contents

CHAPTER I

Cooperation as the central focus of microeconomics

1. Three modes of cooperation

Human beings do not live or work in isolation because cooperation often pays. For economics (and social sciences beyond economics) cooperation is a central concern. While cooperation between individuals is a micro phenomenon, economics is also interested in the consequences for macro phenomena: prices, distribution of income, inflation etc. This is the topic of both micro- and macroeconomics. Cooperation is also influenced by institutions and implicit and explicit norms – the subject matter of institutional economics. We follow Moulin (1995) and consider three different modes of cooperation: decentral mechanism, bargaining, and dictatorship.

1.1. Decentral mechanism. The market, auctions, and elections are decentral mechanisms. Everybody does what he likes without following somebody's order and without prior agreement. The macro result (price, bid, quantity handed over to buyer or bidder, law voted for) follows from the many individual decisions. Microeconomics puts forth models of auctions, perfect competition, monopoly, and oligopoly theory. Political science deals with different voting mechanisms. The analysis of these mechanism uses one or other equilibrium concept (Walras equilibrium for perfect competition, Nash equilibrium or even subgame-perfect equilibrium in oligopoly theory, and others).

Oftentimes, the mechanism is given. Sometimes, economists or political scientists ask the question of which mechanism is best for certain purposes. For example, is a Walras equilibrium of perfect competition always Pareto efficient? Which auction maximizes the auctioneer's expected revenue? This is the field of mechanism design.

The adjective "decentral" can be misleading. From the point of view of mechanism design, a mechanism is put in place by a center, the auctioneer in case of an auction, the election by the political institutions, ... However, from the point of view of individual participants, "decentral" makes sense.

H. Wiese, *Advanced Microeconomics*,
https://doi.org/10.1007/978-3-658-34959-2_1

Every market participant, bidder or voter is an island and makes the buying or selling, the bidding and voting decision for him- or herself.

1.2. Bargaining. In contrast to decentral mechanisms, bargaining is a few-agents affair and often face to face. Bargaining is analyzed by way of noncooperative game theory (e.g., the Rubinstein alternating offer game) or by way of cooperative game theory (Nash bargaining solution, core, Shapley value). Parallel to mechanism design for decentral mechanisms, we can ask the question of which bargaining protocol is best (in a sense to be specified). Also, bargaining is normally preceded by the search for suitable bargaining partners.

Under ideal conditions (no transaction cost, no uncertainty), one might expect that the agents exhaust every possibility for Pareto improvements. Then, by definition, they have achieved a Pareto efficient outcome. For example, Pareto efficiency

- between consumers is characterized by the equality of the marginal rates of substitution (in an Edgeworth box),
- between countries by the equality of the marginal rates of transformation (this is the law of Ricardo).

1.3. Dictatorship. The third mode is called "dictatorship". Here, we are mainly concerned with arrangements or rules enacted by the government. As a theoretical exercise, one may consider the actions taken by a "benevolent dictator". This fictitious being is normally assumed to aim for Pareto efficiency and other noble goals. We can then ask questions like these:

- Are minimum wages good?
- Should cartels be allowed?
- Should the government impose taxes on environmentally harmful activities?
- Should we allow markets for all kinds of goods? How about security, slaves, prostitution, roads?

In some models, the benevolent dictator has not only noble aspirations, but also no restrictions in terms of knowledge. More to the point are models where the government has to gather the information it uses while the private agents may have an incentive to hide the true state of affairs. Public-choice theory argues against these models, too. Governments and bureaucracies consist of agents with selfish interests.

2. This book

2.1. Overview. In this book, we deal with most topics alluded to in the previous section. I finally decided on the following order:

- Part A on decision and preference theory covers decision theory in both strategic and extensive form. It also deals with utility functions for bundles of goods and for lotteries.
- Part B is an application of decision and preference theory on specific groups of deciders, households and firms. In that part, we deal with household optima, perfect substitutes, Slutsky equations, consumers' rent, profit maximization, supply functions and the like.
- Part C is on noncooperative games. Of the many examples we use in that part, most are from industrial organization which analyzes oligopolistic markets.
- Part D comes under the heading of "Bargaining theory and Pareto optimality". In chapter XIV, we have a look at diverse microeconomic models from the point of view of Pareto optimality. Pareto optimality can be considered the most popular concept of cooperative game theory. The two other very famous concepts are the Shapley value and the core – the subject matter of chapter XV. We close this part with a short chapter on the (non-cooperative) Rubinstein bargaining model.
- Turning again to noncooperative games, part E is concerned with game theory under uncertainty. The central concept is the Bayesian game where some payoffs are not known to some players. We define those games, develop the equilibrium concept for them and analyze auctions. We also consider the question of what kind of outcomes are achievable for suitably chosen games – the subject matter of mechanism design.
- The second-to-last part F deals with perfect competition which is often seen as a welfare-theoretic benchmark. We contrast this benchmark with other competition theories due to Hayek, Schumpeter, Kirzner etc. We also comment on competition laws and competition theory.
- Part G deals with contract theory and in particular principal-agent theory. We cover asymmetric information as well as hidden action.

2.2. Concept. This book is written with three main ideas (and some minor ones) in mind. First, apart from differentiation techniques, we introduce (nearly) all the mathematical concepts necessary to understand the

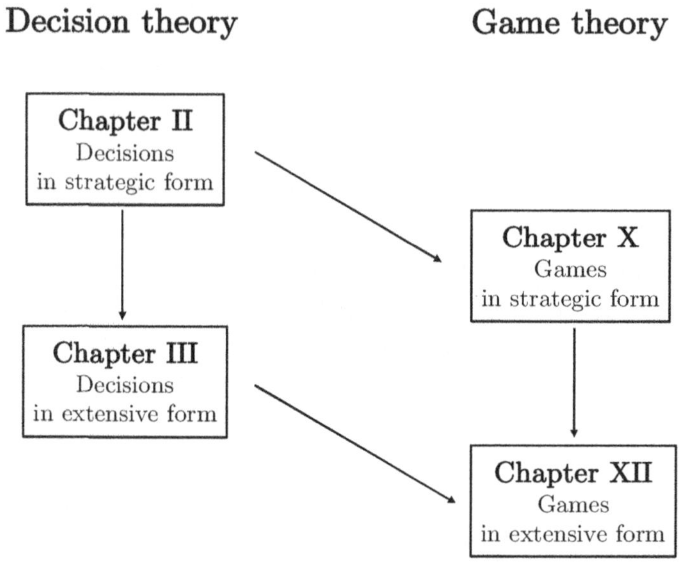

FIGURE I.1. Game theory builds on decision theory

material which is sometimes difficult. Thus, the book is self-contained. Also, we present the mathematical definitions and theorems when and where we need them. We decided against a mathematical appendix. After all, the mathematics has to be presented sometime and students would probably not be amused if a micro course begins with 3 weeks of mathematics.

Second, some basic game theory concepts can already be explained within the simpler decision framework. Therefore, part A prepares for part C by covering

- the strategic-form concepts dominance, best responses, and rationalizability and
- the extensive-form notions actions and strategies, subtree perfection, and backward induction.

The basic chapters on decision and game theory are related in the manner depicted in figure I.1.

Finally, apart from microeconomics in a narrow sense, we also introduce the reader to some basic notions from cooperative game theory. Of course, one can argue that cooperative game theory has no role to play in a microeconomic textbook. After all, the players in cooperative game theory do not act, do not form expectations, do not maximize a profit or utility function, all of which are considered central characteristics of microeconomic models. We do not take such a puristic view. First of all, cooperative concepts do not belong to macroeconomics either and any standard curriculum (being based on micro- and macroeconomics) would leave out these important methods and ways of economic thinking. After all, any economist worth his salt should be familiar with Pareto efficiency, the Shapley value, or the core. Second, analyzing non-cooperative concepts from a cooperative point of view and vice versa, are illuminating ways to gain further insight, compared to a purely non-cooperative or to a purely cooperative approach.

Part A

Basic decision and preference theory

The first part of our course introduces the reader to decision theory. We focus on one agent or one decision maker. The part has four chapters, only. We present some elementary decision theory along with interesting examples in the first two chapters. Chapter II treats the strategic (viz., static) form and chapter III the extensive (viz., dynamic) form. Chapters IV and V deal with preference theory. A central topic of preference theory concerns utility functions that are used to describe preferences. Chapter IV presents the general theory and chapter V treats the special case of preferences for lotteries.

The basic decision and preference theory stops short of explaining household theory and the theory of the firm. This is done in part B.

CHAPTER II

Decisions in strategic form

The first two chapters have two aims. First, they are an introduction to important aspects of decision theory. Second, they help to ease into game theory, the subject matter of the third part of our book. Indeed, some basic game theory concepts can already be explained within the simpler decision framework. In particular, we treat

- dominance, best responses, and rationalizability in this chapter and
- actions and strategies, subtree perfection, and backward induction in the next.

Strategic-form decision theory (this chapter) is concerned with one-time (or once-and-for-all) decisions where the decision maker's outcome (payoff) depends on the decision maker's strategy and also on the so-called state of the world.

1. Introduction and three examples

Assume a firm that produces umbrellas or sunshades. In order to avoid preproduction costs, it decides on the production of either umbrellas or sunshades (in the given time period). The firm's profits depend on the weather. There are two states of the world, good or bad weather. The following payoff matrix indicates the profit as a function of the firm's decision (strategy) and of the state of the world.

		state of the world	
		bad weather	good weather
strategy	production of umbrellas	100	81
	production of sunshades	64	121

FIGURE II.1. Payoff matrix

H. Wiese, *Advanced Microeconomics*,
https://doi.org/10.1007/978-3-658-34959-2_2

The highest profit is obtained if the firm produces sunshades and the weather is good. However, the production of sunshades carries the risk of a very low profit, in case of rain. The payoff matrix examplifies important concepts in our basic decision model: strategies, states of the world, payoffs, and payoff functions.

- The firm has two strategies, producing umbrellas or producing sunshades.
- There are two states of the world, bad and good weather.
- The payoffs are 64, 81, 100, or 121.
- The payoff function determines the payoffs resulting from strategies and states of the world. For example, the firm obtains a profit of 121 if it produces sunshades and it is sunny.

We have the following definition:

DEFINITION II.1 (DECISION SITUATION IN STRATEGIC FORM). *A decision situation in strategic form is a triple*

$$\Delta = (S, W, u),$$

where

- *S is the decision maker's strategy set,*
- *W is the set of states of the world, and*
- *$u : S \times W \to \mathbb{R}$ is the payoff function.*

$\Delta = (S, u : S \to \mathbb{R})$ is called a decision situation in strategic form without uncertainty.

In the umbrella-sunshade decision situation, we have the strategy set $S = \{\text{umbrella, sunshade}\}$ and the payoff function u given by

$$\begin{aligned} u\,(\text{umbrella, bad weather}) &= 100, \\ u\,(\text{umbrella, good weather}) &= 81, \\ u\,(\text{sunshade, bad weather}) &= 64, \\ u\,(\text{sunshade, good weather}) &= 121. \end{aligned}$$

If both the strategy set and the set of states of the world are finite, a payoff matrix is often a good way to write down the decision situation in strategic form. We always assume that (i) S and (ii) W are set up so that (i) the decision maker can choose one and only one strategy from S and so that (ii) one and only one state of the world from W can actually happen.

A decision situation in strategic form without uncertainty is a decision situation in strategic form where $u(s, w_1) = u(s, w_2)$ for all $s \in S$ and all

	prediction: box 2, only	prediction: both boxes
you open box 2, only	1 000 000 Euro	0 Euro
you open both boxes	1 001 000 Euro	1 000 Euro

FIGURE II.2. A payoff matrix for Newcomb's problem

$w_1, w_2 \in W$. For instance, this holds in case of $|W| = 1$ where $|W|$ is called the cardinality of W and denotes the number of elements in W.

Our second example is called Newcomb's problem. Two boxes are shown to an agent (you). In box 1, there are 1000 Euro while box 2 holds no money or 1 million Euro. You have the option of opening box 2, only, or both boxes. Before you jump to the conclusion that both boxes are clearly preferable to one box, consider the following twist to the story. You know that a Higher (and rich) Being has put the money into the boxes depending on a prediction of your choice. If the Higher Being predicts that you open box 2, only, He puts 1 million Euro into box 2. If, however, the Higher Being thinks you will open both boxes, He leaves box 2 empty.

You have to understand that the Higher Being is not perfect. He can make good predictions because He knows the books you read and the classes you attend. The prediction about your choice and the filling of the boxes are done (yesterday) once you are confronted with the two boxes (today). The Higher Being cannot and will not change the content of the boxes today.

EXERCISE II.1. *What would you do?*

The decision depends on how you write down your set of states of the world W. Figure II.2 distinguishes between prediction "box 2, only" and prediction "both boxes". Figure II.3 dissects W differently: Either the predicition is correct or it is wrong.

It seems to me that the first matrix is the correct one. The next section shows how to "solve" this decision problem.

Before turning to that section, we consider a third example. If you know some microeconomics, everything will be clear to you. If you do not understand, don't panic, but wait until chapter XI.

	prediction is correct	prediction is wrong
you open box 2, only	1 000 000 Euro	0 Euro
you open both boxes	1 000 Euro	1 001 000 Euro

FIGURE II.3. A second payoff matrix for Newcomb's problem

DEFINITION II.2 (COURNOT MONOPOLY). *A Cournot monopoly is a decision situation in strategic form without uncertainty* $\Delta = (S, \Pi)$, *where*

- $S = [0, \infty)$ *is the set of output decisions,*
- $\Pi : S \to \mathbb{R}$ *is the payoff function defined by an inverse demand function* $p : S \to [0, \infty)$, *a cost function* $C : S \to [0, \infty)$, *and by* $\Pi(s) = p(s)\, s - C(s)$.

2. Sets, functions, and real numbers

2.1. Sets, tuples, and Cartesian products. The above definition of a decision situation in strategic form contains several important mathematical concepts. In line with the philosophy of this book to explain mathematical concepts wherever they arise for the first time, we offer some comments on sets, tuples, the Cartesian product of sets, functions, and real numbers.

First, a set is any collection of objects that can be distinguished from each other. A set can be empty in which case we use the symbol $\emptyset$. The objects are called elements. In the above definition, we have the sets S, W, $\mathbb{R}$, and $S \times W$ (the latter being a Cartesian product).

DEFINITION II.3 (SET AND SUBSET). *Let* M *be a nonempty set. A set* N *is called a subset of* M *(denoted by* $N \subseteq M$*) if and only if every element from* N *is contained in* M. *We use curly brackets* $\{\}$ *to indicate sets. Two sets* M_1 *and* M_2 *are equal if and only if* M_1 *is a subset of* M_2 *and* M_2 *is a subset of* M_1. *We define strict inclusion* $N \subset M$ *by* $N \subseteq M$ *and* $M \nsubseteq N$.

The reader will note the pedantic use of "if and only if" in the above definition. In definitions (!), it is quite sufficient to write "if" instead of "if and only if" (or the shorter "iff").

Sets need to be distinguished from tuples where the order is important:

DEFINITION II.4 (TUPLE). *Let M be a nonempty set. A tuple on M is an ordered list of elements from M. Elements can appear several times. A tuple consisting of n entries is called an n-tuple. We use round brackets () to denote tuples. Two tuples $(a_1, ..., a_n)$ and $(b_1, ..., b_m)$ are equal if they have the same number of entries, i.e., if $n = m$ holds, and if the respective entries are the same, i.e., if $a_i = b_i$ for all $i = 1, ..., n$.*

Oftentimes, we consider tuples where each entry stems from a particular set. For example, $S \times W$ is the set of tuples (s, w) where s is a strategy from S and w a state of the world from W.

DEFINITION II.5 (CARTESIAN PRODUCT). *Let M_1 and M_2 be nonempty sets. The Cartesian product of M_1 and M_2 is denoted by $M_1 \times M_2$ and defined by*

$$M_1 \times M_2 := \{(m_1, m_2) : m_1 \in M_1, m_2 \in M_2\}.$$

EXERCISE II.2. *Let $M := \{1, 2, 3\}$ and $N := \{2, 3\}$. Find $M \times N$ and depict this set in a two-dimensional figure where M is associated with the abscissa (x-axis) and N with the ordinate (y-axis).*

2.2. Injective and surjective functions. We now turn to the concept of a function. The payoff function $u : S \times W \to \mathbb{R}$ is our first example.

DEFINITION II.6 (FUNCTION). *Let M and N be nonempty sets. A function $f : M \to N$ associates with every $m \in M$ an element from N, denoted by $f(m)$ and called the value of f at m. The set M is called the domain (of f), the set N is the range (of f) and $f(M) := \{f(m) : m \in M\}$ the image (of f). A function is called injective if $f(m) = f(m')$ implies $m = m'$ for all $m, m' \in M$. It is surjective if $f(M) = N$ holds. A function that is both injective and surjective is called bijective.*

EXERCISE II.3. *Let $M := \{1, 2, 3\}$ and $N := \{a, b, c\}$. Define $f : M \to N$ by $f(1) = a$, $f(2) = a$ and $f(3) = c$. Is f surjective or injective?*

When describing a function, we use two different sorts of arrows. First, we have $\to$ in $f : M \to N$ where the domain is left of the arrow and the range to the right. Second, on the level of individual elements of M and N, we use $\mapsto$ to write $m \mapsto f(m)$. For example, a quadratic function may be written as

$$\begin{aligned} f \ : \ & \mathbb{R} \to \mathbb{R}, \\ & x \mapsto x^2. \end{aligned}$$

If the domain and the range are obvious or unimportant, we can also write $f : x \mapsto x^2$. It is also not unusual to talk about the function $f(x)$, but this is not correct and sometimes seriously misleading. Strictly speaking, $f(x)$ is an element from the image of f, i.e., the value the function f takes at the specific element x from the domain.

If a function is injective, we can take an inverse look at it:

DEFINITION II.7 (INVERSE FUNCTION). *Let $f : M \to N$ be an injective function. The function $f^{-1} : f(M) \to M$ defined by*

$$f^{-1}(n) = m \Leftrightarrow f(m) = n$$

is called f's inverse function.

Functions help to find out whether a set is larger than another one. For example, if a function $f : M \to N$ is injective, there are at least as many elements in N as in M. If $f : M \to N$ is bijective, we can say that M and N contain the same number of elements. If M is finite, that is obvious. If M is not finite, it is a matter of definition:

DEFINITION II.8 (CARDINALITY). *Let M and N be nonempty sets and let $f : M \to N$ be a bijective function. We then say that M and N have the same cardinality (denoted by $|M| = |N|$). If a bijective function $f : M \to \{1, 2, ..., n\}$ exists, we say that M is finite and contains n elements. Otherwise M is infinite.*

EXERCISE II.4. *Let $M := \{1, 2, 3\}$ and $N := \{a, b, c\}$. Show $|M| = |N|$.*

2.3. Real numbers. Finally, we want to explain real numbers. They contain natural numbers $(1, 2, 3, ...)$, integers $(..., -2, -1, 0, 1, 2, ...)$, and rational numbers (the numbers gained by dividing an integer by a natural number). These sets are defined in this table:

sets	symbol	elements
natural numbers	$\mathbb{N}$	$\{1, 2, 3, ...\}$
integers	$\mathbb{Z}$	$\{..., -2, -1, 0, 1, 2, ...\}$
rational numbers	$\mathbb{Q}$	$\left\{\frac{p}{q} : p \in \mathbb{Z}, q \in \mathbb{N}\right\}$
irrational real numbers	$\mathbb{R}\backslash\mathbb{Q}$	$\sqrt{2} = 1.4142..., e = 2.7183...$, etc.

Of course, we have $\mathbb{N} \subset \mathbb{Z} \subset \mathbb{Q}$. The set of real numbers contains the other three sets but is much bigger. For example, irrational real numbers are $\sqrt{2}$, e or $\pi := 3.1416...$. The dots point to the fact that the number is never finished and, indeed, there is no pattern that is repeated again and again.

EXERCISE II.5. *$\frac{1}{8}$ and $\frac{4}{7}$ are rational numbers. Write these numbers as* $0.1...$ *and* $0.5...$ *and show that a repeating pattern emerges.*

DEFINITION II.9 (COUNTABLY INFINITE SET). *Let M be a set obeying $|M| = |\mathbb{N}|$. Then M is called a countably infinite set.*

Without proof, we note the following theorem:

THEOREM II.1 (CARDINALITY). *The sets $\mathbb{N}$, $\mathbb{Z}$, and $\mathbb{Q}$ are countably infinite sets, i.e., there exist bijective functions $f : \mathbb{N} \to \mathbb{Z}$ and $g : \mathbb{N} \to \mathbb{Q}$. (The exciting point is that f and g are surjective.) Differently put, their cardinality is the same:*

$$|\mathbb{N}| = |\mathbb{Z}| = |\mathbb{Q}| .$$

However, we have

$$|\mathbb{Q}| < |\mathbb{R}|$$

and even

$$|\mathbb{Q}| < |\{x \in \mathbb{R} : a \leq x \leq b\}|$$

for any numbers a, b with $a < b$.

Thus, there are "more" real numbers in the interval between 0 and 1 (or 0 and $\frac{1}{1000}$) than there are rational numbers.

DEFINITION II.10 (INTERVAL). *Intervals are denoted by*

$$\begin{aligned}
[a, b] &: = \{x \in \mathbb{R} : a \leq x \leq b\}, \\
[a, b) &: = \{x \in \mathbb{R} : a \leq x < b\}, \\
(a, b] &: = \{x \in \mathbb{R} : a < x \leq b\}, \\
(a, b) &: = \{x \in \mathbb{R} : a < x < b\}, \\
[a, \infty) &: = \{x \in \mathbb{R} : a \leq x\} \text{ and} \\
(-\infty, b] &: = \{x \in \mathbb{R} : x \leq b\}.
\end{aligned}$$

EXERCISE II.6. *Given the above definition for intervals, can you find an alternative expression for $\mathbb{R}$?*

3. Dominance and best responses

Dominance means that a strategy is better than the others. We distinguish between (weak) dominance and strict dominance:

DEFINITION II.11 (DOMINANCE). *Let $\Delta = (S, W, u)$ be a decision situation in strategic form. Strategy $s \in S$ (weakly) dominates strategy $s' \in S$ if and only if $u(s, w) \geq u(s', w)$ holds for all $w \in W$ and $u(s, w) > u(s', w)$ is true for at least one $w \in W$. Strategy $s \in S$ strictly dominates strategy $s' \in S$ if and only if $u(s, w) > u(s', w)$ holds for all $w \in W$. Then, strategy s' is called (weakly) dominated or strictly dominated, respectively. A strategy* *that dominates every other strategy is called dominant (weakly or strictly, respectively).*

The decision matrix II.2 depicted in figure II.2 (p. 11) is clearly solvable by strict dominance. Opening both boxes yields an extra amount of Euro 1 000.

There is a simple procedure to find out whether we have dominant strategies. For every state of the world, we find the best strategy and put a "R" into the corresponding field. The letter "R" is reminiscent of response – the decision maker *responds* to a state of the world by choosing the payoff maximizing strategy for that state. Take, for example, the second Newcomb matrix:

	prediction is correct	prediction is wrong
you open box 2, only	1 000 000 Euro [R]	0 Euro
you open both boxes	1 000 Euro	1 001 000 Euro [R]

Since the best strategy (best response) depends on the state of the world, no strategy is dominant. The [R]-procedure needs to be formalized. Before doing so, we familiarize the reader with the notion of a power set and with arg max.

DEFINITION II.12 (POWER SET). *Let M be any set. The set of all subsets of M is called the power set of M and is denoted by 2^M.*

For example, $M := \{1, 2, 3\}$ has the power set

$$2^M = \{\emptyset, \{1\}, \{2\}, \{3\}, \{1, 2\}, \{1, 3\}, \{2, 3\}, \{1, 2, 3\}\}.$$

Note that the empty set $\emptyset$ also belongs to the power set of M, indeed to the power set of any set. $M := \{1, 2, 3\}$ has eight elements which is equal to $2^3 = 2^{|\{1,2,3\}|}$. This is a general rule: For any set M, we have $\left|2^M\right| = 2^{|M|}$. 2 plays a special role in the definition of a power set. The reason is simple – every element m belongs to or does not belong to a given subset.

In order to introduce arg max, consider a firm that tries to maximize its profit Π by choosing the output x optimally. The output x is taken from a set X (for example the interval $[0,\infty)$) and the profit is a real (Euro) number. Then, we have a profit function $\Pi : X \to \mathbb{R}$ and

$$\begin{aligned} \Pi(x) &\in \mathbb{R}: \text{ profit resulting from the output } x, \\ \max_x \Pi(x) &\in \mathbb{R}: \text{ maximal profit by choosing } x \text{ optimally}, \\ \operatorname*{argmax}_x \Pi(x) &\subseteq X: \text{ set of outputs that lead to the maximal profit} \end{aligned}$$

Again: $\max_x \Pi(x)$ is the maximal profit (in Euro), while $\operatorname{argmax}_x \Pi(x)$ is the set of optimal decision variables. Therefore, we have

$$\max_x \Pi(x) = \Pi(x^*) \text{ for all } x^* \text{ from } \operatorname*{argmax}_x \Pi(x).$$

We have three different cases:

- $\operatorname{argmax}_x \Pi(x)$ contains several elements and each of these elements leads to the same profit.
- $\operatorname{argmax}_x \Pi(x)$ contains just one element. We often write $x^* = \operatorname{argmax}_x \Pi(x)$ instead of the very correct $\{x^*\} = \operatorname{argmax}_x \Pi(x)$.
- $\max_x \Pi(x)$ does not exist and $\operatorname{argmax}_x \Pi(x)$ is the empty set. As an example consider $X := [0,1) = \{x \in \mathbb{R} : 0 \leq x < 1\}$ and $\Pi(x) = x$. For every $x \in X$, we have $1 > \frac{1+x}{2} > x \geq 0$ so that no x from X maximizes the profit. The reason is somewhat artificial. We cannot find a greatest number smaller than 1 if we search within the rational or real numbers. (For more on solution theory, consult pp. 151.)

Now, at long last, we can proceed with the main text:

Definition II.13 (best response). *Let*

$$\Delta = (S, W, u)$$

be a decision situation in strategic form. The function $s^R : W \to 2^S$ is called a best-response function (a best response) if s^R is given by

$$s^R(w) := \arg\max_{s \in S} u(s, w)$$

Exercise II.7. *Use best-response functions to characterize s as a dominant strategy. Hint: "characterization" means that you are to find a statement that is equivalent to the definition.*

4. Mixed strategies and beliefs

4.1. Probability distribution. In this section, we introduce probability distributions on the set of pure strategies and on the set of states of the world. This important concept merits a proper definition, where $[0,1]$ is short for $\{x \in \mathbb{R} : 0 \leq x \leq 1\}$:

Definition II.14 (probability distribution). *Let M be a nonempty set. A probability distribution on M is a function*

$$prob : 2^M \to [0,1]$$

such that

- $prob(\emptyset) = 0$,
- $prob(A \cup B) = prob(A) + prob(B)$ *for all* $A, B \in 2^M$ *obeying* $A \cap B = \emptyset$ *and*
- $prob(M) = 1$.

Subsets of M are also called events. For $m \in M$, we often write $prob(m)$ rather than $prob(\{m\})$. If $m \in M$ exists such that $prob(m) = 1$, $prob$ is called a trivial probability distribution and can be identified with m.

The requirement $prob(M) = 1$ is called the summing condition.

Exercise II.8. *Throw a fair dice. What is the probability for the event A, "the number of pips (spots) is 2", and the event B, "the number of pips is odd". Apply the definition to find the probability for the event "the number of pips is 1, 2, 3, or 5".*

Thus, a probability distribution associates a number between 0 and 1 to every subset of M. (This definition is okay for finite sets M, but a problem can arise for sets if M is infinite but not countably infinite. For example, in case of $M = [0,1]$, a probability cannot be defined for every subset of M, but for so-called measurable subsets only. However, it is not easy to find a subset of $[0,1]$ that is not measurable. Therefore, we do not discuss the concept of measurability.)

4.2. Mixing strategies and states of the world. Imagine a decision maker who tosses the dice before making an actual strategy choice. He does not choose between "pure" strategies such as umbrella or sunshade, but between probability distributions on the set of these pure strategies. For example, he produces umbrellas in case of 1, 2, 3, or 5 pips and sunshades otherwise.

Definition II.15 (mixed strategy). *Let S be a finite strategy set. A mixed strategy σ is a probability distribution on S, i.e., we have*

$$\sigma(s) \geq 0 \text{ for all } s \in S$$

and

$$\sum_{s \in S} \sigma(s) = 1 \text{ (summing condition).}$$

The set of mixed strategies is denoted by Σ*. A pure strategy* $s \in S$ *is identified with the (trivial) mixed strategy* σ *obeying* $\sigma(s) = 1$*.* $\sigma \in \Sigma$ *is called a properly mixed strategy if* σ *is not trivial. If there are only finitely many pure strategies and if the order of the strategies is clear, a mixed strategy* σ *can be specified by a vector* $\left(\sigma(s_1), \sigma(s_2), ..., \sigma\left(s_{|S|}\right)\right)$.

DEFINITION II.16 (DECISION SITUATION WITH MIXED STRATEGIES). *If mixed strategies are allowed,* $\Delta = (S, W, u)$ *is called a decision situation in strategic form with mixed strategies.*

We can also consider mixing states of the world:

DEFINITION II.17 (BELIEF). *Let* W *be a set of states of the world. We denote the set of probability distributions on* W *by* Ω*.* $\omega \in \Omega$ *is called a belief. If there are only finitely many states of the world and if their order is clear, a probability distribution on* W *can be specified by a vector* $\left(\omega(w_1), ..., \omega\left(w_{|W|}\right)\right)$.

4.3. Extending payoff definitions.

4.3.1. *... for beliefs (lotteries).* So far, our payoff function $u : S \times W \to \mathbb{R}$ is defined for a specific strategy and a specific state of the world, $(s, w) \in S \times W$. We can now extend this definition so as to take care of probability distributions on S (mixed strategies) and on W (beliefs). We begin with beliefs.

Let us revisit the producer of umbrellas and sunshades whose payoff matrix is given below. According to our belief ω, bad weather occurs with probability $\frac{1}{4}$ and good weather with probability $\frac{3}{4}$.

		state of the world	
		bad weather, $\frac{1}{4}$	good weather, $\frac{3}{4}$
strategy	production of umbrellas	100	81
	production of sunshades	64	121

FIGURE II.4. Umbrellas or sunshades?

The strategy "produce umbrellas" yields the payoff 100 with probability $\frac{1}{4}$ and 81 with probability $\frac{3}{4}$. Thus, the probability distribution on the set of states of the world leads to a probability distribution for payoffs, in this example denoted by

$$L_{\text{umbrella}} = \left[100, 81; \frac{1}{4}, \frac{3}{4}\right].$$

DEFINITION II.18 (SIMPLE LOTTERY). *A tuple*

$$L = [x; p] := [x_1, ..., x_\ell; p_1, ..., p_\ell]$$

is called a simple lottery where

- $x_j \in \mathbb{R}$ *is the payoff accruing with probability* $p_j \geq 0$, $j = 1, ..., \ell$, *and*
- $\sum_{j=1}^{\ell} p_j = 1$ *holds.*

In case of $\ell = 1$, L *is called a trivial lottery. We identify* $L = [x; 1]$ *with* x. *The set of simple lotteries is denoted by* $\mathcal{L}$.

A very important characteristic of a lottery is its expected value:

DEFINITION II.19 (EXPECTED VALUE). *Assume a simple lottery*

$$L = [x_1, ..., x_\ell; p_1, ..., p_\ell].$$

Its expected value is denoted by $E(L)$ *and given by*

$$E(L) = \sum_{j=1}^{\ell} p_j x_j. \tag{II.1}$$

This definition contains the answer to our initial question: How can we extend the payoff function $u : S \times W \to \mathbb{R}$ to a payoff function

$$u : S \times \Omega \to \mathbb{R}?$$

Given a strategy s and a belief $\omega \in \Omega$, the payoff under s and ω is defined by

$$u(s, \omega) := \sum_{w \in W} \omega(w) u(s, w)$$

or, equivalently, by

$$u(s, \omega) := E(L_s) \text{ for } L_s = \left[(u(s, w))_{w \in W}; (\omega(w))_{w \in W}\right].$$

4.3.2. ... *for mixed strategies.* After mixing states of the world, we can now proceed to mix strategies.

EXERCISE II.9. *Consider the umbrella-sunshade decision situation given above and calculate the expected payoff if the firm chooses umbrella with* $\frac{1}{3}$ *and sunshade with probability* $\frac{2}{3}$. *Differentiate between* $w =$ *"bad weather" and* $w =$ *"good weather". Hint: You can write* $u\left(\left(\frac{1}{3}, \frac{2}{3}\right), w\right)$ *where* $\left(\frac{1}{3}, \frac{2}{3}\right)$ *is the mixed strategy.*

If you have worked out the above exercise, the following definition is no surprise to you:

DEFINITION II.20 (PAYOFF FOR MIXED STRATEGY). *Given a mixed strategy* σ *and a state of the world* w, *the payoff under* σ *and* w *is defined by*

$$u(\sigma, w) := \sum_{s \in S} \sigma(s)\, u(s, w) \tag{II.2}$$

Thus, the payoff for a mixed strategy is the mean of the payoffs for the pure strategies. This definition has important consequences for best mixed strategies.

LEMMA II.1. *Best pure and best mixed strategies are related by the following two claims:*

- *Any mixed strategy that puts positive probabilities on best pure strategies, only, is a best strategy.*
- *If a mixed strategy is a best strategy, every pure strategy with positive probability is a best strategy.*

For the time being (until chapter V), we will not worry about risk attitudes. If payoffs are risky (because the agent chooses a mixed strategy or because we have probabilities for states of the world), the decision maker is happy to maximize his expected profit or his expected payoff.

4.3.3. ... *for mixed strategies and beliefs.* Finally, we can mix both strategies and states of the world. Do you know how to define $u(\sigma, \omega)$?

EXERCISE II.10. *Consider again the umbrella-sunshade decision situation in which*

- *the firm chooses umbrella with probability* $\frac{1}{3}$ *and sunshade with probability* $\frac{2}{3}$ *and*
- *the weather is bad with probability* $\frac{1}{4}$ *and good with probability* $\frac{3}{4}$.

Calculate $u\left(\left(\frac{1}{3}, \frac{2}{3}\right), \left(\frac{1}{4}, \frac{3}{4}\right)\right)$!

If you proceed according to the example of the above exercise, you do not need to worry about the summing condition.

DEFINITION II.21 (PAYOFF FOR MIXED STRATEGY AND BELIEF). *Given a mixed strategy* σ *and a belief* ω, *the payoff under* σ *and* ω *is defined by*

$$\begin{aligned} u(\sigma,\omega) &:= \sum_{s\in S}\sum_{w\in W}\sigma(s)\,\omega(w)\,u(s,w) \\ &= \sum_{s\in S}\sigma(s)\,u(s,\omega) \\ &= \sum_{w\in W}\omega(w)\,u(\sigma,w) \end{aligned}$$

4.4. Four different best-response functions. Depending on mixing or not mixing the strategy set and/or the set of states of the world, we modify definition II.13:

DEFINITION II.22 (BEST-RESPONSE FUNCTIONS). *Given* $\Delta = (S, W, u)$, *we distinguish four best-response functions:*

$$\begin{aligned} s^{R,W} &: W \to 2^S, \text{ given by } s^{R,W}(w) := \arg\max_{s\in S} u(s,w), \\ \sigma^{R,W} &: W \to 2^\Sigma, \text{ given by } \sigma^{R,W}(w) := \arg\max_{\sigma\in\Sigma} u(\sigma,w), \\ s^{R,\Omega} &: \Omega \to 2^S, \text{ given by } s^{R,\Omega}(\omega) := \arg\max_{s\in S} u(s,\omega), \text{ and} \\ \sigma^{R,\Omega} &: \Omega \to 2^\Sigma, \text{ given by } \sigma^{R,\Omega}(\omega) := \arg\max_{\sigma\in\Sigma} u(\sigma,\omega) \end{aligned}$$

If there is no danger of confusion, we stick to the simpler s^R *or* σ^R *instead of* $s^{R,W}$ *etc.*

EXERCISE II.11. *Complete the sentence:* $\sigma \in \sigma^{R,W}(w)$ *implies* $\sigma(s) = 0$ *for all ... .*

In line with lemma II.1, we obtain the following results:

THEOREM II.2. *Let* $\Delta = (S, W, u)$ *be a decision situation in strategic form. We have*

- $\sigma \in \Sigma$ *and* $\sum_{s\in s^{R,\Omega}(\omega)} \sigma(s) = 1$ *imply* $\sigma \in \sigma^{R,\Omega}(\omega)$ *and*
- $\sigma \in \sigma^{R,\Omega}(\omega)$ *implies* $s \in s^{R,\Omega}(\omega)$ *for all* $s \in S$ *with* $\sigma(s) > 0$.

These implications continue to hold for W *and* w *rather than* Ω *and* ω.

Best-response functions $\sigma^{R,\Omega}$ can be depicted graphically. Consider, for example, the decision matrix

	w_1	w_2
s_1	4	1
s_2	1	2

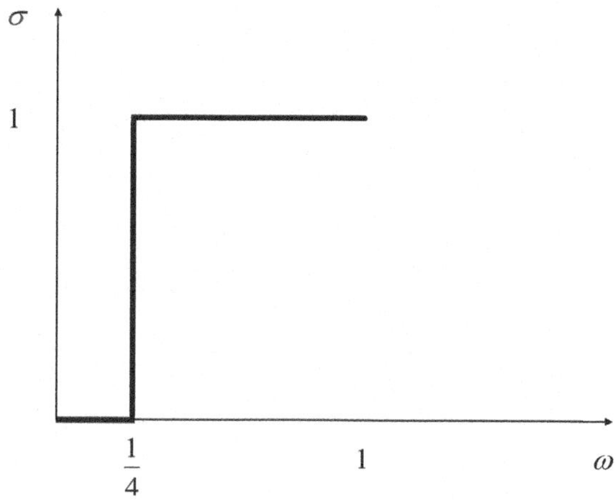

FIGURE II.5. The best-response function

Let $\omega := \omega(w_1)$ be the probability of w_1. We have $s_1 \in s^{R,\Omega}$ in case of

$$\omega \cdot 4 + (1-\omega) \cdot 1 \geq \omega \cdot 1 + (1-\omega) \cdot 2,$$

i.e., if $\omega \geq \frac{1}{4}$ holds. Remember that the best-response function is $\sigma^{R,\Omega} : \Omega \to 2^{\Sigma}$. For $\omega \neq \frac{1}{4}$, there is exactly one best strategy, $\sigma = 0$ (standing for $\sigma = (0,1) = s_2$) or $\sigma = 1$ (standing for $\sigma = (1,0) = s_1$), while $\omega = \frac{1}{4}$ implies that every pure strategy and hence every mixed strategy is best. We obtain

$$\sigma^{R,\Omega}(\omega) = \begin{cases} 1, & \omega > \frac{1}{4} \\ [0,1], & \omega = \frac{1}{4} \\ 0, & \omega < \frac{1}{4} \end{cases}$$

and the graph given in figure II.5.

EXERCISE II.12. *Sketch the best-response function $\sigma^{R,\Omega}$ for*

	w_1	w_2
s_1	1	3
s_2	2	1

5. Rationalizability

There are obvious reasons not to choose a strictly dominated strategy. We now develop a criterion for strategies that a "rational" decision maker might consider. Assume the following payoff matrix:

	w_1	w_2
s_1	4	4
s_2	1	5
s_3	5	1

A rational decision maker may choose s_2 if he thinks that state of the world w_2 will materialize. s_3 also seems a sensible choice. How about s_1? s_1 is a best strategy neither for w_1 nor for w_2. However, a rational decision maker may entertain the belief ω on W with $\omega(w_1) = \omega(w_2) = \frac{1}{2}$. Given this belief, s_1 is a perfectly reasonable strategy:

Exercise II.13. *Show* $s_1 \in s^{R,\Omega}\left(\left(\frac{1}{2}, \frac{1}{2}\right)\right)$!

Definition II.23 (rationalizability). *Let* $\Delta = (S, W, u)$ *be a decision situation in strategic form. A mixed strategy* $\sigma \in \Sigma$ *is called rationalizable with respect to* W *if a* $w \in W$ *exists such that* $\sigma \in \sigma^{R,W}(w)$. *Strategy* $\sigma \in \Sigma$ *is called rationalizable with respect to* Ω *if a belief* $\omega \in \Omega$ *exists such that* $\sigma \in \sigma^{R,\Omega}(\omega)$.

The above example shows that a mixed strategy (such as $(1, 0, 0)$) may be rationalizable with respect to Ω but not with respect to W.

6. Topics and literature

The main topics in this chapter are

- strategic form decision making
- payoff function
- strategy
- state of the world
- dominance
- best response
- mixed strategy
- lottery
- rationalizability
- set, element, interval
- real numbers, rational numbers, natural numbers, integers
- cardinality
- tuple
- Cartesian product
- power set
- function: injective, surjective, bijective
- domain
- range
- image
- probability distribution
- max, arg max

We recommend the mathematical textbooks by de la Fuente (2000) and Chiang & Wainwright (2005).

7. Solutions

Exercise II.1

Are you sure? The other part of humankind makes the opposite choice. See Nozick (1969) and Brams (1983).

Exercise II.2

We find $M \times N = \{(1,2), (1,3), (2,2), (2,3), (3,2), (3,3)\}$ and figure II.6.

Exercise II.3

f is not injective because we have $f(1) = f(2)$ but $1 \neq 2$. f is not surjective because of $b \in N \backslash f(M)$.

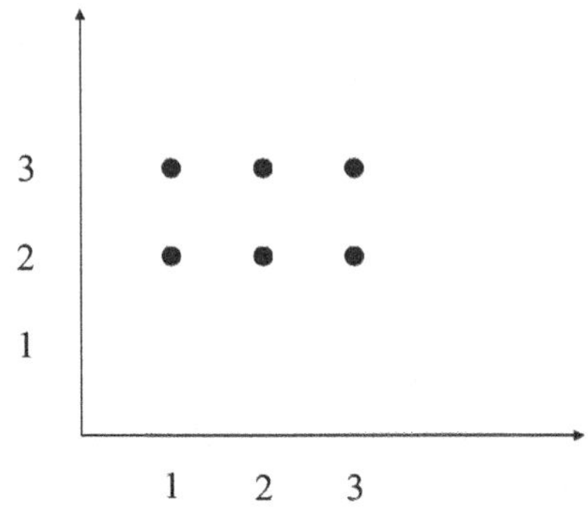

FIGURE II.6. The Cartesian product of M and N

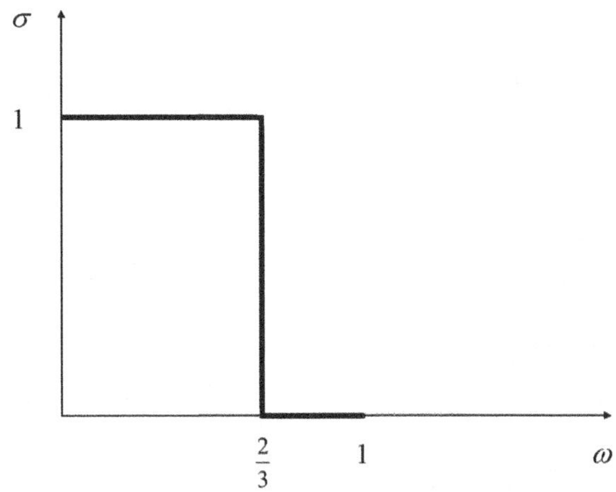

FIGURE II.7. Exercise: the best-reply function

Exercise II.4

Define $f : M \to N$ by $f(1) = a$, $f(2) = b$ and $f(3) = c$. f is surjective (we have $f(M) = N$) and injective (there are no different elements from M that point to the same element from N).

Exercise II.5

We calculate $\frac{1}{8} = 0.125$ and $\frac{4}{7} = 0.571428571428...$ where 571428 repeats itself indefinitely.

Exercise II.6

$\mathbb{R}$ can also be written as $(-\infty, \infty)$.

Exercise II.7

s is a dominant strategy if $s \in s^R(w)$ for all $w \in W$ and if, for every $s' \in S \backslash \{s\}$, we have at least one $w \in W$ such that $s' \notin s^R(w)$.

Exercise II.8

We have $prob(A) = \frac{1}{6}$ and $prob(B) = \frac{1}{2}$ for the two events and, by $A \cap B = \emptyset$, $prob(A \cup B) = prob(A) + prob(B) = \frac{1}{6} + \frac{1}{2} = \frac{4}{6}$.

Exercise II.9

Bad weather yields the payoff

$$\begin{aligned} & u\left(\left(\frac{1}{3}, \frac{2}{3}\right), \text{bad weather}\right) \\ = & \frac{1}{3}u(\text{umbrella, bad weather}) + \frac{2}{3}u(\text{sunshade, bad weather}) \\ = & \frac{1}{3} \cdot 100 + \frac{2}{3} \cdot 64 = 76 \end{aligned}$$

while good weather leads to

$$\begin{aligned} & u\left(\left(\frac{1}{3}, \frac{2}{3}\right), \text{good weather}\right) \\ = & \frac{1}{3}u(\text{umbrella, good weather}) + \frac{2}{3}u(\text{sunshade, good weather}) \\ = & \frac{1}{3} \cdot 81 + \frac{2}{3} \cdot 121 \approx 108 > 76 \end{aligned}$$

Exercise II.10

Using the result of exercise II.9, we obtain

$$u\left(\left(\frac{1}{3}, \frac{2}{3}\right), \left(\frac{1}{4}, \frac{3}{4}\right)\right) = \frac{1}{4} \cdot 76 + \frac{3}{4} \cdot \left(\frac{1}{3} \cdot 81 + \frac{2}{3} \cdot 121\right) = \frac{399}{4} \approx 100.$$

Alternatively, we find

$$\begin{aligned} u(\sigma, \omega) &= \sum_{s \in S} \sum_{w \in W} \sigma(s)\, \omega(w)\, u(s, w) \\ &= \frac{1}{3} \cdot \frac{1}{4} \cdot 100 + \frac{1}{3} \cdot \frac{3}{4} \cdot 81 + \frac{2}{3} \cdot \frac{1}{4} \cdot 64 + \frac{2}{3} \cdot \frac{3}{4} \cdot 121 \\ &= \frac{399}{4} \approx 100 \end{aligned}$$

Exercise II.11

$\sigma \in \sigma^{R,W}(w)$ implies $\sigma(s) = 0$ for all $s \notin s^{R,W}(w)$.

Exercise II.12

By $\omega \cdot 1 + (1-\omega) \cdot 3 \geq \omega \cdot 2 + (1-\omega) \cdot 1$ we find $\omega \leq 2/3$ and hence

$$\sigma^{R,\Omega}(\omega) = \begin{cases} 1, & \omega < \frac{2}{3} \\ [0,1], & \omega = \frac{2}{3} \\ 0, & \omega > \frac{2}{3} \end{cases}$$

This best-response function is depicted in figure II.7.

Exercise II.13

By $u\left(s_1, \left(\frac{1}{2}, \frac{1}{2}\right)\right) = 4$ and $u\left(s_2, \left(\frac{1}{2}, \frac{1}{2}\right)\right) = u\left(s_3, \left(\frac{1}{2}, \frac{1}{2}\right)\right) = \frac{1}{2} \cdot 1 + \frac{1}{2} \cdot 5 = \frac{6}{2} < 4$ we have $s_1 \in s^{R,\Omega}\left(\left(\frac{1}{2}, \frac{1}{2}\right)\right)$.

8. Further exercises without solutions

PROBLEM II.1.

(1) If strategy $s \in S$ strictly dominates strategy $s' \in S$ and strategy s' strictly dominates strategy $s'' \in S$, is it always true that strategy s strictly dominates strategy s''?
(2) If strategy $s \in S$ weakly dominates strategy $s' \in S$ and strategy s' weakly dominates strategy $s'' \in S$, is it always true that strategy s weakly dominates strategy s''?

PROBLEM II.2. Consider the problem of a monopolist faced with the inverse demand function $p(q) = a - b \cdot q$, in which a can either be high, a^h, or low, a^l. The monopolist produces with constant marginal and average cost c. Assume that $a^h > a^l > c$ and $b > 0$. Think of the monopolist as setting the quantity, q, and not the price, p.

(1) Formulate this monopolist's problem as a decision problem in strategic form. Determine $s^{R,W}$!
(2) Assume $a^h = 6, a^l = 4, b = 2, c = 1$ so that you obtain the plot given in figure II.8. Show that any strategy $q \notin \left[\frac{a^l-c}{2\cdot b}, \frac{a^h-c}{2\cdot b}\right]$ is dominated by either $s^{R,W}\left(a^h\right)$ or $s^{R,W}\left(a^l\right)$. Show also that no strategy $q \in \left[\frac{a^l-c}{2\cdot b}, \frac{a^h-c}{2\cdot b}\right]$ dominates any other strategy $q' \in \left[\frac{a^l-c}{2\cdot b}, \frac{a^h-c}{2\cdot b}\right]$.
(3) Determine all rationalizable strategies with respect to W.
(4) Difficult: Determine all rationalizable strategies with respect to Ω. Hint: Show that the optimal output is a linear combination of $s^{R,W}\left(a^h\right)$ and $s^{R,W}\left(a^l\right)$.

PROBLEM II.3. Prove the following assertions or give a counter-example!

(1) If $\sigma \in \Sigma$ is rationalizable with respect to W, then σ is rationalizable with respect to Ω.
(2) If $s \in S$ is a weakly dominant strategy, then it is rationalizable with respect to W.
(3) If $s \in S$ is rationalizable with respect to W, then s is a weakly dominant strategy.

PROBLEM II.4. Let $\Delta^k = \left(S^k, W^k, u^k\right)$ be a decision problem for $k \in \{1,2\}$. Compare the two decision problems Δ^1 and Δ^2, given by

$$\begin{aligned} S^1 &= \{l, r\} \\ W^1 &= \{a, b\} \\ u^1(l,a) &= u^1(r,b) = 1 \\ u^1(r,a) &= u^1(l,b) = 0 \end{aligned}$$

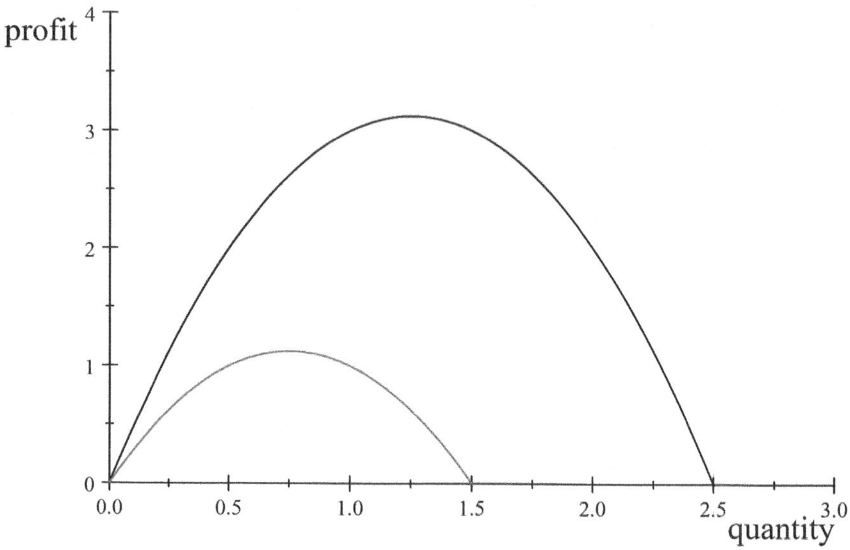

FIGURE II.8. Problem: profits for high demand and for low demand

and

$$\begin{aligned} S^2 &= [0,1] \\ W^2 &= \{a,b\} \\ u^2(1-s,a) &= u^2(s,b) = s. \end{aligned}$$

(1) Are there any dominant strategies?
(2) Calculate $u^1(\sigma,a)$ and $u^1(\sigma,b)$ for mixed strategies σ (where σ is the probability for choosing the pure strategy l). What interpretation of $s \in S^2$ does this suggest?
(3) Can we capture mixed strategies over n strategies in a game without mixed strategies?

PROBLEM II.5. Calculate:

(1) $\arg\max_x \{x+1 \mid x \in [0,1)\}$
(2) $\min \arg\max_y \{(-1)^y \mid y \in \mathbb{N}\}$

CHAPTER III

Decisions in extensive form

Strategic-form decision theory is static. Once a decision maker has chosen his strategy s or σ, he knows what to do and, depending on the state of the world w, he obtains $u(s, w)$ or $u(\sigma, w)$. We now turn to multi-stage decision making, also called extensive-form decision making. We will see later how to reduce extensive-form to strategic-form decision making. We begin with two examples and then go on to discuss more involved concepts such as decision trees, strategies, and subtree perfection. We also introduce the reader to backward induction which was well known to Indian fable tellers:

> The tragedy that follows a wrong plan,
> The Triumph that results from the right plan,
> To the rules of Polity both are linked;
> so the wise can point them out,
> as if displayed in advance.
> (Panchatantra, translated by Olivelle 2006, p. 77)

1. Introduction and two examples

Let us consider a very simple two-stage example. An umbrella-producing firm considers an investment and marketing activities. It has two actions (not strategies!) at his disposal at stage 1: action I (invest) and action nI (not invest). At stage 2, the choice is between actions M (marketing activities) and nM (no marketing activities).

Figure III.1 depicts the corresponding decision tree. It consists of nodes that have predecessors (those to the left) and successors (those to the right). The leftmost node v_0 in that figure is called initial node. Here, the decision maker has to make his first decision, to invest or not to invest. The initial node has no predecessor and the terminal nodes (the rightmost ones) have no successors.

The payoff information is recorded at the terminal nodes after all decisions are made. We assume the payoffs $u(v_3) = 10$, $u(v_4) = 5$, etc. The more formal definitions will follow soon.

H. Wiese, *Advanced Microeconomics*,
https://doi.org/10.1007/978-3-658-34959-2_3

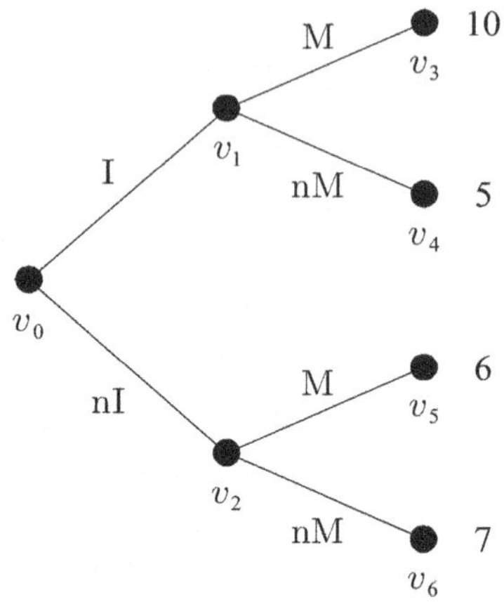

FIGURE III.1. A decision tree

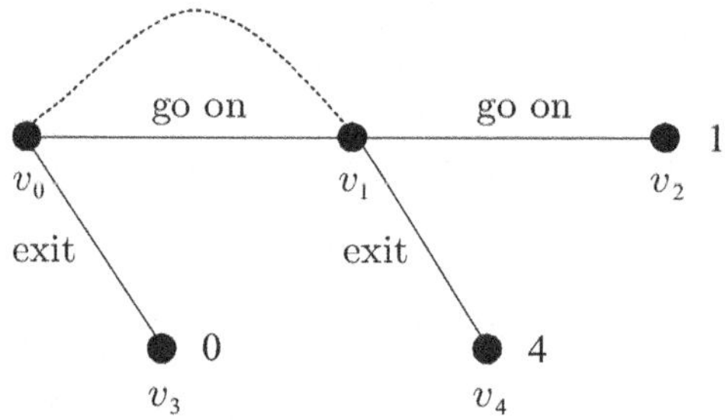

FIGURE III.2. To exit or not to exit?

Our second example is more exciting. It is about an absent-minded driver. He takes a rest near the highway and plans his further route. He knows that it is best to take the second exit. He also knows that he is tired and that he will not know whether the exit he will find himself at is the first or the second. Figure III.2 represents this decision situation. The dotted line linking the first two nodes indicates that the driver cannot distinguish between these nodes.

Thus, the driver has two decision nodes. However, since he cannot distinguish between these two nodes, the actions for the first node, "exit" and "go on", are the actions for the second node, too. What should the driver do?

In the above two examples, we have seen two different kinds of nodes. Nodes indicate that

- I have to make a decision or

- I get something.

Before delving into the next section, the reader should contemplate the difference between the two actions "M" in the upper figure and the two actions "exit" in the lower figure. The decision maker can distinguish the two actions "M" in the investment-marketing decision situation. In a strict sense, they are two different actions and we could have made the difference clear by indicating, M_I and M_{nI}. In contrast, the absent-minded driver cannot distinguish between the actions "exit". Therefore, it is important not to denote them by $\text{exit}_{\text{first}}$ and $\text{exit}_{\text{second}}$.

2. Decision trees and actions

2.1. Perfect information. Before defining a decision situation in extensive form, we need to clarify what we mean by a partition.

DEFINITION III.1 (PARTITION). *Let M be any nonempty set. A partition of M is a subset $\mathcal{P}_M = \{M_1, ..., M_k\}$ of the power set 2^M such that*

$$\begin{aligned} \bigcup_{j=1}^{k} M_j &= M, \\ M_j \cap M_\ell &= \emptyset \text{ for all } j, \ell \in \{1, ..., k\}, j \neq \ell \end{aligned}$$

hold. By $\mathcal{P}_M(m)$ we mean the element of $\mathcal{P}_M$ that contains $m \in M$. The elements of partitions are often called components. A component with one element only is called a singleton.

Most of the time, a partition will not contain the empty set but we allow for this possibility.

EXERCISE III.1. *Write down two partitions of $M := \{1, 2, 3\}$. Find $\mathcal{P}_M(1)$ in each case.*

We begin by describing a decision situation. (This description is not a very formal definition but leans on the investment-marketing decision situation.)

DEFINITION III.2 (EXTENSIVE-FORM DECISION SITUATION). *A decision situation (in extensive form and for perfect information) $\Delta = \left(V, u, (A_d)_{d \in D}\right)$ is given by*

- *a tree with node set V where the nodes are often denoted by v_0, $v_1, ...$ together with*
- *links that connect the nodes, directly or indirectly.*

Additionally:

- *A tree has an initial node v_0 and for every node v there exists exactly one trail (consisting of links) from v_0 to v (see below). Trails are indicated by brackets $\langle\rangle$ that contain at least two nodes.*
- *A trail is called maximal if it cannot be extended.*
- *The length of a trail is defined in the obvious manner (just go from one node to successor nodes and count the number of steps). The length of a tree is defined by its longest trail.*
- *D is the set of non-terminal nodes that are also called decision nodes. A_d is the set of actions that can be chosen at decision node d. Every link at d corresponds to exactly one action. The set of all actions is defined by $A = \cup_{d\in D} A_d$.*
- *E is the set of end nodes where a payoff function $u : E \to \mathbb{R}$ records the payoffs.*

We have three decision nodes and six actions in the investment-marketing decision situation. The link from v_1 to v_3 in figure III.1 corresponds to action M if the firm has chosen I at the first stage. Action M after action nI is a different link (the one from v_2 to v_5). Furthermore, observe $D = \{v_0, v_1, v_2\}$, $E = \{v_3, v_4, v_5, v_6\}$, $A_{v_2} = \{\text{M, nM}\}$, $A = \{\text{I, nI, M, nM}\}$. The trail $\langle v_0, v_3\rangle$ has length 2, while the length of trail $\langle v_1, v_3\rangle$ is 1.

EXERCISE III.2. *What is the length of the investment-marketing tree above? How about the absent minded driver (where you disregard the dotted line)? Indicate all the maximal trails in these two decision situations.*

2.2. Imperfect information. The above definition refers to "perfect information". This means that the decision maker knows the decision node at which he finds himself. In contrast, under "imperfect information" the decision maker does not know exactly the current decision node. We represent imperfect information by information sets that gather decision nodes between which the decision maker cannot distinguish. Therefore, the actions available at different nodes in an information set have to be the same. The absent-minded driver provides an example.

DEFINITION III.3 (INFORMATION PARTITION). *A decision situation (in extensive form and for imperfect information) $\Delta = \left(V, u, I, (A_d)_{d\in D}\right)$ equals the one for perfect information with the following exception: There exists a partition I (called information partition) of the decision nodes D. The elements of I are called information sets (which are components of I). The actions at decision nodes belonging to the same information set have to be identical: $A_d = A_{d'}$ for all $d, d' \in I(d)$.*

For some $d \in D$, we have $I(d) = \{d\}$ and the decision maker knows where he is. In others, we have several nodes d, d' in one information set and obtain $I(d) = I(d') = \{d, d', ...\}$.

EXERCISE III.3. *For the absent-minded driver, specify $I(v_0)$, A_{v_0}, and A_{v_1}!*

3. Strategies and subtrees: perfect information

3.1. Strategies. So far, we did not use the term strategy in extensive-form decision situations. A strategy is a full-fledged plan on how to act at each decision node. We begin with perfect information.

DEFINITION III.4 (STRATEGY). *Let $\Delta = \left(V, u, (A_d)_{d \in D}\right)$ with $A = \cup_{d \in D} A_d$ be a decision situation in extensive form for perfect information. A strategy is a function $s : D \to A$ obeying $s(d) \in A_d$.*

Thus, s is a strategy if it tells us how to act at each decision node. Of course, the actions available at d have to be from A_d. For example,

$$s(v_0) = \text{nI}, s(v_1) = \text{M}, s(v_2) = \text{nM}$$

is a strategy in the investment-marketing decision situation. A more convenient way to write this strategy is

$$\lfloor \text{nI}, \text{M}, \text{nM} \rfloor$$

if the nodes to which M and nM refer are clear.

EXERCISE III.4. *How many strategies do we have in the decision situation of figure III.1 (p. 32)?*

In case of perfect information, we have $|S| = \prod_{d \in D} |A_d|$. Note that some strategies contain information that we do not need, at the moment. For example, if the decision maker chooses not to invest (action nI), he does not need to worry about his action at v_1 which is reached in case he does invest. Would you make detailed plans for a stay in Berlin next Sunday if you plan to visit Munich? Of course, you might end up in Berlin after getting on the wrong train ... .

A strategy tells us how to act at every decision node. Therefore, we can trace the nodes "visited" by a strategy.

DEFINITION III.5 (TRAIL PROVOKED BY STRATEGY). *A node v or a trail $\langle v_0, v_1, ..., v_k = v \rangle$ is provoked or brought about by strategy $s \in S$ if we arrive at v by choosing the actions prescribed by s. The terminal node provoked by strategy s is denoted by v^s. Also, every strategy provokes v_0.*

A strategy s provokes exactly one maximal trail, i.e., the trail from v_0 to a $v^s \in E$. This idea allows us to define a payoff function $S \to \mathbb{R}$ on the basis of the payoff function $E \to \mathbb{R}$.

DEFINITION III.6 (PAYOFF FUNCTION). *In* $\Delta = \left(V, u, (A_d)_{d\in D}\right)$, *we define* $u : S \to \mathbb{R}$ *by*

$$u(s) := u(v^s), s \in S.$$

DEFINITION III.7 (BEST STRATEGIES). *The set of best strategies for* $\Delta = \left(V, u, (A_d)_{d\in D}\right)$ *is defined by*

$$s^R(\Delta) := \arg\max_{s\in S} u(s).$$

EXERCISE III.5. *Indicate all the nodes provoked by the strategy* $\lfloor I, M, M \rfloor$ *in the investment-marketing example. Which strategies are best in the investment-marketing decision situation?*

3.2. Subtrees and subtree perfection. Students often wonder why we define strategies in the very complete manner seen above. First of all, the definition of a strategy is simpler this way. The definition of a more restricted strategy would be rather cumbersome. The second reason is that we want to distinguish between the two best strategies

1. strategy: $s(v_0) = I, s(v_1) = M, s(v_2) = M$
2. strategy: $s(v_0) = I, s(v_1) = M, s(v_2) = nM$

Both of them are optimal, but the first is somewhat peculiar. It advises the decision maker to choose M, should he find himself at v_2. However, at v_2 action nM is better than action M. In order to get rid of this peculiarity, we define subtree perfection. Before doing so, we need the concept of a restriction of a function.

DEFINITION III.8 (RESTRICTION). *Let* $f : X \to Y$ *be a function. For* $X' \subseteq X$, $f|_{X'} : X' \to Y$ *is called the restriction of* f *to* X' *if* $f|_{X'}(x) = f(x)$ *holds for all* $x \in X'$.

Thus, a restriction of a function reduces the domain of a function but stays the same otherwise.

DEFINITION III.9 (SUBTREE). *Let* $\Delta = \left(V, u, (A_d)_{d\in D}\right)$ *be a decision situation in extensive form for perfect information and* $w \in D$. *Let* W *be the set of* w *together with its successor nodes (direct or indirect). Then, we obtain* w*'s decisional subtree of* Δ *called* $\Delta^w = (W, u|_{W\cap E}, A|_W)$. $A|_W$ *is to be understood as* $(A_d)_{d\in D\cap W}$. *We call* $s^w : D \cap W \to A$ *a substrategy of* $s \in S$ *in* Δ^w *if* $s^w = s|_{W\cap D}$ *holds. By* S^w *we denote the set of substrategies in*

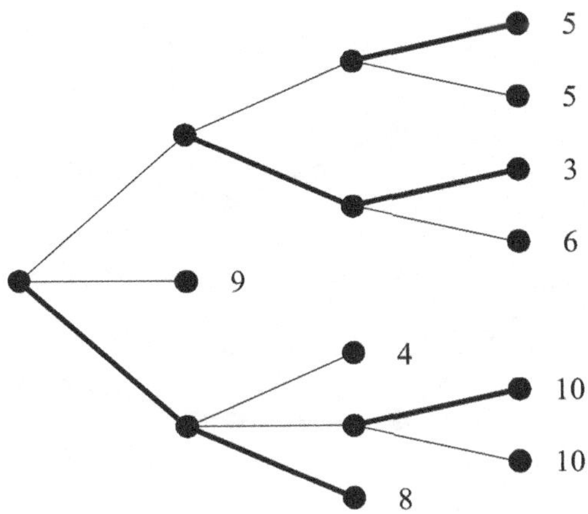

FIGURE III.3. Optimal strategy?

Δ^w. *Δ^w is called a minimal subtree if its length is one. Δ^w is called a proper subtree if $w \neq v_0$.*

Thus, we obtain Δ^w from Δ by choosing a $w \in D$ and restricting the strategies accordingly. Note $\Delta^{v_0} = \Delta$.

DEFINITION III.10 (SUBTREE PERFECTION). *A strategy s is subtree-perfect if, for every $w \in D$, s^w is a best strategy in the decisional subtree Δ^w.*

The strategy $\lfloor I, M, M \rfloor$ noted above is not subtree perfect. At v_2 a subtree Δ^{v_2} begins. It has two actions and also two strategies and M is not the best.

EXERCISE III.6. *Consider the decision trees of figure III.3 and III.4 and check whether the indicated strategies are optimal and/or subtree-perfect. The actions chosen in the respective strategies are indicated by bold lines.*

3.3. Backward induction for perfect information. Backward induction is a very powerful instrument for solving decision situations. The idea is to consider minimal subtrees. Once we know what to do at these "final" decision nodes, we can climb down the tree (climb leftwards).

ALGORITHM III.1. *Let $\Delta = \left(V, u, (A_d)_{d \in D}\right)$ be of finite length. Backward-induction proceeds as follows:*

(1) *Consider the minimal subtrees Δ^w and take note of the best strategies in Δ^w, $s^R(\Delta^w) := \arg\max_{s^w \in S^w} u|_W(s^w)$. If any of these sets are empty (for the reason explained on p. 17), the procedure stops. Otherwise, proceed at point 2.*

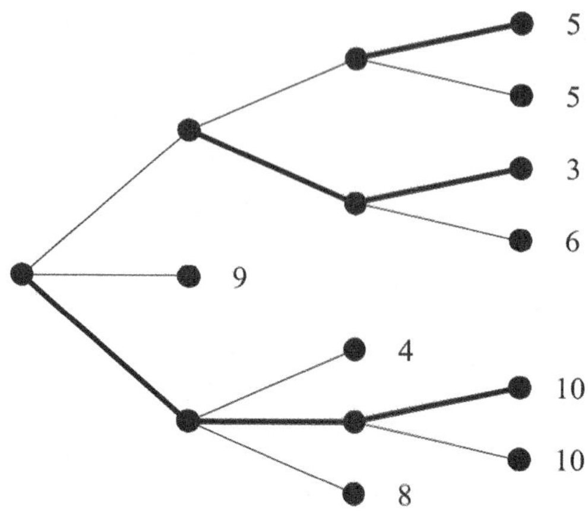

FIGURE III.4. Subgame-perfect strategy?

(2) *Cut the tree by replacing all minimal subtrees* Δ^w *by a terminal node* w *carrying the payoff information obtained at (1). If* $s^R(\Delta^w)$ *contains several best strategies, construct several trees.*
(3) *If the new trees contain minimal subtrees, turn to point (1). Otherwise, the final tree contains (the final trees contain) just one terminal node which is the initial node of the original tree. This tree (all these trees) carries (carry) the same and maximal payoff.*

The maximal trails and the strategies generated by the backward-induction procedure are called backward-induction trails and backward-induction strategies, respectively.

The algorithm is explained in figure III.5 and III.6 by way of example.

If you like to economize on paper, you may prefer another method to find the backward-induction trails in a decision tree. Consider figure III.7. We identify the minimal subtrees (they start at v_1 and v_2) and mark the link leading to the best action. We then consider the subtrees which have v_1 and v_2 as immediate successors. In our simple example, there is only one, the original tree itself. Since 10 is greater than 7, action I is the best action.

EXERCISE III.7. *Solve the decision tree of figure III.8 by applying backward induction. How many backward-induction trails and how many backward-induction strategies can you find?*

Without proof, we note the following theorem:

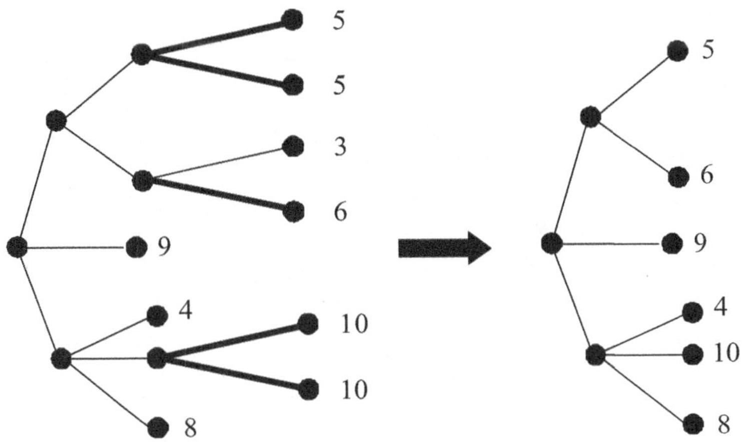

FIGURE III.5. Backward induction, first step

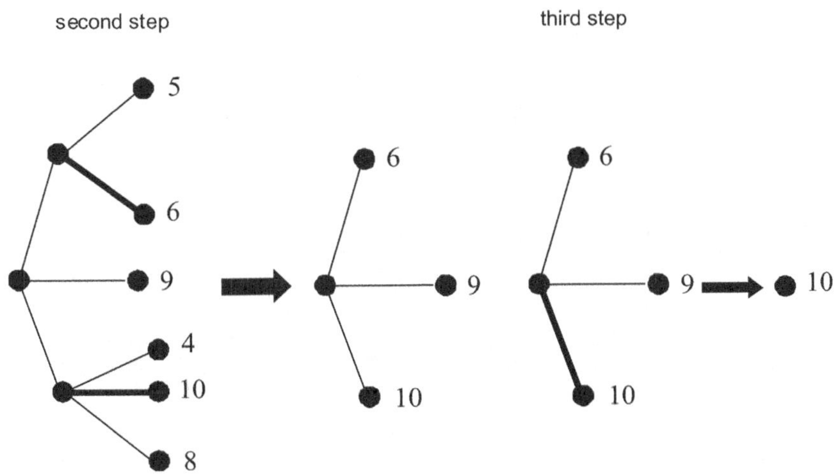

FIGURE III.6. Backward induction, second and third step

THEOREM III.1. *Let* $\Delta = \left(V, u, (A_d)_{d \in D}\right)$ *be of finite length. Then, the set of subtree-perfect strategies and the set of backward-induction strategies coincide. Differently put, a strategy is subtree-perfect iff it is a backward-induction strategy.*

COROLLARY III.1. *Every decision situation* $\Delta = \left(V, u, (A_d)_{d \in D}\right)$ *with* $|V| < \infty$ *has a subtree-perfect strategy.*

Do you see that a finite node set implies a finite tree length, but not vice versa? The corollary (which is Kuhn's theorem) is easy to prove. Since the

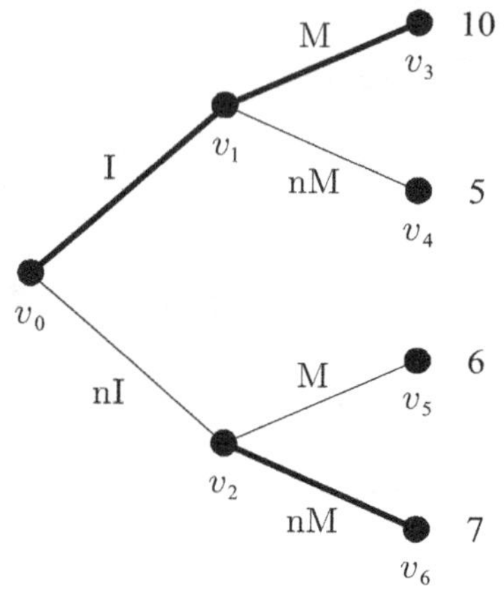

FIGURE III.7. Backward induction, graphically

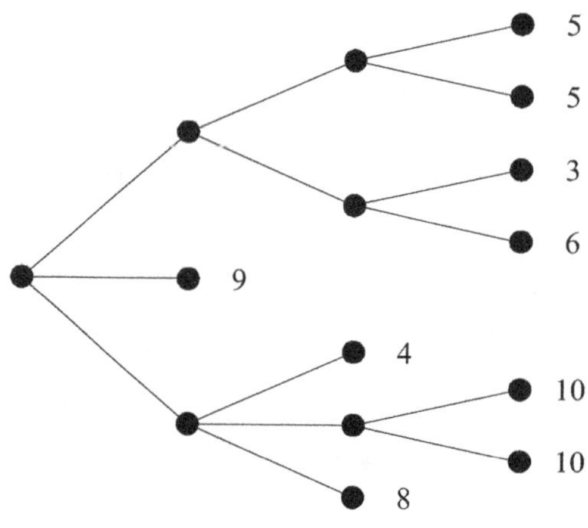

FIGURE III.8. A decision tree

number of nodes is finite, so is the number of actions at each decision node. Therefore, best actions always exist and the backward-induction procedure does not stop halfway. Then, the set of backward-induction strategies is not empty.

3.4. The money pump. A central axiom of utility theory (chapter IV) is transitivity. Transitivity means: if a person prefers z to y and y to x, then she should also prefer z to x. Here x, y, and z are objects, not payoffs. Assume a decision maker whose strict preferences are not transitive:

$$x \prec y \prec z \prec x.$$

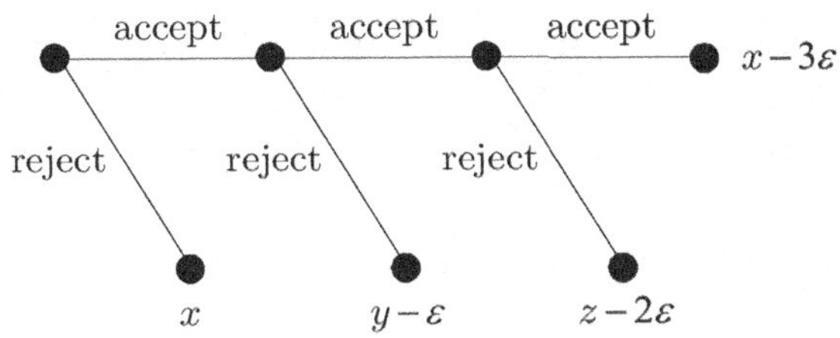

FIGURE III.9. Intransitive preferences and the money pump

$x \prec y$ means that y is strictly preferred to x.

The money-pump argument against intransitive preferences works as follows: Assume, our agent has x. If the agent can exchange x against y, he is happy to do so and is even ready to offer a small amount of money. He now has $y - \varepsilon$ instead of x where $y - \varepsilon$ is shorthand for object y and a reduction of his money stock by ε. Consider figure III.9 to follow the argument. Now, again, an offer is made to exchange z against y. By $y \prec z$ the agent is again willing to accept and also to throw in the very small ε, $y \prec z - \varepsilon$. He now holds $z - 2\varepsilon$. Finally, the agent changes z against $x - \varepsilon$ so that he ends up with $x - 3\varepsilon$.

Of course, that is too bad. The agent starts with x and ends with $x - 3\varepsilon$. It seems as if one could pump money out of the agent again and again. However, we are not forced to accept the argument without further ado. Should the agent not foresee the whole line of transactions and decline immediately? Let us look more closely!

The decision maker can opt for one of the two actions three times. Thus, we have $2^3 = 8$ strategies, among which we find

$$\lfloor \text{accept, accept, accept} \rfloor,$$
$$\lfloor \text{accept, reject, accept} \rfloor \text{ and}$$
$$\lfloor \text{reject, accept, reject} \rfloor$$

EXERCISE III.8. *Write down all strategies that lead to payoff* $y - \varepsilon$.

Let us try backward induction. Assume

$$x \prec y - \varepsilon \prec z - 2\varepsilon \prec x - 3\varepsilon$$

and also

$$x - 3\varepsilon \prec y - \varepsilon$$

which seems reasonable by $x \prec y$. Backward induction is depicted in figure III.10. At the last decision node, the decision maker has to accept by $z -$

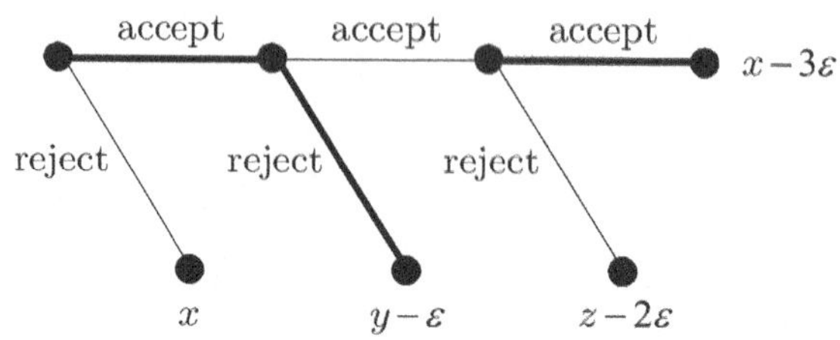

FIGURE III.10. Backward induction for the money pump

$2\varepsilon \prec x - 3\varepsilon$. At the second-to-last node, the above strict preference yields reject. At the first node, the decision maker compares x to $y - \varepsilon$ and will accept. Therefore, backward induction does not support the money-pump argument.

4. Strategies and subtrees: imperfect information

4.1. Strategies and subtrees. We now consider decision situations with information sets as found for the absent-minded driver. The definitions are somewhat more involved because we need to make sure that the decision maker acts in the same fashion at every decision node belonging to the same information set.

DEFINITION III.11 (STRATEGY). *Let* $\Delta = \left(V, u, I, (A_d)_{d \in D}\right)$ *be a decision situation in extensive form for imperfect information. A strategy is a function* $s : D \to A$ *obeying* $s(d) \in A_d$ *and* $s(d) = s(d')$ *for all* $d, d' \in I(d)$.

Consider the absent-minded driver. He has only two strategies, namely $\lfloor$exit, exit$\rfloor$ and $\lfloor$go on, go on$\rfloor$. This manner of writing the strategy conforms to the above definition. However, there is a shorthand manner to indicate his strategy: $\lfloor$exit$\rfloor$ or $\lfloor$go on$\rfloor$. Here, the idea is to report an action for each information set, not for each decision node. Since the actions at different decision nodes within an information set have to be the same, no information is lost with this notation. Therefore, there exists an alternative strategy definition.

DEFINITION III.12 (STRATEGY). *Let* $\Delta = \left(V, u, I, (A_d)_{d \in D}\right)$ *be a decision situation in extensive form for imperfect information. A strategy is a function* $s : I \to A$ *obeying* $s(C_I) \in A_d$ *for every information set* C_I *of* I *(*C_I *is an element and a component of* I*) and any* $d \in I(d) = C_I$.

DEFINITION III.13 (SUBTREE). *Let* Δ *be a decision situation in extensive form for imperfect information. Assume a node* $w \in D$ *with the following* *property: If an information set from I intersects W (w and all its successors), then that information set is included in W. Then, we obtain w's decisional subtree of Δ called $\Delta^w = (W, u|_{W\cap E}, I|_W, A|_W,)$ where $I|_W$ is the subpartition of I whose components are contained in W.*

We call s^w a substrategy of $s \in S$ in Δ^w if $s^w = s|_{W\cap D}$ holds. By S^w we denote the set of substrategies in Δ^w. Δ^w is called a minimal subtree if it does not contain proper subtrees (proper subtrees differ from the tree from which they are derived).

Here, with imperfect information, we have an additional requirement on w : The newly formed subtree must not cut into any information set from I. Rather, any information set from I that intersects with W is wholly contained in W. Therefore, the decision situation of the absent-minded driver has one decisional subtree only, the orginal decision situation.

While a strategy s is a more complicated object under imperfect information than under perfect information, the definitions of

- nodes provoked by s (definition III.5)
- the payoff function $u : S \to \mathbb{R}$ (definition III.6),
- the set of best strategies (definition III.7), and
- subtree perfection (definition III.10)

are the same for imperfect as for perfect information.

EXERCISE III.9. *Consider, again, the absent-minded driver. What is his best strategy? What is his subtree-perfect strategy?*

The driver cannot do any better with a mixed strategy. A mixed strategy yields an average of 1 and 0 which cannot be more than 1. Wait, we did not define a mixed strategy for the extensive form as yet. Here it is:

DEFINITION III.14 (MIXED STRATEGY). *Let $\Delta = \left(V, u, I, (A_d)_{d\in D}\right)$ be a decision situation in extensive form for (perfect or) imperfect information. A mixed strategy is a probability distribution σ on the set of (pure) strategies from S. The set of mixed strategies is denoted by Σ.*

Consider, now, a given terminal node $v \in E$. It can be provoked by different strategies.

DEFINITION III.15 (PROVOKED NODE). *Under a mixed strategy $\sigma \in \Sigma$, the overall probability of reaching $v \in V$ is denoted by*

$$\tau_\sigma(v) := \sum_{\substack{s\in S,\\ s \text{ provokes } v}} \sigma(s).$$

We also say that σ provokes v (or $\{v\}$) with probability $\tau_\sigma(v)$.

EXERCISE III.10. *Let v_0 be the initial node in a decision situation in extensive form. Can you tell $\tau_\sigma(v_0)$?*

EXERCISE III.11. *Consider the investment-marketing decision situation of figure III.7 (p. 40) and the mixed strategy σ given by*

$$\sigma(s) = \begin{cases} \frac{1}{3}, & s = \lfloor I,\ M,\ nM \rfloor \\ \frac{1}{6}, & s = \lfloor nI,\ M,\ nM \rfloor \\ \frac{1}{12}, & s \text{ otherwise} \end{cases}$$

Why is σ well-defined? What is the probability for node v_3?

4.2. Behavioral strategies. As mentioned in the previous section, the absent-minded driver cannot succeed in obtaining the payoff 4 by way of mixed strategies. However, there is one clever alternative to a mixed strategy that allows the driver to fare better than with any pure or mixed strategy. The idea is to mix actions rather than strategies:

DEFINITION III.16 (BEHAVIORAL STRATEGY). *Let $\Delta = (V, u, I, (A_d)_{d\in D})$ be a decision situation in extensive form for (perfect or) imperfect information. A behavioral strategy is a tuple of probability distributions $\beta = (\beta_d)_{d\in D}$ where, for every $d \in D$, β_d is a probability distribution on A_d that obeys $\beta_d = \beta_{d'}$ for all $d, d' \in I(d)$.*

We now show what behavioral strategies can do in the absent-minded driver's case (reconsider figure III.2, p. 32). Since v_0 and v_1 are in the same information set, the probability for exit has to be the same at these two nodes. Let $\beta_{\text{exit}} := \beta_{v_0}(\text{exit})$ be the probability for exit. Then, the driver obtains the expected payoff

$$\begin{aligned} & \underbrace{\beta_{\text{exit}}}_{\substack{\text{exit probability} \\ \text{at } v_0}} \cdot 0 + \underbrace{(1-\beta_{\text{exit}})\,\beta_{\text{exit}}}_{\substack{\text{exit probability} \\ \text{at } v_1}} \cdot 4 + \underbrace{(1-\beta_{\text{exit}})\,(1-\beta_{\text{exit}})}_{\substack{\text{probability} \\ \text{of going on}}} \cdot 1 \\ = \; & -3\beta_{\text{exit}}^2 + 2\beta_{\text{exit}} + 1 \end{aligned}$$

Note that $0 < \beta_{\text{exit}} < 1$ allows payoff 4 with a positive probability. The optimal behavioral strategy is given by

$$\beta_{\text{exit}}^* = \arg\max_{\beta_{\text{exit}}} \left(-3\beta_{\text{exit}}^2 + 2\beta_{\text{exit}} + 1\right) = \frac{1}{3}.$$

Thus, a behavioral strategy (here: β_{exit}^*) can achieve a probability distribution on terminal nodes that no mixed strategy can achieve. The reason for the power of behavioral strategies does not only lie in the fact that we have a decision situation with imperfect information. The Rubinstein decision tree is special in that the driver cannot recall at v_1 that he has been at v_0 before.

4.3. Imperfect recall and strategies. We have imperfect recall when at an information set $\{v, v'\}$ the decision maker's experience at v differs from that at v'.

DEFINITION III.17 (EXPERIENCE). *Let* $\Delta = \left(V, u, I, (A_d)_{d \in D}\right)$ *be a decision situation in extensive form for imperfect information. At* $v \in D$, *the experience* $X(v)$ *is the sequence of information sets and actions at these information sets as they occur from* v_0 *to* v. *An information set is the last element of an experience.*

Thus, an experience is a tuple with an odd number of entries: information set – action – information set – ... – information set. In the absent-minded driver example, we have

- $X(v_0) = (I(v_0))$ and
- $X(v_1) = (I(v_0), \text{ go on}, I(v_1))$.

DEFINITION III.18 (PERFECT RECALL). *Let* $\Delta = \left(V, u, I, (A_d)_{d \in D}\right)$ *be a decision situation in extensive form for imperfect information.* Δ *is characterized by perfect recall if for all* $v, v' \in D$ *with* $I(v) = I(v')$ *we have* $X(v) = X(v')$.

Inversely, if you do not know where you are although your past experience has been different, you suffer from imperfect recall. We do not have perfect recall for the absent-minded driver because v_0 and v_1 belong to the same information set $I(v_0) = I(v_1)$, while the experiences $X(v_0)$ and $X(v_1)$ differ.

EXERCISE III.12. *Can you explain why perfect information implies perfect recall?*

EXERCISE III.13. *Show that the decision situation of figure III.11 exhibits imperfect recall!*

While figure III.11 exhibits imperfect recall, strategies seem to be able to overcome this deficiency. For example, the strategy $\lfloor$I, M$\rfloor$ leads to terminal node v_3, while the strategy $\lfloor$I, nM$\rfloor$ provokes the node v_4. This is somewhat odd given that the decision maker does not know whether he finds himself at v_1 or at v_2 when choosing action M or nM. After all, shouldn't the strategy $\lfloor$I, M$\rfloor$ tell him that he is at v_1?

What do you think about this interpretation for strategy $\lfloor$I, M$\rfloor$? The decision maker has a definite plan on how to act at v_0 and at $I(v_1) =$

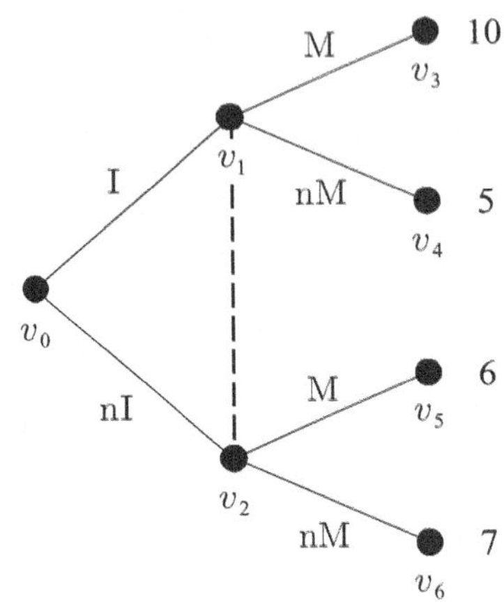

FIGURE III.11. Imperfect recall?

$\{v_1, v_2\}$. After investing at v_0, he forgets this action (it was written with magic ink that disappeared after choosing I), but he still knows that he is to perform marketing activities. Thus, at $\{v_1, v_2\}$ he does not know whether he ends up at v_3 or v_5. (Of course, after receiving the payoff 10, he can surely reconstruct the trail.)

Summarizing, we have

- perfect information if every information set contains one element, only,
- (properly) imperfect information if there is an information set with more than one element,
- perfect recall if all decision nodes in an information set are associated with the same experience and
- imperfect recall if two decision nodes exist that belong to the same information set but result from different experiences.

4.4. Equivalence of mixed and behavioral strategies. The absent-minded driver example prompts the question of whether behavioral strategies are always "more powerful" than mixed strategies. Without giving a proof, we state an important result:

THEOREM III.2 (KUHN'S EQUIVALENCE THEOREM). *Consider a decision situation in extensive form for imperfect information, but perfect recall. A given probability distribution on the set of terminal nodes is achievable by a mixed strategy if and only if it is achievable by a behavioral strategy. We then say that mixed and behavioral strategies are payoff equivalent.*

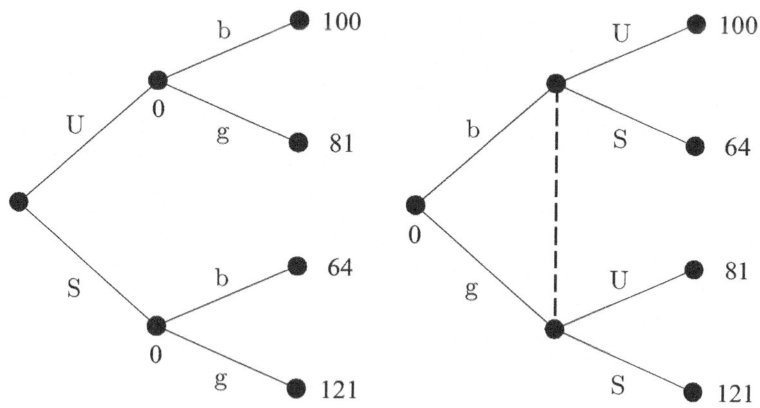

FIGURE III.12. Uncertainty about the weather

The theorem says that behavioral strategies and mixed strategies can achieve the same distribution on the set of terminal nodes in case of perfect recall. For imperfect recall, sometimes, behavioral strategies are more powerful than mixed strategies and sometimes it is the other way around.

5. Moves by nature, imperfect information, and perfect recall

5.1. Decision situation. One origin of imperfect informaton is imperfect recall. A second and maybe more important reason for imperfect information is a move by nature. For example, the weather may be good or bad and the decision maker does not know before he chooses important actions. Revisiting the umbrella-sunshade producer of chapter II, we consider two different trees that reflect the decision maker's uncertainty. The left tree in figure III.12 is one of perfect information. The decision maker moves at the initial node and nature moves at the second stage. The nodes where nature makes its moves are denoted by "0". The right-hand tree is one of imperfect information. Nature moves first and the decision maker does not know nature's "choice". This is indicated by the information set linking the two non-initial decision nodes.

DEFINITION III.19 (DECISION SITUATION). *A decision situation in extensive form for imperfect information with moves by nature is a tuple*

$$\Delta = \left(V, u, \iota, I, (A_d)_{d\in D}, \beta_0\right)$$

that fulfills:

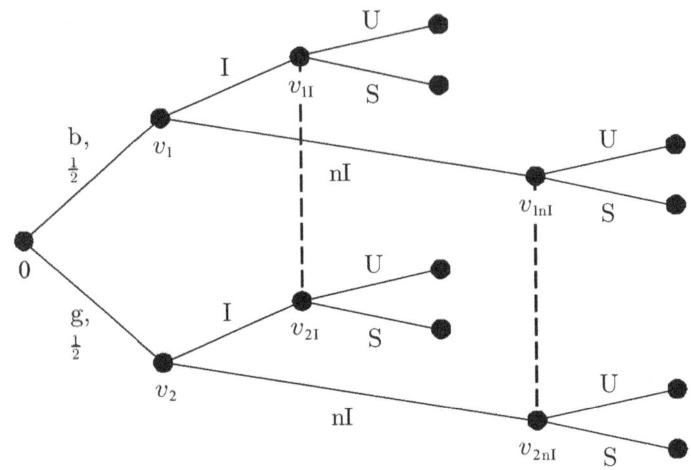

FIGURE III.13. Perfect or imperfect recall?

- $\iota : D \to \{0, 1\}$ *is a player-selection function that yields a partition* $\{D_0, D_1\}$ *of* D*, where* $D_0 := \{d \in D : \iota(d) = 0\}$ *refers to nature and* $D_1 := \{d \in D : \iota(d) = 1\}$ *to the decision maker.*
- I*, a partition of the decision nodes* D_1*, is the information partition.*
- A_d *is the set of actions that can be chosen at decision node* d*, either by the decision maker* $(\iota(d) = 1)$ *or by nature* $(\iota(d) = 0)$*. Every link at* d *corresponds to exactly one action. The decision maker's action sets obey* $A_d = A_{d'}$ *for all* $d, d' \in I(d)$*.*
- β_0 *is a tuple of probability distributions* $(\beta_d)_{d \in D_0}$ *on* A_0*.*

In a sense, we have two players, the decision maker and nature. We model the moves by nature as a behavioral strategy. Consider the two decision trees of figures III.13 and III.14. They report the probability distribution (there is only one) β_0 at the node denoted by 0.

The definition of a strategy is the same as the one on p. 42. For example, in figure III.13, the decision maker's strategies are quadruples such as $\lfloor \text{I, nI, U, U} \rfloor$. Can you find out the probability for nodes provoked by strategies without a formal definition?

EXERCISE III.14. *Reconsider the decision tree of figure III.13. Indicate the probability distributions on the set of terminal nodes provoked by the strategies* $\lfloor I, nI, S, U \rfloor$ *and by* $\lfloor nI, nI, S, S \rfloor$ *by writing the probabilities near these nodes.*

With or without moves by nature, we obtain the following definition.

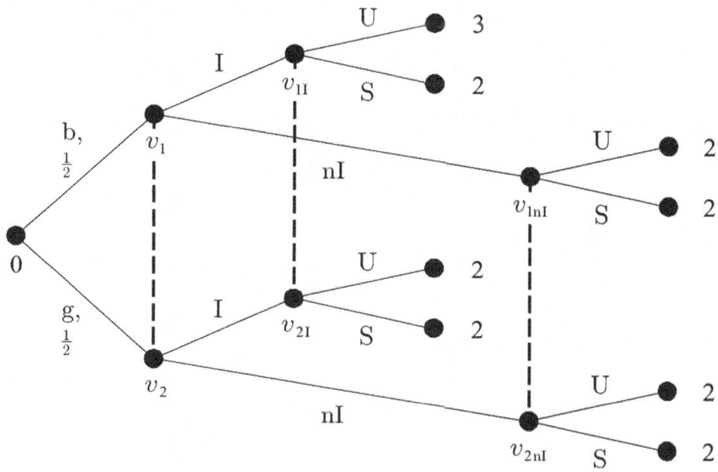

FIGURE III.14. Perfect or imperfect recall?

DEFINITION III.20 (PAYOFF FUNCTION). *Let* $\Delta = \left(V, u, \iota, I, (A_d)_{d \in D}, \beta_0\right)$ *be a decision situation. We define* $u : S \to \mathbb{R}$ *by*

$$u(s) := \sum_{v \in E} \tau_s(v)\, u(v).$$

We feel that it would be overly formalistic to define mixed strategies, subtree perfection, and behavioral strategies for decision situations with moves by nature. Three remarks are important:

- definition of subtrees: Subtrees may start at nodes from $D = D_0 \cup D_1$. In particular, the whole tree is always a subtree of itself.
- definition of experience: If nature moves at v_0 or at any other node along a trail leading to a decision node $v \in D_1$, $X(v)$ does not feature v_0 or other nodes from D_0 and contains the last entry $I(v)$. Of course, the decision maker acts at decision nodes from D_1, only.
- Kuhn's theorem remains valid for moves by nature.

EXERCISE III.15. *Do the decision trees of figures III.13 and III.14 reflect perfect or imperfect recall? How many subtrees can you identify?*

5.2. Backward induction for imperfect information. Although the chances of finding subtrees are slimmer in case of imperfect information, backward induction is still a useful instrument. We need to adjust the procedure (see section 3.3) a little bit because, in general, not every decision

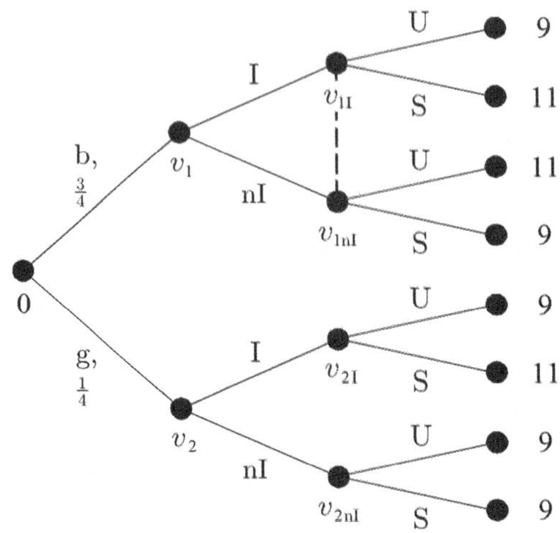

FIGURE III.15. Backward induction?

node gives rise to a subtree. The point is that subtrees are not allowed to cut into information sets (please consult definition III.13 on p. 42). Taking account of this difficulty, backward induction can be applied and yields backward-induction trails and backward-induction strategies as in the case of perfect information.

Without proof, we note the following theorem.

THEOREM III.3. *Let* $\Delta = \left(V, u, \iota, I, (A_d)_{d \in D}, \beta_0\right)$ *be of finite length. Then, the set of subtree-perfect strategies and the set of backward-induction strategies coincide.*

COROLLARY III.2. *Every decision situation* $\Delta = \left(V, u, \iota, I, (A_d)_{d \in D}, \beta_0\right)$ *with* $|V| < \infty$ *has a subtree-perfect strategy.*

EXERCISE III.16. *Consider the decsion tree of figure III.15. Is it characterized by perfect recall? How many subtrees do you find? Apply backward induction! How many subtree-perfect strategies do you find?*

6. Topics

The main topics in this chapter are

- information partition, information set
- perfect versus imperfect information
- strategy, mixed strategy, behavioral strategy
- absent-mindedness
- perfect recall
- subtree perfection
- moves by nature
- player-selection function
- trees and subtrees
- successor function
- predecessor
- decision node
- terminal node
- initial node
- partition
- restriction of a function

7. Solutions

Exercise III.1

The following collections of subsets are partitions of M :

$$\begin{aligned} \{\{1,2\},\{3\}\} \text{ with } \mathcal{P}_M(1) &= \{1,2\}, \\ \{M\} \text{ with } \mathcal{P}_M(1) &= M, \\ \{\{1\},\{2\},\{3\}\} \text{ with } \mathcal{P}_M(1) &= \{1\}, \end{aligned}$$

while these are not:

$$\{\{1\},\{3\}\},\{\{1,2\},\{2,3\}\},\{1,2,3\}.$$

Exercise III.2

In the investment-marketing decision situation we have four maximal trails: $\langle v_0, v_1, v_3\rangle$, $\langle v_0, v_1, v_4\rangle$, $\langle v_0, v_2, v_5\rangle$, and $\langle v_0, v_2, v_6\rangle$. The length of the tree is 2. In the absent-minded driver example, we find three maximal trails: $\langle v_0, v_3\rangle$, $\langle v_0, v_1, v_4\rangle$, and $\langle v_0, v_1, v_2\rangle$. The length of the tree is 2.

Exercise III.3

We find $I(v_0) = I(v_1) = \{v_0, v_1\}$ and $A_{v_0} = A_{v_1} = \{\text{go on, exit}\}$.

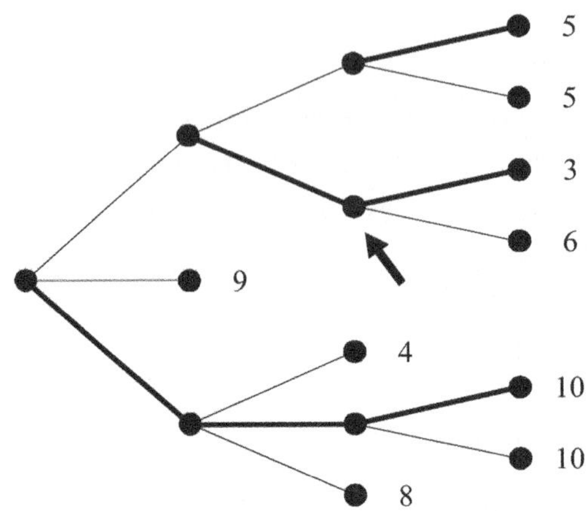

FIGURE III.16. Two profitable one-node deviations

Exercise III.4

We have eight strategies:

1. strategy: $s(v_0) = I, s(v_1) = M, s(v_2) = M$
2. strategy: $s(v_0) = I, s(v_1) = M, s(v_2) = nM$
3. strategy: $s(v_0) = I, s(v_1) = nM, s(v_2) = M$
4. strategy: $s(v_0) = I, s(v_1) = nM, s(v_2) = nM$
5. strategy: $s(v_0) = nI, s(v_1) = M, s(v_2) = M$
6. strategy: $s(v_0) = nI, s(v_1) = M, s(v_2) = nM$
7. strategy: $s(v_0) = nI, s(v_1) = nM, s(v_2) = M$
8. strategy: $s(v_0) = nI, s(v_1) = nM, s(v_2) = nM$

Exercise III.5

The three provoked nodes are v_0, v_1, and v_3. The first two strategies from the previous exercise are best strategies because they lead to the maximal payoff 10.

Exercise III.6

The strategy indicated in figure III.3 is not optimal and hence not subtree-perfect. The strategy of figure III.4 is optimal but not subtree-perfect. The best strategy is not chosen in the subtree highlighted in figure III.16.

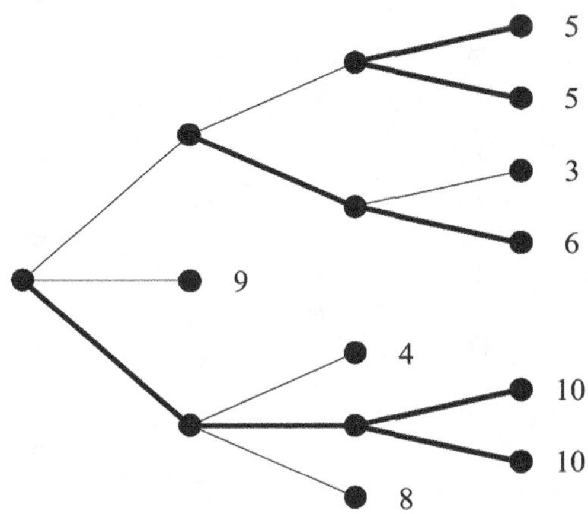

FIGURE III.17. Backward induction

Exercise III.7

Figure III.17 depicts the backward-indcution solution. There are two backward-induction trails. They lead to payoff 10. There are four backward-induction strategies because there are two nodes at which there are two best actions.

Exercise III.8

There are only two strategies compatible with exactly one exchange:

$$\lfloor \text{accept, reject, accept} \rfloor \text{ and}$$
$$\lfloor \text{accept, reject, reject} \rfloor .$$

Exercise III.9

The driver's action choices have to be the same at v_0 and v_1 which are the only decision nodes. If he chooses exit, he obtains 0, while go on yields 1. Therefore, going on is the best strategy. The only decisional subtree is the whole tree itself. Therefore, $\lfloor$go on, go on$\rfloor$ is the subtree-perfect strategy.

Exercise III.10

Since every strategy provokes v_0, we have $\tau_\sigma(v_0) = 1$.

Exercise III.11

In order to be well-defined, the probabilities need to be non-negative and their sum equal to 1. The second requirement is fulfilled by

$$\frac{1}{3} + \frac{1}{6} + 6 \cdot \frac{1}{12} = 1.$$

The probabiliy of reaching v_3 is given by

$$\tau_\sigma(v_3) = \sigma(\lfloor \text{I, M, nM} \rfloor) + \sigma(\lfloor \text{I, M, M} \rfloor) = \frac{1}{3} + \frac{1}{12} = \frac{5}{12}.$$

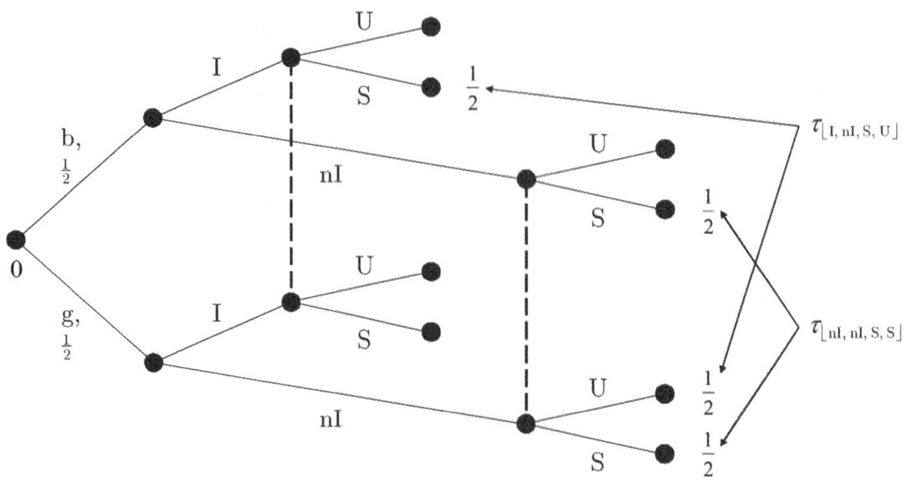

FIGURE III.18. Probabity distributions provoked by strategies

Exercise III.12

Perfect information implies that all information sets are singletons, i.e., $|I(v)| = 1$ for all $v \in D$. Then, there is only one node v in the information set $I(v)$. And the experience at that node is just $X(v)$.

Exercise III.13

We have $I(v_1) = I(v_2)$ but

$$X(v_1) = (\{v_0\}, \text{I}, \{v_1, v_2\}) \neq (\{v_0\}, \text{nI}, \{v_1, v_2\}) = X(v_2).$$

Exercise III.14

Figure III.18 records the two probability distributions on the set of terminal nodes.

Exercise III.15

Both figures reflect imperfect information. Apart from the original trees, there are no further subtrees. In figure III.13, we have imperfect recall, while III.14 reflects perfect recall.

Exercise III.16

The decision maker can not tell $v_{1\text{I}}$ and $v_{1\text{nI}}$ apart, although the experience is different. Therefore, we do not have perfect recall. We have five subtrees, the original tree (where nature moves) and the subtrees originating in v_1, v_2, $v_{2\text{I}}$, and $v_{2\text{nI}}$. Figure III.19 contains the backward-induction solution. Note that there is no subtree at $v_{1\text{I}}$. Every subtree-perfect strategy contains action S at $v_{2\text{I}}$. At v_1, both $\lfloor \text{I, S} \rfloor$ and $\lfloor \text{nI, U} \rfloor$ are optimal. Therefore, in the order of nodes v_1, $v_{1\text{I}}$, $v_{1\text{nI}}$, v_2, $v_{2\text{I}}$, and $v_{2\text{nI}}$, the subtree-perfect

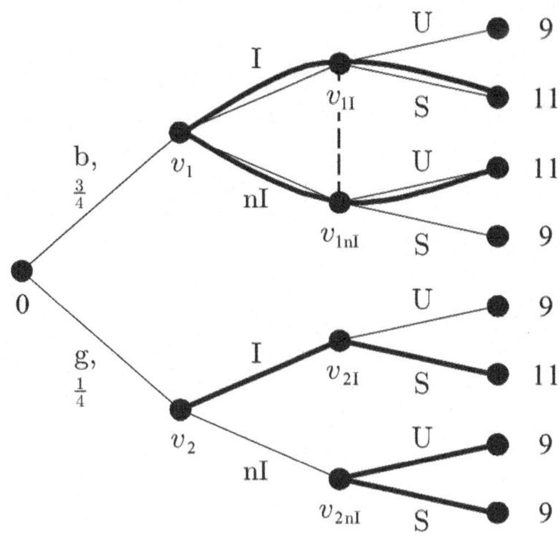

FIGURE III.19. Backward induction!

strategies are

$$\lfloor \text{I, S, S, I, S, U} \rfloor, \ \lfloor \text{I, S, S, I, S, S} \rfloor \text{ (invest at } v_1\text{) and}$$
$$\lfloor \text{nI, U, U, I, S, U} \rfloor, \ \lfloor \text{nI, U, U, I, S, S} \rfloor \text{ (do not invest at } v_1\text{).}$$

8. Further exercises without solutions

PROBLEM III.1. The example of figure III.11 shows that sometimes mixed strategies are more powerful than behavioral strategies. Can you see, why?

PROBLEM III.2. Consider the following decision situation in extensive form:

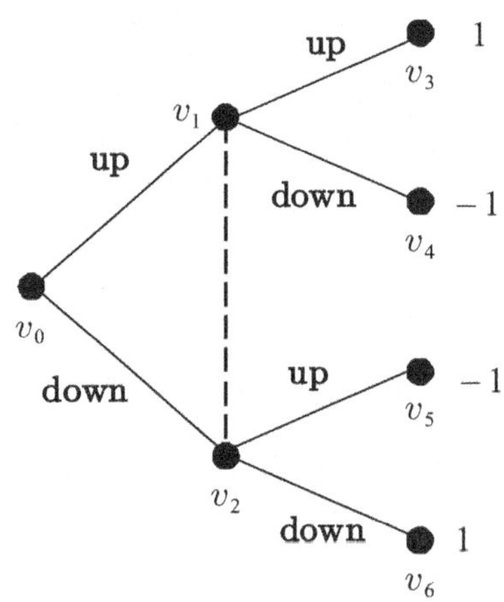

(1) True or false? In this game, any behavioral strategy can be characterized by specifying two probabilities.
(2) Determine the equivalent behavioral strategy for the following mixed strategy! $\sigma(\lfloor \text{up, up} \rfloor) = \frac{1}{5}, \sigma(\lfloor \text{up, down} \rfloor) = \frac{2}{5}, \sigma(\lfloor \text{down, down} \rfloor) = \frac{1}{5}, \sigma(\lfloor \text{down, up} \rfloor) = \frac{1}{5}$.
(3) Is this a game of perfect recall?
(4) Determine the equivalent mixed strategy for the following behavioral strategy! $\beta_{v_0}(\text{up}) = \frac{1}{3}, \beta_{\{v_1,v_2\}}(\text{up}) = \frac{1}{4}$.
(5) Determine the best mixed strategy and the best behavioral strategy!

PROBLEM III.3. Find a decision problem in extensive form with $|V| = |\mathbb{N}|$ that has no subgame-perfect equilibrium.

PROBLEM III.4. Consider the decision situation in extensive form below.

(1) How many subgames can you identify?
(2) How many strategies are there? Which are the best?
(3) Apply backward induction!

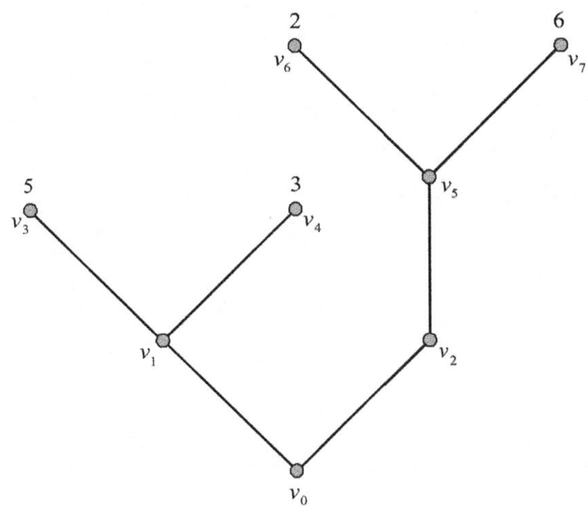
2
6
v_6
v_7
5
3
v_5
v_3
v_4
v_1
v_2
v_0

CHAPTER IV

Ordinal preference theory

Preferences are a relation between objects. Deciders prefer (would rather like to have) one object to another. So far, the preferences of decision makers are depicted by "payoffs". We did not discuss where the payoffs come from and what they mean. We now take a closer look. The objects (over which preferences hold) are bundles of goods that we introduce in section 1. We need to learn some topology for the set of bundles in order to define important preference concepts such as convex or continuous preferences.

The decision maker is often called a household.

1. The vector space of goods and its topology

1.1. The vector space of goods. Our household is confronted with a finite number ℓ of goods. These goods have certain characteristics (for example, apples of a certain weight and class) and are further differentiated according to the region, where they are on offer, and the time interval during which they can be bought. It is also possible to define contingent products, i.e., products distinguished by states of nature. The good in question may be 4 pounds of apple of certain characteristics to be delivered at time t if it does not rain the day before. Payment for such apples is made at point 0. Thus, it may well happen that payments occur but apples cannot be consumed (rain the day before in our example).

Formally, bundles of goods are elements of the vector space $\mathbb{R}^\ell$. The real line depicts the set of real numbers, $\mathbb{R}$. The set of all vectors with two real numbers, $\mathbb{R}^2$, is visualized by the two-dimensional plane, while the set of all vectors with three real numbers, $\mathbb{R}^3$, represents the three-dimensional space. You know how to calculate in a vector space?

EXERCISE IV.1. *Consider the vectors* $x = (x_1, x_2) = (2, 4)$ *and* $y = (y_1, y_2) = (8, 12)$. *Find* $x + y$, $2x$, *and* $\frac{1}{4}x + \frac{3}{4}y$!

$\frac{1}{4}x + \frac{3}{4}y$ is called a convex combination of vectors x and y because the coefficients are non-negative and they sum to 1. It is to be found on the line between x and y. $\frac{1}{4}x + \frac{3}{4}y$ is closer to y because y's coefficient is the highest. An extreme case is $0x + 1y = y$.

H. Wiese, *Advanced Microeconomics*,
https://doi.org/10.1007/978-3-658-34959-2_4

In general (for an arbitrary dimension $\ell \in \mathbb{N}$) we write

$$\mathbb{R}^\ell := \{(x_1, ..., x_\ell) : x_g \in \mathbb{R}, g = 1, ..., \ell\}.$$

$0 \in \mathbb{R}^\ell$ is the null vector $(0, 0, ..., 0)$. The vectors are often called points (in $\mathbb{R}^\ell$).

REMARK IV.1. *For vectors x and y with ℓ entries, we define*

- $x \geq y$ *by* $x_g \geq y_g$ *for all* g *from* $\{1, 2, ..., \ell\}$,
- $x > y$ *by* $x \geq y$ *and* $x \neq y$,
- $x \gg y$ *by* $x_g > y_g$ *for all* g *from* $\{1, 2, ..., \ell\}$.

In household theory, we will work with the goods space

$$\mathbb{R}_+^\ell := \left\{x \in \mathbb{R}^\ell : x \geq 0\right\}$$

rather than $\mathbb{R}^\ell$ (where negative amounts of goods would be possible).

1.2. Distance and balls. We will need some topological concepts and need to learn how to define the distance between points or how to establish that a set is open or closed. While most of our definitions refer to $\mathbb{R}^\ell$, we sometimes use $\mathbb{R}_+^\ell$, the goods space, instead.

DEFINITION IV.1 (DISTANCE). *In $\mathbb{R}^\ell$ the distance between two points x and y is given by the city-block norm (or 1-norm)*

$$\|x - y\|_1 := \sum_{g=1}^{\ell} |x_g - y_g|,$$

by the euclidian (or 2-) norm

$$\|x - y\|_2 := \sqrt{\sum_{g=1}^{\ell} (x_g - y_g)^2}$$

or by the infinity norm

$$\|x - y\|_\infty := \max_{g=1,...,\ell} |x_g - y_g|.$$

Figure IV.1 illustrates these norms.

When we later define certain terms such as open sets, bounded sets, etc., it is not important which of these norms we use. The three norms above are the most common and have a simple geometric interpretation. What is it? We will often just write $\|\cdot\|$ and the reader is free to imagine any of these norms.

EXERCISE IV.2. *What is the distance (in $\mathbb{R}^2$) between $(2, 5)$ and $(7, 1)$, measured by the 2-norm $\|\cdot\|_2$ and by the inifinity norm $\|\cdot\|_\infty$?*

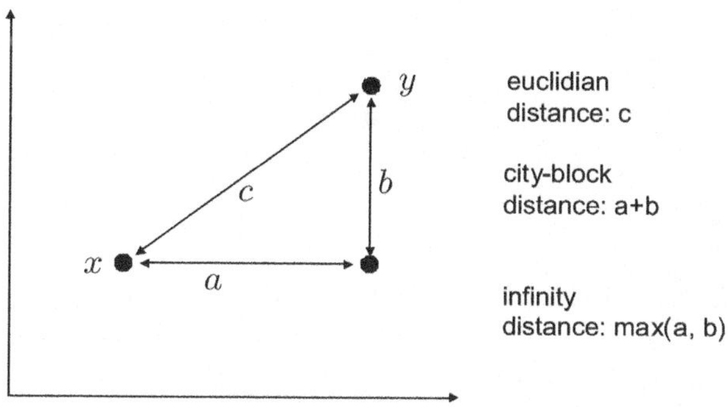

FIGURE IV.1. Different methods to measure distance

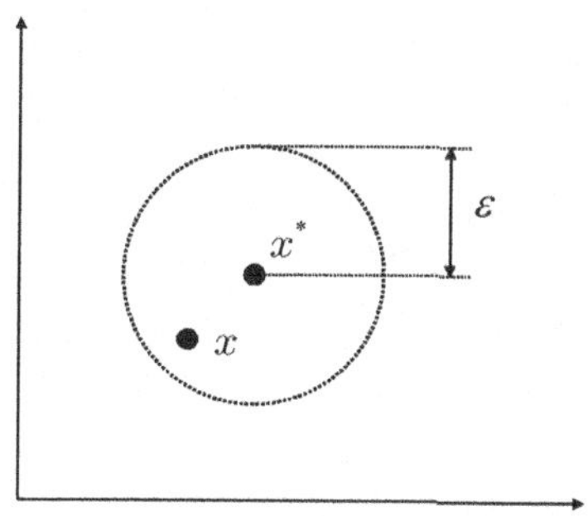

FIGURE IV.2. A point close to the ball's center

DEFINITION IV.2 (BALL). *Let $x^* \in \mathbb{R}^\ell$ and $\varepsilon > 0$.*

$$K = \left\{x \in \mathbb{R}^\ell : \|x - x^*\| < \varepsilon\right\}$$

is called the (open) ε-ball with center x^*. Within the goods space $\mathbb{R}_+^\ell$, the ε-ball with center $x^* \in \mathbb{R}_+^\ell$ is defined by $K = \left\{x \in \mathbb{R}_+^\ell : \|x - x^*\| < \varepsilon\right\}$.

If ε is small, $x \in K$ is "very close to" x^* (see figure IV.2). Note that $\|x - x^*\| = \varepsilon$ holds for all x on the circular line, while K stands for all the points within.

EXERCISE IV.3. *Assuming the goods space $\mathbb{R}_+^2$, sketch three 1-balls with centers $(2,2), (0,0)$, and $(2,0)$, respectively.*

DEFINITION IV.3 (BOUNDEDNESS). *A set M is called bounded if there exists an ε-ball K such that $M \subseteq K$.*

EXAMPLE IV.1. *The set $[0, \infty) = \{x \in \mathbb{R} : x \geq 0\}$ (see p. 15) is not bounded.*

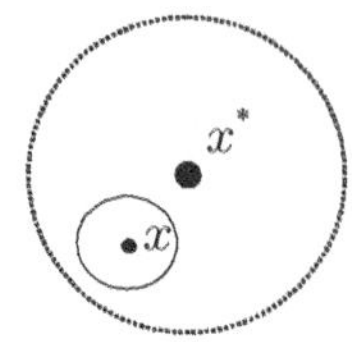

FIGURE IV.3. The open ball is an open set

1.3. Open and closed sets.

DEFINITION IV.4 (INTERIOR POINT). *x^* is called an interior point of some set M if there exists an ε-ball K with center x^* such that $K \subseteq M$.*

EXAMPLE IV.2. *Point 1 is not an interior point of $[0,1]$.*

EXAMPLE IV.3. *x^* is an interior point of the ε-ball K with center x^*.*

DEFINITION IV.5 (OPEN SET). *A set that consists of interior points only is called open. The empty set (symbol $\emptyset$) is open.*

A set M is open if you can take an arbitrary point of this set and find an ε-ball K that is contained in M.

EXAMPLE IV.4. *In $\mathbb{R}^1$, $(0,1) = \{x \in \mathbb{R} : 0 < x < 1\}$ is an open set. $\mathbb{R}^\ell$ is open. Every ε-ball is an open set. You can see this by sketching the ε-ball and by finding a second, smaller one, around every point in the ε-ball (see figure IV.3).*

DEFINITION IV.6 (COMPLEMENT). *$\mathbb{R}^\ell \backslash M = \{x \in \mathbb{R}^\ell : x \notin M\}$ is the complement of M (in $\mathbb{R}^\ell$).*

DEFINITION IV.7 (BOUNDARY POINT). *x^* is called boundary point of a set M if for all ε-balls K with center x^* the following two conditions are fulfilled:*

$$\begin{aligned} K \cap M &\neq \emptyset \text{ and} \\ K \cap \left(\mathbb{R}^\ell \backslash M\right) &\neq \emptyset. \end{aligned}$$

Figure IV.4 illustrates the definition of a boundary point in two-dimensional space.

EXERCISE IV.4. *Find the boundary points of $(0,1) \subset \mathbb{R}$, of $[0,1]$, and of $[0,0]$!*

LEMMA IV.1. *Assume that x^* is contained in M. Then, x^* is an interior point of M if and only if x^* is not a boundary point of M.*

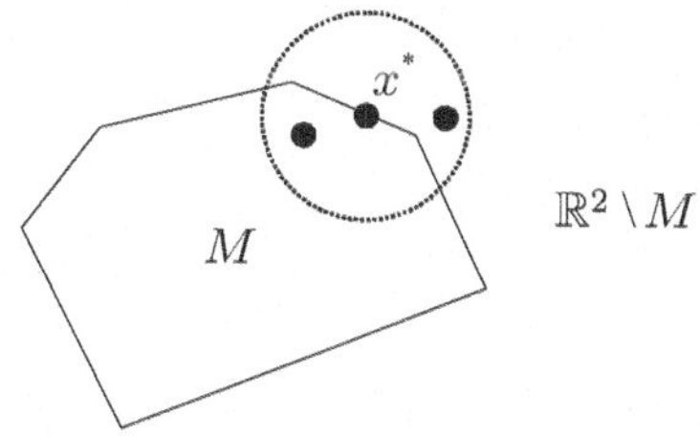

FIGURE IV.4. A boundary point

PROOF. If x^* is an interior point of M, we can find an ε-ball K with center x^* such that $K \subseteq M$. Then, $K \cap (\mathbb{R}^\ell \backslash M) = \emptyset$ and x^* is not a boundary point of M. If x^* is not an interior point of M, every ε-ball K with center x^* fulfills $K \nsubseteq M$ or, differently put, $K \cap (\mathbb{R}^\ell \backslash M) \neq \emptyset$. Also, every ε-balls K with center x^* contains x^*. Therefore, x^* is contained in both K and M so that $K \cap M \neq \emptyset$ holds. Thus, x^* is a boundary point of set M. □

DEFINITION IV.8 (CLOSED SET). *A set containing all its boundary points is called closed.* $\mathbb{R}^\ell$ *itself is also closed.*

EXAMPLE IV.5. *In* $\mathbb{R}^1$, $[0, 1] = \{x \in \mathbb{R} : 0 \leq x \leq 1\}$ *is a closed set.*

LEMMA IV.2. *The complement of an open set is a closed set. The complement of a closed set is an open set.*

It can be the case that a set is neither closed nor open. Consider, for example, the set $\{0\} \cup (1, 2)$. On the other hand, $\mathbb{R}^\ell$ and $\emptyset$ are both open and closed.

DEFINITION IV.9 (COMPACT SET). *A set* $M \subseteq \mathbb{R}^\ell$ *is called compact if it is closed and bounded.*

EXAMPLE IV.6. $[0, 1]$ *is compact.* $\mathbb{R}^\ell$ *is closed but not bounded.* ε*-balls are bounded but not closed. Hence, neither* $\mathbb{R}^\ell$ *nor* ε*-balls are compact.*

1.4. Sequences and convergence. It is often helpful to consider the above concepts from the point of view of sequences in $\mathbb{R}^\ell$.

DEFINITION IV.10 (SEQUENCE). *A sequence* $(x^j)_{j \in \mathbb{N}}$ *in* $\mathbb{R}^\ell$ *is a function* $\mathbb{N} \to \mathbb{R}^\ell$.

EXAMPLE IV.7. $1, 2, 3, 4, ...$ *is a sequence in* $\mathbb{R}$ *that can also be defined by*

$$x^j := j.$$

Examples in $\mathbb{R}^2$ *are*

$$(1,2),(2,3),(3,4),...$$

or

$$\left(1,\frac{1}{2}\right),\left(1,\frac{1}{3}\right),\left(1,\frac{1}{4}\right),\left(1,\frac{1}{5}\right),\ldots.$$

Sometimes, a sequence gets closer and closer to some point:

DEFINITION IV.11 (CONVERGENCE). *A sequence* $(x^j)_{j\in\mathbb{N}}$ *in* $\mathbb{R}^\ell$ *converges towards* $x\in\mathbb{R}^\ell$ *if for every* $\varepsilon>0$ *there is an* $N\in\mathbb{N}$ *such that the distance from* x^j *to* x *is smaller than* ε *for all* $j>N$, *i.e., if*

$$\left\|x^j-x\right\|<\varepsilon \text{ for all } j>N$$

holds. A sequence that converges towards some $x\in\mathbb{R}^\ell$ *is called convergent.*

Convergence towards x means: Given any ε-ball K with center x, nearly all members of the sequence (i.e., all members except a finite number) belong to that ball.

EXAMPLE IV.8. *The sequence*

$$1,2,3,4,...$$

is not convergent towards any $x\in\mathbb{R}$,
while

$$1,1,1,1,...$$

converges towards 1 *and*

$$1,\frac{1}{2},\frac{1}{3},\ldots$$

converges towards zero.

We do not provide a proof for the following lemma, but you can try to confirm it for yourself.

LEMMA IV.3. *Let* $(x^j)_{j\in\mathbb{N}}$ *be a sequence in* $\mathbb{R}^\ell$.

- *If* $(x^j)_{j\in\mathbb{N}}$ *converges towards* x *and* y, *we have* $x=y$.
- $(x^j)_{j\in\mathbb{N}}=\left(x_1^j,...,x_\ell^j\right)_{j\in\mathbb{N}}$ *converges towards* $(x_1,...,x_\ell)$ *if and only if* x_g^j *converges towards* x_g *for every* $g=1,...,\ell$.

EXERCISE IV.5. *Are the sequences*

$$(1,2),(1,3),(1,4),...$$

or

$$\left(1,\frac{1}{2}\right),\left(1,\frac{1}{3}\right),\left(1,\frac{1}{4}\right),\left(1,\frac{1}{5}\right),\ldots.$$

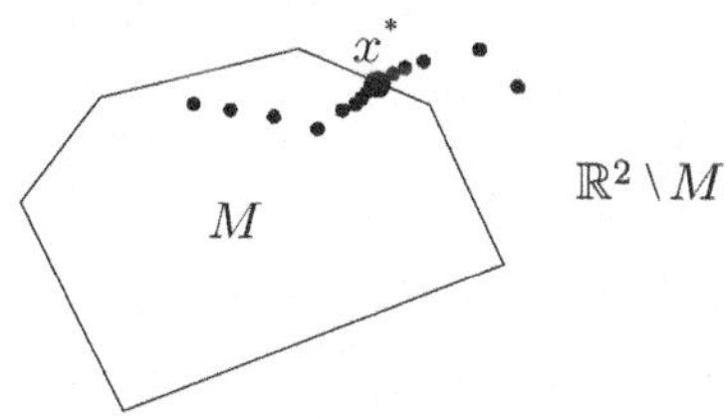

FIGURE IV.5. The sequence definition of a boundary point

convergent?

Now, we can provide alternative definitions for "boundary point" (see figure IV.5) and "closed set".

LEMMA IV.4. *A point $x^* \in \mathbb{R}^\ell$ is a boundary point of $M \subseteq \mathbb{R}^\ell$ if and only if there is a sequence of points in M and another sequence of points in $\mathbb{R}^\ell \setminus M$ so that both converge towards x^*.*

LEMMA IV.5. *A set $M \subseteq \mathbb{R}^\ell$ is closed if and only if every converging sequence in M with convergence point $x \in \mathbb{R}^\ell$ fulfills $x \in M$.*

Differently put, M is not closed if we can find a converging sequence in M with convergence point outside M.

EXERCISE IV.6. *Use the above lemma to show that the sets $\{0\} \cup (1, 2)$ and $\mathbb{R} \setminus [\{0\} \cup (1, 2)]$ are not closed. Can you find out whether they are open with the help of lemma IV.2?*

2. Preference relations

2.1. Relations and equivalence classes. Our aim is to consider relations on the goods space $\mathbb{R}^\ell_+$. However, we begin with three examples from outside preference theory.

EXAMPLE IV.9. *For any two inhabitants from Leipzig, we ask whether*

- *one is the father of the other or*
- *they are of the same sex.*

EXAMPLE IV.10. *For the set of integers $\mathbb{Z}$ (the numbers ..., -2, -1, 0, 1, 2, ...), we consider the difference and examine whether this difference is an even number (i.e., from ..., $-2, 0, 2, 4$, ...).*

All three examples define relations, the first two on the set of the inhabitants from Leipzig, the last on the set of integers. Often, relations are expressed by the symbol $\sim$. To take up the last example on the set of integers, we have $5 \sim -3$ (the difference $5-(-3)=8$ is even) and $5 \nsim 0$ (the difference $5-0=5$ is odd).

DEFINITION IV.12 (RELATION). *A relation on a set* M *is a subset of* $M \times M$. *If a tuple* $(a,b) \in M \times M$ *is an element of this subset, we often write* $a \sim b$.

Relations have, or have not, specific properties:

DEFINITION IV.13 (PROPERTIES OF RELATIONS). *A relation* $\sim$ *on a set* M *is called*

- *reflexive if* $a \sim a$ *holds for all* $a \in M$,
- *transitive if* $a \sim b$ *and* $b \sim c$ *imply* $a \sim c$ *for all* $a,b,c \in M$,
- *symmetric if* $a \sim b$ *implies* $b \sim a$ *for all* $a,b \in M$,
- *antisymmetric if* $a \sim b$ *and* $b \sim a$ *imply* $a = b$ *for all* $a,b \in M$, *and*
- *complete if* $a \sim b$ *or* $b \sim a$ *holds for all* $a,b \in M$.

LEMMA IV.6. *On the set of integers* $\mathbb{Z}$, *the relation* $\sim$ *defined by*

$$a \sim b :\Leftrightarrow a-b \text{ is an even number}$$

is reflexive, transitive, and symmetric, but neither antisymmetric nor complete.

"$:\Leftrightarrow$" means that the expression left of the colon is defined by the expression to the right of the equivalence sign.

PROOF. We have $a-a=0$ for all $a \in \mathbb{Z}$ and hence $a \sim a$; therefore, $\sim$ is reflexive. For transitivity, consider any three integers a,b,c that obey $a \sim b$ and $b \sim c$. Since the sum of two even numbers is even, we find that

$$\begin{aligned} & (a-b)+(b-c) \\ = \;& a-c \end{aligned}$$

is also even. This proves $a \sim c$ and concludes the proof of transitivity. Symmetry follows from the fact that a number is even if and only if its negative is even.

$\sim$ is not complete which can be seen from $0 \nsim 1$ and $1 \nsim 0$. Finally, $\sim$ is not antisymmetric. Just consider the numbers 0 and 2. □

EXERCISE IV.7. *Which properties do the relations "is the father of" and "is of the same sex as" have? Fill in "yes" or "no":*

property	*is the father of*	*is of the same sex as*
reflexive		
transitive		
symmetric		
antisymmetric		
complete		

DEFINITION IV.14 (EQUIVALENCE RELATION). *Let $\sim$ be a relation on a set M which obeys reflexivity, transitivity, and symmetry. Then, any two elements $a, b \in M$ with $a \sim b$ are called equivalent and $\sim$ is called an equivalence relation. By an equivalence class of $a \in M$, we mean the set*

$$[a] := \{b \in M : b \sim a\}.$$

Our relation on the set of integers (even difference) is an equivalence relation. We have two equivalence classes:

$$\begin{aligned} [0] &= \{b \in M : b \sim 0\} = \{..., -2, 0, 2, 4, ...\} \text{ and} \\ [1] &= \{b \in M : b \sim 1\} = \{..., -3, -1, 1, 3, ...\} \end{aligned}$$

EXERCISE IV.8. *Continuing the above example, find the equivalence classes* $[17]$, $[-23]$, *and* $[100]$. *Reconsider the relation "is of the same sex as". Can you describe its equivalence classes?*

Generalizing the above example, $a \sim b$ implies $[a] = [b]$ for every equivalence relation. Here comes the proof. Consider any $a' \in [a]$. We need to show $a' \in [b]$. Now, $a' \in [a]$ means $a' \sim a$. Together with $a \sim b$, transitivity implies $a' \sim b$ and hence $a' \in [b]$. We have shown $[a] \subseteq [b]$. The converse, $[b] \subseteq [a]$, can be shown similarly.

The following lemma uses the above result and the observation $a \in [a]$ which is true by reflexivity.

LEMMA IV.7. *Let $\sim$ be an equivalence relation on a set M. Then, we have*

$$\begin{aligned} \bigcup_{a \in M} [a] &= M \text{ and} \\ [a] &\neq [b] \Rightarrow [a] \cap [b] = \emptyset. \end{aligned}$$

Thus, equivalence classes form a partition of the underlying set.

The other direction holds also: Once we have a partition, we can define an equivalence relation whose equivalence classes are equal to the components of the partition. Just say that two elements are related if they belong to the same component.

2.2. Preference relations and indifference curves. We now assume that every household i has weak preferences (a weak preference relation) on the goods space $\mathbb{R}_+^\ell$, denoted by $\precsim^i$. $x \precsim^i y$ means that household i finds y at least as good as x. If there is no doubt about the household we are talking about, we omit the index.

DEFINITION IV.15 (PREFERENCE RELATION). *A (weak) preference relation $\precsim$ is a relation on $\mathbb{R}_+^\ell$ that is complete, transitive, and reflexive. Given a preference relation $\precsim$, the indifference relation is defined by*

$$x \sim^i y :\Leftrightarrow x \precsim^i y \text{ and } y \precsim^i x$$

and the strict preference by

$$x \prec^i y :\Leftrightarrow x \precsim^i y \text{ and not } y \precsim^i x.$$

While it is hard to imagine preferences without reflexivity, completeness and transitivity are not as innocent as they seem. Completeness means that households can always make up their mind. However, "real" households will sometimes have a hard time to find out what they "really" want. Also, if confronted with many different, but similar bundles, people will often violate transitivity. We discuss the money-pump argument against the violation of transitivity in chapter III, pp. 40.

EXERCISE IV.9. *Is the indifference relation a preference relation or an equivalence relation? How about the strict preference relation? Fill in:*

property	*indifference*	*strict preference*
reflexive		
transitive		
symmetric		
complete		

DEFINITION IV.16 (BETTER SET, INDIFFERENCE SET). *Let $\succsim$ be a preference relation on $\mathbb{R}_+^\ell$. The better set B_y of y is given by*

$$B_y := \left\{x \in \mathbb{R}_+^\ell : x \succsim y\right\}.$$

The worse set W_y of y is

$$W_y := \left\{x \in \mathbb{R}_+^\ell : x \precsim y\right\}.$$

y's indifference set I_y is the intersection of its better and worse set:

$$I_y := B_y \cap W_y = \left\{x \in \mathbb{R}_+^\ell : x \sim y\right\}$$

The geometric locus of an indifference set is called an indifference curve.

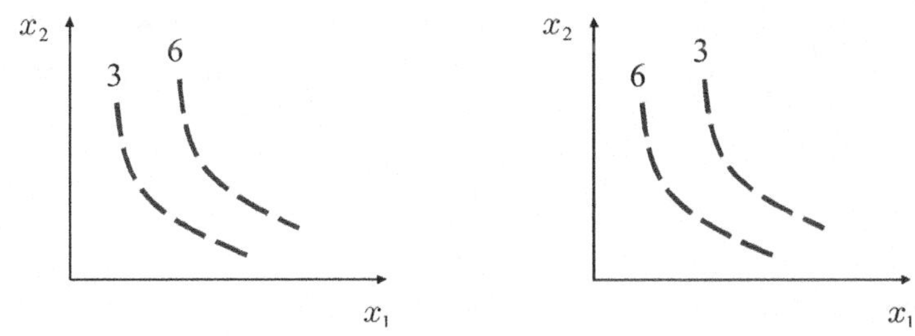

FIGURE IV.6. Numbers associated with indifference curves

The set of indifference sets partitions the goods space. This means that every bundle belongs to one and only one indifference curve. You know from intermediate microeconomics that indifference curves cannot intersect. This follows from the fact that indifference relations are equivalence relations and, in particular, from $[a] \neq [b] \Rightarrow [a] \cap [b] = \emptyset$ in lemma IV.7.

When we draw indifference curves, we often associate them with numbers where a higher number indicates strict preference. Consider figure IV.6. The left-hand graph stands for preferences of so-called goods where the consumer would like to have more of both goods. The right-hand graph represents so-called bads, i.e., the consumer wants as small an amount of both goods as possible. Think of dirt and noise.

EXERCISE IV.10. *Sketch indifference curves for a goods space with just 2 goods and, alternatively,*

- *good* 2 *is a bad (the consumer would like to have less of that good),*
- *good* 1 *represents red matches and good* 2 *blue matches,*
- *good* 1 *stands for right shoes and good* 2 *for left shoes.*

Lexicographic preferences $\precsim_{lex}$ are very interesting preferences. In the two-good case they are defined by

$$x \precsim_{lex} y :\Leftrightarrow x_1 < y_1 \text{ or } (x_1 = y_1 \text{ and } x_2 \leq y_2).$$

Thus, one good (good 1 in our example) is the "most important" good and households look at the amount of this good first when they decide between bundles.

EXERCISE IV.11. *What do the indifference curves for lexicographic preferences look like?*

3. Axioms: convexity, monotonicity, and continuity

3.1. Convex preferences. We will often assume convexity of preferences and monotonicity. We will first need some math.

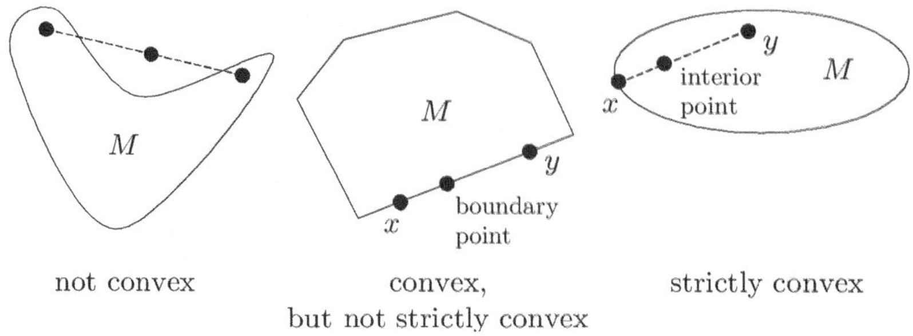

FIGURE IV.7. Convex sets?

DEFINITION IV.17 (CONVEX COMBINATION). *Let x and y be elements of $\mathbb{R}^\ell$. Then,*

$$kx + (1-k)\,y, k \in [0,1]$$

is called the convex combination of x and y.

We have seen a convex combination before, in exercise IV.1 (p. 59). The convex combination of x and y lies on the line connecting x and y. The smaller k, the closer the convex combination to y, which is also clear from

$$kx + (1-k)\,y = y + k\,(x-y)\,.$$

DEFINITION IV.18 (CONVEX SET). *A set $M \subseteq \mathbb{R}^\ell$ is called convex if for any two points x and y from M, their convex combinations are also contained in M.*

Here, convexity is a property of sets and is not to be confused with the convexity of functions. In figure IV.7, the left-hand example shows a set that is not convex, while the other two sets exhibit convexity.

EXERCISE IV.12. *Show that the intersection of two convex sets is also convex.*

DEFINITION IV.19 (STRICTLY CONVEX SET). *A set M is called strictly convex if for any two points x and y from M, $x \neq y$,*

$$kx + (1-k)\,y$$

is an interior point of M for any $k \in (0,1)$.

The right-most set in figure IV.7 is strictly convex, while the middle set is convex but not strictly so. Convince yourself that open ε-balls in $\mathbb{R}^2$ are strictly convex for any norm. How about closed ε-balls?

EXERCISE IV.13. *Are the intervals $(0,\infty)$, $[0,3]$, or $[0,\infty)$ convex or strictly convex?*

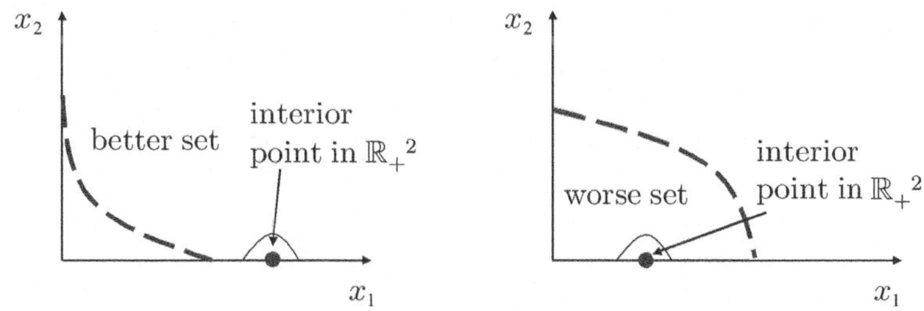

FIGURE IV.8. Two interior points

DEFINITION IV.20 (CONVEX PREFERENCE RELATION). *A preference relation* $\succsim$ *on* $\mathbb{R}_+^\ell$ *is*

- *convex if all its better sets* B_y *are convex,*
- *strictly convex if all its better sets* B_y *are strictly convex,*
- *concave if all its worse sets* W_y *are convex,*
- *strictly concave if all its worse sets* W_y *are strictly convex.*

A rough description of convexity is "mixtures are prefered to extremes". I, for example, would rather have 1 glass of milk and one donut than two glasses of milk or two donuts.

For many preference relations, we have $B_0 = \mathbb{R}_+^\ell$ (where $0 \in \mathbb{R}_+^\ell$ is the null vector). For the goods space $\mathbb{R}_+^\ell$, every point x is an interior point of $\mathbb{R}_+^\ell$. This is no contradiction to our definition of a ball. Consider figure IV.8 where you see two interior points, one within a better set and one within a worse set. The left-hand indifference curve indicates strictly convex preferences, while the right-hand indifference curve points to strictly concave preferences.

EXERCISE IV.14. *Are the preferences depicted in figure IV.9 convex or strictly convex?*

3.2. Monotonicity of preferences. We now have a closer look at monotonicity of preferences. Broadly speaking, monotonicity means "more is better". It comes in three different forms:

DEFINITION IV.21 (MONOTONICITY). *A preference relation* $\succsim$ *obeys*

- *weak monotonicity if* $x > y$ *implies* $x \succsim y$,
- *strict monotonicity if* $x > y$ *implies* $x \succ y$, *and*
- *local non-satiation at* y *if in every* ε*-ball with center* y *a bundle* x *with* $x \succ y$ *can be found.*

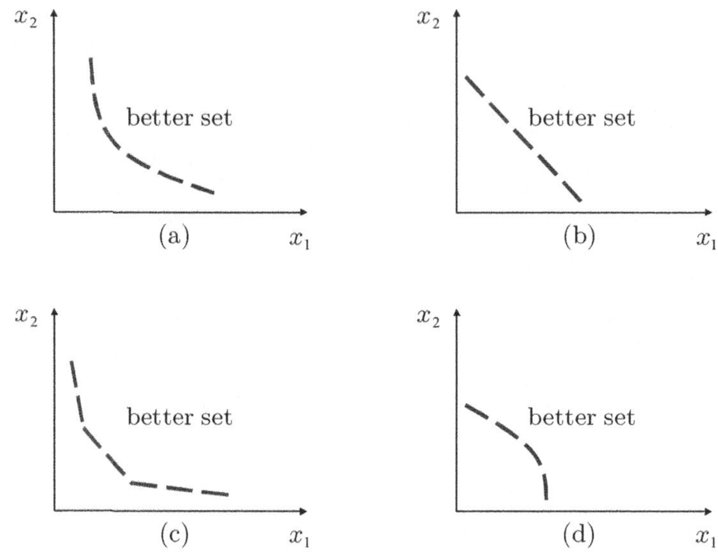

FIGURE IV.9. Convex or strictly convex preferences?

Weak monotonicity excludes the possibility of having "too much of a good thing". Alternatively, it can be seen as the option to throw away unwanted items – this property is often referred to as free disposal. Strict monotonicity is stronger and implies that the agent is strictly better off if he has more of one good and less of no good. Strict monotonicity also implies local non-satiation.

EXERCISE IV.15. *Sketch the better set of $y = (y_1, y_2)$ in case of weak monotonicity!*

Sometimes, consumers prefer a limited amount of goods to more or to less of that good. For example, how many tables would you want in your appartment if you could get them for free? The existence of a so-called bliss point (see figure IV.10) violates all three definitions of monotonicity.

3.3. Continuous preferences. Finally, we consider the property of continuity.

DEFINITION IV.22 (CONTINUOUS PREFERENCES). *A preference relation $\precsim$ is called continuous if for all $y \in \mathbb{R}_+^\ell$ the sets*

$$W_y = \left\{x \in \mathbb{R}_+^\ell : x \precsim y\right\}$$

and

$$B_y = \left\{x \in \mathbb{R}_+^\ell : y \precsim x\right\}$$

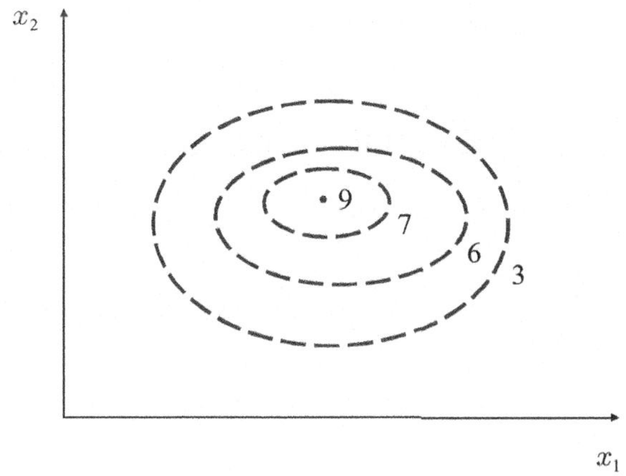

FIGURE IV.10. Bliss point

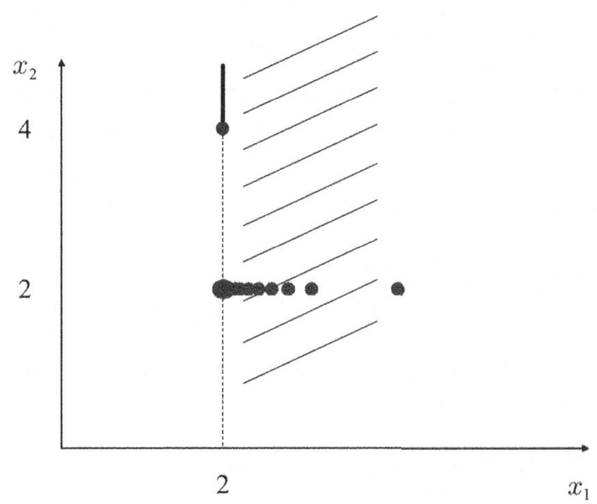

FIGURE IV.11. A better set for lexicographic preferences

are closed.

Lexicographic preferences are not continuous. Have a look at figure IV.11. The hatched area is the "better-set" of point $(2,4)$, i.e., all bundles weakly preferred to $(2,4)$. All points (x_1,x_2) obeying $x_1 > 2$ belong to this better-set and also all bundles with $x_1 = 2$ and $x_2 \geq 4$.

Now, take point $(2,2)$ which is a boundary point of the better-set of $(2,4)$. However, $(2,2)$ does not belong to the better-set. Therefore, the better-set of $(2,4)$ is not closed and lexicographic preferences are not continous. The figure suggests another manner of showing that $B_{(2,4)}$ is not closed.

4. Utility functions

4.1. Definition. Utility functions are used to describe (or "represent") preferences:

DEFINITION IV.23 (UTILITY FUNCTION). *For an agent $i \in N$ with preference relation $\succsim^i$*

$$U^i : \mathbb{R}_+^\ell \mapsto \mathbb{R}$$

is called a utility function if, for all $x, y \in \mathbb{R}_+^\ell$,

$$U^i(x) \geq U^i(y) \Leftrightarrow x \succsim^i y$$

holds. We then say that U^i represents the preferences $\succsim^i$.

This definition implies that we subscribe to ordinal utility theory. That is, the utility function is only used to rank bundles. Therefore, the claims "the utility of bundle x is twice as high as the utility of y" and "the utiliy of bundle x is higher than the utility of y" both express the preference $x \succ y$ and nothing more.

It is natural to ask two questions:

- Existence: Can we find a representation of all sorts of preferences?
- Uniqueness: Can there be several representations?

4.2. Examples. We present a few prominent examples of utility functions.

- Cobb-Douglas utility functions are given by $U(x_1, x_2) = x_1^a x_2^{1-a}$ with $0 < a < 1$. They are weakly monotonic but not strictly so. (Why?) We postpone the question of convexity until after the introduction of the marginal rate of substitution.
- Goods 1 and 2 are called perfect substitutes (the red and blue matches) if the utility function is given by $U(x_1, x_2) = ax_1 + bx_2$ with $a > 0$ and $b > 0$. Draw the indifference curve for $a = 1, b = 4$ and the utility level 5!
- Perfect complements (left and right shoes) are described by utility functions such as $U(x_1, x_2) = \min(ax_1, bx_2)$ with $a > 0$ and $b > 0$. Draw the indifference curve for $a = 1, b = 4$ (a car with four wheels and one engine) and the utility level 5! Does x_1 denote the number of wheels or the number of engines?

4.3. Strictly increasing transformations (uniqueness). We start with the uniqueness question. If a utility function U represents preferences $\precsim$, we can easily find other utility functions standing for the same preferences.

DEFINITION IV.24 (EQUIVALENT UTILITY FUNCTIONS). *Two utility functions U and V are called equivalent if they represent the same preferences.*

Two utility functions U and V represent the same preferences $\precsim$ if there is a strictly increasing function $\tau : \mathbb{R} \to \mathbb{R}$ such that $V = \tau \circ U$ (i.e., V is the composition of τ and U). The proof rests on the equivalence

$$\begin{aligned} & U(x) \geq U(y) \\ \Leftrightarrow \quad & V(x) = \tau(U(x)) \geq \tau(U(y)) = V(y). \end{aligned}$$

For example, multiplying U by 2 or subtracting -17 keeps the ordering intact.

LEMMA IV.8 (EQUIVALENT UTILITY FUNCTIONS). *Two utility functions U and V are called equivalent if there is a strictly increasing function τ : $\mathbb{R} \to \mathbb{R}$ such that $V = \tau \circ U$.*

Importantly, the equivalence that may hold between utility functions is an equivalence relation. If you like, use the above lemma and check for reflexivity, transitivity, and symmetry.

EXERCISE IV.16. *Which of the following utility functions represent the same preferences? Why?*
a) $U_1(x_1, x_2, x_3) = (x_1 + 1)(x_2 + 1)(x_3 + 1)$
b) $U_2(x_1, x_2, x_3) = \ln(x_1 + 1) + \ln(x_2 + 1) + \ln(x_3 + 1)$
c) $U_3(x_1, x_2, x_3) = -(x_1 + 1)(x_2 + 1)(x_3 + 1)$
d) $U_4(x_1, x_2, x_3) = -[(x_1 + 1)(x_2 + 1)(x_3 + 1)]^{-1}$
e) $U_5(x_1, x_2, x_3) = x_1 x_2 x_3$

4.4. Existence.

4.4.1. *Existence is not guaranteed.* The existence of a utility function is not always guaranteed. In particular, lexicographic preferences cannot be represented by a utility function. To see this, consider figure IV.12. We assume that good 1 is the important good. On the x_1-axis, we have real numbers r', r'', and r'''. If (!!) a utility function for lexicographic preferences exists, we have

$$U(A') < U(B') < U(A'') < U(B'') < U(A''') < U(B''')$$

where these utilities are real numbers, not necessarily rational numbers. However, we can be sure to find at least one (indeed, an infinity of) rational

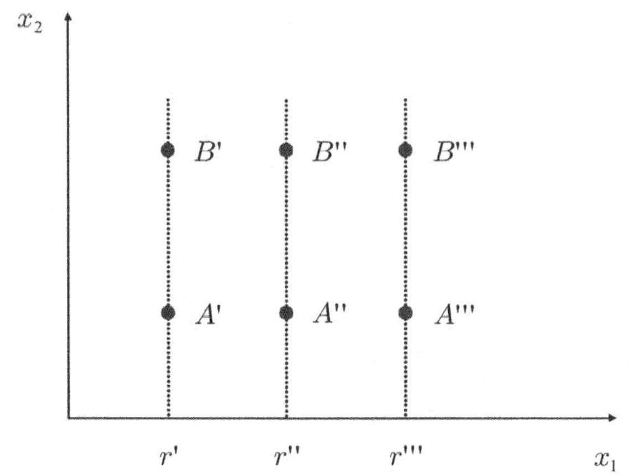

FIGURE IV.12. No utility function for lexicographic preferences

number(s) within the nonempty interval $(U(A'), U(B'))$ which we denote by q'. Similarly, the rational numbers q'' and q''' can be found in the intervals above r'' and r''', respectively. By the above chain of inequalities, we have $q' < q'' < q'''$. In particular, all the rational numbers picked out in this manner are pairwise different. In this fashion, we can define an injective function

$$f : [r', r'''] \to Q$$

that associates with every real number in the interval $[r', r''']$ a rational number. However, by theorem II.1, there are simply not enough rational numbers to make such a function possible. This is the desired contradiction.

We cannot find a utility function for lexicographic preferences because they are not continuous. This is clear from the theorem to be presented shortly.

4.4.2. *Existence of a continuous utility function.* Continuous preferences do not only guarantee the existence of a utility function. We can also be sure that the utility function is continuous itself. We will explain the meaning of a continuous function in a minute. The proof of the following theorem can be found in Debreu (1959, pp. 56).

THEOREM IV.1. *If the preference relation $\precsim^i$ of an agent i is continuous, there is a continuous utility function U^i that represents $\precsim^i$.*

We sketch the idea of a proof. Compare figure IV.13. Consider any bundle $x = (x_1, ..., x_\ell)$ and the indifference curve I_x through x. Let us assume a bundle $x_{45^\circ} := (\bar{x}_1, ..., \bar{x}_\ell)$ such that all the entries are the same and such that $x \sim x_{45^\circ}$ holds. Thus, x_{45° is a bundle on the $45°$-line that

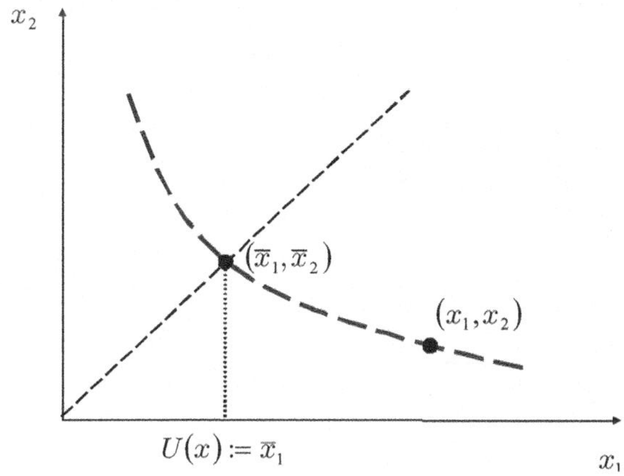

FIGURE IV.13. How to construct a utility curve

is indifferent to x. We can be sure of its existence if $\precsim^i$ is monotonic and continuous. (This claim would need to be shown in a formal manner.) Anyway, if such a bundle exists for every $x \in \mathbb{R}^\ell_+$, we can specify a utility function U by

$$U(x) := \bar{x}_1.$$

EXERCISE IV.17. *Assume a utility function U that represents the preference relation $\precsim$. Can you express weak monotonicity, strict monotonicity, and local non-satiation of $\precsim$ (see definition IV.21, p. 71) through U rather than $\precsim$?*

Belatedly, we now supply the definition of a continuous function. Intuitively, a function f from X to Y is continuous at $x \in X$ if the distance between $f(x)$ and $f(x')$ can be made arbitrarily small (smaller than some given ε) by choosing x' from some δ-ball with center x with a sufficiently small δ.

DEFINITION IV.25 (CONTINUOUS FUNCTION). *Let f be a function $\mathbb{R}^\ell \to \mathbb{R}$. f is called continuous at $x \in \mathbb{R}^\ell$ if for all $\varepsilon > 0$ there is a $\delta > 0$ such that*

$$\left\| f(x) - f(x') \right\| < \varepsilon$$

for every x' obeying

$$\left\| x - x' \right\| < \delta.$$

f is called continuous if it is continuous at every point in $\mathbb{R}^\ell$.

Thus, in order to check continuity at x, we first fix $\varepsilon > 0$ and then find a suitable δ which will in general depend on both ε and x. If x' is

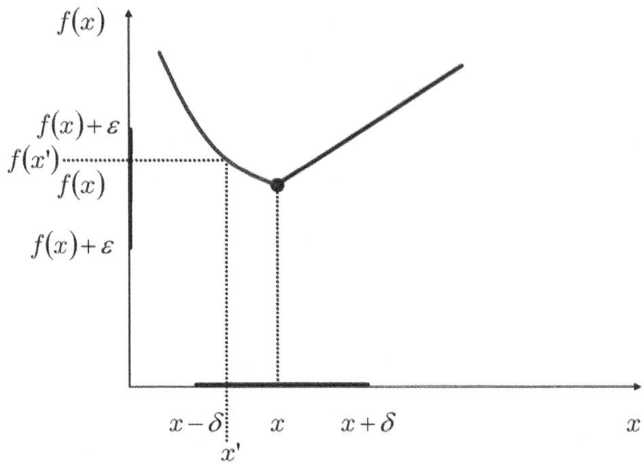

FIGURE IV.14. Continuity at x

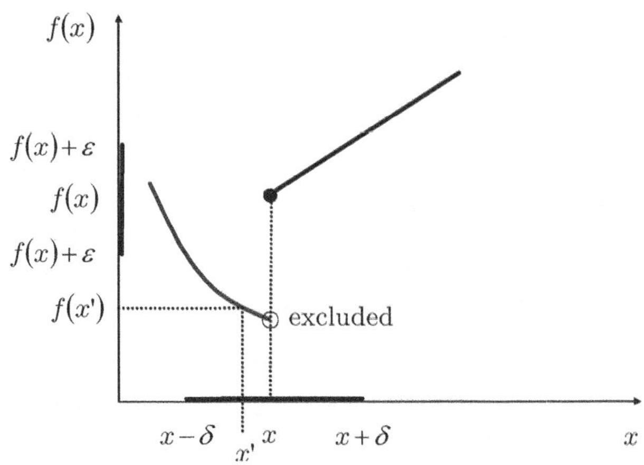

FIGURE IV.15. Discontinuity at x

sufficiently close to x (in terms of δ), $f(x')$ will be near $f(x)$ (defined by ε). Consider figure IV.14 where we have continuity at x. Indeed, δ is sufficiently small so that all the x' in the δ-ball with center x (the open interval $(x-\delta, x+\delta)$) have values $f(x')$ in the ε-ball with center $f(x)$ (the open interval $(f(x)-\varepsilon, f(x)+\varepsilon)$).

Discontinuity can be seen from figure IV.15. For the given ε on the y-axis, the chosen δ on the x-axis or any other delta is not sufficient to guarantee that x' from the δ-interval around x produces a value $f(x')$ inside the ε-interval around $f(x)$.

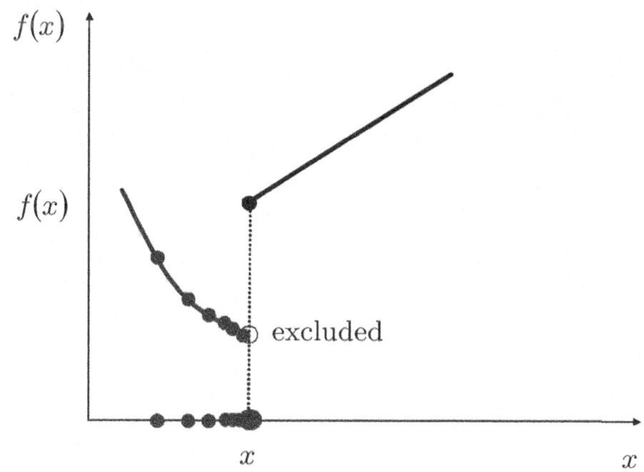

FIGURE IV.16. Discontinuity at x, once more

A utility function U^i is continuous if the difference between two utility levels can be made arbitrarily small by applying U^i to two bundles that are sufficiently close to each other. In other words, there is no jump at x. If we jump at x, we cannot guarantee (by getting very close to x) to get the distance $\|f(x) - f(x')\|$ closer than the jump's height.

Sometimes, you might prefer to use another criterion for continuity which builds on sequences:

THEOREM IV.2. *$f : \mathbb{R}^\ell \to \mathbb{R}$ is continuous at $x \in \mathbb{R}^\ell$ if and only if, for every sequence $\left(x^j\right)_{j\in\mathbb{N}}$ in $\mathbb{R}^\ell$ that converges towards x, the corresponding sequence in the range, $\left(f\left(x^j\right)\right)_{j\in\mathbb{N}}$, converges towards $f(x)$.*

Consider figure IV.16. Can you see that we obtain discontinuity according to this second criterion, also? The sequence of points on the x-axis (domain) converges towards x. The sequence on the graph above (range) does not converge towards $f(x)$.

5. Quasi-concave utility functions and convex preferences

The convexity of preferences is equivalent to a property of utility functions called quasi-concavity:

DEFINITION IV.26 (QUASI-CONCAVITY). *$f : \mathbb{R}^\ell \to \mathbb{R}$ is called quasi-concave if*

$$f(kx + (1-k)y) \geq \min(f(x), f(y))$$

holds for all x, $y \in \mathbb{R}^\ell$ and all $k \in [0,1]$. f is strictly quasi-concave if

$$f(kx + (1-k)y) > \min(f(x), f(y))$$

holds for all x, $y \in \mathbb{R}^\ell$ *with* $x \neq y$ *and all* $k \in (0, 1)$.

Quasi-concavity means that the value of the function (the utility, if f is a utility function) at a point between x and y is at least as high as the lowest of the values $f(x)$ and $f(y)$. You see examples of quasi-concavity in figure IV.17, while figure IV.18 depicts a function that is not quasi-concave.

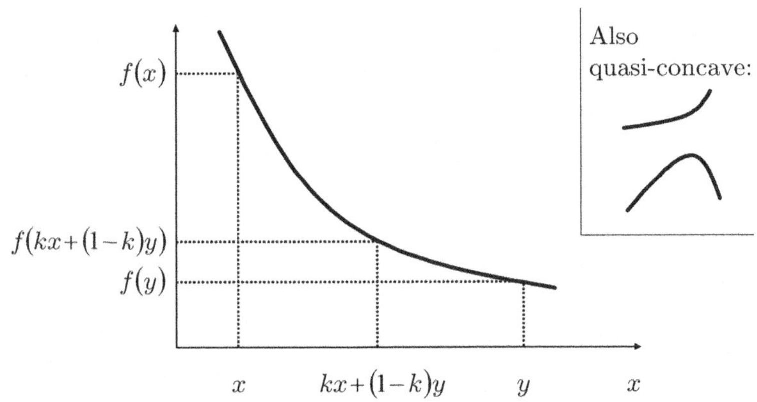

FIGURE IV.17. Strictly quasi-concave functions

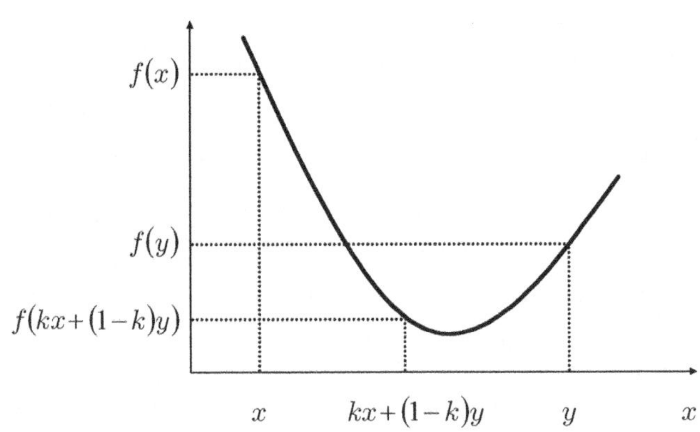

FIGURE IV.18. A function that is not quasi-concave

EXAMPLE IV.11. *Any monotonically increasing or decreasing function* f : $\mathbb{R} \to \mathbb{R}$ *is quasi-concave.*

Better and worse sets and indifference sets are definable for utility functions in the obvious manner.

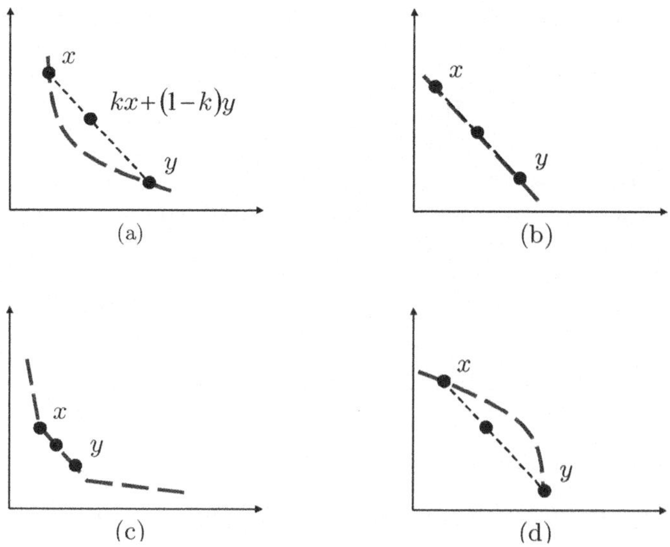

FIGURE IV.19. Concave or strictly concave indifference curves?

DEFINITION IV.27 (BETTER SET). *Let U be a utility function on $\mathbb{R}_+^\ell$. Then, we have the better set B_y of y :*

$$B_{U(y)} := B_y = \left\{x \in \mathbb{R}_+^\ell : U(x) \geq U(y)\right\},$$

the worse set W_y of y :

$$W_{U(y)} := W_y = \left\{x \in \mathbb{R}_+^\ell : U(x) \leq U(y)\right\},$$

and y's indifference set (indifference curve) I_y :

$$I_{U(y)} := I_y = B_y \cap W_y = \left\{x \in \mathbb{R}_+^\ell : U(x) = U(y)\right\}$$

DEFINITION IV.28 (CONCAVE INDIFFERENCE CURVE). *Let U be a utility function on $\mathbb{R}_+^\ell$. The indifference curve I_y is called concave if $U(x) = U(y)$ implies*

$$U(kx + (1-k)y) \geq U(x)$$

for all $x, y \in \mathbb{R}_+^\ell$ and all $k \in [0,1]$. I_y is strictly concave if $U(x) = U(y)$ implies

$$U(kx + (1-k)y) > U(x)$$

for all $x, y \in \mathbb{R}_+^\ell$ with $x \neq y$ and all $k \in (0,1)$.

Figure IV.19 presents four indifference curves where strict concavity holds only in subfigure (a). (d) is an example of a non-concave indifference curve.

U's better sets strictly convex	$\Leftarrow$	U strictly quasi-concave	$\Rightarrow$	U quasi-concave	$\Leftrightarrow$	U's better sets convex
		$\Updownarrow$		$\Updownarrow$		
U's better sets strictly convex and local nonsatiation	$\Rightarrow$	U's indifference curves strictly concave	$\Rightarrow$	U's indifference curves concave		

FIGURE IV.20. Concavity and convexity

We note without proof:

LEMMA IV.9. *Let U be a continuous utility function on $\mathbb{R}_+^\ell$. A preference relation $\succsim$ is convex (in the sense of convex better sets, see definition IV.20) if and only if*

- *all the indifference curves are concave, or*
- *U is quasi-concave.*

These non-strict results and strict results are summarized in figure IV.20.

6. Marginal rate of substitution

6.1. Mathematics: some differentiation rules. We assume that the reader is familiar with the usual differentiation rules for a function $f : M \rightarrow \mathbb{R}$ with open set $M \subseteq \mathbb{R}$ (product rule, quotient rule, chain rule). If M is a subset of $\mathbb{R}^\ell$, the partial derivative of f with respect to x_i is again a real valued function, but the variables other than x_i are held constant. Partial derivates are denoted by

$$\frac{\partial f}{\partial x_i} \text{ rather than } \frac{df}{dx_i}.$$

We can again apply the simple differentiation rules.

We remind the reader of the chain rule of differentiation (which applies to compositions $f \circ g$ of functions f and g):

$$\frac{d(f \circ g)}{dx} = \frac{df}{dg}\frac{dg}{dx}$$

However, differentiation for $M \subseteq \mathbb{R}^\ell$ is not equivalent to partial differentiation. For example, it can happen that a function has partial derivatives everywhere but that the function itself is not differentiable (in a sense not defined by us in this text). We ignore these ugly possibilities and use a definition that will do for all practical purposes:

DEFINITION IV.29 (DIFFERENTIABILITY). *Let $f : M \to \mathbb{R}$ be a real-valued function with open domain $M \subseteq \mathbb{R}^\ell$. f is called differentiable if all the partial derivatives*

$$f_i := \frac{\partial f}{\partial x_i} \ (i = 1, ..., \ell)$$

exist and are continuous. In that case, the column vector

$$f'(x) := \begin{pmatrix} f_1(x) \\ f_2(x) \\ \\ \\ \\ f_\ell(x) \end{pmatrix}$$

is called f's derivative at x.

THEOREM IV.3 (ADDING RULE). *Let $f : \mathbb{R}^\ell \to \mathbb{R}$ be a differentiable function and let $g_1, ..., g_\ell$ be differentiable functions $\mathbb{R} \to \mathbb{R}$. Let $F : \mathbb{R} \to \mathbb{R}$ be defined by*

$$F(x) = f(g_1(x), ..., g_\ell(x)).$$

Then we have

$$\frac{dF}{dx} = \sum_{i=1}^{\ell} \frac{\partial f}{\partial g_i} \frac{dg_i}{dx}.$$

6.2. Economics: the marginal rate of substitution. Consider two goods 1 and 2 (if other goods are present, hold them constant). A bundle $y = (y_1, y_2)$ defines an indifference curve

$$I_y = \left\{(x_1, x_2) \in \mathbb{R}^2_+ : (x_1, x_2) \sim (y_1, y_2)\right\}.$$

We now consider an amount x_1 of good 1 and look for the amount x_2 of good 2 such that (x_1, x_2) is contained in I_y (see figure IV.21 for an illustration). In this manner, we can sometimes define a function

$$I_y : x_1 \mapsto x_2.$$

Note that we have used the symbol I_y in two different ways, as a subset of the goods space and as a function. We do this to economize on symbols and also to make clear that the function I_y is closely related to the indifference curve.

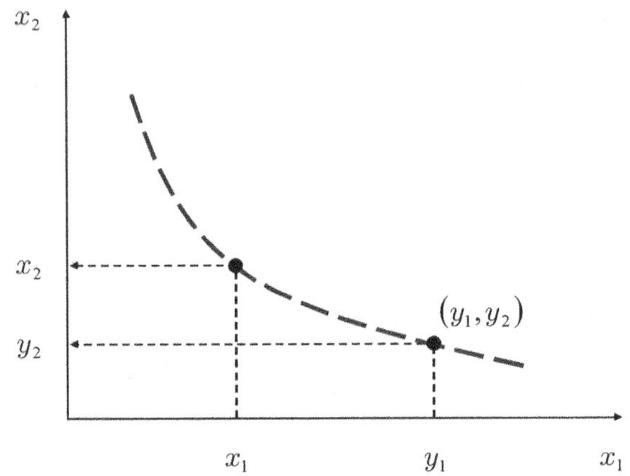

FIGURE IV.21. An implicit function

DEFINITION IV.30 (MARGINAL RATE OF SUBSTITUTION). *If the function I_y is differentiable and if preferences are monotonic, we call*

$$MRS = \left| \frac{dI_y(x_1)}{dx_1} \right|$$

the marginal rate of substitution between good 1 and good 2 (or of good 2 for good 1).

We can readily interpret the marginal rate of substitution: if one additional unit of good 1 is consumed while good 2's consumption is reduced by MRS units, the consumer stays indifferent. We could also say: MRS measures the willingness to pay for one additional unit of good 1 in terms of good 2.

REMARK IV.2. *Note that the above definition "does not work" if one of the goods is a bad. In that case, consuming more of good 1 (nice music) leaves the consumer indifferent if he endures more of good 2 (filthy smoky air). However, since we deal with goods (in the sense of monotonic preferences) most of the time, there is no harm in that definition. Of course, if we are interested in the slope of the indifference curve, we can simply calculate $\frac{dI_y(x_1)}{dx_1}$. In order to avoid tedious repetitions, we will not always point to the fact that we have monotonic preferences.*

REMARK IV.3. *Marginal this and marginal that is standard staple for economists. It is a somewhat peculiar way of saying that we consider the derivative of a function. Apart from the marginal rate of substitution, we will encounter "marginal utility", "marginal cost", "marginal revenue" etc.*

As an example, consider the utility function given by $U(x_1, x_2) = ax_1 + bx_2$, $a > 0$ and $b > 0$, i.e., the case of perfect substitutes. Along an indifference curve, the utility is constant at some level k so that we focus on all goods bundles (x_1, x_2) fulfilling $ax_1 + bx_2 = k$. We find the slope of that indifference curve by

- solving for $x_2 : x_2(x_1) = \frac{k}{b} - \frac{a}{b}x_1$ and
- forming the derivative with respect to $x_1 : \frac{dx_2}{dx_1} = -\frac{a}{b}$.

Therefore, the marginal rate of substitution for perfect substitutes is $\frac{a}{b}$.

So far, we did not make use of a utility function (possibly) representing the preferences. If such a function is available, calculating the marginal rate of substitution is an easy exercise:

LEMMA IV.10. *Let $\succsim$ be a preference relation on $\mathbb{R}^\ell_+$ and let U be the corresponding utility function. If U is differentiable, the marginal rate of substitution between good 1 and good 2 can be obtained by*

$$MRS(x_1) = \left|\frac{dI_y(x_1)}{dx_1}\right| = \frac{\frac{\partial U}{\partial x_1}}{\frac{\partial U}{\partial x_2}}.$$

Here, we make use of the partial derivatives of the utility function, $\frac{\partial U}{\partial x_1}$ and $\frac{\partial U}{\partial x_2}$, for goods 1 and 2, respectively. They are called ... marginal utility.

PROOF. Along an indifference curve, the utility is constant, i.e., we have

$$\text{constant} = U(x_1, I_y(x_1)).$$

By the adding rule (theorem IV.3), differentiating with respect to x_1 yields

$$0 = \frac{\partial U}{\partial x_1} + \frac{\partial U}{\partial x_2}\frac{dI_y(x_1)}{dx_1}.$$

□

Thus, we can find the slope of the function I_y even if I_y is not given explicitly. This is an application of the so-called implicit-function theorem.

Let us return to the case of perfect substitutes considered above. The marginal rate of substitution is found easily:

$$MRS(x_1) = \frac{\frac{\partial(ax_1+bx_2)}{\partial x_1}}{\frac{\partial(ax_1+bx_2)}{\partial x_2}} = \frac{a}{b}$$

We note without proof:

LEMMA IV.11. *Let U be a differentiable utility function and I_y an indifference curve of U. This indifference curve is concave if and only if the marginal rate of substitution is a decreasing function in x_1.*

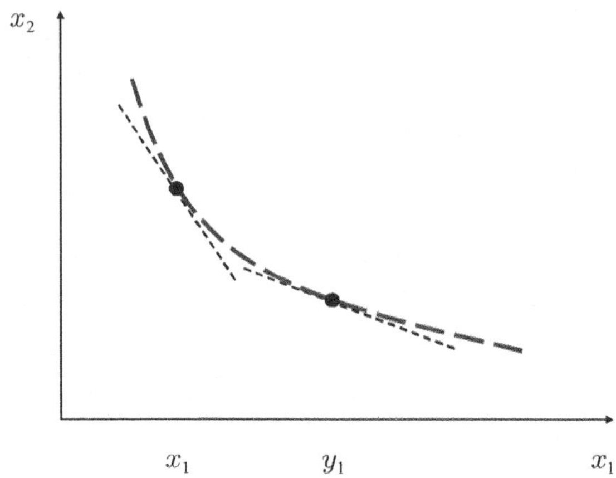

FIGURE IV.22. Concave indifference curve, increasing MRS

This lemma is depicted in figure IV.22. In that figure, we have $x_1 < y_1$ and $MRS(x_1) > MRS(y_1)$.

For example, the MRS of Cobb-Douglas utility functions (which are given by $U(x_1, x_2) = x_1^a x_2^{1-a}$, $0 < a < 1$) is

$$MRS = \frac{\frac{\partial U}{\partial x_1}}{\frac{\partial U}{\partial x_2}} = \frac{a x_1^{a-1} x_2^{1-a}}{(1-a) x_1^a x_2^{-a}} = \frac{a}{1-a} \frac{x_2}{x_1}.$$

If we increase x_1, we need to decrease $x_2 > 0$ along any indifference curve (Cobb-Douglas preferences are monotonic). Therefore, $\frac{x_2}{x_1}$ is a decreasing function of x_1. Thus, the MRS decreases in x_1 and Cobb-Douglas indifference curves are concave (Cobb-Douglas preferences convex or Cobb-Douglas utility functions quasi-concave). Similarly, the utility function $U(x_1, x_2) = x_1 x_2$ with $MRS = \frac{x_2}{x_1}$ is quasi-concave.

7. Topics

The main topics in this chapter are

- preference relation
- indifference
- strict preference
- better set, worse set
- indifference set, indifference curve
- lexicographic preferences
- Cobb-Douglas preferences
- perfect substitutes, perfect complements
- utility function
- the vector space $\mathbb{R}^\ell$
- the first quadrant of $\mathbb{R}^\ell$, $\mathbb{R}_+^\ell$
- $\geq$, $>$, and $\gg$ for vectors
- the distance between points x and y in $\mathbb{R}^\ell$, $\|x-y\|$, in particular
  - the Euclidian norm, $\|x-y\|_2 = \sqrt{\sum_{g=1}^{\ell} (x_g - y_g)^2}$
  - the infinity norm, $\|x-y\|_\infty = \max_{g=1,\ldots,\ell} |x_g - y_g|$
  - the city-block norm, $\|x-y\|_1 = \sum_{g=1}^{\ell} |x_g - y_g|$
- ε-ball with center x^*
- bounded set
- interior point
- boundary point
- open set
- closed set
- sequence
- convergence
- continuity of a function
- relation
- marginal rate of substitution
- equivalence relation
- convex set
- symmetry
- transitivity
- reflexivity
- completeness
- continuity of preferences
- quasi-concave functions

8. Solutions

Exercise IV.1

Adding two vectors reduces to adding the respective entries:

$$\begin{aligned} x+y &= (2,4)+(8,12) \\ &= (2+8,4+12)=(10,16) \end{aligned}$$

Multiplying a vector with a real number is also defined entry-by-entry:

$$2x = 2\,(2,4) = (2\cdot 2, 2\cdot 4) = (4,8)$$

Using both operations, we find

$$\begin{aligned} \frac{1}{4}x+\frac{3}{4}y &= \frac{1}{4}(2,4)+\frac{3}{4}(8,12) \\ &= \left(\frac{1}{4}\cdot 2, \frac{1}{4}\cdot 4\right)+\left(\frac{3}{4}\cdot 8, \frac{3}{4}\cdot 12\right) \\ &= \left(\frac{1}{2},1\right)+(6,9) \\ &= \left(6\frac{1}{2},10\right) \end{aligned}$$

Exercise IV.2

We have

$$\|(2,5)-(7,1)\|_2 = \|(-5,4)\|_2 = \sqrt{25+16} = \sqrt{41} \approx 6,4$$

and

$$\|(2,5)-(7,1)\|_\infty = \|(-5,4)\|_\infty = \max(5,4) = 5.$$

Exercise IV.3

The 1-ball with center $(2,2)$ is a circle with radius 1. The 1-ball with center $(0,0)$ is a quarter-circle with radius 1 that goes through $(1,0)$ and $(0,1)$. The 1-ball with center $(2,0)$ is a half-circle with radius 1 visiting, inter alia, the points $(1,0)$, $(3,0)$ and $(2,1)$.

Exercise IV.4

The boundary points of both $(0,1)$ and $[0,1]$ are 0 and 1, the boundary point of $[0,0]$ ist 0.

Exercise IV.5

The sequence $(1,2),(1,3),(1,4),\ldots$ does not converge because of the second entry. The sequence $\left(1,\frac{1}{2}\right),\left(1,\frac{1}{3}\right),\left(1,\frac{1}{4}\right),\left(1,\frac{1}{5}\right),\ldots$ converges towards $(1,0)$.

Exercise IV.6

We consider the sequence $\left(\frac{3}{2}\right),\left(\frac{4}{3}\right),\left(\frac{5}{4}\right),\left(\frac{6}{5}\right),\ldots$ which is a sequence in $\{0\}\cup(1,2)$. In $\mathbb{R}$, it converges towards $1\notin\{0\}\cup(1,2)$. By lemma

IV.5, $\{0\} \cup (1,2)$ is not closed. The sequence $\left(\frac{1}{2}\right), \left(\frac{1}{3}\right), \left(\frac{1}{4}\right), \left(\frac{1}{5}\right), \ldots$ is contained in $\mathbb{R}\backslash [\{0\} \cup (1,2)]$ and converges towards $0 \notin \mathbb{R}\backslash [\{0\} \cup (1,2)]$. Therefore, $\mathbb{R}\backslash [\{0\} \cup (1,2)]$ is also not closed. By lemma IV.2, we know that $\mathbb{R}\backslash [\{0\} \cup (1,2)]$ is not open because its complement

$$\mathbb{R}\backslash (\mathbb{R}\backslash [\{0\} \cup (1,2)]) = \{0\} \cup (1,2)$$

is not closed. Similarly, $\{0\} \cup (1,2)$ is also not open.

Exercise IV.7

Did you also obtain

property	is the father of	is of the same sex as
reflexive	no	yes
transitive	no	yes
symmetric	no	yes
antisymmetric	yes	no
complete	no	no

Exercise IV.8

We have $[17] = [-23] = [1]$ and $[100] = [0]$. The relation "is of the same sex as" is an equivalence relation (see exercise IV.7). Under gender binarism, the equivalent classes are "the set of all males" and "the set of all females".

Exercise IV.9

The indifference relation is not a preference relation, but an equivalence relation. The strict preference relation inherits transitivity from the transitivity of weak preference:

property	indifference	strict preference
reflexive	yes	no
transitive	yes	yes
symmetric	yes	no
complete	no	no

Exercise IV.10

If you have problems drawing these indifference curves, go back to an intermediate-microeconomics textbook. Your three pictures should look like this:

- If good 2 is a bad, the indifference curve is upward sloping.
- Red and blue matches are perfect substitutes. They are depicted by linear indifference curves with slope -1.
- Left and right shoes are perfect complements. The indifference curves are L-shaped.

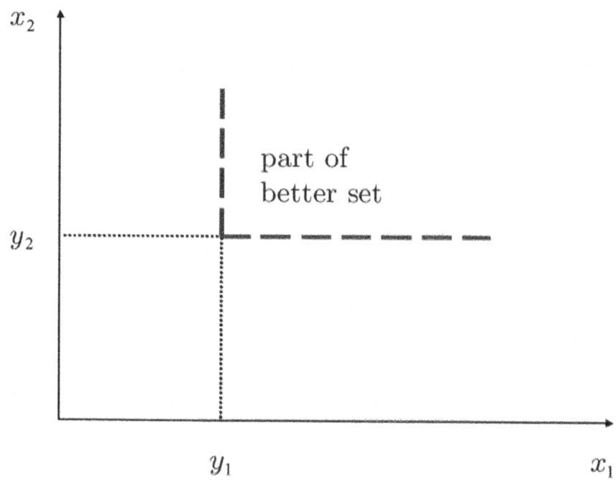

FIGURE IV.23. Weak monotonicity

Exercise IV.11

There is no indifference between two points. Therefore, every point is an indifference curve for itself.

Exercise IV.12

Let M_1 and M_2 be convex sets. $M := M_1 \cap M_2$ is the intersection of these two sets. Two points x and y from M are contained in M_1 and in M_2. By convexity

$$kx + (1 - k)\,y$$

is also contained in both sets and therefore in their intersection, M.

Exercise IV.13

All these sets are strictly convex.

Exercise IV.14

The preferences indicated in (a) are strictly convex, while those in (b) and (c) are convex but not strictly convex. The preferences depicted in (d) are not convex. They are strictly concave.

Exercise IV.15

The set of those bundles $x = (x_1, x_2)$ for which we have $x \geq y$, can be seen in figure IV.23. In case of weak monotonicity all those bundles fulfill $x \succsim y$, i.e., they belong to y's better-set.

Exercise IV.16

We have $\ln U_1(x_1, x_2, x_3) = U_2(x_1, x_2, x_3)$ and $\frac{d\ln y}{dy} = \frac{1}{y} > 0$. Therefore, U_1 and U_2 are equivalent. By $\frac{d(-y)^{-1}}{dy} = (-1)\,(-y)^{-2}\,(-1) = \frac{1}{y^2} > 0$, U_1 and U_4 are also equivalent. Equivalence of utility functions is an equivalence relation (do you see, why?) so that U_1, U_2 and U_4 are equivalent.

U_1 and U_3 are not equivalent because we have

$$\begin{aligned} U_1(0,0,0) &= 1 < 2 = U_1(1,0,0) \text{ but} \\ U_3(0,0,0) &= -1 > -2 = U_3(1,0,0). \end{aligned}$$

U_1 and U_5 are also not equivalent, because of

$$\begin{aligned} U_1(0,0,0) &= 1 < 2 = U_1(1,0,0) \text{ but} \\ U_5(0,0,0) &= 0 = 0 = U_3(1,0,0). \end{aligned}$$

You can also check that U_3 and U_5 are not equivalent.

Exercise IV.17

U obeys

- weak monotonicity iff $x > y$ implies $U(x) \geq U(y)$,
- strict monotonicity iff $x > y$ implies $U(x) > U(y)$, and
- local non-satiation at y iff a bundle x with $U(x) > U(y)$ can be found in every ε-ball with center y.

9. Further exercises without solutions

PROBLEM IV.1. Show that lexicographic preferences are not continuous by applying lemma IV.5. Hint: Consider the better set $B_{(2,4)}$ and the sequence

$$\left(x^j\right)_{j\in\mathbb{N}} = \left(2 + \frac{1}{j}, 2\right).$$

PROBLEM IV.2. Provide a definition of strict anti-monotonicity. In the matrix below, sketch indifference curves for each of the four cases!

	strict monotonicity	strict anti-monotonicity
strict convexity		
strict concavity		

PROBLEM IV.3. Which of the properties (strict) monotonicity, (strict) convexity, and continuity do the following preferences satisfy?

(1) $U(x_1, x_2) = x_1 \cdot x_2$,
(2) $U(x_1, x_2) = \min\{a \cdot x_1, b \cdot x_2\}$ where $a, b > 0$ holds,
(3) $U(x_1, x_2) = a \cdot x_1 + b \cdot x_2$ where $a, b > 0$ holds,
(4) lexicographic preferences

There is no need for mathematical proofs.

PROBLEM IV.4. Let U be a continous utility function representing the preference relation $\precsim$ on $\mathbb{R}^l_+$. Show that $\precsim$ is continous as well. Also, give an example for a continous preference relation that is represented by a discontinous utility function. Hint: Define a function U' that differs from U for $x = 0$, only.

PROBLEM IV.5. Let $\succ$ be a strict preference relation. Show that $\succ$

(1) need not be complete,
(2) is always transitive

PROBLEM IV.6. A preference relation $\succsim$ is called homothetic if $x \succsim y$ implies $\alpha x \succsim \alpha y$ for all $\alpha \geq 0$.

(1) Show that $x \sim y$ implies $\alpha x \sim \alpha y$!
(2) Consider a world with two goods. Sketch two indifference curves.
(3) Show that homogeneous utility functions (defined by $U(\alpha x) = \alpha^\lambda U(x)$ for some $\lambda > 0$) represent homothetic preferences.
(4) Are lexicographic preferences homothetic?

CHAPTER V

Decisions under risk

So far, we have considered decision theory under certainty or for risk neutrality. In this chapter, we focus on the risk attitude of the agents. The model presented by von Neuman and Morgenstern allows to formalize risk aversion and risk loving. The objects of choice are lotteries, i.e., payoffs together with probabilities. We present and discuss the axioms that govern the choice between lotteries. We also introduce important concepts such as the certainty equivalent, the risk premium and the Arrow-Pratt measure of risk averseness. This chapter borrows from Mas-Colell, Whinston & Green (1995, chapter 6).

1. Simple and compound lotteries

1.1. Simple lotteries as bundles and trees. We introduce lotteries on pp. 19. We repeat the definition and show how lotteries can be understood

- as bundles of goods (see chapter VI) or
- as extensive-form decision situations (see chapter III).

We also introduce compound lotteries where the "payoffs" are lotteries themselves. The reader is reminded of our umbrellas-sunshades example:

		state of the world	
		bad weather, $\frac{1}{4}$	good weather, $\frac{3}{4}$
strategy	production of umbrellas	100	81
	production of sunshades	64	121

FIGURE V.1. Umbrellas or sunshades?

H. Wiese, *Advanced Microeconomics*,
https://doi.org/10.1007/978-3-658-34959-2_5

The lotteries

$$L_{\text{umbrella}} = \left[100, 81; \frac{1}{4}, \frac{3}{4}\right] \text{ and}$$
$$L_{\text{sunshade}} = \left[64, 121; \frac{1}{4}, \frac{3}{4}\right]$$

are also depicted in figure V.2 (the probabilities are noted at the axes). Thus, given the probabilities, lotteries are bundles of goods.

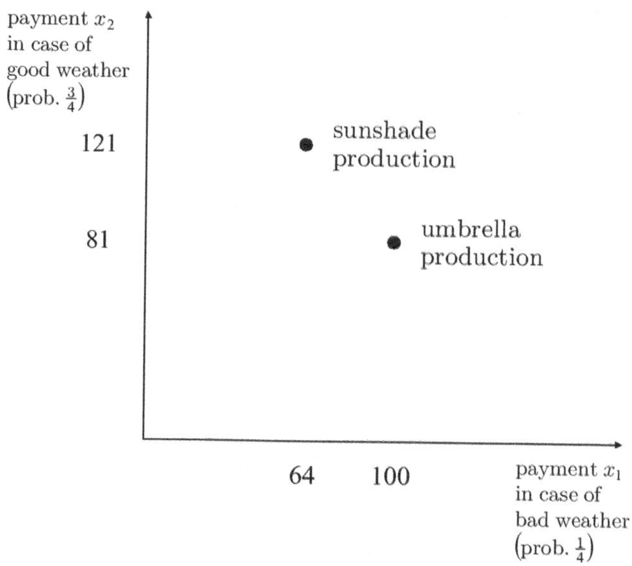

FIGURE V.2. The lotteries resulting from umbrella and sunshade production

We repeat the definition of an expected value:

DEFINITION V.1 (EXPECTED VALUE). *Assume a simple lottery*

$$L = [x_1, ..., x_\ell; p_1, ..., p_\ell].$$

Its expected value is denoted by $E(L)$ and given by

$$E(L) = \sum_{j=1}^{\ell} p_j x_j.$$

In case of two payoffs and two probabilities ($\ell = 2$), the expected value can be derived graphically. Consider the lottery $L = \left[2, 10; \frac{1}{4}, \frac{3}{4}\right]$ which is depicted in figure V.3. Its expected payoff is $\frac{1}{4} \cdot 2 + \frac{3}{4} \cdot 10 = 8$. The line crossing through L is the locus of lotteries (for probabilities 1/4 and 3/4, respectively) that also have this mean. This can be seen from $8 = p_1x_1+p_2x_2$ (which is obviously fulfilled by lottery L) or

$$x_2 = \frac{8}{p_2} - \frac{p_1}{p_2}x_1$$

with slope $-\frac{p_1}{p_2} = -\frac{1}{3}$. Now, the lottery on that line which also lies on the 45°-line (fulfilling $x_1 = x_2$) has expected value

$$\begin{aligned} E(L) &= p_1x_1 + p_2x_2 \\ &= p_1x_1 + p_2x_1 \\ &= x_1. \end{aligned}$$

Therefore, given a lottery L, you draw a line through L with slope $-\frac{p_1}{p_2}$ (the line of constant expected value) and the 45°-line (the line of equal payoffs) and then, either entry of the crossing point equals L's expected value.

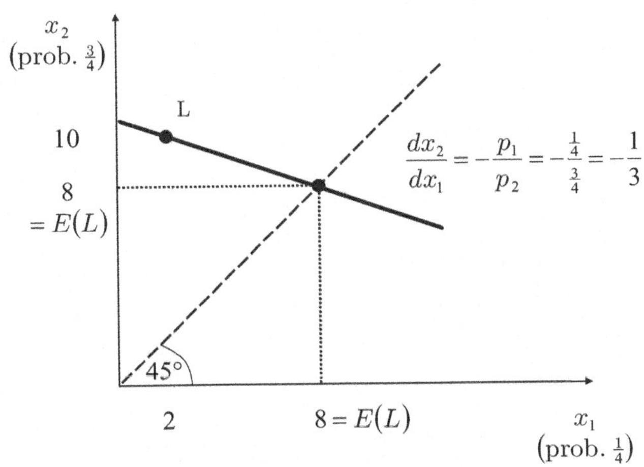

FIGURE V.3. Deriving the expected payoff

As an alternative to the bundle-of-goods interpretation, we can understand a simple lottery as a decision situation in extensive form with moves by nature $\Delta = (V, u, \iota, A_{v_0}, \beta_0)$ with $V_{term} = V \backslash \{v_0\}$, $D_1 = \emptyset$, and $D_0 = \{v_0\}$ (see p. 47). $V_{term} = V \backslash \{v_0\}$ implies that the tree is of length 1. Therefore, lottery $\left[0, 10; \frac{1}{3}, \frac{2}{3}\right]$ can be depicted as in figure V.4 – a decision situation without decision maker.

EXERCISE V.1. *Do you prefer* $L_1 = \left[0, 10; \frac{1}{3}, \frac{2}{3}\right]$ *to* $L_2 = \left[5, 10; \frac{1}{4}, \frac{3}{4}\right]$ *if the payoff numbers are Euro amounts?*

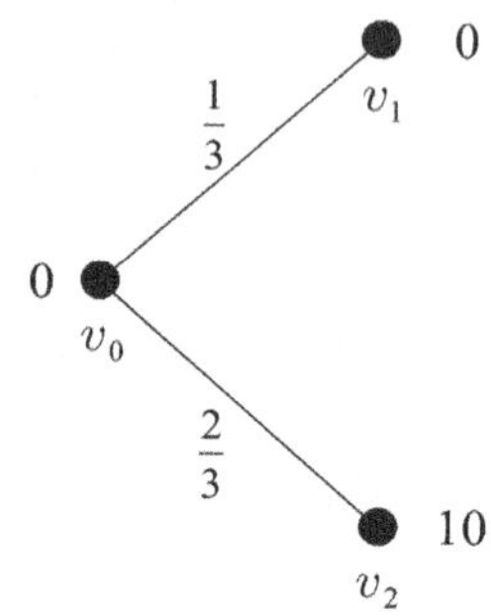

FIGURE V.4. A simple lottery as a decision situation

EXERCISE V.2. *Do you prefer the lottery* $L = \left[95, 105; \frac{1}{2}, \frac{1}{2}\right]$ *to a certain payoff of* 100?

1.2. Compound lotteries. For theoretical purposes, we need to present compound lotteries:

DEFINITION V.2 (COMPOUND LOTTERY). *Let* $L_1, ..., L_\ell$ *be simple lotteries (lotteries in the sense of definition II.18, p. 20). Then*

$$L = [L_1, ..., L_\ell; p_1, ..., p_\ell]$$

is called a compound or two-stage lottery. We also allow for an infinite ℓ.

EXERCISE V.3. *Consider the two simple lotteries* $L_1 = \left[0, 10; \frac{1}{3}, \frac{2}{3}\right]$ *and* $L_2 = \left[5, 10; \frac{1}{4}, \frac{3}{4}\right]$. *Express the compound lottery* $L = \left[L_1, L_2; \frac{1}{2}, \frac{1}{2}\right]$ *as a simple lottery! Can you draw the appropriate trees, one of length* 2 *and one of length* 1*?*

2. The St. Petersburg lottery

2.1. The paradox. Imagine Peter throwing a fair coin j times until "head" occurs for the first time. Head (H) rather than tail (T) occurs at the first coin toss (sequence H) with probability $\frac{1}{2}$, at the second coin toss (sequence TH) with probability $\frac{1}{4}$, and at the jth toss (sequence T...TH) with probability $\frac{1}{2^j}$. Peter pays 2^j to Paul if "head" occurs for the first time at the jth toss. Thus, the lottery is an infinite one and defined by

$$L = \left[2, 4, 8, ..., 2^j, ...; \frac{1}{2}, \frac{1}{4}, \frac{1}{8}, ..., \frac{1}{2^j}, ...\right].$$

The probabilities are positive. However, do they sum up to 1? Note that any (but the first) probability is the product of the previous probability and $\frac{1}{2}$. In general, if the absolute value of the factor q (here: $\frac{1}{2}$) is smaller than 1,

$|q| < 1$, the infinite geometric series $\sum_{j=0}^{\infty} cq^j = c + cq + cq^2 + ...$ converges and we have

$$\text{infinite geometric series} = \frac{\text{first term}}{1 - \text{factor}} = \frac{c}{1-q}.$$

EXERCISE V.4. *Apply the above rule to the sum of the probabilities* $\frac{1}{2} + \frac{1}{4} + \frac{1}{8} + ...$ *!*

EXERCISE V.5. *How much are you prepared to pay for the St. Petersburg lottery?*

Lottery L is known as the St. Petersburg lottery. Its expected payoff is infinite:

$$E(L) = \sum_{j=1}^{\infty} \frac{1}{2^j} \cdot 2^j = \infty$$

In any case, your willingness to pay is well below the expected value of the lottery. This discrepancy between the infinite expected value of the lottery and the very low willingness to pay for it is called the St. Petersburg paradox.

2.2. Limited resources as a resolution? Can we make sense of the paradox? One argument refers to Peter's limited resources. Assume Peter is a millionaire who possesses one million Euro which he is willing to pay to Paul. Then, rather than the infinite lottery above, we have the resource-restricted lottery

$$L_{\text{millionaire}} = \left[2, 4, 8, ..., 2^{18}, 2^{19}; \frac{1}{2}, \frac{1}{4}, \frac{1}{8}, ..., \frac{1}{2^{18}}, \frac{1}{2^{18}}\right].$$

(It is easy to remember $2^{10} = 1024 \approx 10^3$. Therefore 2^{20} is slightly above 10^6 so that the maximum payment Peter can afford is 2^{19}.) The expected value of this lottery is

$$\begin{aligned} E(L_{\text{millionaire}}) &= \sum_{j=1}^{18} \frac{1}{2^j} \cdot 2^j + \frac{1}{2^{18}} \cdot 2^{19} \\ &= 20. \end{aligned}$$

Thus, the resource restriction puts a serious limit to an expected-value maximizer's willingness to pay.

Aumann (1977) thinks that this is a cheap way out of the difficulty posed by the St. Petersburg paradox. He points to the possibility that the utility arises from "religious, aesthetic, or emotional experiences, like entering a monastery, climbing a mountain, or engaging in research with possibly spectacular results. It seems reasonable to suppose that before engaging in such an activity, Paul would perceive the utility of the resulting sensation as a random variable and there is no particular reason to assume that this random variable is bounded." Thus, we can construct a St. Petersburg lottery that does not refer to Peter's limited resources.

2.3. Expected utility as a resolution? A second argument challenges the rationale of maximizing the expected value. Instead, decision makers maximize (should maximize) the expected utility as Daniel Bernoulli proposed in 1738. The idea is to transform the payoffs x_j by a utility function u which is defined on $\mathbb{R}$.

DEFINITION V.3 (EXPECTED UTILITY). *Assume a decision maker who is faced with a lottery* $L = [x_1, ..., x_\ell; p_1, ..., p_\ell]$ *and has the utility function* $u : \mathbb{R} \to \mathbb{R}$. *The expected utility is denoted by* $E_u(L)$ *and given by*

$$E_u(L) = \sum_{j=1}^{\ell} p_j u(x_j).$$

For example, assume an agent with the utility function $u = \ln$.

EXERCISE V.6. *Do you know how to rewrite or calculate* $\ln(1)$ *and, for* $x, y > 0$, $\ln\left(\frac{x}{y}\right)$, $\ln(xy)$, $\ln x^b$, *and* $\frac{d\ln x}{dx}$*?*

Without resource restriction, we obtain the expected utility of the lottery

$$\begin{aligned} E_{\ln}(L) &= \sum_{j=1}^{\infty} \frac{1}{2^j} \ln\left(2^j\right) = \sum_{j=1}^{\infty} \frac{1}{2^j} j \ln 2 \\ &= \ln 2 \sum_{j=1}^{\infty} \frac{1}{2^j} j \\ &= 2 \ln 2 \end{aligned}$$

where you do not need to check the last equality. What amount of money that could be obtained for sure is as valuable to the agent as the lottery itself? Later in the chapter, we call this amount the certainty equivalent CE. It can be calculated from

$$E_{\ln}([CE; 1]) \stackrel{!}{=} E_{\ln}(L) = 2 \ln 2$$

and is given by

$$CE = e^{\ln(CE)} = e^{E_{\ln}([CE;1])} \stackrel{!}{=} e^{2\ln 2} = \left(e^{\ln 2}\right)^2 = 2^2 = 4.$$

2.4. Bounded utility. This expected-utility argument is fine for the original St. Petersburg lottery. However, given the ln utility function, we can construct a lottery that leads to the same paradox:

$$L_{\ln} = \left[4, 16, 256, ..., 2^{(2^j)}, ...; \frac{1}{2}, \frac{1}{4}, \frac{1}{8}, ..., \frac{1}{2^j}, ...\right].$$

EXERCISE V.7. *Calculate the expected utility of the above lottery with respect to the utility function* ln, *i.e., find* $E_{\ln}(L_{\ln})$!

Thus, it seems that we cannot escape the St. Petersburg paradox! As it turns out, the culprit is a utility function that is not bounded.

DEFINITION V.4 (BOUNDED FUNCTION). *A real-valued function* f *is bounded if a* $K \in \mathbb{R}$ *exists such that* $|f(x)| \leq K$ *for all* x *from* f'*s domain.*

EXERCISE V.8. *Consider the utility functions on the domain* $\mathbb{R}_+$:

- $u(x) = \ln x$
- $u(x) = x$
- $u(x) = \sqrt{x}$
- $u(x) = 1 - \frac{1}{1+x}$

Which of these are bounded?

LEMMA V.1. *Let* u *be an unbounded utility function. Then, a St. Petersburg lottery* L_u *can be found such that* $E_u(L_u)$ *is infinite.*

This lemma is due to Menger (1967). The idea is to associate the probability $\frac{1}{2^j}$ with a payoff x such that $\frac{1}{2^j}u(x) \geq 1$. By the unboundedness of u, such an x can be found. That's it. The inverse is even simpler:

EXERCISE V.9. *Show the following proposition: For any bounded utility function obeying* $u(x) \leq K$ *for all* $x \geq 0$ *and for any lottery* L, *we have* $E_u(L) \leq K$.

Aumann (1977) suggests to look here for a resolution to the St. Petersburg paradox: It is simply not true that utility functions are unbounded. Aumann corroborates the claim of boundedness by a thought experiment. Assume Paul has an unbounded utility function. Let x be a very good life ("long, happy, and useful", in Aumann's words) and let y be a very miserable life for Paul. By unboundedness there is another very, very good life for Paul, z, such that

$$\frac{1}{10^{100}}u(z) + \left(1 - \frac{1}{10^{100}}\right)u(y) > u(x)$$

holds. Aumann thinks that the existence of such z is implausible.

Although we find Aumann's argument for bounded utilities convincing, we will often work with unbounded utility functions, just for the sake of convenience. Of course, so far, we do not even know whether the procedure of calculating the expected utility makes sense. And what is the meaning of u and its shape? This is the topic of the next section.

3. Preference axioms for lotteries

3.1. Preference axioms. We assume a weak preference relation on the set of lotteries. Given a weak preference relation $\succsim$, we can define indifference $\sim$ and strict preference $\succ$. The first two axioms are old acquaintances:

Completeness axiom: For two lotteries L_1 and L_2, we have $L_1 \succsim L_2$ or $L_2 \succsim L_1$.

Transitivity axiom: Assume three lotteries L_1, L_2, and L_3 obeying $L_1 \succsim L_2$ and $L_2 \succsim L_3$. Then, we have $L_1 \succsim L_3$.

The third axiom uses the special structure of lotteries:

Continuity axiom: For any three lotteries L_1, L_2, and L_3 with $L_1 \succsim L_2 \succsim L_3$ a probability $p \in [0,1]$ exists such that

$$L_2 \sim [L_1, L_3; p, 1-p]$$

holds.

Assume three real-life lotteries L_1, L_2, and L_3 where L_3 means certain death, L_1 a payoff of 10 Euros, and L_2 a zero payoff. Hopefully, you are not suicidal and your preferences are given by $L_1 \succ L_2 \succ L_3$. Determine your personal p so that you are indifferent between L_2 (obtaining nothing) and the lottery $[L_1, L_3; p, 1-p]$ where you risk your life with some probability $1-p$ and obtain 10 Euros with probability p.

Now the following problem arises. Most people in rich countries point to a probability close to 1. In fact, many say that any probability below 1 makes them stricly prefer the zero payoff. The probability $p = 1$ leads to $[L_1, L_3; 1, 0] = L_1 \succ L_2$ and does not fulfill the continuity axiom's requirement. Therefore, one has a reason to be critical towards the continuity axiom. Note, however, that many people are willing to cross the road to pick up a 10 Euro bill. Since crossing the road is a dangerous activity which leads to death with a nonzero probability, these people should be able to name a p below 1.

Finally, we turn to the independence axiom which is of central importance for decisions between lotteries. It claims that the decision between compound lotteries depends on the differences between these lotteries:

Independence axiom: Assume three lotteries L_1, L_2, and L_3 and assume a probability $p > 0$. We have

$$[L_1, L_3; p, 1-p] \precsim [L_2, L_3; p, 1-p] \Leftrightarrow L_1 \precsim L_2.$$

EXERCISE V.10. *Assume a decision maker who is indifferent between*

$$L_1 = \left[0, 100; \frac{1}{2}, \frac{1}{2}\right] \text{ and } L_2 = \left[16, 25; \frac{1}{4}, \frac{3}{4}\right].$$

Can you show the indifference between $L_3 = \left[0, 50, 100; \frac{1}{4}, \frac{1}{2}, \frac{1}{4}\right]$ *and* $L_4 = \left[16, 25, 50; \frac{1}{8}, \frac{3}{8}, \frac{1}{2}\right]$ *by verifying*

$$L_3 = \left[L_1, 50; \frac{1}{2}, \frac{1}{2}\right] \text{ and } L_4 = \left[L_2, 50; \frac{1}{2}, \frac{1}{2}\right].$$

The independence axiom has been heavily critized. Consider the following four lotteries:

$$\begin{aligned}
L_1 &= \left[12 \cdot 10^6, 0; \frac{10}{100}, \frac{90}{100}\right], \\
L_2 &= \left[1 \cdot 10^6, 0; \frac{11}{100}, \frac{89}{100}\right], \\
L_3 &= \left[1 \cdot 10^6; 1\right], \\
L_4 &= \left[12 \cdot 10^6, 1 \cdot 10^6, 0; \frac{10}{100}, \frac{89}{100}, \frac{1}{100}\right].
\end{aligned}$$

Pause for a moment to consider your preferences between L_1 and L_2 on the one hand and between L_3 and L_4 on the other hand.

Many people prefer L_1 to L_2 and L_3 to L_4. However, these preferences are in violation of the independence axiom. By $L_1 \succ L_2$, this axiom implies $\left[L_1, L_3; \frac{1}{2}, \frac{1}{2}\right] \succ \left[L_2, L_3; \frac{1}{2}, \frac{1}{2}\right]$, while $L_3 \succ L_4$ leads to $\left[L_2, L_3; \frac{1}{2}, \frac{1}{2}\right] \succ \left[L_2, L_4; \frac{1}{2}, \frac{1}{2}\right]$. Transitivity implies $\left[L_1, L_3; \frac{1}{2}, \frac{1}{2}\right] \succ \left[L_2, L_4; \frac{1}{2}, \frac{1}{2}\right]$.

EXERCISE V.11. *Reduce* $\left[L_1, L_3; \frac{1}{2}, \frac{1}{2}\right]$ *and* $\left[L_2, L_4; \frac{1}{2}, \frac{1}{2}\right]$ *to simple lotteries!*

The result reported in the exercise shows that the preferences $L_1 \succ L_2$ and $L_3 \succ L_4$ do not conform with the independence and transitivity axioms. Too bad for the axioms! Or too bad for your preferences?

Indeed, how do you react when somebody points out the violation of these axioms? You may say: "Well, I made a mistake", and then you change your preference between L_1 and L_2 (or between L_3 and L_4). After all, we all make mistakes in complicated situations in the same way as we sometimes do not manage to add up numbers correctly. Others take these examples as an argument against the independence axiom (rarely against the transitivity axiom).

Whatever your position is, from now on we assume decision makers who adhere to all four axioms explained above. Apart from the intrinsic appeal of the independence axiom, the preference theory built on these axioms is very intriguing and helpful in the analysis of decisions under risk.

3.2. A utility function for lotteries. We begin with a startling and far-reaching statement:

THEOREM V.1. *Preferences between lotteries obey the four axioms mentioned in the previous subsection if and only if there is a utility function* $u : \mathbb{R}_+ \to \mathbb{R}$ *such that*

$$L_1 \succsim L_2 \Leftrightarrow E_u(L_1) \geq E_u(L_2)$$

holds for all $L_1, L_2 \in \mathcal{L}$. *In that case, the utility function* u *is said to represent the preferences* $\succsim$ *on the set of lotteries* $\mathcal{L}$. u *is called a von Neumann Morgenstern utility function (vNM utility function).*

The theorem makes two claims. First, if a decision maker's preferences are represented by expected utility for a suitably chosen utility function $u : \mathbb{R}_+ \to \mathbb{R}$, these preferences obey the four axioms. Second, if preferences on the set of lotteries do not violate any of the four axioms, we can find a utility function u (that depends on the preferences) such that the expected utility with respect to u delivers all the preference information.

It is important to distinguish two utility functions very strictly:

- $u : \mathbb{R}_+ \to \mathbb{R}$ is the vNM (von Neumann and Morgenstern) utility function and has payoffs as its domain,
- $E_u : \mathcal{L} \to \mathbb{R}$ is the expected utility and has lotteries as its domain.

The preferences that an expected utility function represents are not changed by strictly monotonic transformations (for a more detailed discussion, see pp. 75). For example, we can multiply E_u by 4 or $\frac{1}{5}$ or we can apply the ln or square-root function if $E_u(L)$ is nonnegative for all $L \in \mathcal{L}$. However, we have no reason to do that. It is more interesting that we can change u in certain ways without affecting the preferences to be represented:

LEMMA V.2. *If* u *represents the preferences on the set of lotteries, so does any utility function* v *that obeys* $v(x) = a + bu(x)$ *for* $a \in \mathbb{R}$ *and* $b > 0$.

The lemma says that we can subject u to an affine transformation (adding a constant and multiplying by a positive constant) without changing the preferences u represents.

EXERCISE V.12. *Find a vNM utility function that is simpler than* $u(x) = 100 + 3x + 9x^2$ *while representing the same preferences.*

The proof of the lemma is not difficult:

EXERCISE V.13. *Consider two lotteries* $L^A := \left[x_1^A, ..., x_{\ell_A}^A; p_1^A, ..., p_{\ell_A}^A\right]$ *and* $L^B := \left[x_1^B, ..., x_{\ell_B}^B; p_1^B, ..., p_{\ell_B}^B\right]$. *Let* v *be an affine transformation of* u. *Show*

$$E_u\left(L^A\right) \geq E_u\left(L^B\right) \Leftrightarrow E_v\left(L^A\right) \geq E_v\left(L^B\right).$$

3.3. The construction of the vNM utility function. We still do not know how to construct u and how to interpret it. We use the continuity axiom for the construction of u. Consider a very bad lottery L_{bad} and a very good lottery L_{good} ($L_{good} \succ L_{bad}$) and consider an in-between lottery L obeying $L_{good} \succsim L \succsim L_{bad}$. By the continuity axiom, there exists a probability $p(L)$ such that $L \sim [L_{good}, L_{bad}; p(L), 1 - p(L)]$ holds.

EXERCISE V.14. *Find* $p(L_{good})$ *and* $p(L_{bad})$! *Hint: Translate*

$$L \sim [L_{good}, L_{bad}; p(L), 1 - p(L)]$$

into a statement on expected utilities.

For a payment x we consider the trivial lottery $L := [x; 1]$ and define a vNM utility function u by

$$u(x) := p(L).$$

Then, we have indifference between x and $[L_{good}, L_{bad}; u(x), 1 - u(x)]$. $u(x)$ is a value between 0 (the probability for L_{bad}) and 1 (the probability for L_{good}) and u represents the preferences of the decision maker as shown by Myerson (1991, pp. 12). We assume that u is strictly increasing but allow for convex or concave shapes.

4. Risk attitudes

4.1. Concave and convex functions. We will see that risk preference is closely related to the concavity or convexity of vNM utility functions.

DEFINITION V.5 (CONCAVE FUNCTION). *Let* $f : M \to \mathbb{R}$ *be a function on a convex domain* $M \subseteq \mathbb{R}$. f *is called concave if we have*

$$f(kx + (1-k)y) \geq kf(x) + (1-k)f(y)$$

for all $x, y \in M$ *and for all* $k \in [0,1]$. f *is called strictly concave if*

$$f(kx + (1-k)y) > kf(x) + (1-k)f(y)$$

holds for all $x, y \in M$ *with* $x \neq y$ *and for all* $k \in (0,1)$. *If the inequality signs are the other way around,* f *is convex or strictly convex, respectively.*

Figure V.5 illustrates concave functions. You see that the value at the point in between x and y is higher than the average of the values at x and y. Graphically speaking, concavity means that the straight line connecting $(x, f(x))$ and $(y, f(y))$ lies below the graph of f.

If $f : \mathbb{R} \to \mathbb{R}$ is differentiable, we have another way to characterize concavity. Consider the slopes (derivatives of f) at points A and B. Both slopes are negative but the slope is lower at B than at A. This means that the first derivative declines, or, differently put, that the second derivative is negative. Just think of a mountain walk. At first, the slope may be positive but you reach a plateau and go downhill afterwards. The slope gets steeper and steeper (until you fall off).

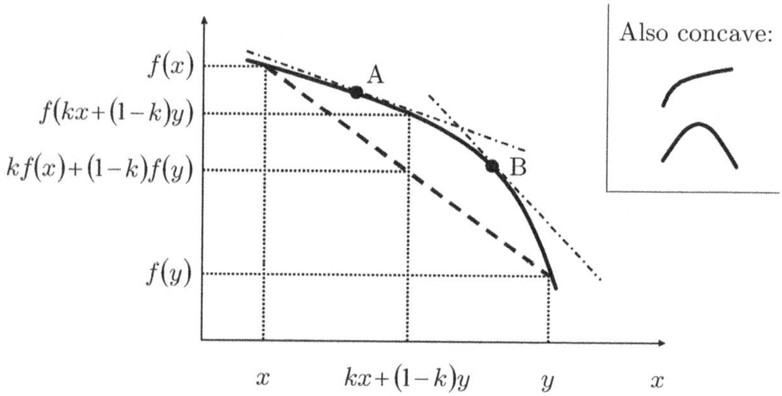

FIGURE V.5. Concave functions and negative second derivative

LEMMA V.3. *Let $f : M \to \mathbb{R}$ with convex domain $M \subseteq \mathbb{R}$ be twice differentiable. f is concave if and only if*

$$f''(x) \leq 0$$

holds for all $x \in M$. f is convex if and only if

$$f''(x) \geq 0$$

holds for all $x \in M$.

Convexity is the opposite of concavity (see figure V.6). If you march down a hill, the slope gets less steep; if you march upwards, the slope increases steadily.

EXERCISE V.15. *Comment: If a function $f : \mathbb{R} \to \mathbb{R}$ is not concave, it is convex.*

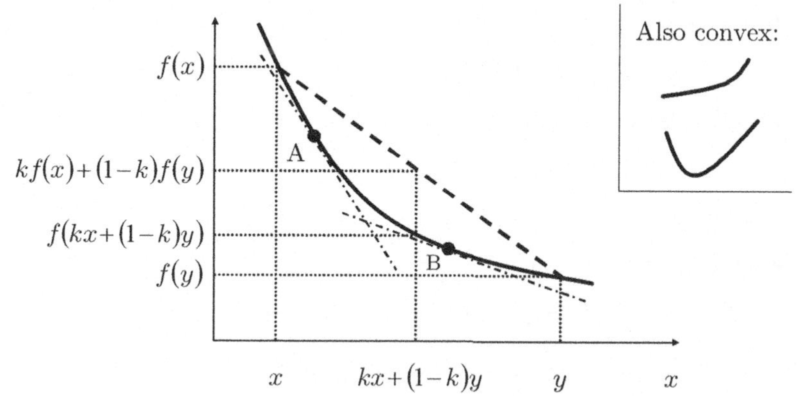

FIGURE V.6. Convex functions and positive second derivative

In chapter IV, we introduce quasi-concavity. In chapter VIII (exercise VIII.2, p. 221), you will be asked to show that concavity implies quasi-concavity in a more general setting. At this point, it may suffice to present the four different possibilities – see figure V.7.

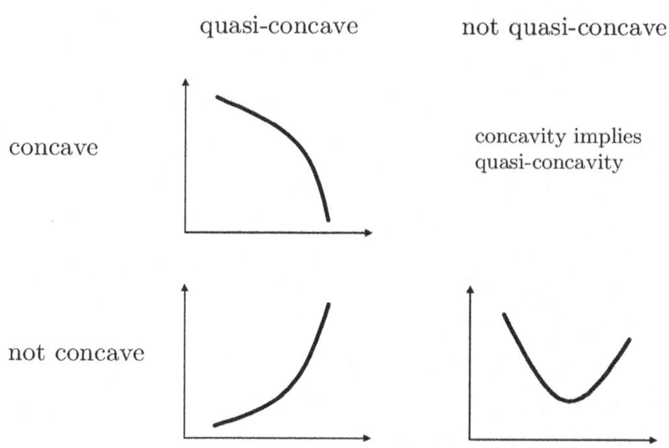

FIGURE V.7. Concavity versus quasi concavity

4.2. Risk aversion and risk loving. We now show that the curvature of u reflects the decision maker's attitude towards risk. But first of all, we need the appropriate definitions:

DEFINITION V.6 (RISK PREFERENCES). *Assume preferences* $\succsim$ *on* $\mathcal{L}$. *A decision maker is called*

- *risk neutral in case of*

$$L \sim [E(L);1] \text{ or } E_u(L) = u(E(L));$$

- *risk-averse in case of*

$$L \precsim [E(L);1] \text{ or } E_u(L) \leq u(E(L));$$

and

- *risk-loving in case of*

$$L \succsim [E(L);1] \text{ or } E_u(L) \geq u(E(L)).$$

Thus, a risk-loving decision maker prefers a lottery over the expected value of that lottery. A risk neutral person cares only for the expected value of the lottery. Of course, we can also express the three attitudes towards risk by way of expected utility. For a risk-averse decision maker, the expected utility of a lottery L is not higher than the utility of the lottery $[E(L);1]$ which equals $u(E(L))$. Thus, risk-averseness can also be expressed by $E_u(L) \leq u(E(L))$.

Consider the lottery $\left[95, 105; \frac{1}{2}, \frac{1}{2}\right]$. Its expected value is 100 and its expected utility $\frac{1}{2}u(95) + \frac{1}{2}u(105)$. Have a look at figure V.8 and observe the concave shape of the vNM utility function u. Apparently, we have

$$u(100) = u(E(L)) > E_u(L) = \frac{1}{2}u(95) + \frac{1}{2}u(105),$$

i.e., the utility of the expected value is greater than the expected utility. In other words, the decision maker prefers the (sure) expected value to the (uncertain) lottery. He is risk averse. This is true in general and the inverse holds, too:

LEMMA V.4. *Assume preferences $\succsim$ on $\mathcal{L}$ and an associated vNM utility function u. A decision maker is*

- *risk neutral iff u is an affine function (i.e., $u(x) = ax + b$, $a > 0$),*
- *risk-averse iff u is concave, and*
- *risk-loving iff u is convex.*

EXERCISE V.16. *Do the preferences characterized by the following utility functions exhibit risk-averseness?*

- $u_1(x) = x^2$, $x > 0$
- $u_2(x) = 2x + 3$
- $u_3(x) = ln(x)$, $x > 0$
- $u_4(x) = -e^{-x}$
- $u_5(x) = \frac{x^{1-\theta}}{1-\theta}, \theta > 0, \theta \neq 1$

Hint: apply lemma V.3.

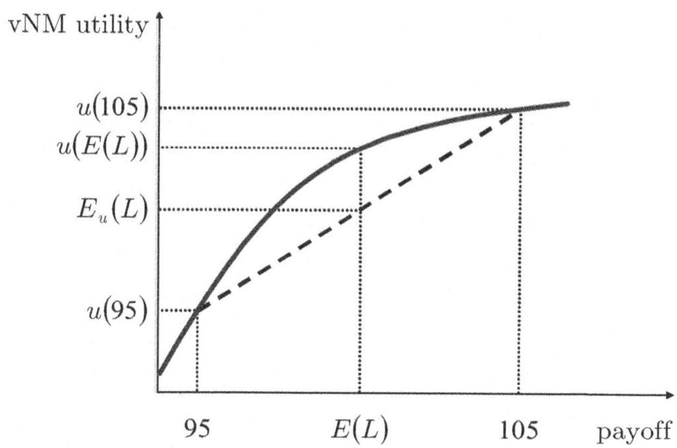

FIGURE V.8. Risk-averseness

4.3. Certainty equivalent and risk premium. For a risk-averse decision maker, the expected value of a lottery is worth more than the lottery itself. An amount smaller than the expected value may yield indifference:

DEFINITION V.7 (CERTAINTY EQUIVALENT). *Assume preferences $\succsim$ on $\mathcal{L}$. For any lottery $L \in \mathcal{L}$, the payoff $CE(L)$ is called certainty equivalent of L if*

$$L \sim [CE(L); 1]$$

holds.

Figure V.9 shows a concave vNM utility function and the certainty equivalent which is definable by $E_u(L) = u(CE(L))$ (why?). If the risk-averse decision maker has the chance to obtain $E(L)$ in exchange for L, he is happy to do so and to throw in at most the difference between the expected value and the certainty equivalent. The decision maker's willingness to pay for exchanging L against $E(L)$ is called the risk premium.

DEFINITION V.8 (RISK PREMIUM). *Assume preferences $\succsim$ on $\mathcal{L}$ and an associated vNM utility function u. For any lottery $L \in \mathcal{L}$, the risk premium is denoted by $RP(L)$ and defined by*

$$RP(L) := E(L) - CE(L).$$

Thus, the risk premium is the amount the individual is willing to pay for being relieved of the risk.

4.4. Arrow-Pratt measure of risk aversion. Sometimnes, it is helpful to have a measure of risk aversion.

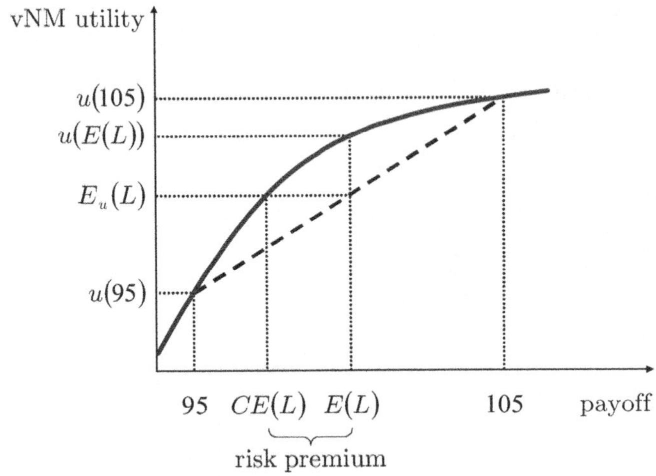

FIGURE V.9. Certainty equivalent and risk premium

DEFINITION V.9 (ARROW-PRATT MEASURE). *Let u be a twice differentiable vNM utility function. We denote the Arrow-Pratt measure of absolute risk aversion by*

$$ARA_u(x) := -\frac{u''(x)}{u'(x)}$$

and the Arrow-Pratt measure of relative risk aversion by

$$RRA_u(x) := -x\frac{u''(x)}{u'(x)}.$$

The Arrow-Pratt measures are positive for risk-aversion.

EXERCISE V.17. *Calculate the two Arrow-Pratt measures for the following utility functions:*

- $u_1(x) = x^2$, $x > 0$
- $u_2(x) = 2x + 3$
- $u_3(x) = ln(x)$, $x > 0$
- $u_4(x) = -e^{-x}$
- $u_5(x) = \frac{x^{1-\theta}}{1-\theta}, \theta \geq 0, \theta \neq 1$

The relative Arrow-Pratt measure has an elasticity interpretation. Do you remember the definition of the price elasticity of demand at price p for a demand function $x(p)$? In chapter VI, we will denote this elasticity by $\varepsilon_{x,p}$ and write

$$\varepsilon_{x,p} := \frac{\frac{dx}{x}}{\frac{dp}{p}} = \frac{dx}{dp}\frac{p}{x}.$$

Indeed, we obtain

$$x\frac{u''(x)}{u'(x)} = x\frac{\frac{du'}{dx}}{u'} = \frac{\frac{du'}{u'}}{\frac{dx}{x}} =: \varepsilon_{u',x}.$$

Thus, if x is increased by 1 percent, marginal vNM utility decreases by $RRA(x)$ percent in case of risk-averseness. Roughly speaking, the higher $RRA(x)$ the more concave the vNM utility function.

4.5. Risk aversion and risk loving in an x_1-x_2-diagram. If a decision maker's preferences on lotteries are determined by a vNM utility function u, the utility function E_u represents his preferences on these lotteries. Fixing the probabilities $p = (p_1, ..., p_\ell)$ for the ℓ outcomes $x = (x_1, ..., x_\ell)$, we are in the framework developed in the previous chapter. We define the utility function E_u^p by

$$\begin{aligned} E_u^p &: \mathbb{R}_+^\ell \to \mathbb{R} \\ &(x_1, ..., x_\ell) \mapsto E_u^p(x) = E_u(x,p). \end{aligned}$$

In the case of $\ell = 2$, we can then describe the agent's preferences by indifference curves. Consider figure V.10. It depicts an agent's indifference curves for given probabilities p_1 and p_2. The agent is risk averse. He would rather have the lottery's mean than the lottery itself.

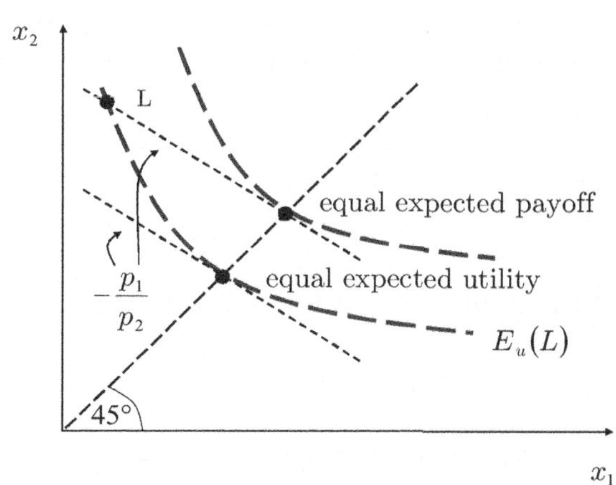

FIGURE V.10. The indifference curve of a risk-averse agent

Note that the indifference curve touches a constant-mean curve at the 45°-line (slope $-\frac{p_1}{p_2}$, compare figure V.3, p. 95). This is no coincidence.

Indeed, the absolute value of the slope of the indifference curve is

$$\begin{aligned} MRS &= \frac{\frac{\partial E_u^p}{\partial x_1}}{\frac{\partial E_u^p}{\partial x_2}} \\ &= \frac{\frac{\partial [p_1 u(x_1) + p_2 u(x_2)]}{\partial x_1}}{\frac{\partial [p_1 u(x_1) + p_2 u(x_2)]}{\partial x_2}} \\ &= \frac{p_1 \frac{du(x_1)}{dx_1}}{p_2 \frac{du(x_2)}{dx_2}} \end{aligned}$$

which reduces to

$$MRS = \frac{p_1}{p_2} \text{ for } x_1 = x_2.$$

Thus, the slopes are identical along the 45°-line.

For a risk-averse agent, an increase in x_1 leads to a decrease in $\frac{du(x_1)}{dx_1}$ by lemma V.3 (p. 104). Along an indifference curve, x_2 decreases when x_1 increases. Hence, by the same lemma, the increase of x_1 is associated with an increase in $\frac{du(x_2)}{dx_2}$. Therefore, the marginal rate of substitution

$$MRS(x_1) = \frac{p_1 \frac{du(x_1)}{dx_1}}{p_2 \frac{du(x_2)}{dx_2}}$$

decreases when x_1 increases. By lemma IV.11 (p. 85), risk aversion is equivalent to concave indifference curves and therefore to convex better sets.

In case of risk neutrality, the vNM utility function is given by $u(x) = ax + b$, $a > 0$. Then, we have

$$MRS(x_1) = \frac{p_1 \frac{du(x_1)}{dx_1}}{p_2 \frac{du(x_2)}{dx_2}} = \frac{p_1 a}{p_2 a} = \frac{p_1}{p_2}$$

so that the slope of the indifference curve is constant and equal to the slope of a constant-mean curve.

5. Stochastic dominance

5.1. Distribution and density functions. We introduce probability distributions on a set M on p. 18. We now let $M := \mathbb{R}_+$ and consider events $[0, x]$ for $x \geq 0$. Then, distribution functions may be used to describe probabilities. We present this alternative because it allows handy characterizations of an agent's preferences over lotteries.

DEFINITION V.10 (DISTRIBUTION FUNCTION). *A monotonically increasing, but not necessarily continuous, function* $F : \mathbb{R} \rightarrow [0,1]$ *obeying*

$$\begin{aligned} \lim_{x \rightarrow -\infty} F(x) &= 0 \text{ and} \\ \lim_{x \rightarrow \infty} F(x) &= 1 \end{aligned}$$

is called a distribution function. If F *is constant and equal to zero on some interval* $(-\infty, \underline{x}]$ *or constant and equal to* 1 *on interval* $[\bar{x}, \infty)$, *we can focus on the distribution functions* $F : [\underline{x}, \infty) \rightarrow [0,1]$, $F : (-\infty, \bar{x}] \rightarrow [0,1]$, *or* $F : [\underline{x}, \bar{x}] \rightarrow [0,1]$. *If there is a function* $f : [\underline{x}, \bar{x}] \rightarrow [0,1]$ *(allow for* $\underline{x} = -\infty$ *and* $\bar{x} = \infty$*) with*

$$F(x) = \int_{\underline{x}}^{x} f(t)\, dt, \tag{V.1}$$

f *is called* F*'s density function.*

We understand $F(x)$ as the probability of having payoff x or below. Thus, for $x < y$,

$$F(y) - F(x) \text{ or } \int_{x}^{y} f(t)\, dt$$

is the probability for values between x and y.

F does not need to be continuous as you can see in figure V.11 which reflects the lottery

$$L_{\text{umbrella}} = \left[100, 81; \frac{1}{4}, \frac{3}{4}\right].$$

Here, the umbrella producer obtains 81 or less with probability $\frac{3}{4}$. He obtains 100 or less with probability 1.

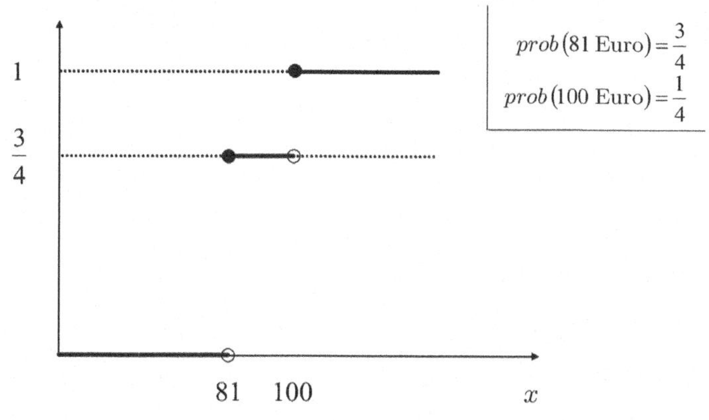

FIGURE V.11. The distribution function for two payoffs

DEFINITION V.11. *Let $F : [\underline{x}, \bar{x}] \to [0,1]$ be a discrete distribution function with ℓ different values $x_1, ..., x_\ell$. It represents the lottery*

$$L_F = [x_1, ..., x_\ell; p_1, ..., p_\ell]$$

given by

$$\begin{aligned} F(x_0) &:= 0 \text{ and} \\ p_j &:= F(x_j) - F(x_{j-1}), j = 1, ..., \ell. \end{aligned}$$

The relationship between a continuous (!) distribution function F and its density function f is governed by the "Fundamental Theorem of Calculus".

DEFINITION V.12 (ANTIDERIVATIVE). *Let $f : [\underline{x}, \bar{x}] \to \mathbb{R}$ be a real-valued function (allow for $\underline{x} = -\infty$ and $\bar{x} = \infty$). A continuous function $F : [\underline{x}, \bar{x}] \to \mathbb{R}$ is called the antiderivative of f if it obeys $F' = f$.*

THEOREM V.2 (FUNDAMENTAL THEOREM OF CALCULUS). *Let $f : [\underline{x}, \bar{x}] \to \mathbb{R}$ be a continuous function.*

- *Define $F : [\underline{x}, \bar{x}] \to \mathbb{R}$ by*

$$F(x) := \int_{\underline{x}}^{x} f(t)\, dt.$$

Then, F is an antiderivative of f.

- *Let F be any antiderivative of f. Then, we have, for $\underline{x} \leq a \leq b \leq \bar{x}$,*

$$\int_a^b f(t)\, dt = F(b) - F(a) =: F(x)|_a^b.$$

As an example of a density function, consult figure V.12 with the payoff interval $[0, \bar{x}]$. The probability for the small payoff interval $\Delta x = [a, b]$ is the corresponding area under the density function, i.e.,

$$\int_a^b f(x)\, dx.$$

By definition V.10, the area under the whole interval $[0, \bar{x}]$ is,

$$\int_0^{\bar{x}} f(x)\, dx = 1.$$

EXERCISE V.18. *Figure V.12 is not complete. Find a constant (!) $f(x)$ for $x \in \left[\frac{3}{4}\bar{x}, \bar{x}\right]$ Calculate and draw the corresponding distribution function! Hint: You need to distinguish four cases!*

The advantage of working with F rather than f is that f cannot be used to describe discrete lotteries. Indeed, any F obeying eq. V.1 is continuous.

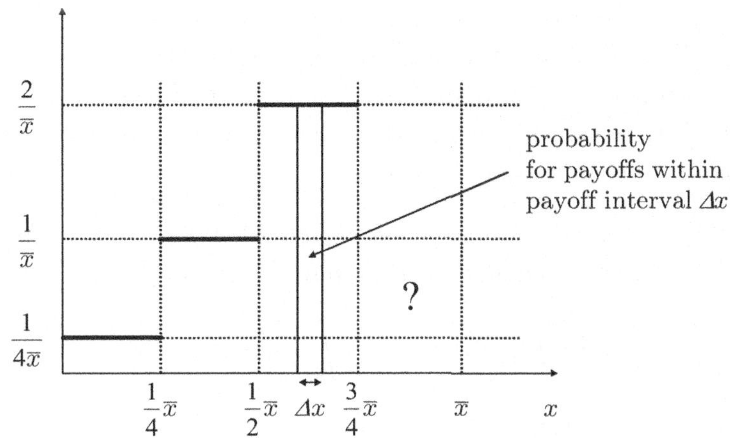

FIGURE V.12. A density function (to be completed)

5.2. First-order stochastic dominance. Consider figure V.13 which is taken from Mas-Colell et al. (1995, Figure 6.D.1, p. 196). Tell which distribution function you prefer. You may be tempted to point to G which lies above F everywhere, $G(x) \geq F(x)$ for all x.

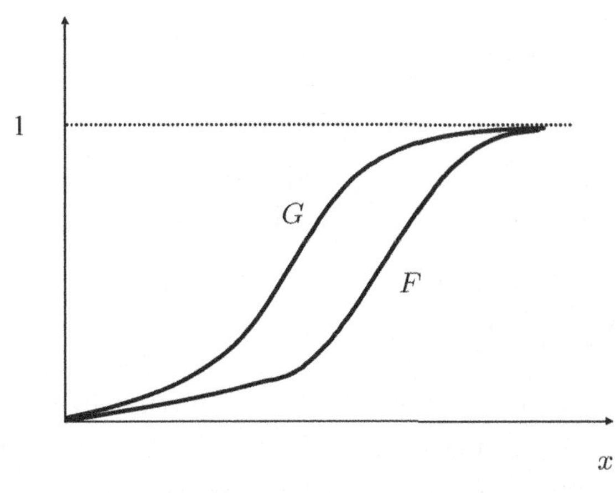

FIGURE V.13. F first-order stochastically dominates G

DEFINITION V.13 (FIRST-ORDER STOCHASTIC DOMINANCE). *Consider two distribution functions F and $G : [\underline{x}, \overline{x}] \to [0,1]$. F is said to first-order stochastically dominate G if*

$$F(x) \leq G(x)$$

for all $x \in [\underline{x}, \overline{x}]$.

F is better than G. After all, $G(100) \geq F(100)$ means that the probability of obtaining 100 or less is greater for G than for F. This implies that the probability of obtaining more than 100 is larger for F than for G. This is the intuitive reason behind the following two theorems, the first for continuous distribution functions, the second for discrete ones.

If density functions exist, we obtain

THEOREM V.3. *Distribution function F (with density function f) first-order stochastically dominates distribution function G (with density function g) iff*

$$\int_{\underline{x}}^{\overline{x}} f(x)\, u(x)\, dx \geq \int_{\underline{x}}^{\overline{x}} g(x)\, u(x)\, dx$$

holds for every vNM utility function u.

If we have no density function but discrete lotteries, we have the discrete counterpart of the above theorem:

THEOREM V.4. *The discrete distribution function F (with lottery $L_F = \left[x_1^F, ..., x_{\ell_F}^F; p_1^F, ..., p_{\ell_F}^F\right]$) first-order stochastically dominates the discrete distribution function G (with lottery $L_G = \left[x_1^G, ..., x_{\ell_G}^G; p_1^G, ..., p_{\ell_G}^G\right]$) iff*

$$\begin{aligned} E_u(L_F) &= \sum_{j=1}^{\ell_F} p_j^F u\left(x_j^F\right) \\ &\geq \sum_{j=1}^{\ell_G} p_j^G u\left(x_j^G\right) \\ &= E_u(L_G) \end{aligned}$$

holds for every vNM utility function u.

Read the two theorems carefully. They say that every agent (irrespective of his vNM utility function) prefers F to G whenever F first-order stochastically dominates G. Of course, it may well be the case that there is no first-order stochastic dominance between F and G one way or the other. Then, some agents might prefer F, while others prefer G.

5.3. Second-order stochastic dominance. Remember from the previous subsection that first-order stochastic dominance is equivalent to every agent preferring the dominating distribution. More restrictedly, second-order stochastic dominance means that every risk-averse agent prefers the dominating distribution. In case of first-order stochastic domination, the dominating distribution function lies below the dominated one. Second-order stochastic domination means that the area below the dominating distribution function is smaller than the area below the dominated one, where the integral is calculated from $\underline{x}$ up to any larger b.

DEFINITION V.14 (SECOND-ORDER STOCHASTIC DOMINANCE). *Consider two distribution functions F and $G : [\underline{x}, \bar{x}] \to [0, 1]$. F is said to second-order stochastically dominate G if*

$$\int_{\underline{x}}^{b} F(x)\, dx \leq \int_{\underline{x}}^{b} G(x)\, dx$$

for all $b \in [\underline{x}, \bar{x}]$.

Note that this definition applies to both discrete and continuous distribution functions.

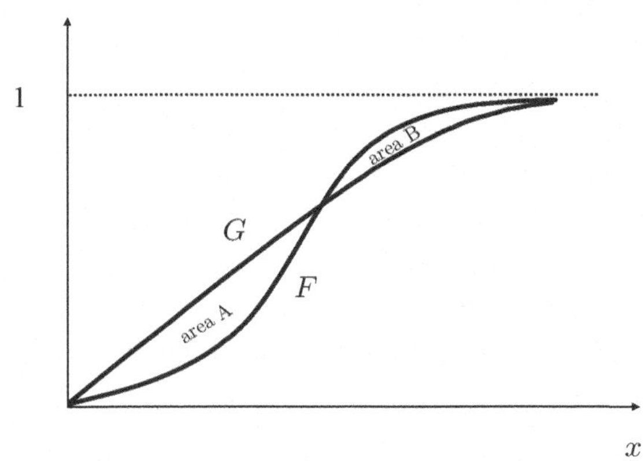

FIGURE V.14. F second-order stochastically dominates G

Of course, first-order stochastic dominance implies second-order stochastic dominance. Just have a look at figure V.13 and compare the area below the two curves between $\underline{x}$ and $b \geq \underline{x}$. Consider figure V.14 where F second-order stochastically dominates G. We see this from the fact that area A is larger than area B. Also, the means of G and F differ. Generally, two distribution functions have the same mean if the areas under the curves are the same. In the figure, the mean of F is larger than the mean of G because the area under F is smaller than the area under G. Figure V.15 (taken from Mas-Colell et al. 1995, p. 199) provides an illustration where F second-order stochastically dominates G and still has the same mean.

The best way to approach second-order stochastic dominance is by mean-preserving spreads. For example, $\left[95, 105; \frac{1}{2}, \frac{1}{2}\right]$ is a mean-preserving

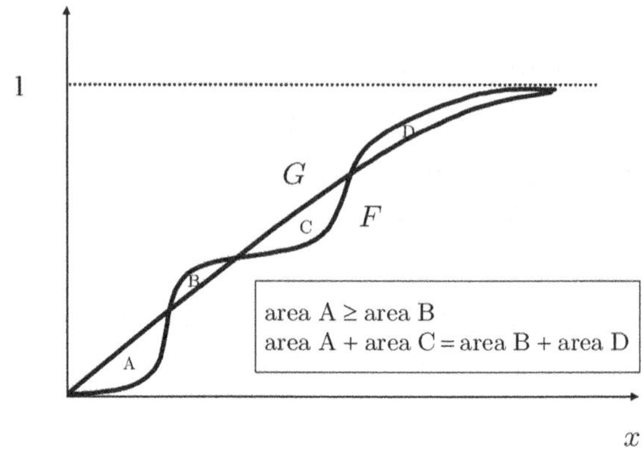

FIGURE V.15. Second-order stochastic dominance and same mean

spread of [100; 1] (compare exercise V.2, p. 96). The mean is the same while the payoffs are spread out. A risk-averse agent prefers the non-spread lottery to the spread-out one. In this fashion, the risk-averse agent gets to worse and worse lotteries as the payoffs are spread out more as in figure V.16. Mean-preserving spreads can also be shown with density functions. In figure V.17, the payoffs between $\frac{1}{2}\bar{x}$ and $\frac{3}{4}\bar{x}$ have density $\frac{2}{\bar{x}}$ in the upper subfigure. In the lower subfigure this density is reduced to $\frac{3}{2\bar{x}}$, half of which goes to interval $\left[\frac{1}{4}\bar{x}, \frac{1}{2}\bar{x}\right]$ and the other half to interval $\left[\frac{3}{4}\bar{x}, \bar{x}\right]$.

Spreading payoffs means increasing risk. This fact provides an intuitive reason for the following theorem.

THEOREM V.5. *Consider two lotteries L and $\hat{L}$ that have the same expected value. Lottery $L = [x_1, ..., x_\ell; p_1, ..., p_\ell]$ is a mean preserving spread of lottery $\hat{L} = [\hat{x}_1, ..., \hat{x}_{\hat{\ell}}; \hat{p}_1, ..., \hat{p}_{\hat{\ell}}]$ if*

$$E_u(L) \leq E_u\left(\hat{L}\right)$$

holds for every concave vNM utility function u.

Beautifully, this theorem can be connected to second-order stochastic dominance.

THEOREM V.6. *Assume two distribution functions F (with density function f) and G (with density function g) with the same mean, i.e., obeying*

$$\int_{\underline{x}}^{\bar{x}} f(x)\, x dx = \int_{\underline{x}}^{\bar{x}} g(x)\, x dx.$$

We have

- *F second-order stochastically dominates G iff*

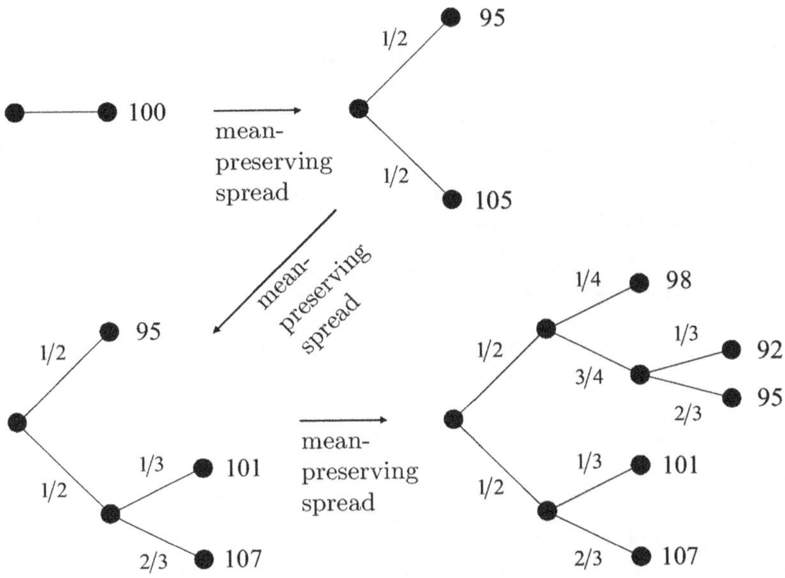

FIGURE V.16. Mean-preserving spreads

- *every risk-averse agent prefers F over G, i.e.,*

$$\int_{\underline{x}}^{\overline{x}} f(x)\, u(x)\, dx \geq \int_{\underline{x}}^{\overline{x}} g(x)\, u(x)\, dx$$

holds for every concave vNM utility function u iff

- *G is a mean-preserving spread of F.*

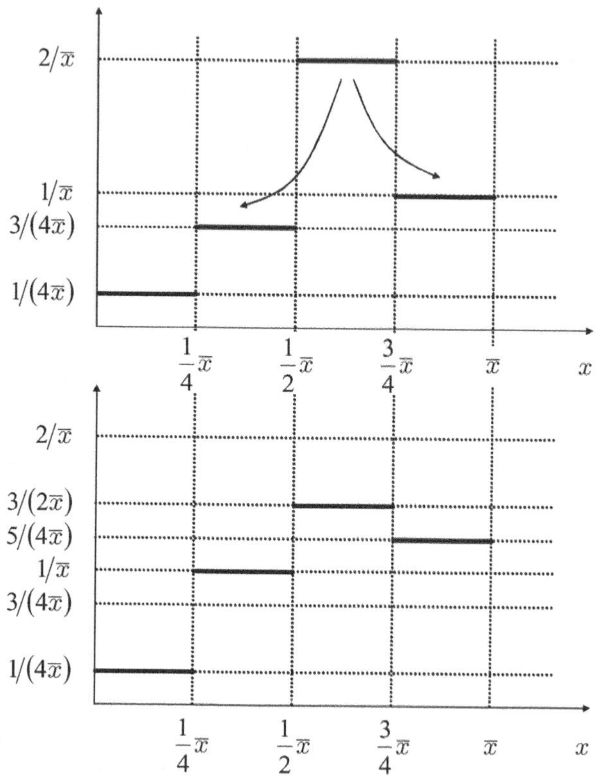

FIGURE V.17. Mean-preserving spread

6. Topics

The main topics in this chapter are

- compound lottery
- expected value
- expected utility
- Bernoulli principle
- independence axiom
- continuity axiom
- risk aversion, risk neutrality and risk loving
- St. Petersburg lottery
- Arrow-Pratt measure
- risk premium
- certainty equivalent
- first-order stochastic dominance
- second-order stochastic dominance

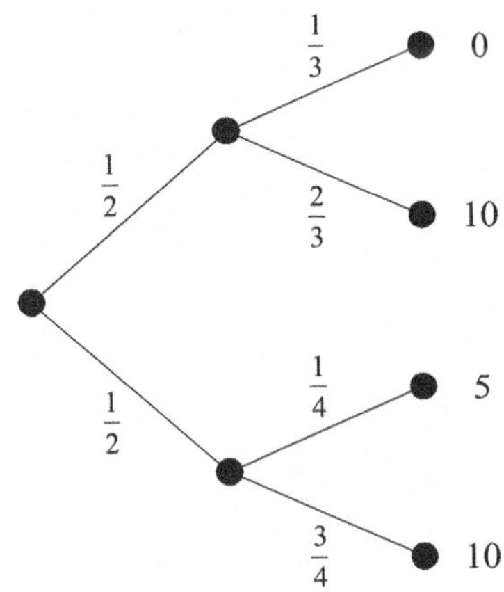

FIGURE V.18. The two-stage lottery

- concave and convex function
- Fundamental Theory of Calculus

7. Solutions

Exercise V.1

If L_1 is your preferred lottery, you do not understand the concept of a lottery, the concept of probability, or you hate money.

Exercise V.2

There is no correct answer here. However, if you prefer the lottery to the expected value of the lottery, you are a risk-loving person.

Exercise V.3

We find

$$\begin{aligned} L &= \left[L_1, L_2; \frac{1}{2}, \frac{1}{2}\right] \\ &= \left[0, 5, 10; \frac{1}{2} \cdot \frac{1}{3}, \frac{1}{2} \cdot \frac{1}{4}, \frac{1}{2} \cdot \frac{2}{3} + \frac{1}{2} \cdot \frac{3}{4}\right] \\ &= \left[0, 5, 10; \frac{1}{6}, \frac{1}{8}, \frac{17}{24}\right] \end{aligned}$$

The two-stage lottery $\left[L_1, L_2; \frac{1}{2}, \frac{1}{2}\right]$ is depicted in figure V.18 and the one-stage lottery $\left[0, 5, 10; \frac{1}{6}, \frac{1}{8}, \frac{17}{24}\right]$ in figure V.19.

Exercise V.4

We obtain

$$\frac{\frac{1}{2}}{1 - \frac{1}{2}} = 1.$$

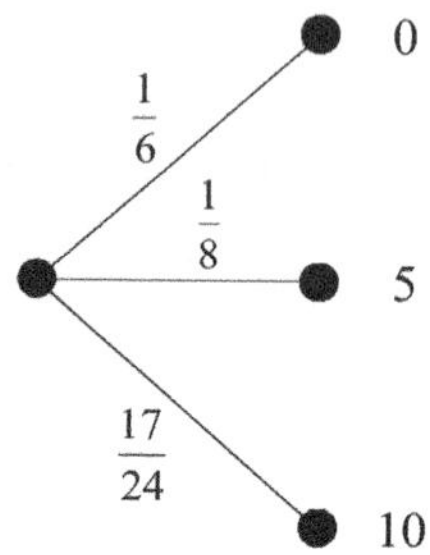

FIGURE V.19. The one-stage lottery

Exercise V.5

Again, no correct answer. However, your willingness to pay should be greater than 2.

Exercise V.6

You may remember (otherwise commit to your memory!)

- $\ln(1) = 0$,
- $\ln\left(\frac{x}{y}\right) = \ln x - \ln y$
- $\ln(xy) = \ln x + \ln y$
- $\ln x^b = b \ln x$
- $\frac{d \ln x}{dx} = \frac{1}{x}$.

Exercise V.7

For the ln utility function, the lottery $L_{\ln}$ yields an infinite expected utility:

$$\begin{aligned} E_{\ln}(L_{\ln}) &= \sum_{j=1}^{\infty} \frac{1}{2^j} \ln\left(2^{(2^j)}\right) = \sum_{j=1}^{\infty} \frac{1}{2^j} 2^j \ln 2 \\ &= \ln 2 \sum_{j=1}^{\infty} \frac{1}{2^j} 2^j = \ln 2 (1 + 1 + \ldots) = \infty. \end{aligned}$$

Exercise V.8

Only the last one is bounded, by 1.

Exercise V.9

Here is the proof of the simple proposition:

$$E_u(L) = \sum_{j=1}^{\ell} p_j \cdot u(x_j) \leq \sum_{j=1}^{\ell} p_j \cdot K = K \sum_{j=1}^{\ell} p_j = K$$

Exercise V.10

We find indifference between L_3 and L_4 :

$$\begin{aligned}
& \left[0, 50, 100; \frac{1}{4}, \frac{1}{2}, \frac{1}{4}\right] \\
= & \left[L_1, 50; \frac{1}{2}, \frac{1}{2}\right] \text{ (compound lottery)} \\
\sim & \left[L_2, 50; \frac{1}{2}, \frac{1}{2}\right] \text{ (independence axiom, } L_1 \sim L_2\text{)} \\
= & \left[16, 25, 50; \frac{1}{8}, \frac{3}{8}, \frac{1}{2}\right] \text{ (compound lottery)}
\end{aligned}$$

Exercise V.11

We find

$$\begin{aligned}
\left[L_1, L_3; \frac{1}{2}, \frac{1}{2}\right] & = \left[12 \cdot 10^6, 1 \cdot 10^6, 0; \frac{1}{2} \cdot \frac{10}{100}, \frac{1}{2} \cdot 1, \frac{1}{2} \cdot \frac{90}{100}\right] \\
& = \left[12 \cdot 10^6, 1 \cdot 10^6, 0; \frac{5}{100}, \frac{50}{100}, \frac{45}{100}\right]
\end{aligned}$$

and

$$\begin{aligned}
& \left[L_2, L_4; \frac{1}{2}, \frac{1}{2}\right] \\
= & \left[12 \cdot 10^6, 1 \cdot 10^6, 0; \frac{1}{2} \cdot \frac{10}{100}, \frac{1}{2} \cdot \frac{11}{100} + \frac{1}{2} \cdot \frac{89}{100}, \frac{1}{2} \cdot \frac{89}{100} + \frac{1}{2} \cdot \frac{1}{100}\right] \\
= & \left[12 \cdot 10^6, 1 \cdot 10^6, 0; \frac{5}{100}, \frac{50}{100}, \frac{45}{100}\right].
\end{aligned}$$

Oops.

Exercise V.12

An obvious choice is $v(x) = \frac{-100}{3} + \frac{1}{3} u(x) = x + 3x^2$.

Exercise V.13

Let v be defined by $v(x) = a + bu(x)$ for $a \in \mathbb{R}$ and $b > 0$. Then, we obtain the desired chain of equivalences

$$\begin{aligned}
& E_v\left(L^A\right) \geq E_v\left(L^B\right) \\
\Leftrightarrow & \sum_{j=1}^{\ell_A} p_j^A v\left(x_j^A\right) \geq \sum_{j=1}^{\ell_B} p_j^B v\left(x_j^B\right) \\
\Leftrightarrow & \sum_{j=1}^{\ell_A} p_j^A \left[a + bu\left(x_j^A\right)\right] \geq \sum_{j=1}^{\ell_B} p_j^B \left[a + bu\left(x_j^B\right)\right]
\end{aligned}$$

$$
\begin{aligned}
\Leftrightarrow \quad & a\sum_{j=1}^{\ell_A} p_j^A + b\sum_{j=1}^{\ell_A} p_j^A u\left(x_j^A\right) \geq a\sum_{j=1}^{\ell_B} p_j^B + b\sum_{j=1}^{\ell_B} p_j^B u\left(x_j^B\right) \\
\Leftrightarrow \quad & \sum_{j=1}^{\ell_A} p_j^A u\left(x_j^A\right) \geq \sum_{j=1}^{\ell_B} p_j^B u\left(x_j^B\right) \text{ (b>0)} \\
\Leftrightarrow \quad & E_u\left(L^A\right) \geq E_u\left(L^B\right).
\end{aligned}
$$

Exercise V.14

$L \sim [L_{good}, L_{bad}; p(L), 1-p(L)]$ means $E_u(L) = p(L)\, E_u(L_{good}) + [1-p(L)]\, E_u(L_{bad})$ for the appropriate vNM utility u. For $L := L_{good}$ we obtain

$$[1-p(L_{good})]\,[E_u(L_{good}) - E_u(L_{bad})] = 0.$$

By $E_u(L_{good}) > E_u(L_{bad})$, we find $p(L_{good}) = 1$ and $p(L_{bad}) = 0$.

Exercise V.15

Consider the function $f : M \to \mathbb{R}$ with convex domain $M \subseteq \mathbb{R}$ and defined by $f(x) = \begin{cases} 1, & x \in Q \\ 0, & x \notin Q \end{cases}$. It is neither concave, nor convex. If f is twice differentiable, it is easy to find a function that is concave on some subset of M and convex on the other.

Exercise V.16

By

- $u_1''(x) = 2 > 0$,
- $u_2''(x) = 0$,
- $u_3''(x) = -\frac{1}{x^2} < 0$, and
- $u_4''(x) = -e^{-x} < 0$ (chain rule)
- $u_5''(x) = -\theta x^{-\theta-1} < 0$

we have a risk-lover in case of u_1, a risk-neutral decision maker in case of u_2 and risk-averse decision maker in case of u_3, u_4, and u_5.

Exercise V.17

We obtain

- $ARA_{u_1}(x) = -\frac{2}{2x} = -\frac{1}{x}$ and $RRA_{u_1}(x) = -1$
- $ARA_{u_2}(x) = 0$ and $RRA_{u_2}(x) = 0$
- $ARA_{u_3}(x) = -\frac{-\frac{1}{x^2}}{\frac{1}{x}} = \frac{1}{x}$ and $RRA_{u_3}(x) = 1$
- $ARA_{u_4}(x) = -\frac{-e^{-x}}{e^{-x}} = 1$ and $RRA_{u_4}(x) = x$
- $ARA_{u_5}(x) = -\frac{-\theta x^{-\theta-1}}{\frac{(1-\theta)x^{-\theta}}{1-\theta}} = \frac{\theta}{x}$ and $RRA_{u_5}(x) = \theta$

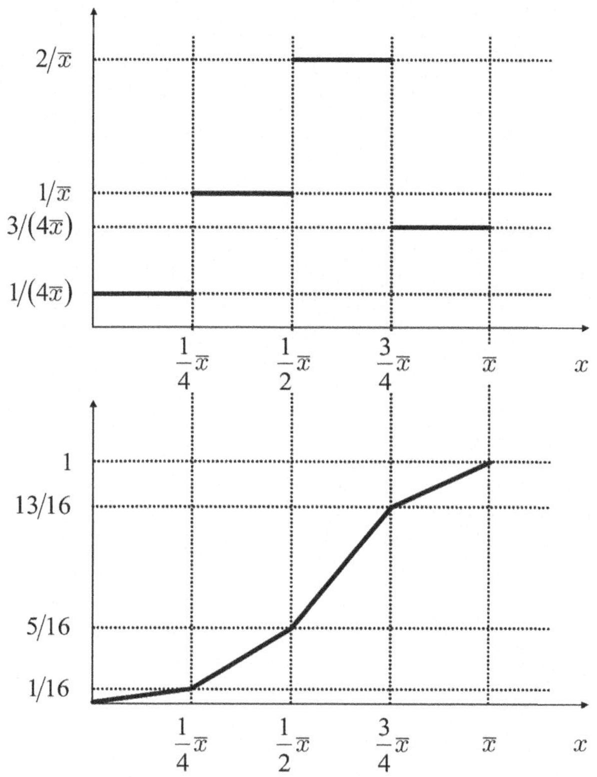

FIGURE V.20. Deriving the distribution function from the density function

Exercise V.18

For $x \in \left[\frac{3}{4}\bar{x}, \bar{x}\right]$, we have

$$1 = \frac{1}{4\bar{x}} \cdot \frac{1}{4}\bar{x} + \frac{1}{\bar{x}} \cdot \frac{1}{4}\bar{x} + \frac{2}{\bar{x}} \cdot \frac{1}{4}\bar{x} + f(x) \cdot \frac{1}{4}\bar{x}$$

and hence

$$f(x) = \frac{3}{4\bar{x}}.$$

We now calculate the corresponding distribution function F and distinguish four cases (see figure V.20):

- $0 \le b \le \frac{1}{4}\bar{x}$:

$$F(b) = \int_0^b \frac{1}{4\bar{x}} dx = \frac{1}{4\bar{x}} x \Big|_0^b = \frac{1}{4\bar{x}} b$$

- $\frac{1}{4}\bar{x} \leq b \leq \frac{1}{2}\bar{x}$:

$$\begin{aligned} F(b) &= \frac{1}{4\bar{x}}\frac{1}{4}\bar{x} + \int_{\frac{1}{4}\bar{x}}^{b} \frac{1}{\bar{x}}dx = \frac{1}{16} + \frac{1}{\bar{x}}x\Big|_{\frac{1}{4}\bar{x}}^{b} \\ &= \frac{1}{16} + \frac{1}{\bar{x}}b - \frac{1}{\bar{x}}\frac{1}{4}\bar{x} = \frac{1}{\bar{x}}b - \frac{3}{16} \end{aligned}$$

- $\frac{1}{2}\bar{x} \leq b \leq \frac{3}{4}\bar{x}$:

$$\begin{aligned} F(b) &= \frac{1}{\bar{x}}\frac{1}{2}\bar{x} - \frac{3}{16} + \int_{\frac{1}{2}\bar{x}}^{b} \frac{2}{\bar{x}}dx = \frac{1}{2} - \frac{3}{16} + \frac{2}{\bar{x}}x\Big|_{\frac{1}{2}\bar{x}}^{b} \\ &= \frac{1}{2} - \frac{3}{16} + \frac{2}{\bar{x}}b - \frac{2}{\bar{x}}\frac{1}{2}\bar{x} = \frac{2}{\bar{x}}b - \frac{11}{16} \end{aligned}$$

- $\frac{3}{4}\bar{x} \leq b \leq \bar{x}$:

$$\begin{aligned} F(b) &= \frac{2}{\bar{x}}\frac{3}{4}\bar{x} - \frac{11}{16} + \int_{\frac{3}{4}\bar{x}}^{b} \frac{3}{4\bar{x}}dx = \frac{3}{2} - \frac{11}{16} + \frac{3}{4\bar{x}}x\Big|_{\frac{3}{4}\bar{x}}^{b} \\ &= \frac{3}{2} - \frac{11}{16} + \frac{3}{4\bar{x}}b - \frac{3}{4\bar{x}}\frac{3}{4}\bar{x} = \frac{3}{4\bar{x}}b + \frac{1}{4} \end{aligned}$$

8. Further exercises without solutions

PROBLEM V.1. Socrates has an endowment of 225 million Euro most of which is invested in a luxury cruise ship worth 200 million Euro. The ship sinks with a probability of $\frac{1}{5}$. Socrates vNM utility function is given by $u(x) = \sqrt{x}$. What is his willingness to pay for full insurance?

PROBLEM V.2. Identify the certainty equivalent and the risk premium in figure V.10 (p. 109).

PROBLEM V.3. Let $W = \{w_1, w_2\}$ be a set of 2 states of the world. The contingent good 1 that pays one Euro in case of state of the world w_1 and nothing in the other state is called an Arrow security. Determine this Arrow security in an x_1-x_2-diagram.

PROBLEM V.4. Sarah may become a paediatrician or a clerk in an insurance company. She expects to earn 40 000 Euro as a clerk every year. Her income as paediatrician depends on the number of children that will be born. In case of a baby boom, her yearly income will be 100 000 Euro, otherwise 20 000 Euro. She estimates the probability of a babyboom at $\frac{1}{2}$. Sarah's vNM utility function is given by $u(x) = 300 + \frac{4}{5}x$.

- Formulate Sarah's choices as lotteries!
- What is Sarah's best choice?
- The Institute of Advanced Demography (IAD) has developed a secret, but reliable, method of predicting a baby boom. Sarah can buy the information for constant yearly rates. What is the maximum yearly willingness to pay?
- Sketch Sarah's decision problem in x_1-x_2 space where income without babyboom is noted at the x_1-axis and income with babyboom at the x_2-axis.

PROBLEM V.5. Consider figure V.9, p. 108, and draw a corresponding figure for risk neutral and risk-loving preferences.

PROBLEM V.6. André's von Neumann and Morgenstern utility function satisfies $u(0) = 0$ and $u(1) = 1$. In addition, André is risk averse.

(1) Show that $u(x) \geq x$ holds for all $x \in [0,1]$. Hint: Begin with $x = (1-x) \cdot u(0) + x \cdot u(1)$!

(2) André has to choose between the following two lotteries:

$$L_A = \left[0, 0.3, 0.6, 1; \frac{4}{10}, \frac{3}{10}, \frac{2}{10}, \frac{1}{10}\right] \text{ and}$$

$$L_B = \left[0, 0.3, 0.6, 1; \frac{29}{50}, \frac{1}{10}, \frac{1}{10}, \frac{11}{50}\right].$$

Can we infer something about his preferences over these two lotteries?

- Hint: Just calculate the expected utilities and use (1)!
- Alternative hint: Apply second-order stochastic dominance twice:
  - replace 0.6 by a lottery involving the payoffs 0 and 1,
  - reduce the probability for 0.3 from $\frac{3}{10}$ to $\frac{1}{10}$ and introduce a lottery involving the payoffs 0 and 0.6 instead.

PROBLEM V.7. Show that we have first-order domination between the lotteries

$$L_1 = \left[\frac{1}{8}, \frac{1}{6}, \frac{1}{4}, \frac{1}{2}; \frac{1}{4}, \frac{1}{4}, \frac{1}{4}, \frac{1}{4}\right]$$

and

$$L_2 = \left[\frac{1}{8}, \frac{1}{6}, \frac{1}{4}, \frac{1}{2}; \frac{1}{4}, \frac{1}{6}, \frac{1}{6}, \frac{5}{12}\right].$$

Can you find vNM utility functions u and v such that $E_u(L_1) > E_u(L_2)$ and $E_v(L_1) < E_v(L_2)$?

Part B

Household theory and theory of the firm

We now turn to the second part of the course where the deciders are households or firms. We begin with household theory. We need to describe the bundles of goods a household can afford (the budget) and the preferences of the household. The relationship between preferences and utility functions has been worked out in the two previous chapters. The budget and the optimal goods bundle chosen by the household are considered in chapter VI. Finally, in chapter VII, we examine how the household's decision depends on the parameters of the model, i.e., we turn to comparative statics.

We then deal with the very simple theory of the firm. It is simple in that we assume a single decider who tries to maximize his profits. We cover production theory in chapter VIII and both cost minimization and profit maximization in chapter IX. For the time being, we leave aside principal-agent problems. The simple theory of the firm benefits from household theory (or vice versa) – many concepts can be transferred from one setting to the other. It also lays the groundwork for oligopoly theory treated in part C and for general equilibrium theory (part F, chapter XIX).

A simplifying assumption assumed throughout this part is price takership. That is, households and firms consider prices of goods and factors as given. If households and firms are small relative to the market, this assumption is not too weird. Of course, in later chapters, we will do without it.

CHAPTER VI

The household optimum

We now consider decisions of households facing prices for each of the ℓ goods. Assuming price takership, the households buy a best bundle within their budget. Therefore, we analyze the budget first and then derive a best bundle on the basis of budget and preferences.

1. Budget

1.1. Money budget. We first assume that the household has some monetary amount m at his disposal. The budget is the set of bundles that the household can afford, i.e., the set of bundles whose expenditure is not above m. The expenditure for a bundle of goods $x = (x_1, x_2, ..., x_\ell)$ at a vector of prices $p = (p_1, p_2, ..., p_\ell)$ is the dot product (or the scalar product) of the two vectors:

$$p \cdot x := \sum_{g=1}^{\ell} p_g x_g.$$

DEFINITION VI.1 (MONEY BUDGET). *For a price vector $p \in \mathbb{R}^\ell$ and monetary income $m \in \mathbb{R}_+$, the money budget is defined by*

$$B(p, m) := \left\{ x \in \mathbb{R}_+^\ell : p \cdot x \leq m \right\}$$

where

$$\left\{ x \in \mathbb{R}_+^\ell : p \cdot x = m \right\}$$

is called the budget line.

For example, in case of two goods, the budget is the set of bundles fulfilling $p_1 x_1 + p_2 x_2 \leq m$. If the household does not consume good 1 ($x_1 = 0$), he can consume up to m/p_2 units of good 2. (Just solve the inequality for x_2.). In figure VI.1, the household can afford bundles A and B, but not C.

The following theorem should be no surprise to you. If you double all prices and income, your budget remains unchanged:

LEMMA VI.1. *For any number $\alpha > 0$, we have $B(\alpha p, \alpha m) = B(p, m)$.*

H. Wiese, *Advanced Microeconomics*,
https://doi.org/10.1007/978-3-658-34959-2_6

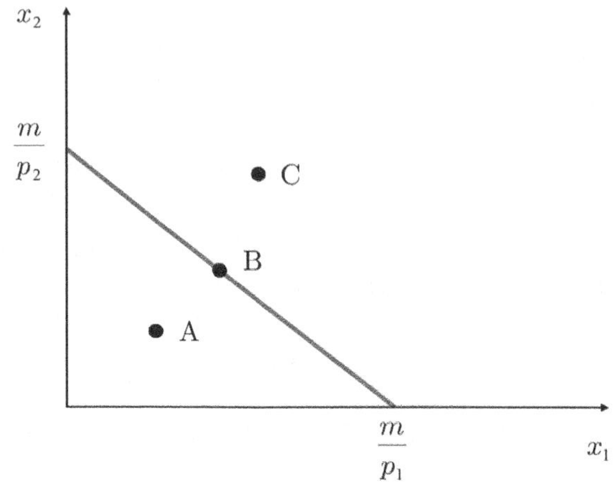

FIGURE VI.1. The budget for two goods

EXERCISE VI.1. *Fill in: For any number* $\alpha > 0$, *we have* $B(\alpha p, m) = B(p, ?)$.

EXERCISE VI.2. *Assume that the household consumes bundle A in figure VI.1. Identify the "left-over" in terms of good* 1, *in terms of good* 2, *and in money terms.*

LEMMA VI.2. *The money budget is nonempty, closed, and convex. If* $p >> 0$ *holds, the budget is bounded.*

PROOF. The budget is nonempty because we have $(0, ..., 0) \in \mathbb{R}_+^\ell$ and $0 \cdot p = 0 \leq m$. It is closed because it is defined by way of weak inequalities ($x_g \geq 0$, $g = 1, ..., \ell$, $x \cdot p \leq m$). We now show convexity. Consider two bundles x and x' from $B(p, m)$ and a number $k \in [0, 1]$. Then $x \cdot p \leq m$ and $x' \cdot p \leq m$ hold and we have $(kx + (1-k)x') \cdot p = kx \cdot p + (1-k)x' \cdot p \leq km + (1-k)m = m$ so that $kx + (1-k)x'$ is also contained in $B(p, m)$. Therefore, the budget is convex. Finally, the budget is bounded in case of $p >> 0$ because every bundle x in the budget fulfills $0 \leq x \leq \left(\frac{m}{p_1}, ..., \frac{m}{p_\ell}\right)$. □

EXERCISE VI.3. *Verify that the budget line's slope is given by* $-\frac{p_1}{p_2}$ *in case of two goods and* $p_2 \neq 0$.

If both prices are positive, the budget line is negatively sloped.

DEFINITION VI.2 (MARGINAL OPPORTUNITY COST). *If all the prices are non-negative and the price of good* 2 *is positive, the marginal opportunity*

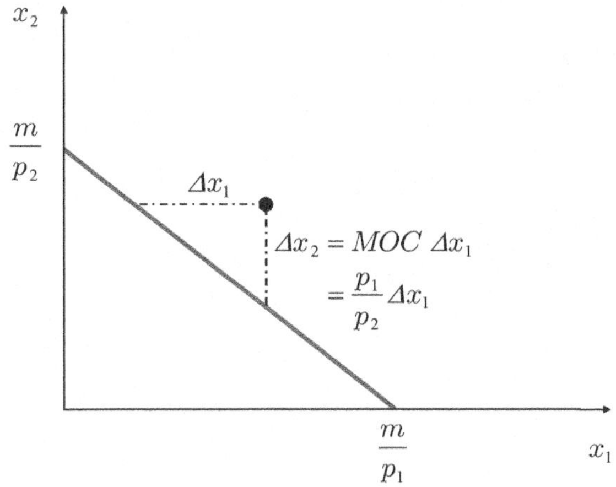

FIGURE VI.2. The opportunity cost of one additional unit of good 1

cost of consuming one unit of good 1 *in terms of good* 2 *is denoted by* $MOC(x_1)$ *and given by*

$$MOC(x_1) = \left|\frac{dx_2}{dx_1}\right| = \frac{p_1}{p_2}.$$

Thus, if the household wants to consume one additional unit of good 1, he needs to forgo MOC units of good 2 (see also figure VI.2). Note that we use the absolute value of the budget line's slope – very similar to the definition of the marginal rate of substitution on p. 84.

1.2. Endowment budget.

1.2.1. *Definition.* In the previous section, the budget is defined by some monetary income m. We now assume that the household has some endowment $\omega \in \mathbb{R}^\ell_+$, which he can consume or, at the prevailing prices, use to buy another bundle. In any case, we obtain the following definition:

DEFINITION VI.3 (ENDOWMENT BUDGET). *For a price vector* $p \in \mathbb{R}^\ell$ *and an endowment* $\omega \in \mathbb{R}^\ell_+$, *the endowment budget is defined by*

$$B(p,\omega) := \left\{x \in \mathbb{R}^\ell_+ : p \cdot x \leq p \cdot \omega\right\}.$$

Again, equality defines the budget line.

By $m := \omega \cdot p$, endowment budgets turn into money budgets. Therefore, lemma VI.2 holds for an endowment budget as well.

In case of two goods, the budget line is written as

$$p_1x_1 + p_2x_2 = p_1\omega_1 + p_2\omega_2$$

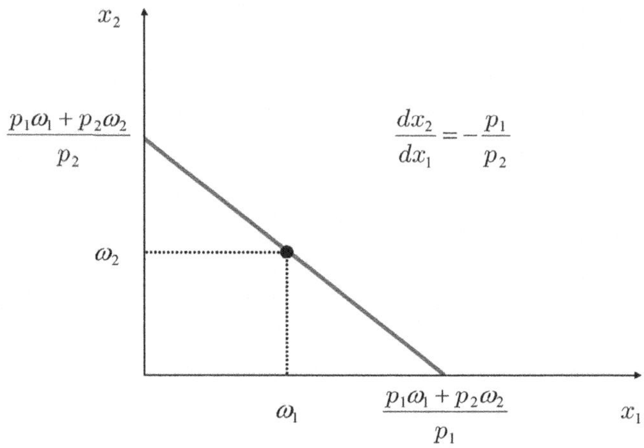

FIGURE VI.3. The endowment budget

and depicted in figure VI.3.

1.2.2. *Application: consumption today versus consumption tomorrow* . We now present three very important examples of endowment budgets. Our first example deals with intertemporal consumption. Consider a household whose monetary income in periods 1 and 2 is ω_1 and ω_2, respectively. His consumption is denoted by x_1 and x_2. We assume that he can borrow ($x_1 > \omega_1$) or lend ($x_1 < \omega_1$). Of course, he can also decide to just consume what he earns ($x_1 = \omega_1$). In either case, he has to break even at the end of the second period. At a given rate of interest r, his second-period consumption is

$$\begin{aligned} x_2 &= \underbrace{\omega_2}_{\substack{\text{second-period} \\ \text{income}}} + \underbrace{(\omega_1 - x_1)}_{\substack{\text{amount borrowed } (<0) \\ \text{or lended } (>0)}} + \underbrace{r(\omega_1 - x_1)}_{\substack{\text{interest payed } (<0) \\ \text{or earned } (>0)}} \\ &= \omega_2 + (1+r)(\omega_1 - x_1) \end{aligned}$$

We can rewrite the break-even condition (the budget equation) in two different fashions.

- Equalizing the future values of consumption and income yields

$$(1+r)x_1 + x_2 = (1+r)\omega_1 + \omega_2,$$

while

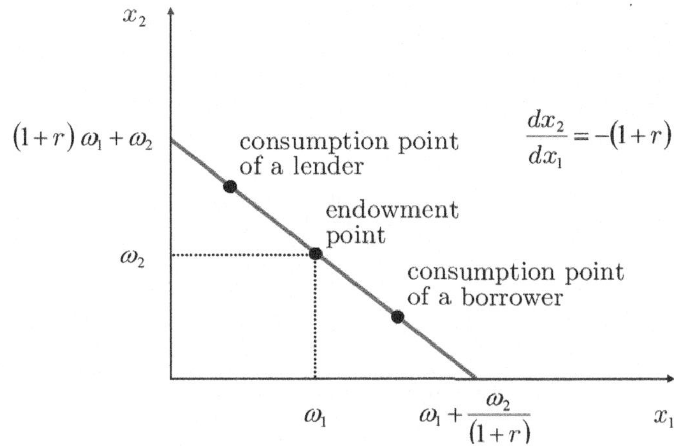

FIGURE VI.4. Save or borrow?

- the equality of the present values of consumption and income is behind the budget equation

$$x_1 + \frac{x_2}{1+r} = \omega_1 + \frac{\omega_2}{1+r}.$$

Consider also figure VI.4 where the present value of the income stream (ω_1, ω_2) is found at the x_1-axis and the future value at the x_2-axis. The marginal opportunity cost of one additional unit of consumption in period 1 is

$$MOC = \left|\frac{dx_2}{dx_1}\right| = 1 + r$$

units of consumption in period 2.

1.2.3. *Application: leisure versus consumption.* A second application concerns the demand for leisure or, differently put, the supply of labor. We depict the budget line in figure VI.5. Recreational hours are denoted by x_R. By definition, the household works $24 - x_R$ hours. For obvious reasons, we have $0 \le x_R \le 24 = \omega_R$. Recreational time is good 1, the second good is real consumption x_C. x_C may stand for the only consumption good (bread) bought and sold at price p. Alternatively, you can think of a bundle of goods x_C and an aggregate price (index) p.

At a wage rate w, the household earns $w\,(24 - x_R)$. He may also obtain some non-labor income $p\omega_C$ where p is the price index and ω_C the real non-labor income. Thus, the household's consumption in nominal terms is

$$px_C = p\omega_C + w\,(24 - x_R)$$

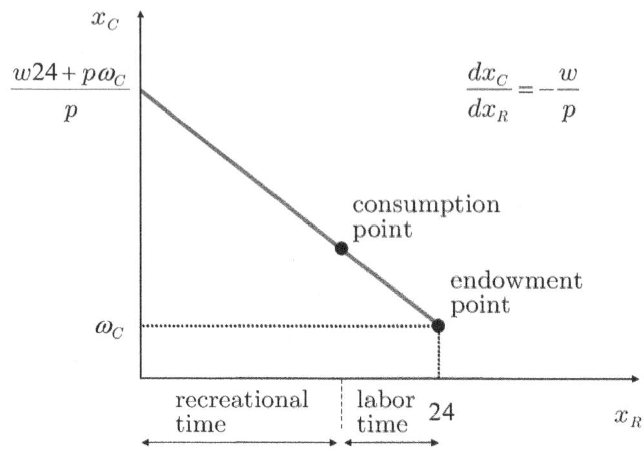

FIGURE VI.5. Recreational versus labor time

which can also be rewritten in endowment-budget form

$$wx_R + px_C = w24 + p\omega_C$$

where $(24, \omega_C)$ is the endowment point. Thus, the price of leisure is the wage rate. Indeed, if a household chooses to increase its recreational time by one unit, he foregoes w (in monetary consumption terms) or $\frac{w}{p}$ (in real consumption terms). The marginal opportunity cost of one unit of recreational time is

$$MOC = \left|\frac{dx_C}{dx_R}\right| = \frac{w}{p}$$

units of real consumption.

1.2.4. *Application: contingent consumption.* Our last application deals with contingent consumption and insurance. (You may want to revisit figures V.3 (p. 95) and V.10 (p. 109).) We consider a household whose wealth A may be hit by some calamity resulting in a damage D. Let p be the probability of this bad event. Then, the household is confronted with the lottery

$$L = [A - D, A; p, 1 - p].$$

For a given damage probability p, this lottery is the no-insurance point in x_1-x_2-space as shown in figure VI.6. We now assume the possibility to insure against the damage. The household's decision is the insurance sum K which is to be paid to the household in case of loss. The insurance premium (to be paid to the insurance company) is γK. Thus, the household has

$$x_1 = A - D + K - \gamma K = A - D + (1 - \gamma) K \qquad \text{(VI.1)}$$

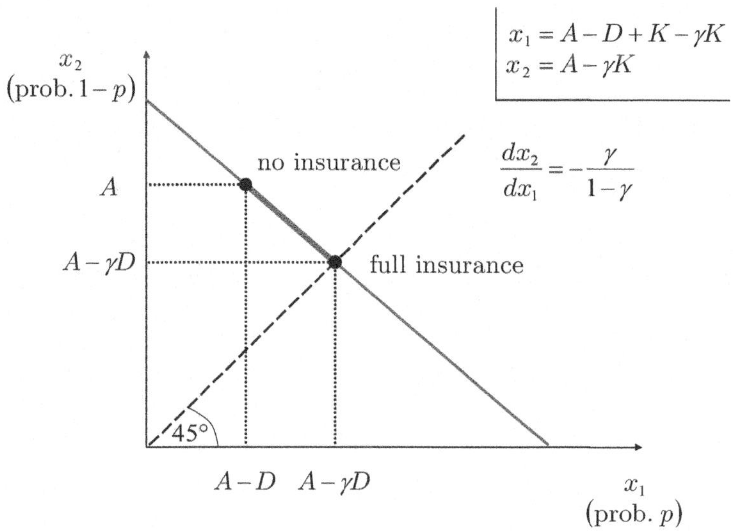

FIGURE VI.6. Insure or not insure?

in case of damage or

$$x_2 = A - \gamma K \tag{VI.2}$$

if no damage occurs.

The special case of full insurance is defined by $K := D$ and leads to the payoffs

$$x_1 = x_2 = A - \gamma D.$$

Equations VI.1 and VI.2 can be rewritten into a single equation by solving the second for K and substituting into the first. After rearranging the terms appropriately, we obtain

$$x_1 + \frac{1-\gamma}{\gamma} x_2 = (A - D) + \frac{1-\gamma}{\gamma} A$$

which has the usual form of an endowment-budget equation.

In figure VI.6, the part of the budget line left of the no-insurance point results from $K < 0$. This means that the premium γK is paid to the household who pays K to the insurance company if the damage occurs. This is a negative insurance.

EXERCISE VI.4. *Interpret the part of the budget line right of the full-insurance point.*

2. The household optimum

2.1. The household's decision situation and problem. In chapter II (p. 10), a decision situation in strategic form without uncertainty is denoted by $\Delta = (S, u : S \to \mathbb{R})$. A household's strategy set is the budget set B. Therefore, we propose the following definition.

Definition VI.4 (a household's decision situation). *A household's decision situation is a tuple*

$$\begin{aligned} \Delta &= (B, \precsim) \text{ with} \\ B &= B(p, m) \subseteq \mathbb{R}_+^\ell \text{ or } B = B(p, \omega) \subseteq \mathbb{R}_+^\ell \end{aligned}$$

where $p \in \mathbb{R}^\ell$ is a vector of prices and $\precsim$ a preference relation on $\mathbb{R}_+^\ell$. The household's problem is to find the best-response function given by

$$x^R(B) := \{x \in B\text{: there is no } x' \in B \text{ with } x' \succ x\}$$

If $\precsim$ is representable by a utility function U on $\mathbb{R}_+^\ell$, we have the decision situation $\Delta = (B, U)$ and the best-response function

$$x^R(B) := \arg\max_{x \in B} U(x).$$

Any x^* from $x^R(B)$ is called a household optimum. We often write $x^R(p, m)$ or $x(p, m)$.

Thus, the household aims to find the highest indifference curve attainable with his budget. As a very obvious corollary from lemma VI.1, we obtain the following lemma.

Lemma VI.3. *For any number $\alpha > 0$, we have $x^R(\alpha p, \alpha m) = x^R(p, m)$.*

Exercise VI.5. *Look at the household situations depicted in figure VI.7. Assume monotonicity of preferences. Are the highlighted points A or B optima?*

Exercise VI.6. *Assume a household's decision problem with endowment $\Delta = (B(p, \omega), \precsim)$. $x^R(B)$ consists of the bundles x that fulfill the two conditions:*

(1) *The household can afford x, i.e.,*

$$p \cdot x \leq p \cdot \omega$$

(2) *There is no other bundle y that the household can afford and that he prefers to x, i.e.,*

$$y \succ x \Rightarrow ??$$

Substitute the question marks by an inequality.

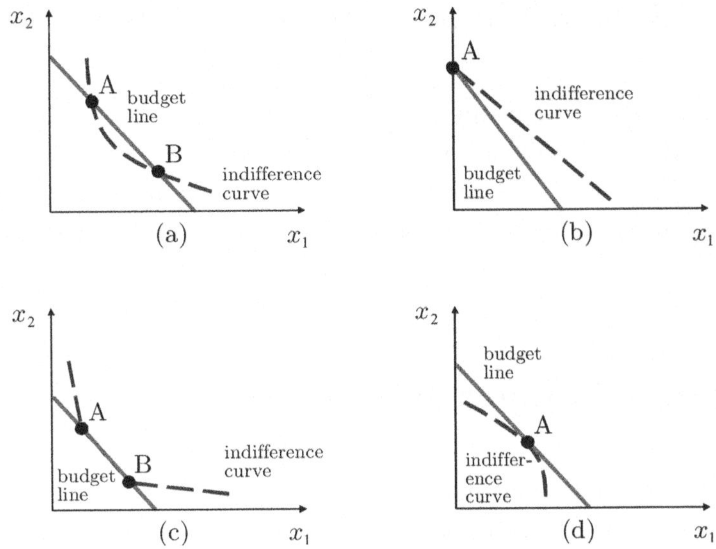

FIGURE VI.7. Household optima?

2.2. MRS versus MOC. A good part of household theory can be couched in terms of the marginal rate of substitution and the marginal opportunity cost. Consider figure VI.8. We can ask two questions:

- What is the household's willingness to pay for one additional unit of good 1 in terms of units of good 2? The answer is MRS units of good 2.
- What is the household's cost for one additional unit of good 1 in terms of units of good 2? The answer: MOC units of good 2.

Now, the interplay of the marginal rate of substitution MRS and marginal opportunity cost MOC helps to find the household optimum. Consider the inequality

$$MRS = \underbrace{\left|\frac{dx_2}{dx_1}\right|}_{\substack{\text{absolute value}\\ \text{of the slope of}\\ \text{the indifference}\\ \text{curve}}} > \underbrace{\left|\frac{dx_2}{dx_1}\right|}_{\substack{\text{absolute value}\\ \text{of the slope of}\\ \text{the budget line}}} = MOC.$$

If, now, the household increases his consumption of good 1 by one unit, he can decrease his consumption of good 2 by MRS units and still stay on the same indifference curve. Compare figure VI.9. The increase of good 1

Marginal willingness to pay: $MRS = \left|\frac{dx_2}{dx_1}\right|$

If the household consumes one additional unit of good 1, how many units of good 2 can he forgo so as to remain indifferent. — movement on the indifference curve

Marginal opportunity cost: $MOC = \left|\frac{dx_2}{dx_1}\right|$

If the household consumes one additional unit of good 1, how many units of good 2 does he have to forgo so as to remain within his budget. — movement on the budget line

FIGURE VI.8. Willingness to pay and opportunity cost

necessitates a decrease of only $MOC < MRS$ units of good 2. Therefore, the household needs to give up less than he would be prepared to. In case of strict monotonicity, increasing the consumption of good 1 leads to a higher indifference curve. Thus, we cannot have $MRS > MOC$ at the optimal bundle unless it is impossible to further increase the consumption of good 1. This latter possibility is depicted in figure VI.10.

Thus, if the household consumes both goods in positive quantities, we can derive the optimality condition

$$MRS \stackrel{!}{=} MOC$$

(if both terms are defined).

Alternatively, we can derive this first-order condition with the help of a utility function (if we have one). The household tries to maximize

$$U\left(x_1, \frac{m}{p_2} - \frac{p_1}{p_2}x_1\right).$$

If the household increases the consumption of good 1 by one unit, we have two effects. First, his utility increases by $\frac{\partial U}{\partial x_1}$. Second, an increase in x_1 leads

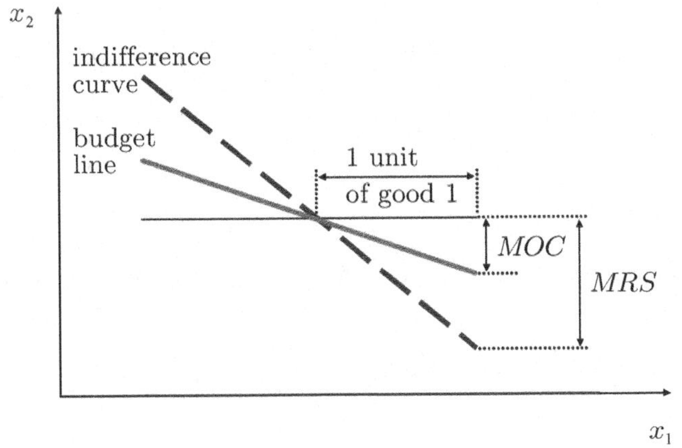

FIGURE VI.9. Not optimal

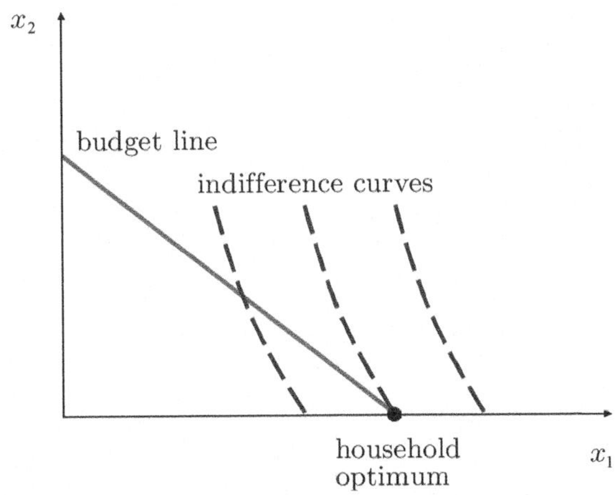

FIGURE VI.10. The willingness to pay can be higher than the cost.

to a reduction in x_2 by $MOC = \left|\frac{dx_2}{dx_1}\right| = \frac{p_1}{p_2}$ and this reduced consumption of good 2 decreases utility (chain rule). Therefore, the household increases x_1 as long as

$$\underbrace{\frac{\partial U}{\partial x_1}}_{\substack{\text{marginal benefit}\\ \text{of increasing } x_1}} > \underbrace{\frac{\partial U}{\partial x_2}\left|\frac{dx_2}{dx_1}\right|}_{\substack{\text{marginal cost}\\ \text{of increasing } x_1}}$$

holds. Dividing by $\frac{\partial U}{\partial x_2}$, an increase in x_1 leads to an increase in utility if

$$\begin{aligned} MRS &= \frac{\frac{\partial U}{\partial x_1}}{\frac{\partial U}{\partial x_2}} \text{ (chapter IV on p. 82)} \\ &> \left|\frac{dx_2}{dx_1}\right| = MOC \end{aligned}$$

holds.

The MRS-versus-MOC rule can help to derive the household optimum in some cases:

- Cobb-Douglas utility functions $U(x_1, x_2) = x_1^a x_2^{1-a}$ with $0 < a < 1$ lead to

$$MRS = \frac{\frac{\partial U}{\partial x_1}}{\frac{\partial U}{\partial x_2}} = \frac{a}{1-a}\frac{x_2}{x_1} \stackrel{!}{=} \frac{p_1}{p_2}$$

  and, together with the budget line, the household optimum

$$\begin{aligned} x_1(m,p) &= a\frac{m}{p_1}, \\ x_2(m,p) &= (1-a)\frac{m}{p_2}. \end{aligned}$$

- Goods 1 and 2 are perfect substitutes if the utility function is given by $U(x_1, x_2) = ax_1 + bx_2$ with $a > 0$ and $b > 0$. An increase of good 1 enhances utility if

$$\frac{a}{b} = MRS > MOC = \frac{p_1}{p_2}$$

  holds so that we find the household optimum

$$x(m,p) = \begin{cases} \left(\frac{m}{p_1}, 0\right), & \frac{a}{b} > \frac{p_1}{p_2} \\ \left\{\left(x_1, \frac{m}{p_2} - \frac{p_1}{p_2}x_1\right) \in \mathbb{R}_+^2 : x_1 \in \left[0, \frac{m}{p_1}\right]\right\} & \frac{a}{b} = \frac{p_1}{p_2} \\ \left(0, \frac{m}{p_2}\right) & \frac{a}{b} < \frac{p_1}{p_2} \end{cases}$$

- Preferences are concave with utility function $U(x_1, x_2) = x_1^2 + x_2^2$. We have the marginal rate of substitution

$$MRS = \frac{\frac{\partial U}{\partial x_1}}{\frac{\partial U}{\partial x_2}} = \frac{2x_1}{2x_2} = \frac{x_1}{x_2}$$

  so that

$$\frac{x_1}{x_2} = MRS > MOC = \frac{p_1}{p_2}$$

  holds for sufficiently large x_1 which calls for an increase of x_1. Inversely,

$$\frac{x_1}{x_2} = MRS < MOC = \frac{p_1}{p_2}$$

holds for sufficiently large x_2 so that an increase of x_2 seems a good idea. Therefore, we need to compare the extreme bundles $\left(\frac{m}{p_1}, 0\right)$ and $\left(0, \frac{m}{p_2}\right)$ and obtain

$$\begin{aligned} \left(\frac{m}{p_1}\right)^2 + 0^2 &\geq 0^2 + \left(\frac{m}{p_2}\right)^2 \text{ and} \\ p_1 &\leq p_2 \end{aligned}$$

and finally

$$x(m,p) = \begin{cases} \left(\frac{m}{p_1}, 0\right), & p_1 < p_2 \\ \left\{\left(\frac{m}{p_1}, 0\right), \left(0, \frac{m}{p_2}\right)\right\} & p_1 = p_2 \\ \left(0, \frac{m}{p_2}\right) & p_1 > p_2 \end{cases}$$

2.3. Household optimum and monotonicity. We now turn to specific implications that can be drawn from the fact that some x^* is a household optimum and that some sort of monotonicity holds.

LEMMA VI.4. *Let x^* be a household optimum of the decision situation $\Delta = (B(p,m), \precsim)$. Then, we have the following implications:*

- *Walras' law: Local nonsatiation implies $p \cdot x^* = m$.*
- *Strict monotonicity implies $p >> 0$.*
- *Local nonsatiation and weak monotonicity imply $p \geq 0$.*

PROOF. We use proofs by contradiction for each statement:

- Because of $x^* \in B$, we can exclude $p \cdot x^* > m$. Assume $p \cdot x^* < m$. Then, the household can afford bundles sufficiently close to x^*. By local nonsatiation, within the set of those affordable bundles, a bundle y exists that the household strictly prefers to x^* (see figure VI.11). This is a contradiction to x^* being a household optimum.
- Turning to the second implication, assume a household optimum and a price p_g which is zero or negative. Then, the household can afford more of good g at bundle x^*. By strict monotonicity, the household is better off than at x^*, which yields the desired contradiction.
- Assume a negative price for some good g. By weak monotonicity the household can "buy" additional units of that good without being worse off. Since the price is negative, the household has additional funding for preferred bundles which exist by nonsatiation. Proceed as shown at the first bullet.

□

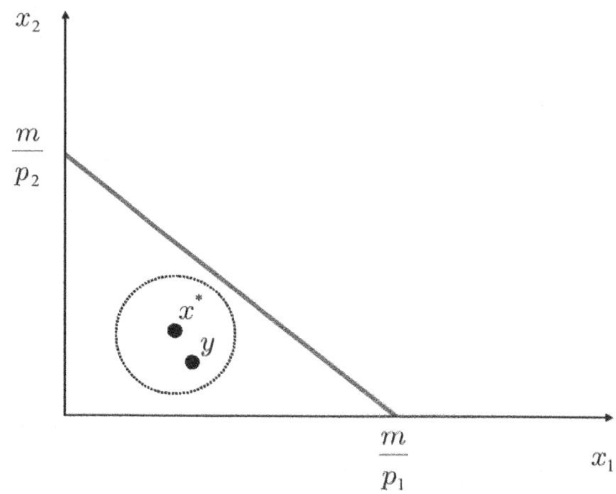

FIGURE VI.11. Proving Walras' law

3. Comparative statics and vocabulary

3.1. Vocabulary. In household theory, we carefully distinguish parameters and variables. If we focus on the price p_g of good x_g, we treat the other prices and the income or endowment as parameters. That means, they are fixed for the time being. If we plot $x_g^R(p_g, ...)$ as a function of p_g, we obtain a demand curve. A change in p_g results in a movement along the demand curve, while a parameter change shifts the whole demand curve.

DEFINITION VI.5. *In household theory, we omit the R and often write $x(p, m)$ or $x(p, \omega)$ instead of $x^R(p, m, \precsim)$ or the like. Holding some of the parameters constant, we distinguish*

- *the (Marshallian) demand function for good g, denoted by $x_g(p_g)$,*
- *the Engel function for good g, denoted by $x_g(m)$, and*
- *the cross demand function for good g with respect to the price p_k of some other good $k \neq g$, denoted by $x_g(p_k)$.*

In case of an endowment budget, the household is called a net supplier of good g in case of $x_g(p, \omega) < \omega_g$ and a net demander in case of $x_g(p, \omega) > \omega_g$.

Once we know the three functions defined in the above definition, we can form the derivatives:

DEFINITION VI.6. *We call a good g*

- *ordinary if*

$$\frac{\partial x_g}{\partial p_g} \leq 0$$

holds and non-ordinary for the inverse inequality (demand function),

- *normal if*

$$\frac{\partial x_g}{\partial m} \geq 0$$

holds and inferior for the inverse inequality (Engel function),

- *a substitute of good k if*

$$\frac{\partial x_g}{\partial p_k} \geq 0$$

holds and a complement of good k for the inverse inequality (cross demand function).

EXERCISE VI.7. *Consider the Cobb-Douglas demand functions* $x_1 = a\frac{m}{p_1}$ *and* $x_2 = (1-a)\frac{m}{p_2}$, $0 < a < 1$, *in case of a money budget and find out whether*

- *good* 1 *is an ordinary good?*
- *good* 1 *is normal?*
- *good* 1 *is a substitute or a complement of good* 2*?*

3.2. Price-consumption curve and demand curve. The demand function defined in the previous section is sometimes qualified as "Marshallian" in order to differentiate between Marshallian demand and Hicksian demand, the latter becoming a central topic in the next chapter. In this section, we show how to derive demand curves.

Assume a money budget for two goods 1 and 2. For fixed values p_2 and m, we vary the price p_1 of good 1 in x_1-x_2-space (prices p_1^B, p_1^C and p_1^D with $p_1^B > p_1^C > p_1^D$) and obtain the price-consumption curve which is the geometric locus of household optima (see the upper part of figure VI.12). We then associate all the different prices of good 1 with the demand for that good. The graph obtained (the lower part of our figure) is called the demand curve where – normally – the ordinate is the price axis.

EXERCISE VI.8. *Assuming that good* 1 *and good* 2 *are complements, sketch a price-consumption curve and the associated demand curve for good* 1.

Assuming the utility function $U(x_1, x_2) = x_1^{\frac{1}{3}} \cdot x_2^{\frac{2}{3}}$, we now calculate the price-consumption curve and the demand curve for good 1. You remember that the household optimum (x_1^*, x_2^*) is given by

$$x_1^* = \frac{1}{3}\frac{m}{p_1},\ x_2^* = \frac{2}{3}\frac{m}{p_2}.$$

Thus, the demand curve for good 1 is $x_1^* = f(p_1) = \frac{1}{3}\frac{m}{p_1}$.

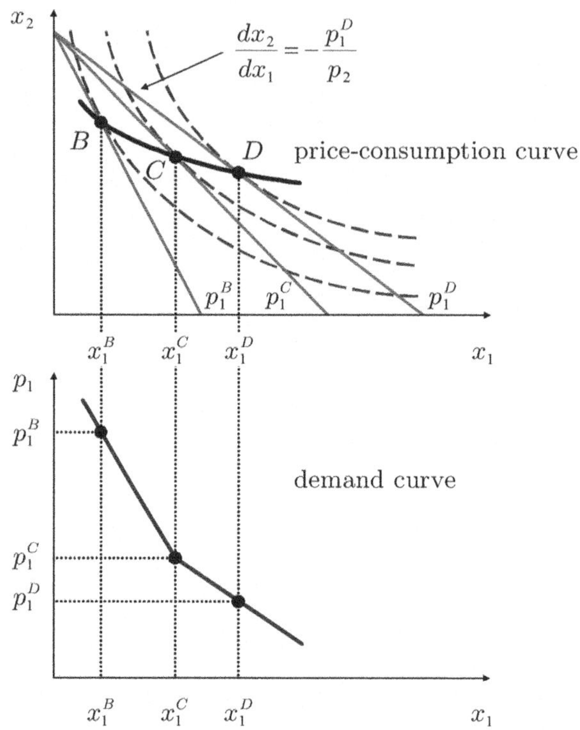

FIGURE VI.12. The price-consumption curve and the demand curve

In general, the price-consumption curve for variing p_1 is determined by the following procedure:

- We associate each p_1 with the household optimum $(x_1^*(p_1), x_2^*(p_1))$.
- We look for the geometric locus of these optima and express it as a function $x_2 = h(x_1)$. It is important that h is *not* a function of p_1.

In our example, $x_2^* = h(x_1) = \frac{2}{3}\frac{m}{p_2}$ is already the price-consumption curve – the price of good 1 affects the demand of good 1, but not of good 2. Therefore, the price-consumption curve is a horizontal line.

EXERCISE VI.9. *Determine the demand function for good 2 in case of the utility function* $U(x_1, x_2) = \min(x_1, 2x_2)$! *Be careful and check for zero and negative prices; you can use the case distinction given by*

$$x_2(p_1, p_2, m) = \begin{cases} ?, & p_1 > 0, p_2 > 0 \\ ?, & p_1 \leq 0 \text{ and } p_2 \leq 0 \\ ?, & p_1 > 0, p_2 = 0 \\ ?, & p_1 = 0, p_2 > 0 \end{cases}$$

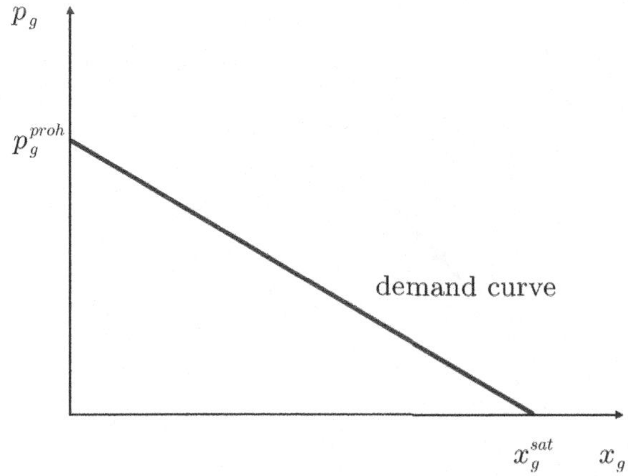

FIGURE VI.13. Saturation quantity and prohibitive price

Determine also the price-consumption curve for varying p_1!

Sometimes, demand curves hit the axes (see figure VI.13):

DEFINITION VI.7. *Let* $x_g(p_g)$ *be the quantity demanded for any price* $p_g \geq 0$. *We call*

$$x_g^{sat} := x_g(0)$$

the saturation quantity and

$$p_1^{proh} := \min\{p_g \geq 0 : x_g(p_g) = 0\}$$

the prohibitive price.

3.3. Income-consumption curve and Engel curve. Figure VI.14 shows how to derive the Engel curve from the income-consumption curve. The latter one connects the household optima in our x_1-x_2 diagram.

EXERCISE VI.10. *Assuming that good* 1 *and good* 2 *are complements, sketch an income-consumption curve and the associated Engel curve for good* 1!

Assuming the same Cobb-Douglas utility function as in the previous section, we determine the income-consumption curve and the Engel curve. Algebraically, the Engel curve can be obtained from the household optimum and is given by

$$x_1^* = q(m) = \frac{1}{3}\frac{m}{p_1}.$$

In order to express the income-consumption curve algebraically, we have to write x_2 as a function of x_1, but not of income m (which takes on all values).

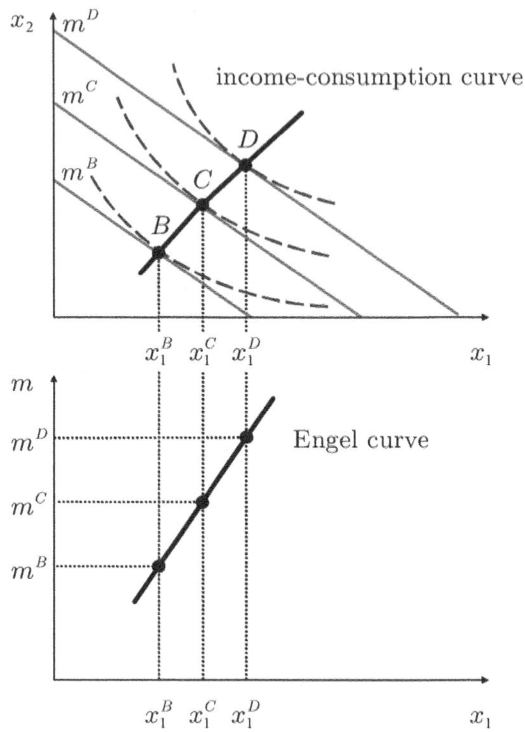

FIGURE VI.14. Deriving the Engel curve

We solve good 1's demand for m and obtain $m = 3p_1x_1^*$. Substituting in x_2^* yields

$$\begin{aligned} x_2^* &= \frac{2}{3}\frac{m}{p_2} \\ &= \frac{2}{3}\frac{3p_1x_1^*}{p_2} \\ &= 2\frac{p_1}{p_2}x_1^* \end{aligned}$$

and hence the income-consumption curve $x_2^* = g(x_1^*) = 2\frac{p_1}{p_2}x_1^*$.

EXERCISE VI.11. *Determine the income-consumption curve for the case of perfect complements,* $U(x_1, x_2) = \min(x_1, 2x_2)$! *Can you also find the Engel-curve function for good* 2?

3.4. Defining substitutes and complements. The definitions of substitutes and complements seem innocuous. However, they are highly problematic as you will realize when you solve the following exercise:

EXERCISE VI.12. *Determine* $\frac{\partial x_1(p,m)}{\partial p_2}$ *and* $\frac{\partial x_2(p,m)}{\partial p_1}$ *for the quasi-linear utility function given by*

$$U(x_1, x_2) = \ln x_1 + x_2 \qquad (x_1 > 0)!$$

Assume positive prices and $\frac{m}{p_2} > 1$ *(in order to avoid a corner solution)!*

Thus, good g can be the substitute of good k while k is not a (strict) substitute of g. We will see how to avoid this problem in the next chapter.

3.5. Price elasticities of demand. An important characteristic of demand is $\frac{dx_g}{dp_g}$, i.e., the question of how demand changes if the price of a good changes. However, this slope of the demand curve depends on the units of measurement – do we have Euros or dollars? This problem can be avoided by using relative quantities. By how many percent does demand change if price is changed by one percent?

DEFINITION VI.8 (PRICE ELASTICITY). *Let* $x_g(p_g)$ *be the demand at price* p_g *(and other prices which are hold constant). The price elasticity (of demand) is denoted by* ε_{x_g,p_g} *and given by*

$$\varepsilon_{x_g,p_g} := \underbrace{\frac{\frac{dx_g}{x_g}}{\frac{dp_g}{p_g}}}_{\substack{\text{mathematically}\\ \text{doubtful, but}\\ \text{easily interpretable}}} = \frac{dx_g}{dp_g}\frac{p_g}{x_g}.$$

EXERCISE VI.13. *Calculate the price elasticities of demand for the demand function (individual or aggregate) given by*

$$x_g(p_g) = 100 - p_g \text{ and } x_k(p_k) = \frac{1}{p_k}.$$

If we know that we are dealing with ordinary goods, we can consider the absolute value of the price elasticity. Then $|\varepsilon_{x,p}| < 1$ and $\varepsilon_{x,p} > -1$ are equivalent. The price elasticity can help to assess the effect of a price change on expenditure, i.e., we are interested in

$$\frac{d(px(p))}{dp}.$$

If the absolute value of the price elasticity is smaller than 1, the expenditure increases if the price increases. You can see this from

$$\begin{aligned}\frac{d(px(p))}{dp} &= x + p\frac{dx}{dp} \\ &= x\left(1 + \frac{p}{x}\frac{dx}{dp}\right) \\ &= x(1 + \varepsilon_{x,p}) \\ &= x(1 - |\varepsilon_{x,p}|) > 0.\end{aligned}$$

This result can be used as an argument for a liberal drug policy. It is plausible that demand for drugs is inelastic, $|\varepsilon_{x,p}| < 1$. Assume the government increases the price of drugs by taxing them or by criminalizing selling or buying. Then the expenditure of drug users increases and so does drug-related crime (stealing money in order to finance the addiction).

3.6. Income elasticity of demand. In a similar fashion, we can define the income elasticity. Income increases by one percent. By how many percent does demand increase?

DEFINITION VI.9 (INCOME ELASTICITY). *Let $x_g(m)$ be the demand at income m. The income elasticity (of demand) is denoted by $\varepsilon_{x_g,m}$ and given by*

$$\varepsilon_{x_g,m} := \frac{\frac{dx_g}{x_g}}{\frac{dm}{m}} = \frac{dx_g}{dm}\frac{m}{x_g}.$$

If a good is normal, its income elasticity is positive. We can subdivide normal goods in

- luxury goods such as Kaviar: your consumption increases more strongly than your income and
- necessity goods such as housing: your consumption in terms of square meters increases with your income, but in a less than proportional manner.

DEFINITION VI.10 (LUXURY GOOD). *We call a good g*

- *a luxury good if*

$$\varepsilon_{x_g,m} \geq 1$$

  holds
- *a necessity good if*

$$0 \leq \varepsilon_{x_g,m} \leq 1$$

  holds.

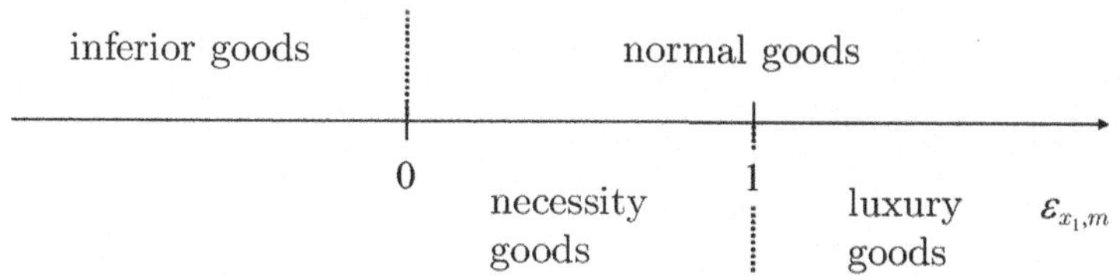

FIGURE VI.15. Inferior and normal, luxury and necessity goods

EXERCISE VI.14. *Calculate the income elasticity of demand for the Cobb-Douglas utility function* $U(x_1, x_2) = x_1^{\frac{1}{3}} \cdot x_2^{\frac{2}{3}}$*! How do you classify (demand for) good* 1*?*

In a sense, an income elasticity of 1 is very normal.

LEMMA VI.5. *Assume local nonsatiation and the household optimum* x^**. Then the average income elasticity is* 1*:*

$$\sum_{g=1}^{\ell} s_g \varepsilon_{x_g,m} = 1$$

where the weights are the relative expenditures, $s_g := \frac{p_g x_g}{m}$.

According to Walras' law (lemma VI.4, p. 141), the household chooses x^* on the budget line, $p \cdot x^*(m) = m$. Of course, if there is only one good, $\ell = 1$, we have $x = \frac{m}{p}$ and the income elasticity is

$$\varepsilon_{x,m} = \frac{dx}{dm}\frac{m}{x} = \frac{1}{p}\frac{m}{\frac{m}{p}} = 1.$$

To show the lemma in the general case, we form the derivative of the budget equation $m = \sum_{g=1}^{\ell} p_g x_g^*(m)$ with respect to m to obtain

$$1 = \sum_{g=1}^{\ell} p_g \frac{dx_g^*}{dm}$$

and, by multiplying the summands with $\frac{x_g^*}{m}\frac{m}{x_g^*} = 1$,

$$1 = \sum_{g=1}^{\ell} p_g \frac{dx_g^*}{dm}\frac{x_g^*}{m}\frac{m}{x_g^*} = \sum_{g=1}^{\ell} \frac{p_g x_g^*}{m}\frac{dx_g^*}{dm}\frac{m}{x_g^*} = \sum_{g=1}^{\ell} s_g \varepsilon_{x_g,m}$$

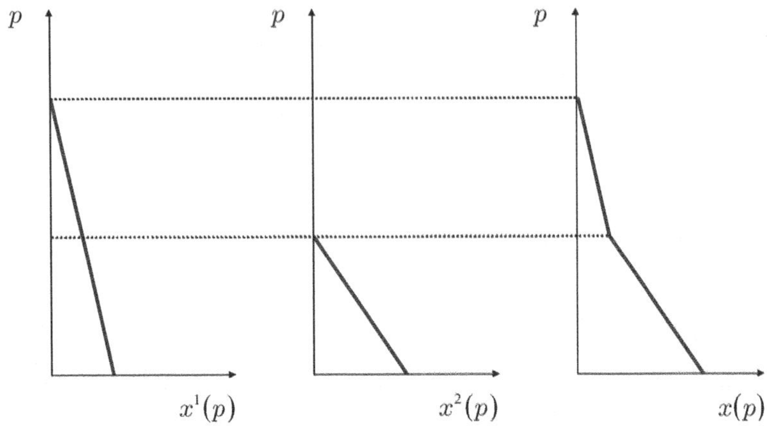

FIGURE VI.16. Aggregation of individual demand curves

3.7. Aggregation of individual demand curves. Household theory shows how to derive individual demand curves. We now aggregate several individual demand curves in order to arrive at an aggregate demand curve.

DEFINITION VI.11 (AGGREGATE DEMAND). *Let $x^i(p)$ be the demand functions of individuals $i = 1, ..., n$. Aggregate demand is then given by*

$$x(p) := \sum_{i=1}^{n} x^i(p).$$

Consider figure VI.16 to see how "horizontal aggregation" works. For every price, the quantities demanded by households A and B are added.

EXERCISE VI.15. *Consider the individual demand functions for good g given by*

$$\begin{aligned} x_g^1(p_g) &= \max(0, 100 - p_g), \\ x_g^2(p_g) &= \max(0, 50 - 2p_g) \text{ and} \\ x_g^3(p_g) &= \max(0, 60 - 3p_g). \end{aligned}$$

Find the aggregate demand function. Hint: Find the prohibitive prices first!

Individual and aggregate demand functions can sometimes be inversed: Have a look at figure VI.17. It can be considered as a graphical representation of the demand curve. At price $\hat{p}_g$ (ordinate) we obtain the quantity demanded $x_g(\hat{p}_g)$. Inversely, for a given quantity $\hat{x}_g$ of good g, we can ask for the price $p_g(\hat{x}_g)$ that is just sufficient to yield the quantity $\hat{x}_g$. The resulting function is called the inverse demand function.

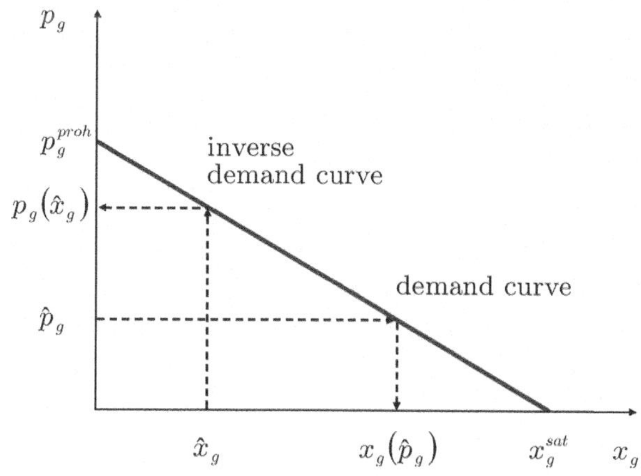

FIGURE VI.17. Demand curve and inverse demand curve

DEFINITION VI.12 (INVERSE DEMAND FUNCTION). *Let* $x_1 : \left[0, p_1^{proh}\right] \to \left[0, x_1^{sat}\right]$ *be a bijective (individual or market) demand function. The inverse of this function is called the inverse (individual or market) demand function and given by*

$$\begin{aligned} p_1 &= x_1^{-1} : \left[0, x_1^{sat}\right] \to \left[0, p_1^{proh}\right] \\ & x_1 \mapsto p_1(x_1) \text{ where } p_1(x_1) \text{ is the unique price} \\ & \text{resulting in } x_1. \end{aligned}$$

4. Solution theory

4.1. The general setup of an optimization problem. We now want to take a broader, more general view of the problem facing a household. (We present solution theory in this section and in a later chapter on pp. 243.) Usually, an optimization problem consists of a feasible set and a preference relation on that set. If the preference relation is expressed by a function, this function is often called an objective function. A firm's profit function or a household's utility function are examples of objective functions.

Feasible sets can be strategy sets as in chapter II or budgets as in the present chapter. In chapter VIII, we introduce production sets which describe the production possibilities open to a firm. A production set is yet another instance of a feasible set.

Summarizing these examples,

- in decision theory (and similarly in game theory), we have
  - feasible set = strategy set and

  - preference relation $\succsim^{\omega}$ on the set of strategies S for a belief ω on W defined by $s \succsim^{\omega} s' :\Leftrightarrow u(s,\omega) \geq u(s',\omega)$
- in household theory, we find
  - feasible set = budget and
  - preference relation on the set of bundles, possibly definable by a utility function
- in the theory of the firm, we encounter
  - feasible set = production set and
  - preference relation given by a profit function

DEFINITION VI.13 (MAXIMIZATION OR MINIMIZATION PROBLEM). *Let $f : \mathbb{R}^{\ell} \to \mathbb{R}$ be an objective function and $M \subseteq \mathbb{R}^{\ell}$ the feasible set. (f, M) is called an optimization problem if it is a maximization or a minimization problem, i.e., if we look for a maximal or a minimal element in $f(M)$, respectively. $x^* \in M$ is called a solution to the*

- *maximization problem if $f(x^*) \geq f(x)$ for all $x \in M$ holds,*
- *minimization problem if $f(x^*) \leq f(x)$ for all $x \in M$ holds.*

$x^* \in M$ is called a solution if it is a solution to the maximization or the minimization problem.

Mathematically, the above definition of the optimization problem (f, M) is equivalent to looking for a maximum or minimum of the function $f|_M$.

Solution theory deals with two questions:

- Existence: Can we be sure that a solution to the optimization problem exists?
- Uniqueness: Are there several solutions or is there just one solution?

In the best of all worlds, we have exactly one solution.

4.2. Existence. Not every maximization (or minimization) problem needs to have a solution. The two most relevant examples can be shown with the help of figure VI.18. In both subfigures, the value of the objective function can be increased by increasing the variable x. In (a), the domain is $\mathbb{R}_+$ (and hence not bounded) and for any x from the domain, $x+1$ yields a higher value. In (b), for every $x \in [0,1)$ (which is not closed), we have $1 > \frac{1+x}{2} > x \geq 0$. That is, we cannot find a greatest number smaller than 1.

In household theory, these problems do not need to bother us a lot. The reason lies in the following theorem and its corollary:

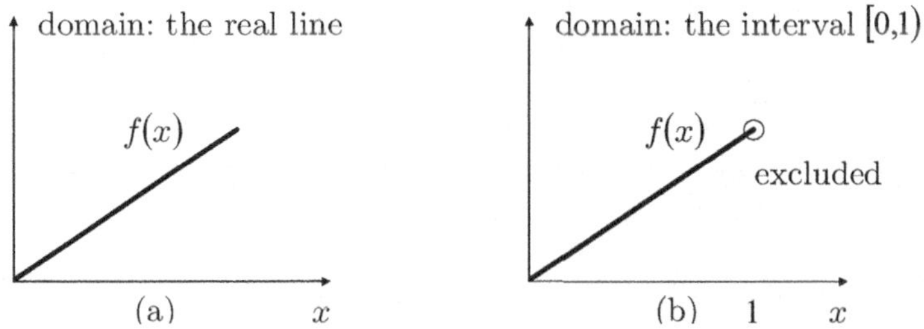

FIGURE VI.18. No solutions

THEOREM VI.1. *Let $f : M \to \mathbb{R}$ be a continuous function where $M \subseteq \mathbb{R}^\ell$ is nonempty, closed, and bounded. Then, f adopts a maximum and a minimum on M.*

We only hint at a proof. By the continuity of f, the image $f(M)$ is also closed and bounded. Every closed and bounded subset of $\mathbb{R}$ takes on a maximum and a minimum.

COROLLARY VI.1. *Let $\precsim$ be a continuous preference relation and let $p \in \mathbb{R}^\ell_+$ obey $p >> 0$. Then, the household's decision problem has a solution, i.e., $x(p,m)$ or $x(p,\omega)$ is nonempty for every $m \geq 0$ and $\omega \in \mathbb{R}^\ell_+$.*

The corollary follows from the above theorem together with theorem IV.1 (p. 76) and lemma VI.2 (p. 130).

4.3. Uniqueness. Optimization problems can have several solutions as you have seen in figure VI.7 (c), p. 137. Subfigure (a) hints at the importance of convexity:

THEOREM VI.2. *Let $f : M \to \mathbb{R}$ be a strictly quasi-concave function on a convex domain $M \subseteq \mathbb{R}^\ell$. Then, we cannot have two different solutions to a maximization problem.*
Let $f : M \to \mathbb{R}$ be a quasi-concave function that obeys strict monotonicity or local nonsatiation (see exercise IV.17, p. 77) and let M be a strictly convex domain $M \subseteq \mathbb{R}^\ell$. Then, we cannot have two different solutions to a maximization problem.

PROOF. We provide a proof (by contradiction) of the first assertion. Let $x, y \in M$ be two solutions with $x \neq y$. By the convexity of M, we have

$$kx + (1-k)\,y \in M$$

for all $k \in (0,1)$. Strict quasi-concavity implies

$$f(kx + (1-k)\,y) > \min(f(x), f(y)) = f(x) = f(y).$$

Thus, we have found elements from our feasible set M that yield a higher value than the assumed solutions. This desired contradiction shows that two different solutions are impossible. □

COROLLARY VI.2. *Let U be a strictly quasi-concave utility function. Then, the household's decision problem has at most one solution.*

Returning to figure VI.7 (c) on p. 137, note that

- the utility function is quasi-concave but not strictly quasi-concave and
- the budget is convex but not strictly convex.

Therefore, this example does fit neither the first nor the second part of the theorem.

The corollary is derived from the first part of the theorem. With respect to the second part, just reconsider figure VI.7 (d) and exchange the indifference and budget curves.

Uniqueness in the above theorem is possible from strict quasi-concavity of the objective function or strict convexity of the domain. If we have neither, we cannot hope for uniqueness, but we can still ensure that the solutions form a convex set:

THEOREM VI.3. *Let $f : M \to \mathbb{R}$ be a quasi-concave function on a convex domain $M \subseteq \mathbb{R}^\ell$. Then the set of solutions is convex.*

Therefore, if you have two solutions, every convex combination of these solutions is also a solution. A good example is provided by perfect substitutes where $x(p, m)$ is an interval if the budget line's slope happens to coincide with the indifference curves' slopes.

4.4. Local solutions and global solutions.

DEFINITION VI.14 (LOCAL SOLUTION). *A solution $x^* \in M$ is also called a global solution. A solution $x^* \in M$ is called a local solution if we have an ε-ball K with center x^* such that x^* is a solution to $(f, M \cap K)$.*

Of course, every global solution is a local one (consider figure VI.19). The inverse is more interesting.

THEOREM VI.4. *Let $f : M \to \mathbb{R}$ be a strictly quasi-concave function on a convex domain $M \subseteq \mathbb{R}^\ell$. Then, a local maximum of f on M is a global one.*

PROOF. Assume a local maximum at $x^* \in M$ and hence an ε-ball K with center x^* such that $f(x^*) \geq f(x)$ for all $x \in M \cap K$ (see figure VI.20).

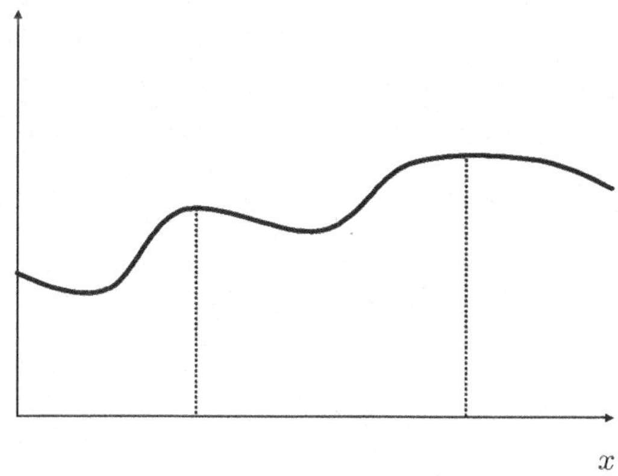

FIGURE VI.19. Every global solution is a local one.

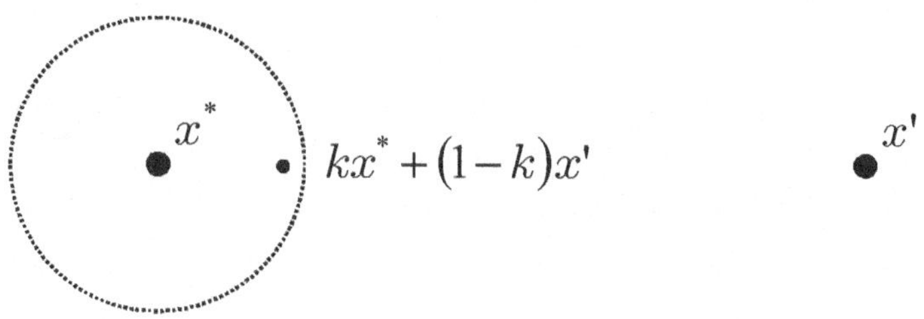

FIGURE VI.20. A local solution is global.

Assume further that x^* is not a global solution. Then there exists an $x' \in M$ obeying $f(x^*) < f(x')$. By convexity of M, we have

$$kx^* + (1-k)\,x' \in M$$

for every $k \in (0,1)$. Strict quasi-concavity yields

$$f\left(kx^* + (1-k)\,x'\right) > f(x^*).$$

Now, we can choose k so close to 1 that $kx^* + (1-k)\,x'$ is included in the ε-ball K with center x^*. This is the desired contradiction to the local maximum at x^*. □

COROLLARY VI.3. *Let U be a strictly quasi-concave utility function. Then, a local solution to the household's decision problem is already a global one.*

4.5. Interior solutions, boundary solutions, and corner solutions. Assume an optimization problem (f, M). We distinguish between different kinds of solutions:

Definition VI.15 (interior solution). *A solution $x^* \in M$ is called an interior solution if it is an interior point of M. A solution $x^* \in M$ is called a boundary solution if it is a boundary point of M. If all solutions are boundary solutions, the feasibility constraint is called binding.*

For interior solutions, we have an obvious necessary condition:

Theorem VI.5. *Let $x^* \in M$ be an interior solution to an optimization problem (f, M) and let f be differentiable. Then,*

$$\left.\frac{\partial f}{\partial x_g}\right|_{x^*} = 0$$

for all $g = 1, .., \ell$.

The reason is obvious. If the derivative with respect to any x_g were not zero, we could increase or decrease x_g and increase or decrease the value of our objective function.

By lemma IV.1 (p. 62), a solution $x^* \in M$ is an interior solution if and only if it is not a boundary solution. In contrast to figure IV.8 on p. 71, we consider M (the budget set) a subset of $\mathbb{R}^\ell$ rather than $\mathbb{R}^\ell_+$. Of course, in household theory, we are especially interested in the boundary solutions defined by the budget line (see figure VI.21) because this line gets displaced by changes in income (endowment) and prices.

There are two sorts of boundary solutions on the budget line. A corner solution means that at least one good is not consumed at all. Non-corner solutions are those where every good is consumed with a non-zero quantity:

Definition VI.16 (corner solution). *A solution $x^* \in M$ is called a non-corner solution if $x^* >> 0$ holds. A solution is called a corner solution if we have $x_g^* = 0$ for at least one $g \in \{1, .., \ell\}$.*

4.6. Boundary non-corner solutions and the Lagrange method.

4.6.1. *The Lagrange theorem.* In a previous section (see p. 138), we argue that the household maximizing

$$U\left(x_1, \frac{m}{p_2} - \frac{p_1}{p_2}x_1\right)$$

pays the additional utility from consuming another unit of good 1 with a reduction of utility stemming from consuming less of good 2. The Lagrange method helps to find the optimum by merging the feasibility constraint with the objective function.

Theorem VI.6 (Lagrange theorem). *Let f and g be differentiable functions $\mathbb{R}^\ell \rightarrow \mathbb{R}$ where f is the obejctive function and $g(x) = 0$ specifies the*

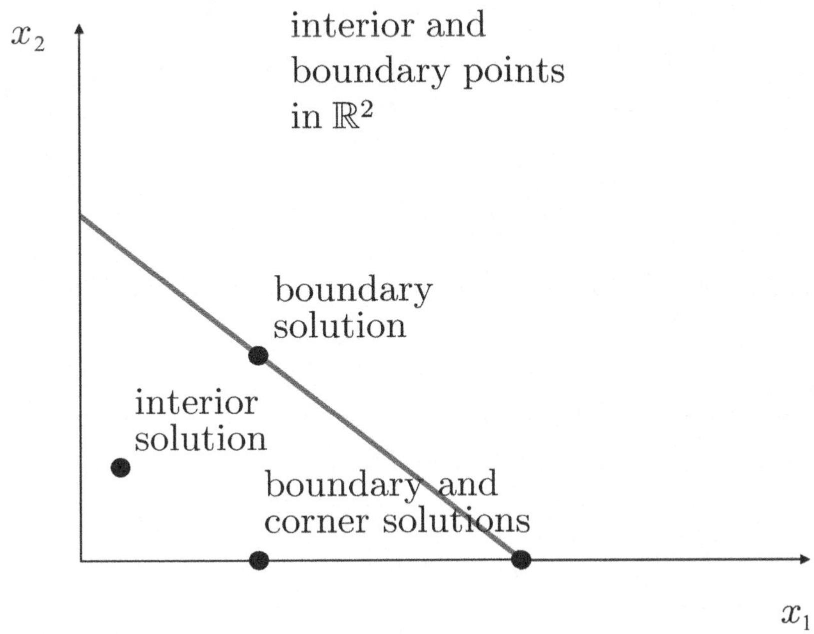

FIGURE VI.21. Solutions: interior, boundary, corner

side condition. If the optimization problem $\left(f, \left\{x \in \mathbb{R}^\ell : g(x) = 0\right\}\right)$ *has a local non-corner solution* x^*, *there exists a* $\lambda \in \mathbb{R}$ *such that*

$$f'(x^*) = -\lambda g'(x^*).$$

Note that the Lagrange theorem does not tell whether a solution exists. If (!) we have a local non-corner solution, it fulfills the above equation. In the next subsection, we learn how to apply the Lagrange method to our household problem.

4.6.2. *Applying the Lagrange method.* Assume strict quasi-concavity of the utility function U and positive prices, $p >> 0$. Furthermore, assume strict monotonicity of U. The Lagrange function is given by

$$L(x, \lambda) = U(x) + \lambda \left[m - \sum_{g=1}^{\ell} p_g x_g \right].$$

For the time being, think of λ, the so-called Lagrange multiplier, as a positive parameter. It translates a budget surplus (in case of $m > \sum_{g=1}^{\ell} p_g x_g$) into additional utility. Thus, increasing consumption has a positive direct effect via the utility function U and negative indirect effect via a decreasing budget surplus and λ. Therefore (we do not give a proof), maximizing L with respect to x and λ is equivalent to maximizing U on $B(p, m)$.

Indeed, we differentiate L with respect to x_g to obtain

$$\frac{\partial L\left(x_1, x_2, ..., \lambda\right)}{\partial x_g} = \frac{\partial U\left(x_1, x_2, ..., x_\ell\right)}{\partial x_g} - \lambda p_g \overset{!}{=} 0 \tag{VI.3}$$

or

$$\frac{\partial U\left(x_1, x_2, ..., x_\ell\right)}{\partial x_g} \overset{!}{=} \lambda p_g. \tag{VI.4}$$

We want to check whether these conditions are those claimed in theorem VI.6. Letting

- objective function $f := U$ and
- side condition $g\left(x\right) := m - \sum_{g=1}^{\ell} p_g x_g = 0$

the theorem claims – in the presence of a local non-corner solution x^* – the existence of $\lambda \in \mathbb{R}$ such that $f'\left(x^*\right) = -\lambda g'\left(x^*\right)$ holds, i.e., such that

$$\frac{\partial U\left(x_1, x_2, ..., x_\ell\right)}{\partial x_g} = -\lambda \frac{\partial\left(m - \sum_{g=1}^{\ell} p_g x_g\right)}{\partial x_g} = -\lambda\left(-p_g\right) = \lambda p_g$$

is true for all $g = 1, ..., \ell$ (see definition IV.29, p. 83). Thus, our recipe is in accordance with the above theorem.

Now, dividing condition VI.4 for a good g by the same condition for good k yields

$$\frac{\frac{\partial U(x_1,x_2,...,x_\ell)}{\partial x_g}}{\frac{\partial U(x_1,x_2,...,x_\ell)}{\partial x_k}} \overset{!}{=} \frac{p_g}{p_k}$$

which, again, is our familiar condition $MRS \overset{!}{=} MOC$.

EXERCISE VI.16. *Set the derivative of L with respect to λ equal to* 0. *What do you find?*

λ is called the shadow price of the restriction (the budget equation). It can be shown that λ is equal to the additional utility accruing to the household if the restriction is eased by one unit. One might be tempted to express this by

$$\lambda = \frac{dU}{dm}.$$

Then, λ would be the marginal utility of income. However, this is not quite correct. After all, U does not have income m as an argument but only bundles of goods. One purpose of the next section is to deal with this problem.

function	**arguments**	**optimal bundles**
utility function	good bundles $x \in \mathbb{R}_+^\ell$	$x(p,m)$
indirect utility function	income m and prices $p \in \mathbb{R}^\ell$	$x(p,m)$

FIGURE VI.22. Direct versus indirect utility function

5. Indirect utility function

5.1. Definition. If the household problem has a solution, the maximum utility attainable can be expressed as a function of the budget:

DEFINITION VI.17 (INDIRECT UTILITY FUNCTION). *Consider a household with utility function U. The function (if it exists)*

$$\begin{aligned} V &: \mathbb{R}^\ell \times \mathbb{R}_+ \to \mathbb{R}, \\ &(p,m) \mapsto V(p,m) := U(x(p,m)) \end{aligned}$$

is called indirect utility function.

Figure VI.22 compares the utility function and the indirect utility function.

EXERCISE VI.17. *Determine the indirect utility function for the Cobb-Douglas utility function $U(x_1,x_2) = x_1^a x_2^{1-a}$ $(0<a<1)$!*

5.2. Revisiting the Lagrange multiplier. The aim of this section is to substitute the somewhat incorrect $\lambda = \frac{dU}{dm}$ by $\lambda = \frac{dV}{dm}$. This is not difficult. We differentiate both the budget equation and the indirect utility function with respect to income m. For the budget equation $\sum_{g=1}^\ell p_g x_g(p,m) = m$, we obtain

$$\sum_{g=1}^{\ell} p_g \frac{\partial x_g}{\partial m} = \frac{dm}{dm} = 1. \tag{VI.5}$$

The indirect utility function is defined by

$$V(p,m) = U(x(p,m)).$$

Forming the derivative with respect to m yields

$$\frac{\partial V}{\partial m} = \sum_{g=1}^{\ell} \frac{\partial U}{\partial x_g}\frac{\partial x_g}{\partial m}.$$

where the derivative on the right-hand side uses the adding rule (p. 83). Now, we obtain the desired result:

$$\begin{aligned}\frac{\partial V}{\partial m} &= \sum_{g=1}^{\ell} \lambda p_g \frac{\partial x_g}{\partial m} \text{ (eq. VI.4, p. 158)} \\ &= \lambda \sum_{g=1}^{\ell} p_g \frac{\partial x_g}{\partial m} \text{ (distributivity)} \qquad \text{(VI.6)} \\ &= \lambda \text{ (Gl. VI.5).} \qquad \text{(VI.7)}\end{aligned}$$

We can now rewrite the above optimization condition VI.4 in this fashion:

$$\underbrace{\frac{\partial U(x_1, x_2, ..., x_\ell)}{\partial x_g}}_{\text{marginal utility}} \stackrel{!}{=} \lambda p_g = \underbrace{\frac{\partial V}{\partial m} p_g}_{\text{marginal cost}}$$

That is, the household consumes every good so that the marginal utility of his consumption (left side) is equal to the marginal cost (also in terms of utility) of consumption (right side) where consuming one additional unit of good g means that the expenditure increases by p_g so that the income left for the consumption of other goods decreases by the same amount and hence the utility decreases by $\frac{dV}{dm} p_g$.

5.3. The indirect utility function is quasi-convex in prices and income. We plan to show that the indirect utility function is quasi-convex in both prices and income. Reversing the inequality signs and turning the min into the max operator in definition IV.26 (p. 79), we obtain the following

DEFINITION VI.18 (QUASI-CONVEXITY). $f : \mathbb{R}^\ell \to \mathbb{R}$ *is called quasi-convex if*

$$f(kx + (1-k)y) \le \max(f(x), f(y))$$

holds for all $x, y \in \mathbb{R}^\ell$ *and all* $k \in [0,1]$. f *is strictly quasi-convex if*

$$f(kx + (1-k)y) < \max(f(x), f(y))$$

holds for all $x, y \in \mathbb{R}^\ell$ *with* $x \neq y$ *and all* $k \in (0,1)$.

Thus, the value of the function f at a point between x and y is smaller than or equal to the largest of the values $f(x)$ and $f(y)$. Consider the examples of quasi-convexity in figureVI.23, while figure VI.24 depicts a function that is not quasi-convex.

EXAMPLE VI.1. *Any monotonically increasing or decreasing function* $f : \mathbb{R} \to \mathbb{R}$ *is quasi-convex.*

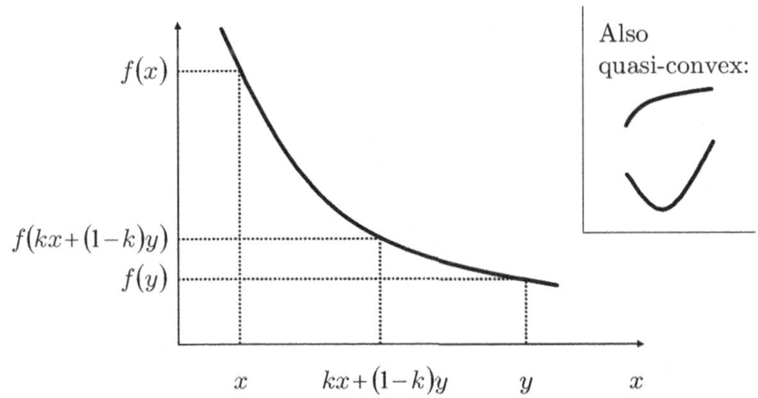

FIGURE VI.23. Strictly quasi-convex functions

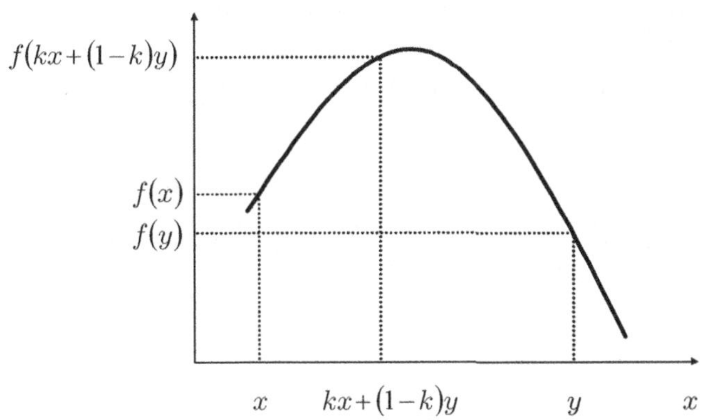

FIGURE VI.24. A function that is not quasi-convex

Similar to figure V.7 on p. 105, we contrast convexity (see the definition on p. 103 and also figure V.6 on p. 105) and quasi-convexity in figure VI.25.

THEOREM VI.7. *The indirect utility function is quasi-convex in both prices and income.*

PROOF. We assume two vectors

$$\begin{gathered}(p', m'), \\ (p'', m'')\end{gathered}$$

from $\mathbb{R}^\ell \times \mathbb{R}_+$ and consider the linear combination

$$(p^*, m^*) := k\,(p', m') + (1-k)\,(p'', m'') . \tag{VI.8}$$

We want to show

$$V\,(p^*, m^*) \leq \max\left(V\,(p', m'), V\,(p'', m'')\right) .$$

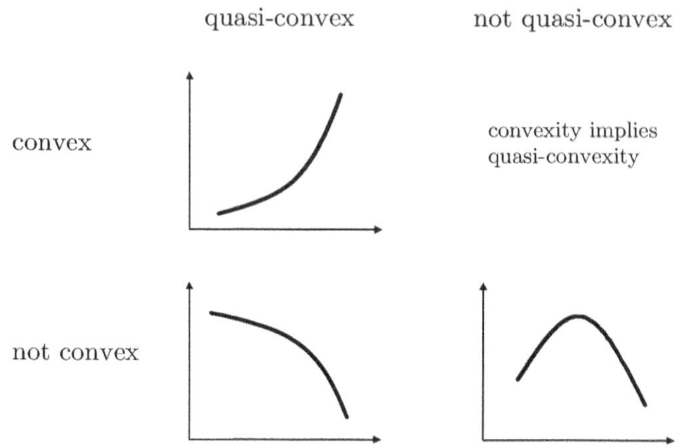

FIGURE VI.25. Convexity versus quasi convexity

Let $x^* := x(p^*, m^*)$ be the household optimum at prices p^* and income m^*. We find

$$\begin{aligned} kp'x^* + (1-k)\,p''x^* &= p^*x^* \text{ (eq. VI.8, distributivity)} \\ &\leq m^* \text{ (budget inequality)} \\ &= km' + (1-k)\,m'' \text{ (eq. VI.8)} \end{aligned}$$

Therefore, we obtain

$$\begin{aligned} p'x^* &\leq m' \text{ or} \\ p''x^* &\leq m'' \end{aligned}$$

and hence the hoped-for inequality:

$$\begin{aligned} & V\left(k\left(p', m'\right) + (1-k)\left(p'', m''\right)\right) \\ = \;& V\left((p^*, m^*)\right) \text{ (eq. VI.8)} \\ = \;& U(x^*) \;\; (x^* = x(p^*, m^*) \text{ is the household optimum}) \\ \leq \;& \max\left(V\left(p', m'\right), V\left(p'', m''\right)\right) \;\; (x^* \text{ affordable at } \left(p', m'\right) \text{ or } \left(p'', m''\right)) \end{aligned}$$

Do you see why the last inequality holds? Since x^* is affordable at (p', m') or (p'', m''), the utility $V(p', m')$ or $V(p'', m'')$ obtainable at these price-income vectors is at least as high as $U(x^*)$ so that we have $U(x^*) \leq V(p', m')$ or $U(x^*) \leq V(p'', m'')$. □

6. Consumer's rent and Marshallian demand

One can use the Marshallian demand curve $x_1(p_1)$ to derive an important welfare-theoretical concept, consumer's rent. We take the first-order

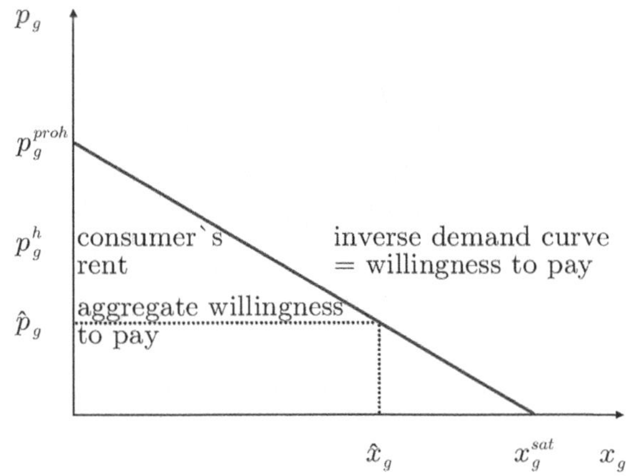

FIGURE VI.26. Willingness to pay and consumer's rent

condition

$$MRS \overset{!}{=} \frac{p_1}{p_2}$$

as a starting point. We concentrate on good 1 and assume that good 2 stands for "all the other goods" (also called money) and has price 1. Then, the willingness to pay for one extra unit of good 1 measured in terms of good 2 is p_1. Thus, the price is a good indication for the consumer's willingness to pay.

Have a look at the inverse demand function in figure VI.26. From the point of view of the above discussion, $p_g(x_g)$ can be addressed as the marginal willingness to pay for one extra unit of good g.

Then, we can distinguish several aggregate concepts:

DEFINITION VI.19 (MARSHALLIAN WILLINGNESS TO PAY). *Let p_g be an inverse demand function. The Marshallian willingness to pay for the quantity $\hat{x}_g$ is defined by*

$$\int_0^{\hat{x}_g} p_g(x_g)\, dx_g.$$

The Marshallian willingness to pay for a price decrease from p_g^h to $\hat{p}_g < p_g^h$ is defined by

$$\int_{\hat{p}_g}^{p_g^h} x_g(p_g)\, dp_g.$$

The Marshallian consumer's rent at price $\hat{p}_g \leq p_g^{proh}$ *is*

$$\begin{aligned} CR^{Marshall}(\hat{p}_g) &= \int_0^{x_g(\hat{p}_g)} p_g(x_g)\,dx_g - \hat{p}_g x_g(\hat{p}_g) \\ &= \int_0^{x_g(\hat{p}_g)} (p_g(x_g) - \hat{p}_g)\,dx_g \\ &= \int_{\hat{p}_g}^{p_g^{proh}} x_g(p_g)\,dp_g. \end{aligned}$$

Sketch the different integrals to check on our definitions. According to the last equality, the Marshallian consumer's rent at price $\hat{p}_g$ is the Marshallian willingness to pay for a price decrease from p_g^{proh} to $\hat{p}_g < p_g^{proh}$.

The aggregate Marshallian concepts are somewhat incorrect. Consider the Marshallian willingness to pay for the quantity $\hat{x}_g$, $\int_0^{\hat{x}_g} p_g(x_g)\,dx_g$. We have to "sum" the prices for all the units from 0 up to $\hat{x}_g$. Unfortunately, this procedure is at odds with the construction of the Marshallian demand curve:

- According to Marshallian demand, the consumers pay one price for all the units.
- In contrast, the integral $\int_0^{\hat{x}_g} p_g(x_g)\,dx_g$ presupposes that the consumer pays (in general) different prices for the first, the second, and so on, units.

Now, if higher prices are to be paid for the first units, the household has less income available to spend on the additional units so that we cannot expect that the "real" consumer's rent is as high as Marshallian demand has us believe. You will have to wait until the next chapter where we can present a correct graphical representation of the consumer's willingness to pay for a price decrease.

7. Topics

The main topics in this chapter are

- money budget
- endowment budget
- marginal opportunity cost
- feasibility
- objective function
- indirect utility function
- marginal utility of income
- labor supply
- intertemporal consumption
- contingent consumption
- demand function
- cross demand function
- Engel function
- substitutes and complements
- ordinary and nonordinary goods
- normal and inferior goods
- Marshallian consumer's rent
- local solution
- global solution
- interior solutions
- boundary solutions
- corner solutions
- Lagrange method
- Lagrange multiplier

8. Solutions

Exercise VI.1

For any number $\alpha > 0$, we have $B(\alpha p, m) = B\left(p, \frac{m}{\alpha}\right)$.

Exercise VI.2

Figure VI.27 shows the left-over in terms of good 1 and good 2. In order to obtain the left-over in money terms, we need to multiply the left-over of good 1 with p_1 (or the left-over of good 2 with p_2).

Exercise VI.3

Solving $p_1x_1 + p_2x_2 = m$ for x_2 yields $x_2 = \frac{m}{p_2} - \frac{p_1}{p_2}x_1$ so that the derivative of x_2 (as a function of x_1) is $-\frac{p_1}{p_2}$. Do you remember the line of constant expected value in the previous chapter?

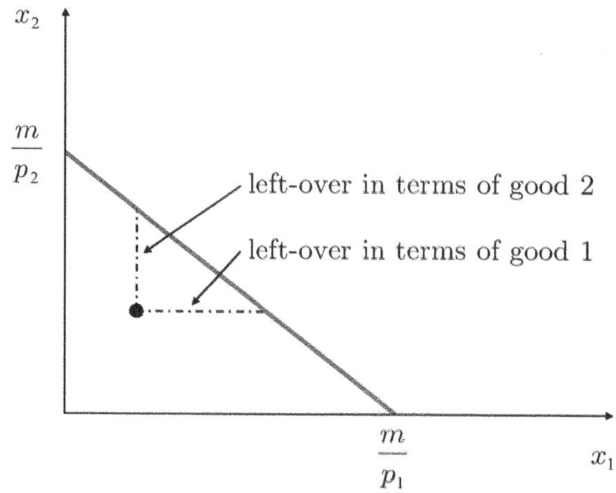

FIGURE VI.27. The left-over

Exercise VI.4

In figure VI.6, the part of the budget line right of the full-insurance point is associated with insurance sums $K > D$. In case of a loss, the household is refunded the damage D plus some extra. Then, the household benefits from the damage. This is called over-insurance.

Exercise VI.5

In subfigure (a), points A and B do not correspond to an optimum. The preferences are strictly convex and every point between A and B is better than A or B. Subfigure (b) depicts perfect substitutes. Point A is the household optimum. In subfigure (c), points A and B are optima but so are all the points in between. Turning to subfigure (d), the point of tangency A is the worst bundle of all the bundles on the budget line. There are two candidates for household optima in this case of concave preferences, the two extreme bundles $\left(\frac{m}{p_1}, 0\right)$ and $\left(0, \frac{m}{p_2}\right)$.

Exercise VI.6

If y is better than x, the household optimum, the household cannot afford y :

$$y \succ x \Rightarrow p \cdot y > p \cdot \omega$$

Exercise VI.7

Good 1 is an ordinary good by $\frac{\partial x_1}{\partial p_1} < 0$, it is normal because of $\frac{\partial x_1}{\partial m} > 0$ and is it both a substitute and a complement of good 2 by $\frac{\partial x_1}{\partial p_2} = 0$.

Exercise VI.8

Consider figure VI.28.

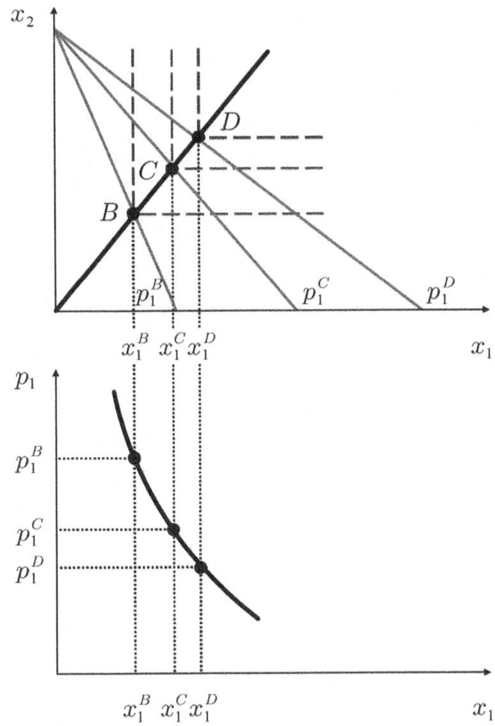

FIGURE VI.28. The price-consumption curve and the demand curve for perfect complements

Exercise VI.9

The household optima can be found via $x_1 = 2x_2$ and the budget line. They are given by $\left(x_1^* = \frac{m}{p_1+\frac{1}{2}p_2}, \; x_2^* = \frac{m}{2p_1+p_2}\right)$. Hence good 2's demand curve is given by $x_2 = f(p_2) = \frac{m}{2p_1+p_2}$ for positive prices p_1 and p_2. The complete demand curve is given by

$$x_2(p_1, p_2, m) = \begin{cases} \frac{m}{2p_1+p_2}, & p_1 > 0, p_2 > 0 \\ \emptyset, & p_1 \leq 0 \text{ and } p_2 \leq 0 \\ \left[\frac{m}{2p_1}, \infty\right) & p_1 > 0, p_2 = 0 \\ \frac{m}{p_2}, & p_1 = 0, p_2 > 0 \end{cases}$$

The price-consumption curve for varying p_1 seems to be $x_2 = h(x_1) = \frac{1}{2}x_1$. However, for $p_2 < 0$, no household optimum and hence no price-consumption curve exists. For $p_2 = 0$, optima exist for $p_1 > 0$ only (see the left-hand graph in figure VI.29). For $p_2 > 0$, the price-consumption curve consists of

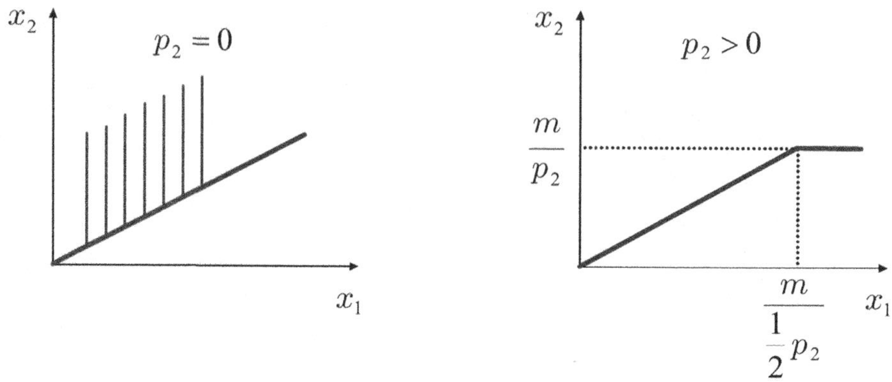

FIGURE VI.29. The price-consumption curves in case of perfect complements

two parts which are depicted in the right-hand graph:

$$x_2 = \begin{cases} \frac{1}{2}x_1, & x_1 \le \frac{m}{\frac{1}{2}p_2} \ (p_1 > 0) \\ \frac{m}{p_2}, & x_1 > \frac{m}{\frac{1}{2}p_2} \ (p_1 = 0) \end{cases}$$

Exercise VI.10

See figure VI.30.

Exercise VI.11

Again, we use $x_1 = 2x_2$ and $x_2^* = \frac{m}{2p_1+p_2}$. The income-consumption curve is given by $x_2 = g(x_1) = \frac{1}{2}x_1$. Good 2's Engel curve is given by $x_2 = q(m) = \frac{m}{2p_1+p_2}$.

Exercise VI.12

The first-order condition

$$\frac{\frac{\partial U}{\partial x_1}}{\frac{\partial U}{\partial x_2}} = \frac{1}{x_1} \overset{!}{=} \frac{p_1}{p_2}$$

and the budget constraint $p_1x_1 + p_2x_2 = m$ yield

$$x(m,p) = \left(\frac{p_2}{p_1}, \frac{m}{p_2} - 1\right).$$

Therefore, we have

$$\begin{aligned} \frac{\partial x_1(p,m)}{\partial p_2} &= \frac{1}{p_1} > 0 \text{ and} \\ \frac{\partial x_2(p,m)}{\partial p_1} &= 0. \end{aligned}$$

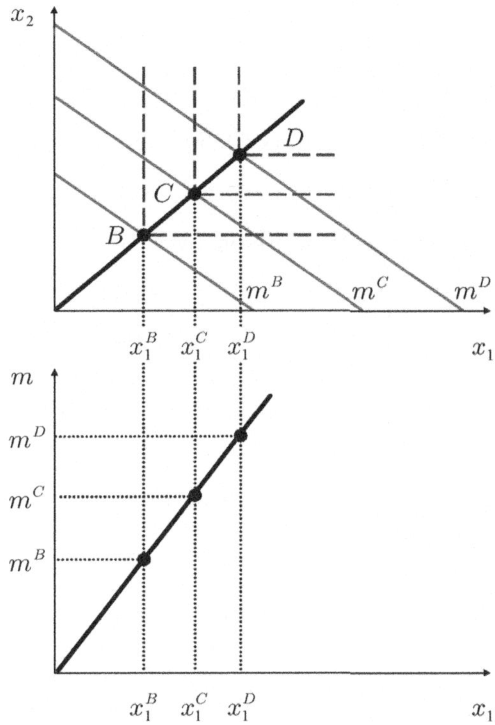

FIGURE VI.30. The income-consumption curve and the Engel curve for perfect complements

Exercise VI.13

We find

$$\begin{aligned} \varepsilon_{x_g,p_g} &= \frac{dx_g}{dp_g}\frac{p_g}{x_g} = (-1)\frac{p_g}{100-p_g} \text{ and} \\ \varepsilon_{x_k,p_k} &= \frac{dx_k}{dp_k}\frac{p_k}{x_k} = (-1)\frac{1}{p_k^2}\frac{p_k}{\frac{1}{p_k}} = -1. \end{aligned}$$

Exercise VI.14

We have found $x_1^* = q(m) = \frac{1}{3}\frac{m}{p_1}$. The income elasticity is

$$\varepsilon_{x_1,m} = \frac{\partial x_1}{\partial m}\frac{m}{x_1} = \frac{\frac{1}{3}}{p_1}\frac{m}{\frac{1}{3}\frac{m}{p_1}} = 1$$

and the good is normal, just in between luxury and necessity.

Exercise VI.15

We find the prohibitive prices

$$\begin{aligned} &100 \text{ for } x_g^1 \ \max(0, 100 - p_g), \\ &25 \text{ for } x_g^2 \ \max(0, 50 - 2p_g) \text{ and} \\ &20 \text{ for } x_g^3 \ \max(0, 60 - 3p_g) \end{aligned}$$

and therefore aggregate demand x_g given by

$$x_g(p) = \begin{cases} 0, & p_g \geq 100 \\ 100 - p_g, & 25 \leq p_g < 100 \\ 150 - 3p_g & 20 \leq p_g < 25 \\ 210 - 6p_g & 0 \leq p_g < 20 \end{cases}$$

Exercise VI.16

Setting the derivative of L with respect to λ equal to 0 is just a reformulation of the budget equality.

Exercise VI.17

We know from p. 140 the household optimum

$$\begin{aligned} x_1(m,p) &= a\frac{m}{p_1}, \\ x_2(m,p) &= (1-a)\frac{m}{p_2}. \end{aligned}$$

Therefore, we obtain the indirect utility function as

$$\begin{aligned} V(p,m) &= U(x(p,m)) \\ &= \left(a\frac{m}{p_1}\right)^a \left((1-a)\frac{m}{p_2}\right)^{1-a} \\ &= \left(\frac{a}{p_1}\right)^a \left(\frac{1-a}{p_2}\right)^{1-a} m. \end{aligned}$$

9. Further exercises without solutions

PROBLEM VI.1. Discuss the units in which to measure price, quantity, expenditure. If you are right, expenditure should be measured in the same units as the product of price and quantity.

PROBLEM VI.2. Sketch budget lines or the displacements of budget lines for the following examples:

- Time $T = 18$ and money $m = 50$ for football F (good 1) or basket ball B (good 2) with prices
  - $p_F = 5$, $p_B = 10$ in monetary terms
  - $t_F = 3$, $t_B = 2$ in temporary terms
- Two goods, bread (good 1) and other goods (good 2). Transfer in kind with and without prohibition to sell:
  - $m = 20$, $p_B = 2$, $p_{\text{other}} = 1$
  - Transfer in kind: $B = 5$

PROBLEM VI.3. Assume two goods 1 and 2. Abba faces a price p for good 1 in terms of good 2. Think of good 2 as the numeraire good with price 1. Abba's utility functions U is given by $U(x_1, x_2) = \sqrt{x_1} + x_2$. His endowment is $\omega = (25, 0)$. Find Abba's optimal bundle. *Hint: Distinguish* $p \geq \frac{1}{10}$ *and* $p < \frac{1}{10}$*!*

PROBLEM VI.4. Show by way of example that $\frac{\partial x_1(p,m)}{\partial p_1} < 0$ and $\frac{\partial x_1(p,\omega)}{\partial p_1} > 0$ may well happen. *Hint: use perfect complements.*

PROBLEM VI.5. Derive the indirect utility functions of the following utility functions:

(1) $U(x_1, x_2) = x_1 \cdot x_2$
(2) $U(x_1, x_2) = \min\{a \cdot x_1, b \cdot x_2\}$ where $a, b > 0$ holds
(3) $U(x_1, x_2) = a \cdot x_1 + b \cdot x_2$ where $a, b > 0$ holds.

PROBLEM VI.6. Discuss the following claim: The lexicographic preference relation on $\mathbb{R}^2_+$, where good 1 is the important good, gives rise to the indirect utility function given by $V(p_1, p_2, m) = \frac{m}{p_1}$ in case of $p >> 0$. How about $p_1 = 0$ or $p_2 = 0$?

PROBLEM VI.7. Show: If $\precsim$ is strictly monotonic and homothetic (compare Problem 6 in Chapter IV), then $x^R(p, \lambda\omega) = \lambda x^R(p, \omega)$ holds for all $\lambda > 0$. *Hint:*

- the two bundles cost the same (how much?) and hence household optimality implies $x^R(p, \lambda\omega) \succsim \lambda \cdot x^R(p, \omega)$ for all bundles of both sets,

- the two bundles $x^R(p,\omega)$ and $\frac{1}{\lambda}x^R(p,\lambda\omega)$ cost the same and hence household optimality implies $x^R(p,\omega) \succsim \frac{1}{\lambda}x^R(p,\lambda\omega)$ for all bundles of both sets,
- homotheticity then implies $\lambda x^R(p,\omega) \succsim x^R(p,\lambda\omega)$ for all bundles of both sets
- the first and the third bullets imply $x^R(p,\lambda\omega) \sim \lambda x^R(p,\omega)$ for all bundles of both sets and hence $\lambda x^R(p,\omega) \subseteq x^R(p,\lambda\omega)$.
- apply the same procedure as above with $\frac{1}{\lambda}$ (rather than λ) and $\lambda\omega$ (replacing ω) to find $\frac{1}{\lambda}x^R(p,\lambda\omega) \subseteq x^R(p,\omega)$ and hence $x^R(p,\lambda\omega) \subseteq \lambda x^R(p,\omega)$
- since inclusion holds in both directions, the desired equality of the two sets is established

PROBLEM VI.8. Consider the preferences given by $U(x_1,x_2) = x_1 + 2x_2$. Find $x(p,m)$. Sketch the demand functionn for good 1. Sketch the Engel curve for good 1 in case of $p_1 < \frac{1}{2}p_2$ while observing the usual convention that the x_1-axis is the abscissa!

CHAPTER VII

Comparative statics and duality theory

In this chapter, we deal with the effects that prices or income have on the household optimum. Differently put, we discuss demand functions, cross demand functions, and Engel functions. We also discuss monetary measures for changes affecting the household.

1. The duality approach

1.1. Maximization problem and minimization problem. In the previous chapter, we have expressed the problem facing a household as a maximization problem. In good-natured problems, there is basically no difference between the following two problems:

- Maximization problem: Find the bundle that maximizes the utility for a given budget line.
- Minimization problem: Find the bundle that minimizes the expenditure needed to achieve a given utility level.

This fact is illustrated by figure VII.1. Given the bold budget line (the lower one) with income m, the household maximizes utility by choosing bundle B rather than bundle C which he can also afford. The household then achieves the utility level $\bar{U}$. This is the utility-maximization problem. Turning to the minimization problem, we fix the utility level $\bar{U}$ and try to find the minimal expenditure necessary. The expenditure-minimizing bundle is (again!) bundle B rather than bundle A.

We say that the maximization and the minimization problem are dual problems. They are basically identical. We do not, however, define "dual" or "basically" formally. In this chapter we spell out the duality approach to household theory in some detail and present some important theoretical results.

1.2. The expenditure function. From chapter VI, we are familiar with the indirect utility function

$$\begin{aligned} V &: \mathbb{R}^\ell \times \mathbb{R}_+ \to \mathbb{R}, \\ & (p,m) \mapsto V(p,m) := \max_{x \in B(p,m)} U(x). \end{aligned}$$

H. Wiese, *Advanced Microeconomics*,
https://doi.org/10.1007/978-3-658-34959-2_7

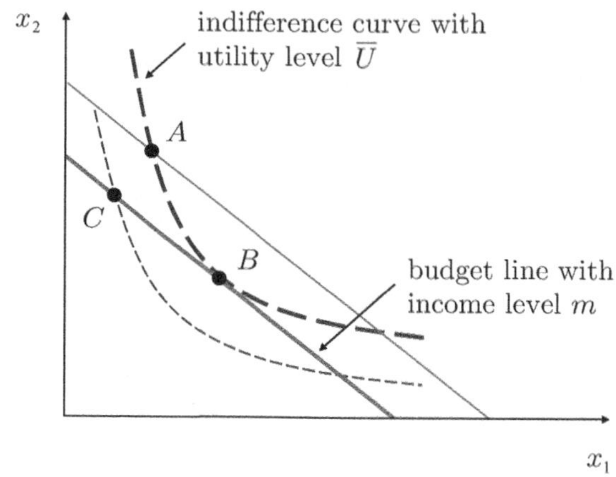

FIGURE VII.1. Maximizing utility corresponds to minimizing expenditure

It tells us the maximum utility achievable with a given income. We now turn this question on its head and ask for the minimum expenditure necessary to achieve a specific utility level (or a specific indifference curve) or a better one:

DEFINITION VII.1 (EXPENDITURE FUNCTION). *Consider a household with utility function U. The function (if it exists)*

$$\begin{aligned} e &: \mathbb{R}^\ell \times \mathbb{R} \to \mathbb{R}, \\ &(p, \bar{U}) \mapsto e(p, \bar{U}) := \min_{\substack{x \text{ with} \\ U(x) \geq \bar{U}}} px \end{aligned}$$

is called expenditure function. The solution to this minimization problem (if there is any) is the function

$$\begin{aligned} \chi &: \mathbb{R}^\ell \times \mathbb{R} \to 2^{\mathbb{R}^\ell_+}, \\ &(p, \bar{U}) \mapsto \chi(p, \bar{U}) := \arg \min_{\substack{x \text{ with} \\ U(x) \geq \bar{U}}} px \end{aligned}$$

χ is called the Hicksian demand function. The Greek letter χ (to be pronounced chi) resembles the Roman letter x and also hints to the H in Hicks.

EXERCISE VII.1. *Express*

- *e in terms of χ and*
- *V in terms of the household optima!*

For household maximization and minimization problems, one optimality condition depends on the specific utility function. For example, one considers a corner solution in case of perfect substitutes, the equality of MRS and MOC if we have strictly convex preferences, etc. For Marshallian demand (maximization problem), the second optimality condition is the budget line in case of monotonic preferences. In order to find Hicksian demand (minimization problem), the second optimality conditions is the utility function, namely $\bar{U} = U(x)$. However, sometimes, one can take Marshallian demand to derive Hicksian demand (and vice versa).

EXERCISE VII.2. *Determine the expenditure function and the Hicksian demand function for the Cobb-Douglas utility function $U(x_1, x_2) = x_1^a x_2^{1-a}$ with $0 < a < 1$! Hint: From exercise VI.17 (p. 159) we know that the indirect utility function V is given by*

$$\begin{aligned} V(p, m) &= U(x(p, m)) \\ &= \left(a\frac{m}{p_1}\right)^a \left((1-a)\frac{m}{p_2}\right)^{1-a} \\ &= \left(\frac{a}{p_1}\right)^a \left(\frac{1-a}{p_2}\right)^{1-a} m. \end{aligned}$$

Use this equation to derive the expenditure function. Hicksian demand can be derived from $\chi_1(p, \bar{U}) = x_1(p, e(p, \bar{U}))$.

From chapter VI, we know $x(\alpha p, \alpha m) = x(p, m)$ and hence $V(\alpha p, \alpha m) = V(p, m)$ for any number $\alpha > 0$. We have a similar result for Hicksian demand and the expenditure function:

LEMMA VII.1. *For any $\alpha > 0$, we have $\chi(\alpha p, \bar{U}) = \chi(p, \bar{U})$ and $e(\alpha p, \bar{U}) = \alpha e(p, \bar{U})$.*

PROOF. For $\alpha > 0$, we have

$$\chi(\alpha p, \bar{U}) = \arg \min_{\substack{x \text{ with} \\ U(x) \geq \bar{U}}} \alpha p x = \arg \min_{\substack{x \text{ with} \\ U(x) \geq \bar{U}}} p x = \chi(p, \bar{U});$$

the bundle x with $U(x) \geq \bar{U}$ that minimizes px also minimizes $\alpha p x$. Now, if prices are changed by a factor $\alpha > 0$, the expenditure necessary to buy $\chi(p, \bar{U})$ is also changed by the same factor α:

$$e(\alpha p, \bar{U}) = (\alpha p) \cdot \chi(p, \bar{U}) = \alpha\left[p \cdot \chi(p, \bar{U})\right] = \alpha e(p, \bar{U}).$$

□

The Hicksian demand function is also called compensating demand function. It tells how to change demand in response to price changes while

	utility maximization	expenditure minimization
objective function	utility	expenditure
parameters	prices p, income m	prices p, utility $\bar{U}$
notation for best bundle(s)	$x(p,m)$	$\chi(p,\bar{U})$
name of demand function	Marshallian	Hicksian
value of objective function	$V(p,m)$ $= U(x(p,m))$	$e(p,\bar{U})$ $= p\cdot\chi(p,\bar{U})$

FIGURE VII.2. Duality

income is adapted in order to keep the utility constant. In contrast, the usual demand function x (which is also called Marshallian demand function) holds income constant. Figure VII.2 juxtaposes utility maximization and expenditure minimization.

1.3. Applying the Lagrange method to expenditure minimization. As in chapter VI (pp. 156), we assume strict quasi-concavity of the utility function U, strict monotonicity of U, and positive prices $p >> 0$. The Lagrange function for expenditure minimization is given by

$$L(x,\mu) = \sum_{g=1}^{\ell} p_g x_g + \mu\left[\bar{U} - U(x)\right].$$

Differentiating with respect to x_g and μ yields

$$\frac{\partial L(x_1, x_2, ..., \mu)}{\partial x_g} = p_g - \mu \frac{\partial U(x_1, x_2, ..., x_\ell)}{\partial x_g} \stackrel{!}{=} 0$$

and the side condition

$$\frac{\partial L(x,\mu)}{\partial \mu} = \bar{U} - U(x) \stackrel{!}{=} 0.$$

The Lagrange multiplier μ has an economic interpretation. It translates a utility surplus (in case of $U(x) > \bar{U}$) into a possible reduction of expenditure. Thus, increasing consumption has a positive direct effect on expenditure (expenditure increases) and a negative indirect effect via a utility surplus (scope for expenditure reduction). Therefore (we do not give a proof), maximizing L with respect to x and μ is equivalent to minimizing expenditure for p and $\bar{U}$.

A comparison with equation VI.3 (p. 158) shows that expenditure minimization also leads to the $MRS \stackrel{!}{=} MOC$ condition. The comparison of the Lagrange multipliers is also instructive:

- λ is the shadow price for utility maximization. It tells how additional income leads to higher utility: $\lambda = \frac{\partial V}{\partial m}$.
- μ is the shadow price for expenditure minimization. Increasing $\bar{U}$ by one unit leads to an increase of expenditure by $\mu = \frac{\partial e(p,\bar{U})}{\partial \bar{U}}$.

This comparison does not prove but makes plausible the (correct) equation

$$\mu = \frac{1}{\lambda}.$$

1.4. The duality theorem. The duality of utility maximization and expenditure minimization means that we can use the solution of one problem to find the solution of the other. However, this cannot always work. Consider, for example, point B of figure VII.3, which is an interior point of the budget with income m. This household optimum is sometimes called "bliss point". The household cannot do any better even if his income would increase by a lot. In order to achieve the utility of the bliss point ($V(p,m) = 9$), the budget indicated by the budget line is not necessary, a smaller budget is sufficient to afford the bliss point. Thus, in this case, we have

$$e(p, V(p,m)) < m.$$

We now present a theorem that states conditions under which duality does work:

THEOREM VII.1. *Let $U : \mathbb{R}^\ell_+ \to \mathbb{R}$ be a continuous utility function that obeys local nonsatiation and let $p >> 0$ be a price vector. We then obtain duality in both directions:*

- *If $x(p,m)$ is the household optimum for $m > 0$, we have*

$$\chi(p, V(p,m)) = x(p,m) \qquad \text{(VII.1)}$$

and

$$e(p, V(p,m)) = m. \qquad \text{(VII.2)}$$

- *If $\chi(p,\bar{U})$ is the expenditure-minimizing bundle for $\bar{U} > U(0)$, we have*

$$x(p, e(p,\bar{U})) = \chi(p,\bar{U}) \qquad \text{(VII.3)}$$

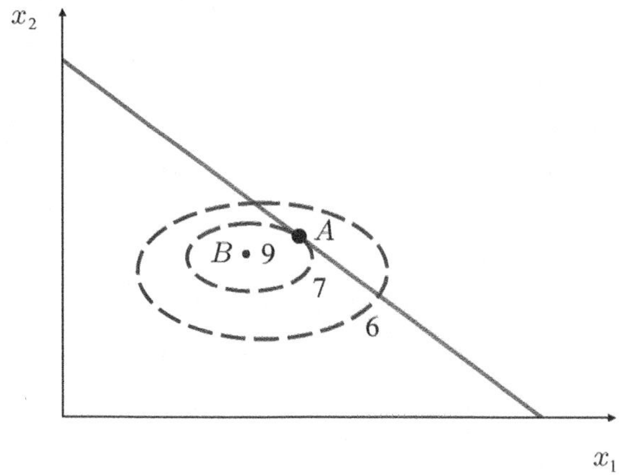

FIGURE VII.3. Duality does not work here.

and

$$V\left(p, e\left(p, \bar{U}\right)\right) = \bar{U}. \tag{VII.4}$$

The duality theorem implies that expenditure is an increasing function of the utility level:

LEMMA VII.2. *Let $U : \mathbb{R}_+^\ell \to \mathbb{R}$ be a continuous utility function that obeys local nonsatiation and let $p >> 0$ be a price vector. We have*

$$\frac{\partial e\left(p, \bar{U}\right)}{\partial \bar{U}} > 0.$$

PROOF. Assume two utility levels $\bar{U}^l$ and $\bar{U}^h$ with $\bar{U}^l < \bar{U}^h$. By the definition of the expenditure function, we have

$$\begin{aligned} e\left(p, \bar{U}^l\right) &= \min_{\substack{x \text{ with} \\ U(x) \geq \bar{U}^l}} px \\ &\leq \min_{\substack{x \text{ with} \\ U(x) \geq \bar{U}^h}} px \text{ (we have less bundles to choose from)} \\ &= e\left(p, \bar{U}^h\right) \end{aligned}$$

We now assume $e\left(p, \bar{U}^l\right) = e\left(p, \bar{U}^h\right)$, i.e., $p \cdot \chi\left(p, \bar{U}^l\right) = p \cdot \chi\left(p, \bar{U}^h\right)$. This means $e\left(p, \bar{U}^l\right)$ is the minimum expenditure necessary for the achievement of utility level $\bar{U}^l$, but the higher utility level $\bar{U}^h$ can be obtained without additional expenditure. By duality (equation VII.3), the above equality can be written as

$$p \cdot x\left(p, e\left(p, \bar{U}^l\right)\right) = p \cdot x\left(p, e\left(p, \bar{U}^h\right)\right).$$

Also, we find

$$\begin{aligned}
U\left(x\left(p,e\left(p,\bar{U}^l\right)\right)\right) &= V\left(p,e\left(p,\bar{U}^l\right)\right) \text{ (definition of indirect utility)} \\
&= \bar{U}^l \text{ (duality theorem, equation VII.4)} \\
&< \bar{U}^h \text{ (assumption)} \\
&= V\left(p,e\left(p,\bar{U}^h\right)\right) \text{ (duality theorem, eq. VII.4)} \\
&= U\left(x\left(p,e\left(p,\bar{U}^h\right)\right)\right) \text{ (def. of indirect utility).}
\end{aligned}$$

This means that the bundles $x\left(p,e\left(p,\bar{U}^l\right)\right)$ and $x\left(p,e\left(p,\bar{U}^h\right)\right)$ are equally expensive, but the first one leads to a lower utility than the second one. This is a contradiction to utility maximization. Thus, the strict inequality $e\left(p,\bar{U}^l\right) < e\left(p,\bar{U}^h\right)$ holds, proving the lemma. □

1.5. Main results. In this last subsection, we present some important results about Hicksian demand and the expenditure function. Some of these results are derived with the help of the duality approach. The rest of this chapter is devoted to proving these propositions and discussing important conclusions.

THEOREM VII.2. *Consider a household with a continuous utility function U, Hicksian demand function χ, and expenditure function e. We have the following results:*

- *Shephard's lemma (p. 183): The price increase of good g by one small unit increases the expenditure necessary to uphold the utility level by χ_g.*
- *Monotonicity of expenditure function (p. 178): In case of local non-satiation and strictly positive prices, the expenditure function is monotonic in utility.*
- *Hicksian law of demand (pp. 185 and 189): If the price of a good g increases, the Hicksian demand χ_g does not increase.*
- *Concavity of expenditure function (p. 187): The expenditure function is concave in its prices.*
- *The Hicksian cross demands are symmetric (p. 189): $\frac{\partial \chi_g(p,\bar{U})}{\partial p_k} = \frac{\partial \chi_k(p,\bar{U})}{\partial p_g}$.*
- *Roy's identity (p. 184): A price increase of good g by one small unit decreases*
  - *the budget available for the other goods by χ_g and*
  - *indirect utility by the product of the marginal utility of income $\frac{\partial V}{\partial m}$ and χ_g.*

- *Convexity of indirect utility function (pp. 160): The indirect utility function is convex in prices and income.*
- *Slutsky equations (pp. 193): Marshallian demand and Hicksian demand are related by the money-budget Slutsky equation*

$$\frac{\partial x_g}{\partial p_g} = \frac{\partial \chi_g}{\partial p_g} - \frac{\partial x_g}{\partial m}\chi_g$$

and the endowment Slutsky equation

$$\frac{\partial x_g^{endowment}}{\partial p_g} = \frac{\partial \chi_g}{\partial p_g} + \frac{\partial x_g^{money}}{\partial m}\left(\omega_g - \chi_g\right)$$

2. Envelope theorems and Shephard's lemma

2.1. The example of the expenditure function. This section deals with a set of useful mathematical results that we will need several times. The application at hand concerns the expenditure function. Assume a price increase of one good g, from p_g to p_g'. If the household consumes the same bundle as before (and therefore achieves the same utility level), he has to increase expenditure by

$$\left(p_g' - p_g\right)\chi_g$$

where χ_g is the quantity consumed at the old price. Therefore, we have

$$e\left(p', \bar{U}\right) \leq e\left(p, \bar{U}\right) + \left(p_g' - p_g\right)\chi_g$$

where the entries in the price vectors p' and p are the same for all goods except good g.

The inequality may be strict if the household reshuffles his expenditure-minimizing bundle in response to the price increase. An implication of the envelope theorem is that reshuffling does not pay in case of a very small price increase. Then, $e\left(p', \bar{U}\right) = e\left(p, \bar{U}\right) + \left(p_g' - p_g\right)\chi_g$ implies

$$\frac{\partial e\left(p, \bar{U}\right)}{\partial p_g} = \lim_{p_g' \to p_g} \frac{e\left(p', \bar{U}\right) - e\left(p, \bar{U}\right)}{p_g' - p_g} = \lim_{p_g' \to p_g} \frac{\left(p_g' - p_g\right)\chi_g}{p_g' - p_g} = \chi_g.$$

2.2. Envelope theorem without constraints. There are two envelope theorems: one without, and one with, constraints. We begin with the simpler case and assume a function f with values $f(a, x)$ where a is a parameter and x a variable. For example, x may denote an output decision undertaken by a firm whose profit is influenced by some weather or demand parameter a. In contrast to problem II.2 (p. 29), we assume that both x and a are taken from the real line (or perhaps some interval of the real line).

The problem is to choose x so as to maximize $f(a,x)$. The optimal value of x (we assume that there is exactly one) is a function of a, written as

$$x^R(a).$$

This is our well-known best-response function. On the basis of f, we define another function:

DEFINITION VII.2 (REDUCED-FORM). *Let $f:\mathbb{R}\times\mathbb{R}\to\mathbb{R}$ be a differentiable function with values $f(a,x)$. We call a a parameter and x a variable. Then,*

$$\begin{aligned} \hat{f} &: \mathbb{R}\to\mathbb{R}, \\ & a\mapsto \hat{f}(a) := f\left(a, x^R(a)\right) \end{aligned}$$

is a well-defined function, often called reduced function. We assume that x^R is a differentiable function of a.

Now, a change in a influences the value $\hat{f}(a)$ directly (through a) and indirectly (through $x^R(a)$). The envelope theorem claims that we can forget about the indirect effect if we consider derivatives:

THEOREM VII.3. *Let f and $\hat{f}$ be given as in the preceding definition. Assume that $x^R(a)$ is an interior solution for every $a\in\mathbb{R}$. Then, we have*

$$\frac{d\hat{f}}{da} = \left.\frac{\partial f}{\partial a}\right|_{x^R(a)}.$$

PROOF. The claim follows from

$$\begin{aligned} \frac{d\hat{f}}{da} &= \frac{df\left(a,x^R(a)\right)}{da} \text{ (this is not the partial derivative of } f!) \\ &= \left.\frac{\partial f}{\partial a}\right|_{x^R(a)} + \left.\frac{\partial f}{\partial x}\right|_{x^R(a)}\frac{dx^R}{da} \text{ (theorem IV.3, p. 83)} \\ &= \left.\frac{\partial f}{\partial a}\right|_{x^R(a)} + \underbrace{\left.\frac{\partial f}{\partial x}\right|_{x^R(a)}}_{0}\frac{dx^R}{da} \text{ (theorem VI.5, p. 156)} \\ &= \left.\frac{\partial f}{\partial a}\right|_{x^R(a)}. \end{aligned}$$

□

2.3. Envelope theorem with equality constraints. We now assume an optimization problem with equality constraint:

DEFINITION VII.3. *Let $f : \mathbb{R} \times \mathbb{R} \to \mathbb{R}$ be a differentiable function with values $f(a, x)$ where a is a parameter and x a variable. x has to obey an equality constraint*

$$g(a, x) = 0$$

where $g : \mathbb{R} \times \mathbb{R} \to \mathbb{R}$ is also differentiable. Assume $\left|x^R(a)\right| = 1$ for all $a \in \mathbb{R}$ and that x^R is a differentiable function of a. Then,

$$\begin{aligned} \hat{f} &: \mathbb{R} \to \mathbb{R}, \\ &a \mapsto \hat{f}(a) := f\left(a, x^R(a)\right) \end{aligned}$$

is a well-defined function.

Again, a change in a influences the value $\hat{f}(a)$ directly (through a) and indirectly (through $x^R(a)$). The envelope theorem claims that we are allowed to ignore the indirect effect if we consider derivatives. However, we need to take the constraint into account:

THEOREM VII.4. *Let f and $\hat{f}$ be given as in the preceding definition. Assume f's Lagrange function L given by*

$$L(a, x, \lambda) := f(a, x) + \lambda g(a, x).$$

Assume that $x^R(a)$ is the solution obtained by the Lagrange method for every $a \in \mathbb{R}$. Then, we have

$$\frac{d\hat{f}}{da} = \left.\frac{\partial f(a, x)}{\partial a}\right|_{x^R(a)} + \lambda \left.\frac{\partial g(a, x)}{\partial a}\right|_{x^R(a)}.$$

Thus, in order to find the effect of the parameter a on the optimal value, we can ignore the change of the optimal response to a.

PROOF. The Lagrange function is given by

$$L(a, x, \lambda) = f(a, x) + \lambda g(a, x).$$

By theorem VI.6 (p. 156), we find

$$\left.\frac{\partial L}{\partial x}\right|_{x^R(a)} = \left.\frac{\partial f}{\partial x}\right|_{x^R(a)} + \lambda \left.\frac{\partial g}{\partial x}\right|_{x^R(a)} = 0.$$

The proof now follows from

$$\begin{aligned}
\frac{d\hat{f}}{da} &= \frac{dL\left(a, x^R(a), \lambda\right)}{da} \text{ (this is not the partial derivative of } f\text{!)} \\
&= \left.\frac{\partial L}{\partial a}\right|_{x^R(a)} + \left.\frac{\partial L}{\partial x}\right|_{x^R(a)} \frac{dx^R}{da} \text{ (theorem IV.3 (p. 83))} \\
&= \left.\frac{\partial L}{\partial a}\right|_{x^R(a)} + \underbrace{\left.\frac{\partial L}{\partial x}\right|_{x^R(a)}}_{0} \frac{dx^R}{da} \text{ (just shown)} \\
&= \left.\frac{\partial L}{\partial a}\right|_{x^R(a)} \\
&= \left.\frac{\partial f(a,x)}{\partial a}\right|_{x^R(a)} + \lambda \left.\frac{\partial g(a,x)}{\partial a}\right|_{x^R(a)}.
\end{aligned}$$

□

2.4. Application: Shephard's lemma. We apply definition VII.3 and theorem VII.4 to the question of how the minimal expenditure varies with a price change of a good p_g. We take the other prices p_{-g} as given and use the following correspondances:

- the role of $f(a,x)$ is taken over by the expenditure $e(p_g, x) = p \cdot x$,
- $\hat{f}(a) = f\left(a, x^R(a)\right)$ is translated into the minimal expenditure $\hat{e}(p_g) := p \cdot \chi(p_g)$ and
- the equality constraint is $U(x) - \bar{U} = 0$.

Note that the equality constraint does not depend on the parameter p_g and that theorem VII.4 says: Forget about indirect effects (working through χ). Thus, we obtain

$$\begin{aligned}
&\frac{d\hat{e}}{dp_g} \text{ (definition of } \hat{e}) \\
&= \left.\frac{\partial e(p_g, x)}{\partial p_g}\right|_{\chi(p_g)} + \lambda \left.\frac{\partial\left(U(x) - \bar{U}\right)}{\partial p_g}\right|_{\chi(p_g)} \text{ (theorem VII.4)} \\
&= \left.\frac{\partial e(p_g, x)}{\partial p_g}\right|_{\chi(p_g)} \text{ } (U(x) \text{ does not depend on } p_g) \\
&= \frac{\partial \sum_{k=1}^{\ell} p_k \chi_k}{\partial p_g} \text{ (definition of dot product, partial (!) derivative)} \\
&= \chi_g, \qquad \text{(VII.5)}
\end{aligned}$$

a result known as Shephard's lemma. (There are several lemmata due to this name.)

2.5. Application of the application: Roy's identity. In the last subsection, we apply Shephard's lemma in order to find out how a price change affects the achievable utility, i.e., to examine

$$\frac{\partial V}{\partial p_g}$$

for some good g. We use the duality equation

$$\bar{U} = V\left(p, e\left(p, \bar{U}\right)\right)$$

known from p. 178. We differentiate both sides with respect to p_g and obtain

$$\begin{aligned} 0 &= \frac{\partial V}{\partial p_g} + \frac{\partial V}{\partial m}\frac{\partial e}{\partial p_g} \\ &= \frac{\partial V}{\partial p_g} + \frac{\partial V}{\partial m}\chi_g \text{ (Shephard's Lemma)} \end{aligned}$$

which can be rewritten as Roy's identity:

$$\frac{\partial V}{\partial p_g} = \frac{\partial V}{\partial m}\left(-\chi_g\right). \qquad \text{(VII.6)}$$

A price increase increases necessary expenditure (necessary to keep the utility level constant) by χ_g (this is Shephard's lemma). If, however, the budget is given, the budget for the other good decreases by χ_g. The marginal utility of income $\frac{\partial V}{\partial m}$ (see also p. 160 in chapter VI) translates this budget reduction into a reduction of utility.

3. Concavity, the Hesse matrix, and the Hicksian law of demand

3.1. Compensated (Hicksian) law of demand. We now continue our examination of the expenditure function and the Hicksian demand function. The Hicksian law of demand states that prices and quantities move inversely.

Let p and p' be two price vectors from $\mathbb{R}^\ell$ and let $\chi\left(p, \bar{U}\right) \in \mathbb{R}^\ell_+$ and $\chi\left(p', \bar{U}\right) \in \mathbb{R}^\ell_+$ be the respective expenditure-minimizing bundles necessary to achieve a utility of at least $\bar{U}$. By definition, the expenditure-minimizing bundle at p' cannot lead to a higher expenditure than another bundle (here: the expenditure-minimizing bundle at p):

$$p' \cdot \underbrace{\chi\left(\bar{U}, p'\right)}_{\substack{\text{expenditure-}\\ \text{minimizing}\\ \text{bundle at } p'}} \leq p' \cdot \underbrace{\chi\left(\bar{U}, p\right)}_{\substack{\text{expenditure-}\\ \text{minimizing}\\ \text{bundle at } p}}.$$

Analogously, we have

$$p \cdot \chi(\bar{U}, p') \geq p \cdot \chi(\bar{U}, p)$$

which we then multiply by -1 to obtain

$$-p \cdot \chi(\bar{U}, p') \leq -p \cdot \chi(\bar{U}, p).$$

Adding the first and the third inequalities yields

$$p' \cdot \chi(\bar{U}, p') - p \cdot \chi(\bar{U}, p') \leq p' \cdot \chi(\bar{U}, p) - p \cdot \chi(\bar{U}, p)$$

and hence

$$[p' - p] \cdot \chi(\bar{U}, p') \leq [p' - p] \cdot \chi(\bar{U}, p)$$

and finally

$$[p' - p] \cdot [\chi(\bar{U}, p') - \chi(\bar{U}, p)] \leq 0.$$

This inequality can be applied to a price change of only one good $\hat{g}$. In that case, we have

$$\begin{aligned} 0 &\geq [p' - p] \cdot [\chi(\bar{U}, p') - \chi(\bar{U}, p)] \\ &= \sum_{g=1}^{\ell} [p'_g - p_g] \cdot [\chi_g(\bar{U}, p') - \chi_g(\bar{U}, p)] \\ &= [p'_{\hat{g}} - p_{\hat{g}}] \cdot [\chi_{\hat{g}}(\bar{U}, p') - \chi_{\hat{g}}(\bar{U}, p)]. \end{aligned}$$

Thus, if the price of one good increases, the demand of that good cannot increase:

$$\frac{\partial \chi_g}{\partial p_g} \leq 0. \tag{VII.7}$$

This result is a special case of a more general result that we can derive from the fact that the expenditure function is concave in the prices.

3.2. Concavity and the Hesse matrix. Generalizing definition V.5 (p. 103) to a multi-dimensional domain, we obtain:

DEFINITION VII.4 (CONCAVE FUNCTION). *Let $f : M \to \mathbb{R}$ be a function on a convex domain $M \subseteq \mathbb{R}^\ell$. f is called concave if we have*

$$f(kx + (1-k)y) \geq kf(x) + (1-k)f(y)$$

for all $x, y \in M$ and for all $k \in [0,1]$. f is called strictly concave if

$$f(kx + (1-k)y) > kf(x) + (1-k)f(y)$$

holds for all $x, y \in M$ with $x \neq y$ and for all $k \in (0,1)$. If the inequality signs are the other way around, f is convex or strictly convex, respectively.

Concavity means that the value at the convex combination $kx+(1-k)\,y$ is greater than the convex combination of the values $f(x)$ and $f(y)$. If $f: M \to \mathbb{R}$ with $M \subseteq \mathbb{R}$ is twice differentiable, we can characterize concavity by the second derivative, $f''(x) \leq 0$ for all $x \in M$. The second derivative is called the Hesse matrix. It is an $\ell \times \ell$ matrix whose entries are the second-order partial derivatives of a function.

DEFINITION VII.5 (HESSE MATRIX). *Let $f: \mathbb{R}^\ell \to \mathbb{R}$ be a function. The second-order partial derivative of f with respect to x_i and x_j (if it exists) is given by*

$$f_{ij}(x) := \frac{\partial \frac{\partial f(x)}{\partial x_i}}{\partial x_j}.$$

If all the second-order partial derivatives exist, the Hesse matrix of f is given by

$$f''(x) = \begin{pmatrix} f_{11}(x) & f_{12}(x) & & f_{1\ell}(x) \\ f_{21}(x) & f_{22}(x) & & \\ & & & \\ & & & \\ f_{\ell 1}(x) & f_{\ell 2}(x) & & f_{\ell\ell}(x) \end{pmatrix}.$$

EXERCISE VII.3. *Determine the Hesse matrix for the function f given by $f(x,y) = x^2 y + y^2$. How about the off-diagonal elements?*

The Hesse matrix is symmetric if all the second-order partial derivatives are continuous:

LEMMA VII.3. *Let $f: \mathbb{R}^\ell \to \mathbb{R}$ be twice differentiable. If f_{ij} is continuous, so is f_{ji}, and we have*

$$f_{ij}(x) = f_{ji}(x).$$

DEFINITION VII.6. *Let $T = (t_{ij})_{\substack{i=1,..,\ell \\ j=1,...,\ell}}$ be a matrix with real entries. For any column vector with ℓ entries $z = \begin{pmatrix} z_1 \\ z_2 \\ \\ \\ z_\ell \end{pmatrix}$,*

$$z^t T z$$

is a real number. T is called

- *negative-semidefinite if we obtain $z^t T z \leq 0$ for all $z \in \mathbb{R}^\ell$,*
- *negative-definite if we obtain $z^t T z < 0$ for all $z \in \mathbb{R}^\ell \setminus \{0\}$,*

- *positive-semidefinite if we obtain* $z^t T z \geq 0$ *for all* $z \in \mathbb{R}^\ell$ *and*
- *positive-definite if we obtain* $z^t T z > 0$ *for all* $z \in \mathbb{R}^\ell \backslash \{0\}$.

EXERCISE VII.4. *Show that the diagonal entries of a negative-semidefinite matrix are nonpositive. Hint: Apply the vectors* $e_i \in \mathbb{R}^\ell$ *whose ith entry is* 1 *and* 0 *otherwise,* $e_i = \left(0, ..., 0, \underbrace{1}_{ith\ entry}, 0, ..., 0\right)^t$.

Now, at long last, we come to the lemma that allows to check the concavity or convexity of a function:

LEMMA VII.4. *A function* $f : M \rightarrow \mathbb{R}$ *with convex domain* $M \subseteq \mathbb{R}^\ell$ *that is twice differentiable with continuous derivative is (strictly) concave if and only if its Hesse matrix is negative-semidefinite (negative-definite). It is (strictly) convex if and only if its Hesse matrix is positive-semidefinite (positive-definite).*

We do not provide a proof. Instead, we show that the function given by $f(x, y) = x^2 + y^2$ is convex. We first calculate its Hessian

$$f''(x, y) = \begin{pmatrix} 2 & 0 \\ 0 & 2 \end{pmatrix}.$$

We then find

$$(z_1, z_2) \begin{pmatrix} 2 & 0 \\ 0 & 2 \end{pmatrix} \begin{pmatrix} z_1 \\ z_2 \end{pmatrix} = (z_1, z_2) \begin{pmatrix} 2z_1 \\ 2z_2 \end{pmatrix} = 2z_1^2 + 2z_2^2 > 0 \text{ for } z \in \mathbb{R}^2 \backslash \{0\}.$$

Therefore, f is strictly convex by lemma VII.4.

EXERCISE VII.5. *Show that the function given by* $f(x, y) = xy$ *is neither convex or concave by identifying vectors* $z, \hat{z} \in \mathbb{R}^\ell$ *such that* $z^t T z < 0$ *and* $\hat{z}^t T \hat{z} > 0$.

The reader might be curious to learn the relationship between quasi-concavity and concavity. However, since we have no special interest in quasi-concave expenditure functions, we postpone this discussion until chapter VIII where the concavity of a production function carries a certain meaning absent in the concavity of a utility function.

3.3. The expenditure function is concave. In order to show the concavity of the expenditure function, assume price vectors p, p' and their convex combination

$$\bar{p} = kp + (1 - k)\, p', k \in [0, 1].$$

Using a similar argument as before, we have

$$\begin{aligned} e\left(p,\bar{U}\right) &= p\cdot\chi\left(p,\bar{U}\right)\leq p\cdot\chi\left(\bar{p},\bar{U}\right) \text{ and} \\ e\left(p',\bar{U}\right) &= p'\cdot\chi\left(p',\bar{U}\right)\leq p'\cdot\chi\left(\bar{p},\bar{U}\right) \end{aligned}$$

for any given utility level $\bar{U}$. Therefore, we obtain

$$\begin{aligned} & e\left(kp+(1-k)p',\bar{U}\right) \\ = & \left[kp+(1-k)p'\right]\cdot\chi\left(\bar{p},\bar{U}\right) \text{ (definition of } \chi) \\ = & kp\cdot\chi\left(\bar{p},\bar{U}\right)+(1-k)p'\cdot\chi\left(\bar{p},\bar{U}\right) \text{ (distributivity)} \\ \geq & ke\left(p,\bar{U}\right)+(1-k)e\left(p',\bar{U}\right) \text{ (above inequalities)} \end{aligned}$$

which shows the concavity of e with respect to prices (see definition VII.4 on p. 185).

3.4. Application: the diagonal entries of the expenditure function's Hesse matrix. The concavity of the expenditure function has two useful implications. First, we can analyze the effect of a price change on Hicksian demand,

$$\frac{\partial\chi_g\left(p,\bar{U}\right)}{\partial p_k}.$$

Second, we can define substitutes and complements.

By Shephard's lemma, we have

$$\frac{\partial e\left(p,\bar{U}\right)}{\partial p_g}=\chi_g\left(p,\bar{U}\right)$$

and thus

$$\frac{\partial\chi_g\left(p,\bar{U}\right)}{\partial p_k}=\frac{\partial\frac{\partial e\left(p,\bar{U}\right)}{\partial p_g}}{\partial p_k}=\frac{\partial^2 e\left(p,\bar{U}\right)}{\partial p_g\partial p_k}.$$

Now, since the expenditure function is concave in its prices, lemma VII.4 (p. 187) implies that the Hesse matrix of a (twice differentiable) expenditure function e

$$e''\left(p,\bar{U}\right)=\begin{pmatrix} \frac{\partial^2 e(p,\bar{U})}{(\partial p_1)^2} & \frac{\partial^2 e(p,\bar{U})}{\partial p_1\partial p_2} & & \frac{\partial^2 e(p,\bar{U})}{\partial p_1\partial p_\ell} \\ \frac{\partial^2 e(p,\bar{U})}{\partial p_2\partial p_1} & \frac{\partial^2 e(p,\bar{U})}{(\partial p_2)^2} & & \\ & & & \\ & & & \\ \frac{\partial^2 e(p,\bar{U})}{\partial p_\ell\partial p_1} & \frac{\partial^2 e(p,\bar{U})}{\partial p_\ell\partial p_2} & & \frac{\partial^2 e(p,\bar{U})}{(\partial p_\ell)^2} \end{pmatrix}.$$

is negative-semidefinite so that its diagonal elements obey

$$\frac{\partial \chi_g (p, \overline{U})}{\partial p_g} = \frac{\partial \frac{\partial e(p,\bar{U})}{\partial p_g}}{\partial p_g} = \frac{\partial^2 e (p, \bar{U})}{(\partial p_g)^2} \leq 0 \qquad \text{(VII.8)}$$

by exercise VII.4. This result has already been obtained (by different means) in subsection 3.1.

3.5. Application: the off-diagonal entries of the expenditure function's Hesse matrix. We now turn to the off-diagonal elements of the expenditure function's Hesse Matrix. If all the entries of this matrix are continuous, lemma VII.3 (p. 186) claims the equality

$$\frac{\partial \chi_g (p, \bar{U})}{\partial p_k} = \frac{\partial \chi_k (p, \bar{U})}{\partial p_g}.$$

This important fact means that the following definition is okay:

Definition VII.7 (substitutes, complements). *Let e be a twice continuously differentiable function in its prices. Goods g and k are called substitutes if*

$$\frac{\partial \chi_g (p, \bar{U})}{\partial p_k} = \frac{\partial \chi_k (p, \bar{U})}{\partial p_g} \geq 0$$

holds and complements in case of

$$\frac{\partial \chi_g (p, \bar{U})}{\partial p_k} = \frac{\partial \chi_k (p, \bar{U})}{\partial p_g} \leq 0.$$

Compare the definition of substitutes and complements via Marshallian demand and exercise VI.12 on p. 147.

One final remark: We know from lemma VII.1 (p. 175) that

$$\chi_g (\alpha p, \bar{U}) = \chi_g (p, \bar{U})$$

holds for all $\alpha > 0$. Doubling all prices does not change the bundle that needs to be consumed for utility level $\bar{U}$. Differentiating this equality with respect to α yields

$$\sum_{k=1}^{\ell} \frac{\partial \chi_g (\alpha p, \bar{U})}{\partial (\alpha p_k)} \cdot p_k = \frac{\partial \chi_g (\alpha p, \bar{U})}{\partial \alpha} = \frac{\partial \chi_g (p, \bar{U})}{\partial \alpha} = 0$$

and hence

$$\sum_{\substack{k=1,\\ k \neq g}}^{\ell} \frac{\partial \chi_g (\alpha p, \bar{U})}{\partial (\alpha p_k)} \cdot p_k = -\underbrace{\frac{\partial \chi_g (\alpha p, \bar{U})}{\partial (\alpha p_k)}}_{\leq 0} \cdot p_g \geq 0.$$

This inequality implies

LEMMA VII.5. *Assume* $\ell \geq 2$ *and* $p >> 0$. *Every good has at least one substitute.*

4. Slutsky equations

4.1. Three effects of a price increase. The Hicksian demand curve is downward sloping. How about Marshallian demand, i.e., can we expect

$$\frac{\partial x_g(p,m)}{\partial p_g} \leq 0?$$

No.

We can dissect the overall effect of a price increase on Marshallian demand into three components:

(1) Substitution effect or opportunity-cost effect:
A price increase of good 1 increases the marginal opportunity cost p_1/p_2 of good 1 in terms of good 2. For this reason alone, the household is inclined to consume less of good 1, i.e., to substitute x_2 for x_1. However, in order to separate this opportunity-cost effect from the income effect (see below), we adjust income to hold purchasing power constant (see next subsection).

(2) Consumption-income effect:
A price increase reduces the consumption possibilities of a household. The household's budget buys less and the household will therefore consume less of all normal goods. Of course, this effect is the stronger, the greater the household's consumption of the good in question. If the household's budget is a money budget, we are done. However, in case of an endowment budget, we have to consider a third component.

(3) Endowment-income effect:
A price increase enhances the value of the household's endowment in line with the amount of the specific good. The household will consume more, in general and in particular more of the more expensive good (if it is normal). Expenditure effect and endowment effect are opposing effects. Indeed, the difference between the household's endowment of the more expensive good and his consumption of this good will turn out to be crucial.

4.2. Two different substitution effects. In response to a price change, there are two different ways to keep real income constant:

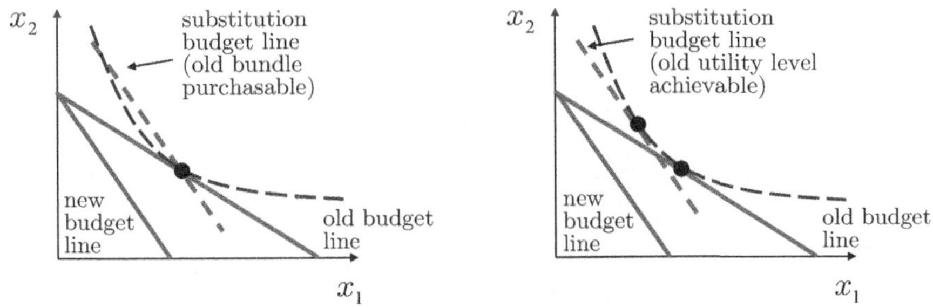

FIGURE VII.4. Two substitution effects

- Old-household-optimum substitution effect: The household's income is increased so as to enable the household to buy the old bundle (chosen before the price increase).
- Old-utility-level substitution effect: The household's income is increased so as to keep the household on his old indifference curve.

Figure VII.4 illustrates these two methods. Imagine a price increase of good 1 so that the bundles of affordable goods diminish. In the left-hand graph you see the "the old bundle can still be purchased" method, while the right-hand graph depicts the "the old utility level can still be achieved" method. Either you pivot the old budget line around the old household optimum or you slide along the indifference curve. In any case, one arrives at a new, at the substitution budget line.

Depending on the application, either of these two substitution effects can be relevant. Imagine spending your money on train rides T (abscissa) and other goods G (ordinate). A train ride costs 20 cents per kilometer, other goods 1 Euro. The "Bahncard 50" allows to buy one train kilometer with 10 cents rather than 20 cents. Find your willingness to pay for the "Bahncard 50" in terms of the other goods! Which substitution effect do you need?

We look for the willingness to pay for the improvement "lower price for good T". Therefore, we use the old-utility-level substitution effect. Figure VII.5 shows how to determine this willingness to pay in terms of good G.

Sometimes, the old-household-optimum substitution effect may be relevant. Assume that the government wants to push back the consumption of a good (energy, cigarettes etc.) without, however, eating into the purchasing power of the consumers. A tax-and-rebate scheme could be employed for that purpose. Turning to a concrete example, assume you are a smoker –

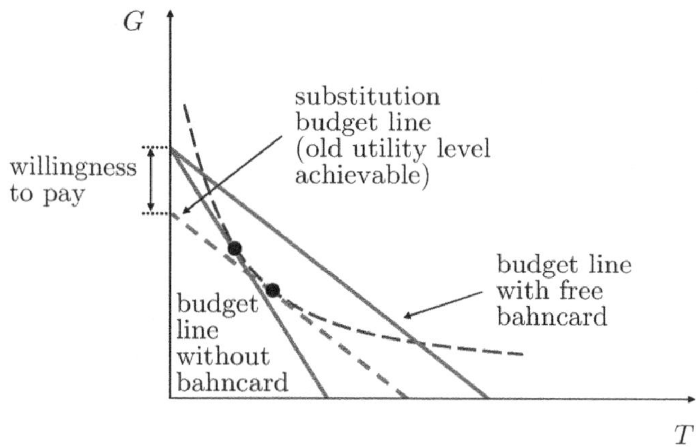

FIGURE VII.5. The willingness to pay for a Bahncard 50

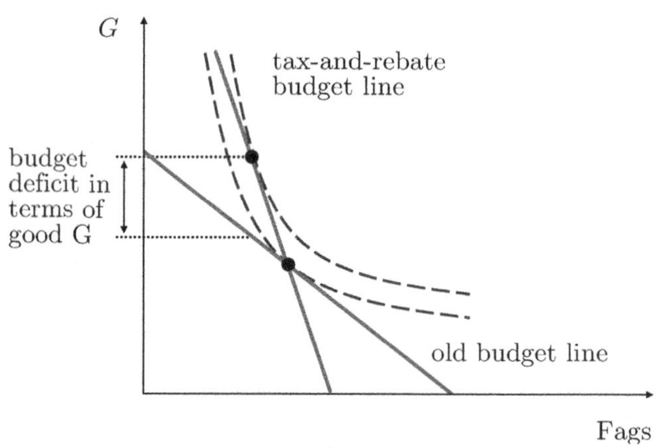

FIGURE VII.6. Tax and rebate

10 cigarettes per day. The government imposes a quantity tax of 10 cents on cigarettes but pays you an additional income (rebate) of 1 Euro per day.

Consult figure VII.6. You could afford to continue smoking as before. However, smoking has become more expensive and you will (in most cases) choose another bundle with a smaller amount of cigarettes. (A higher amount of cigarettes had been available before and you did not choose it!) Then, the government suffers a budget deficit, while you are better off.

4.3. The Slutsky equation for the money budget. From p. 177, we know the duality equation

$$\chi_g\left(p,\bar{U}\right) = x_g\left(p, e\left(p,\bar{U}\right)\right).$$

It is the starting point for the derivation of the so-called Slutsky equations. In case of a money budget, we differentiate with respect to p_k to obtain

$$\begin{aligned}\frac{\partial \chi_g}{\partial p_k} &= \frac{\partial x_g}{\partial p_k} + \frac{\partial x_g}{\partial m}\frac{\partial e}{\partial p_k} \\ &= \frac{\partial x_g}{\partial p_k} + \frac{\partial x_g}{\partial m}\chi_k \text{ (Shephard's Lemma)} \end{aligned} \tag{VII.9}$$

If we focus on the own-price effect ($g = k$) and on normal goods, we find the money-budget Slutsky equation

$$\frac{\partial x_g}{\partial p_g} = \underbrace{\frac{\partial \chi_g}{\partial p_g}}_{\leq 0} - \underbrace{\frac{\partial x_g}{\partial m}}_{\substack{>0 \\ \text{for normal goods}}} \chi_g \tag{VII.10}$$

and the following lemma.

Lemma VII.6. *If good g is normal, it is also ordinary. In that case, the effect of a price increase is stronger on Marshallian demand than on Hicksian demand. If the income effect is zero ($\frac{\partial x_g}{\partial m} = 0$), Hicksian demand does not depend on the utility level to be attained.*

The intuitive reason for this lemma lies in the consumption-income effect. A price increase reduces the consumption possibilities so that the opportunity-cost effect and the consumption-income effect work together achiving a strong effect on Marshallian demand. In case of Hicksian demand, we compensate for the price increase so as to hold the utility constant. Figure VII.7 serves as an illustration. Consider

- a utility level $\bar{U}$,
- a price $\hat{p}_g$ (the other prices are constant) and
- the corresponding Hicksian demand $\chi_g\left(\hat{p}_g,\bar{U}\right)$ where the Hicksian demand function is given by the right-hand dotted line in the figure.

By the duality equation

$$\chi_g\left(\hat{p}_g,\bar{U}\right) = x_g\left(\hat{p}_g, e\left(\hat{p}_g,\bar{U}\right)\right),$$

the Hicksian demand equals the Marshallian demand for income $e\left(\hat{p}_g,\bar{U}\right)$. If we increase (or decrease) the price p_g, the Marshallian reaction is stronger than the Hicksian reaction because of the income effect.

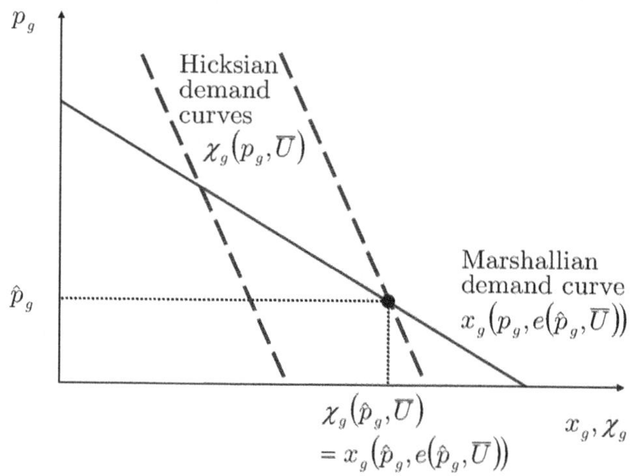

FIGURE VII.7. Marshallian versus Hicksian demand

Of course, in case of inferior goods, the Marshallian demand curve is steeper than the Hicksian demand curves. In fact, the Marshallian demand may be so steep that the good in question becomes non-ordinary. In that case, we have so-called Giffen goods.

4.4. The Slutsky equation for the endowment budget.

4.4.1. *Derivation.* The derivation of the Slutsky equation for the endowment budget is also not difficult. Note

$$x_g^{\text{endowment}}(p,\omega) = x_g^{\text{money}}(p, p\cdot\omega)$$

where the budget type (money or endowment) is explicitly noted. Differentiation with respect to p_k yields

$$\begin{aligned}
\frac{\partial x_g^{\text{endowment}}}{\partial p_k} &= \frac{\partial x_g^{\text{money}}}{\partial p_k} + \frac{\partial x_g^{\text{money}}}{\partial m}\frac{\partial (p\cdot\omega)}{\partial p_k} \\
&= \frac{\partial x_g^{\text{money}}}{\partial p_k} + \frac{\partial x_g^{\text{money}}}{\partial m}\omega_k \text{ (definition of dot product)} \\
&= \frac{\partial \chi_g}{\partial p_k} - \frac{\partial x_g^{\text{money}}}{\partial m}\chi_k + \frac{\partial x_g^{\text{money}}}{\partial m}\omega_k \text{ (eq. VII.10)} \\
&= \frac{\partial \chi_g}{\partial p_k} + \frac{\partial x_g^{\text{money}}}{\partial m}(\omega_k - \chi_k) \qquad \text{(VII.11)}
\end{aligned}$$

and hence the endowment Slutsky equation for the own-price effect ($g = k$)

$$\frac{\partial x_g^{\text{endowment}}}{\partial p_g} = \underbrace{\frac{\partial \chi_g}{\partial p_g}}_{\leq 0} + \underbrace{\frac{\partial x_g^{\text{money}}}{\partial m}}_{\substack{>0 \\ \text{for a normal} \\ \text{good } g}} \underbrace{(\omega_g - \chi_g)}_{\substack{<0 \\ \text{for net demander}}} . \qquad \text{(VII.12)}$$

Lemma VII.7. *If good g is normal and the household consumes more than his endowment, it is also ordinary. A normal good g may be non-ordinary if the household is a net supplier.*

4.4.2. *Application: consumption today versus consumption tomorrow* . The intertemporal budget equation in terms of the future value (if necessary, consult p. 132) is

$$(1 + r)\, x_1 + x_2 = (1 + r)\, \omega_1 + \omega_2.$$

It leads to the Slutsky equation

$$\frac{\partial x_1^{\text{endowment}}}{\partial (1+r)} = \underbrace{\frac{\partial \chi_1}{\partial (1+r)}}_{\leq 0} + \underbrace{\frac{\partial x_1^{\text{money}}}{\partial m}}_{\substack{>0 \\ \text{for normal good} \\ \text{first-period consumption}}} \underbrace{(\omega_1 - \chi_1)}_{\substack{>0 \\ \text{for lender}}}$$

where $\frac{\partial x_1^{\text{money}}}{\partial m}$ stands for the effect of an outward shift of the budget line on consumption in the first period.

We can have a non-ordinary price reaction if the household is a lender ($\omega_1 > \chi_1$) and if first-period consumption is a normal good. In that case, the income effects are positive and work against the negative substitution effect.

4.4.3. *Application: leisure versus consumption.* On p. 133, we derive the budget equation for the leisure-consumption model

$$wx_R + px_C = w24 + p\omega_C.$$

The associated Slutsky equation is

$$\frac{\partial x_R^{\text{endowment}}}{\partial w} = \underbrace{\frac{\partial \chi_R}{\partial w}}_{\leq 0} + \underbrace{\frac{\partial x_R^{\text{money}}}{\partial m}}_{\substack{>0 \\ \text{for normal} \\ \text{good recreation}}} \underbrace{(24 - \chi_R)}_{\substack{\geq 0 \\ \text{by definition}}} .$$

According to the substitution effect, a wage increase leads to a decrease of demand for leisure and hence to an increase of labor supply. The income effects might outweigh the substitution effect. Then a wage increase might lead to a decrease in labor supply. A higher wage allows the household to consume more of the consumption good and also more of leisure.

4.4.4. *Application: contingent consumption.* Contingent consumption is introduced on pp. 134. We obtain the budget equation

$$x_1 + \frac{1-\gamma}{\gamma} x_2 = (A - D) + \frac{1-\gamma}{\gamma} A$$

where γK is the payment to the insurance if K is to be paid to the insuree in case of damage D. After multiplying this equation by $\frac{\gamma}{1-\gamma}$, we can write down the Slutsky equation for the first good, i.e., for consumption in case of damage:

$$\frac{\partial x_1^{\text{endowment}}}{\partial \frac{\gamma}{1-\gamma}} = \underbrace{\frac{\partial \chi_1}{\partial \frac{\gamma}{1-\gamma}}}_{\leq 0} + \underbrace{\frac{\partial x_1^{\text{money}}}{\partial m}}_{\substack{>0 \\ \text{for normal} \\ \text{good consumption} \\ \text{in case of damage}}} \underbrace{(A - D - \chi_1)}_{\substack{\leq 0 \\ \text{in case of} \\ \text{a nonnegative insurance}}} .$$

Usually, the insuree aspires to a higher consumption level in case of damage than the endowment level $A - D$, i.e., the insurance sum K obeys $K \geq 0$. Then, the income effects are negative and an increase of $\frac{\gamma}{1-\gamma}$ (which is equivalent to an increase in γ) leads to a reduction in consumption in case of damage $A-D+(1-\gamma)K$. Note that this does not help in telling whether K decreases or increases.

5. Compensating and equivalent variations

5.1. The case of good air quality. We link the theory developed so far to the problem of putting a monetary value on changes of all different kinds. For example, the air may have become better or worse, a price has increased or decreased. These changes are also called events. We distinguish equivalent variations from a compensating ones (see also figure VII.8).

DEFINITION VII.8 (VARIATIONS). *A variation (a sum of money) is called equivalent to an event, if both (the event or the variation) lead to the same indifference curve. A variation is compensating if it restores the individual to its old indifference curve (prior to the event). Variations are denoted by EV (event) or CV (event), respectively. Both variations are defined as non-negative amounts of money.*

Equivalent variation	**Compensating variation**
in lieu of an event monetary variation is equivalent (i.e., achieving the same utility)	**because of an event** monetary variation compensates for event (i.e., holding utility constant)

FIGURE VII.8. Variations are monetary values of events

Consider figure VII.9. It depicts the variations for an improvement (from low air quality q_1 to high air quality q_2) or a degradation (from high to low quality) of air. Consider, for example, point b. A degradation of air quality (the movement from b to a) leads to the lower indifference curve. The equivalent variation of this event is the money $EV(\text{b} \to \text{a}) \geq 0$ taken from the individual that also leads to the lower indifference curve, i.e., $EV(\text{b} \to \text{a}) = m_2 - m_1$. The compensating variation of this event is the money $CV(\text{b} \to \text{a}) = m_3 - m_2$ given to the individual that keeps him on the original high indifference curve, i.e., that compensates for the worse air quality.

If a utility function $U(m, q)$ is given (with income m and air quality q), the other two variations are implicitly defined by

$$\begin{aligned} U(m_2, q_1) &= U(m_2 - CV(\text{a} \to \text{b}), q_2) \text{ and} \\ U(m_2 + EV(\text{a} \to \text{b}), q_1) &= U(m_2, q_2) \end{aligned}$$

Do you see why we have $EV(\text{a} \to \text{b}) = CV(\text{b} \to \text{a})$?

We can link up the variations with the terms "willingness to pay" and "compensation money" (or "loss compensation"):

- If some amount of money is given to the individual, it can also be addressed as compensation money. For example, CV (degradation) is the compensation money for the degradation of the air quality.
- If money is taken from the individual, we talk about the willingness to pay. EV (degradation) is the willingness to pay for the prevention of the degradation.

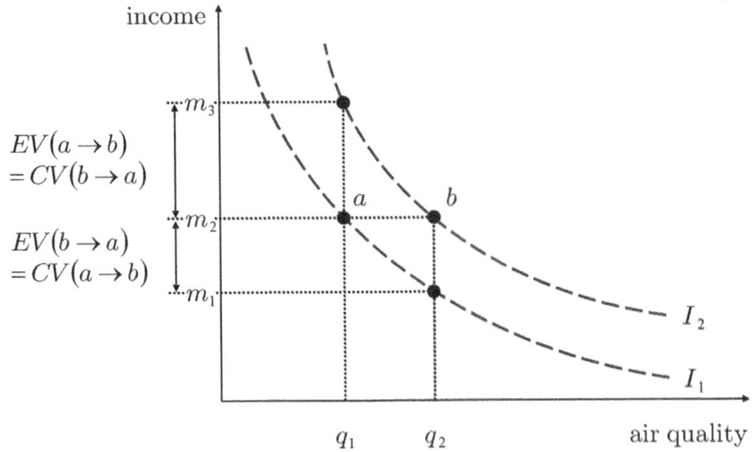

FIGURE VII.9. Variations for better (or worse) air

If your variation turns out to be negative, you should exchange $-EV$ for EV or EV for $-EV$ (similarly for CV).

5.2. Compensating or equivalent variation? Is there a way to decide which variation (compensating or equivalent) is the "correct" one? Not in general. However, in the framework of markets and usual property rights, we will certainly focus on compensating variations:

- A consumer asks himself how much he is prepared to pay for a good. That is the compensating variation for obtaining the good (the event). The corresponding equivalent variation is the compensation payment for not getting the good. You go into a shop and ask for compensation for not taking (stealing?) the good.
- A producer's compensating variation is the compensation money he gets for selling a good. Framed in terms of an equivalent variation, the producer asks himself how much he would be willing to pay if the good were not taken away from him.

Summarizing, exchange is governed by "quid pro quo" so that we will usually deal with compensating variations. This does not say that equivalent variations are useless. For example, the inhabitants of a region may have the impression that they have a right to good air quality. They may then ask themselves how much they should get as a compensation for forgoing the better quality which is nothing but EV (improvement). Also, every compensating variation for an event can be expressed by way of an equivalent variation of the opposite event.

5.3. Price changes. The air-quality example is easily transferred to price changes. Let p_g^h be a higher price than $p_g^l < p_g^h$. Using the intuitive notation "$p_g^l \to p_g^h$" and "$p_g^h \to p_g^l$", we have the willingness to pay for the price decrease of good g

$$CV\left(p_g^h \to p_g^l\right) = EV\left(p_g^l \to p_g^h\right)$$

and the compensation money for the price increase of good g

$$EV\left(p_g^h \to p_g^l\right) = CV\left(p_g^l \to p_g^h\right).$$

We now link these terms to the theory learned in this chapter. Sometimes, one may not be sure whether a change is a good or a bad thing. Then, we use small-letter symbols for the variations, cv and ev.

LEMMA VII.8. *Consider the event of a price change from p^{old} to p^{new} (in general, these are price vectors). Iff CV and EV are the compensating and equivalent variations of this event, respectively, we find*

$$U^{old} \quad : \quad = V\left(p^{old}, m\right) = V\left(p^{new}, m + cv\right), \ CV = |cv|, \quad \text{(VII.13)}$$

$$U^{new} \quad : \quad = V\left(p^{new}, m\right) = V\left(p^{old}, m + ev\right), \ EV = |ev|, \quad \text{(VII.14)}$$

respectively.

The above lemma provides an implicit definition of variations for price changes.

EXERCISE VII.6. *Tell the sign of cv and ev for a price increase of all goods.*

We consider the situation of a household who is confronted with a price increase of good 1 (see figure VII.10). The old optimum is at point O, the new one at point A. How much is the household affected by the price increase? The compensating variation is the amount of money necessary to keep the household on the old (high) indifference curve. By a parallel upward movement of the new (steeper) budget line, we finally reach point C and find CV_2, the compensating variation in terms of good-2 units (or CV_1, the compensating variation in terms of good-1 units). Multiplying by p_2 (or p_1^{new}) yields the compensating variation

$$CV = p_1^{new} \cdot CV_1 = p_2 \cdot CV_2.$$

Consider, for example, Cobb-Douglas preferences given by

$$U(x_1, x_2) = x_1^a x_2^{1-a} \qquad (0 < a < 1)$$

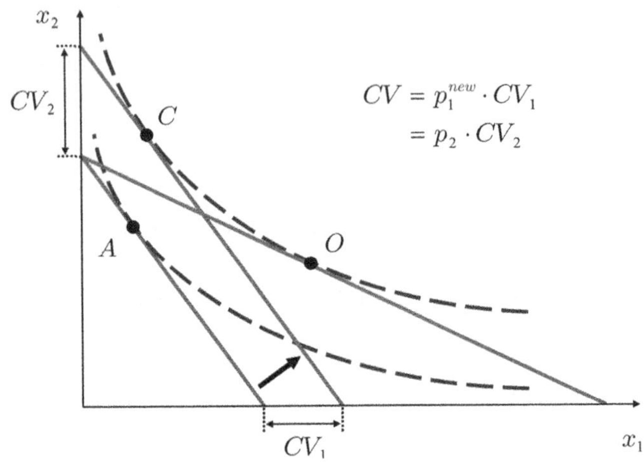

FIGURE VII.10. Compensating variation for price increase

and a price decrease from p_1^h to $p_1^l < p_1^h$. Equation VII.13 translates into

$$\underbrace{\left(a\frac{m}{p_1^h}\right)^a \left((1-a)\frac{m}{p_2}\right)^{1-a}}_{\text{utility at the old, high price}}$$

$$= \underbrace{\left(a\frac{m + cv\left(p_1^h \to p_1^l\right)}{p_1^l}\right)^a \left((1-a)\frac{m + cv\left(p_1^h \to p_1^l\right)}{p_2}\right)^{1-a}}_{\substack{\text{utility at the new, lower price} \\ \text{and compensating variation}}}.$$

We find

$$cv\left(p_1^h \to p_1^l\right) = -\left(1 - \left(\frac{p_1^l}{p_1^h}\right)^a\right) m < 0$$

which means that the household has to pay for the price decrease.

EXERCISE VII.7. *Determine the equivalent variation for a price decrease in case of Cobb-Douglas utility preferences.*

It can be shown (by some algebraic manipulations) that, in absolute terms, the willingness to pay for the price decrease is smaller than the compensation money: $CV\left(p_1^h \to p_1^l\right) < EV\left(p_1^h \to p_1^l\right)$. We will soon learn the reason for this discrepancy (see p. 202).

EXERCISE VII.8. *Determine the compensating variation and the equivalent variation for the price decrease from* p_1^h *to* $p_1^l < p_1^h$ *and the quasi-linear utility function given by*

$$U(x_1, x_2) = \ln x_1 + x_2 \qquad (x_1 > 0)!$$

Assume $\frac{m}{p_2} > 1$*! Hint: the household optimum is* $x(m,p) = \left(\frac{p_2}{p_1}, \frac{m}{p_2} - 1\right)$.

The exercise shows that the willingness to pay and the compensation money can have the same absolute magnitude in case of quasi-linear utility functions.

5.4. Applying duality. Duality theory now helps to find the compensating and the equivalent variations. We begin with the duality equation $e(p, V(p,m)) = m$ which can be written as

$$\begin{aligned} e\left(p^{old}, V\left(p^{old}, m\right)\right) &= m \text{ and} && \text{(VII.15)} \\ e\left(p^{new}, V\left(p^{new}, m + cv\right)\right) &= m + cv && \text{(VII.16)} \end{aligned}$$

where the price-income vector is $\left(p^{old}, m\right)$ in the first and $\left(p^{new}, m + cv\right)$ in the second equation. The explicit definition of the compensating variation is given by

$$\begin{aligned} cv &= e\left(p^{new}, V\left(p^{new}, m + cv\right)\right) - m \text{ (eq. VII.16)} \\ &= e\left(p^{new}, U^{old}\right) - e\left(p^{old}, U^{old}\right) \text{ (eq. VII.13, VII.15)} && \text{(VII.17)} \end{aligned}$$

Thus, the compensating variation cv is the expenditure at the new prices minus the expenditure at the old prices where the utility level stays at the original level U^{old}. The household is given, or is relieved of, the money necessary to uphold the old utility level.

Similarly, the duality equation $e(p, V(p,m)) = m$ leads to

$$\begin{aligned} e\left(p^{new}, V\left(p^{new}, m\right)\right) &= m \text{ and} && \text{(VII.18)} \\ e\left(p^{old}, V\left(p^{old}, m + ev\right)\right) &= m + ev. && \text{(VII.19)} \end{aligned}$$

so that we find

$$\begin{aligned} ev &= e\left(p^{old}, V\left(p^{old}, m + ev\right)\right) - m \text{ (eq. VII.19)} \\ &= e\left(p^{old}, U^{new}\right) - e\left(p^{new}, U^{new}\right) \text{ (eq. VII.14, VII.18)} \end{aligned}$$

The equivalent variation is the expenditure at the old prices minus the expenditure at the new prices where the utility is $U^{new} = V\left(p^{new}, m\right)$. Assume that the new prices would make the household better off, $V\left(p^{new}, m\right) > V\left(p^{old}, m\right)$. The equivalent variation is the amount of money necessary to increase the household's income from $m = e\left(p^{new}, U^{new}\right)$ to $e\left(p^{old}, U^{new}\right)$

so that the household can achieve the new utility level U^{new} because of the income change rather than the price change.

5.5. Variations for a price change and Hicksian demand.

5.5.1. *Applying the fundamental theorem of calculus.* The definitions for both the compensating and the equivalent variations build on a given utility level. Since the Hicksian demand holds utility constant, we can try to use Hicksian demand for an alternative characterization. An important ingredient is the "Fundamental Theorem of Calculus" that we know from p. 112.

We obtain

$$\begin{aligned}
& cv\left(p_g^h \to p_g^l\right) \\
= \; & e\left(p_g^l, V\left(p_g^h, m\right)\right) - e\left(p_g^h, V\left(p_g^h, m\right)\right) \text{ (eq. VII.17)} \\
= \; & -\left[e\left(p_g^h, V\left(p_g^h, m\right)\right) - e\left(p_g^l, V\left(p_g^h, m\right)\right)\right] \\
= \; & -e\left(p_g, V\left(p_g^h, m\right)\right)\Big|_{p_g^l}^{p_g^h} \text{ (Fundamental Theorem)} \\
= \; & -\int_{p_g^l}^{p_g^h} \frac{\partial e\left(p, V\left(p_g^h, m\right)\right)}{\partial p_g} dp_g \; \left(e \text{ is } \frac{\partial e}{\partial p_g}\text{'s antiderivative}\right) \\
= \; & -\int_{p_g^l}^{p_g^h} \chi_g\left(p_g, V\left(p_g^h, m\right)\right) dp_g \text{ (Shephard's lemma)}
\end{aligned}$$

Thus, we have a graphical expression for the compensating variation for a price decrease: the area to the left of the Hicksian demand curve associated with the utility level $V\left(p_g^h, m\right)$ (see figure VII.11). The minus sign reminds us of the fact that we are dealing with willingness to pay, not with loss compensation.

5.5.2. *Comparisons.* The willingness to pay for a price decrease from p_g^h to p_g^l is $CV\left(p_g^h \to p_g^l\right)$, which is the small area in figure VII.12. Following a price decrease, this payment by the consumer keeps him on the low utility level $V\left(p_g^h, m\right)$. The compensation money for the price increase $CV\left(p_g^l \to p_g^h\right)$ is calculated so as to hold the higher utility level $V\left(p_g^l, m\right)$ constant. The higher utility level is associated with higher expenditure (see lemma VII.2, p. 178) and, for a normal good g, with a higher demand of good g. Thus, in case of a normal good g, the willingness to pay is lower than the compensation money.

Figure VII.12 also shows that (again for normal goods) the willingness to pay for a price decrease (which we can measure with the Hicksian demand curve) is smaller, in absolute terms, than the area to the left of the

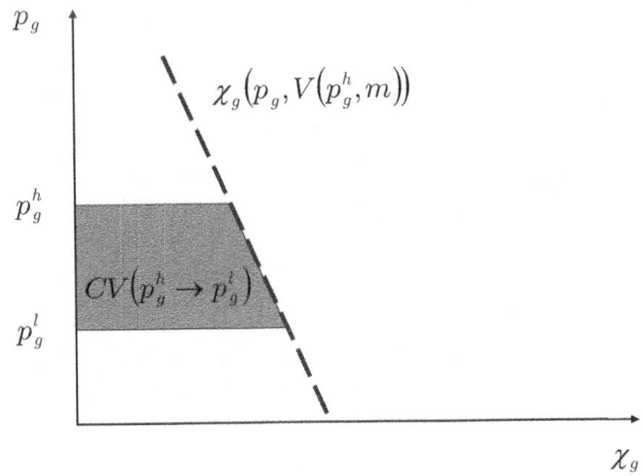

FIGURE VII.11. Compensating variation for price decrease

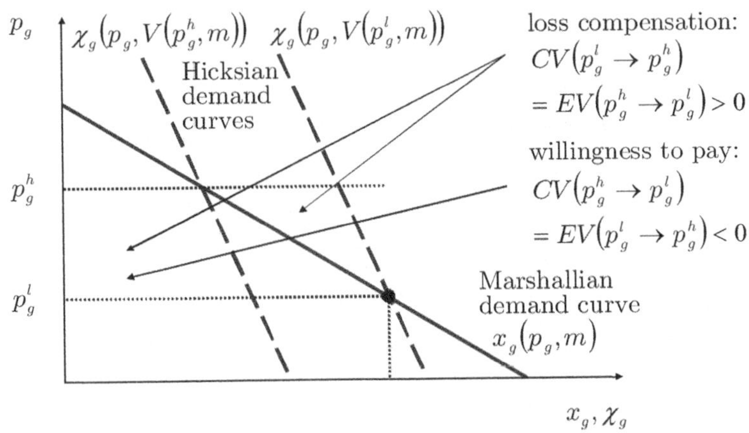

FIGURE VII.12. Variations for a price change

Marshallian demand curve (compare p. 163 in the previous chapter). The reason is explained above, with the help of figure VII.7, p. 194.

We summarize these results:

THEOREM VII.5. *Consider any good g and any price decrease from p_g^h to $p_g^l < p_g^h$. We find*

$$cv\left(p_g^h \to p_g^l\right) = -\int_{p_g^l}^{p_g^h} \chi_g\left(p_g, V\left(p_g^h, m\right)\right) dp_g.$$

If g is a normal good, we obtain

$$\underbrace{CV\left(p_g^h \to p_g^l\right)}_{\substack{\text{(Hicksian)} \\ \text{willingness to pay}}} \leq \underbrace{\int_{p_g^l}^{p_g^h} x_g\left(p_g\right) dp_g}_{\substack{\text{Marshallian} \\ \text{willingness to pay}}} \leq \underbrace{CV\left(p_g^l \to p_g^h\right)}_{\substack{\text{(Hicksian)} \\ \text{loss compensation}}} .$$

5.5.3. *Consumers' rent.* We remind the reader of the Marshallian consumers' rent (pp. 162). We can now define the Hicksian concept of a consumer's rent. Remember that the prohibitive price p^{proh} is the smallest price for which demand is choked off.

DEFINITION VII.9 (CONSUMER'S RENT). *The Hicksian consumer's rent at price $\hat{p}_g < p_g^{proh}$ is given by*

$$\begin{aligned} CR^{Hicks}\left(\hat{p}_g\right) &:= CV\left(p_g^{proh} \to \hat{p}_g\right) \\ &= \int_{\hat{p}_g}^{p_g^{proh}} \chi_g\left(p_g, V\left(p_g^{proh}, m\right)\right) dp_g. \end{aligned}$$

6. Topics

The main topics in this chapter are

- duality theory
- expenditure function
- Hicksian demand
- Marshallian demand
- Shephard's lemma
- Roy's identity
- Slutsky equations
- substitution effect
- income effect
- compensating variation
- equivalent variation
- willingness to pay
- compensation money (loss compensation)
- Hicksian consumer's rent
- the vector space $\mathbb{R}^\ell$
- the first quadrant of $\mathbb{R}^\ell$, $\mathbb{R}^\ell_+$
- envelope theorem
- Hesse matrix
- concavity

7. Solutions

Exercise VII.1

We have

$$e\left(p, \bar{U}\right) = p\chi\left(p, \bar{U}\right)$$

as well as

$$V\left(p, m\right) = U\left(x\left(p, m\right)\right).$$

Exercise VII.2

In order to achieve utility level $\bar{U}$, the household needs to spend

$$e\left(p, \bar{U}\right) := \frac{\bar{U}}{\left(\frac{a}{p_1}\right)^a \left(\frac{1-a}{p_2}\right)^{1-a}}.$$

This defines the expenditure function $e : \mathbb{R}^2 \times \mathbb{R} \to \mathbb{R}$. The solution to the minimization problem is

$$\begin{aligned}\chi_1\left(p,\bar{U}\right) &= x_1\left(p,e\left(p,\bar{U}\right)\right)\\ &= a\frac{e\left(p,\bar{U}\right)}{p_1}\\ &= a\frac{\frac{\bar{U}}{\left(\frac{a}{p_1}\right)^a\left(\frac{1-a}{p_2}\right)^{1-a}}}{p_1}\end{aligned}$$

and

$$\chi_2\left(p,\bar{U}\right) = (1-a)\,\frac{e\left(p,\bar{U}\right)}{p_2}.$$

Exercise VII.3

The Hesse matrix is given by

$$f''(x,y) = \begin{pmatrix} 2y & 2x \\ 2x & 2 \end{pmatrix}.$$

Exercise VII.4

The claim follows from $e_i^t T e_i = t_{ii}$.

Exercise VII.5

We obtain $f''(x,y) = \begin{pmatrix} 0 & 1 \\ 1 & 0 \end{pmatrix}$ and

$$(z_1,z_2)\begin{pmatrix} 0 & 1 \\ 1 & 0 \end{pmatrix}\begin{pmatrix} z_1 \\ z_2 \end{pmatrix} = (z_1,z_2)\begin{pmatrix} z_2 \\ z_1 \end{pmatrix} = 2z_1z_2.$$

Therefore, we have $z^tTz < 0$ for $z_1 = 1$, $z_2 = -1$, but $\hat{z}^tT\hat{z} > 0$ for $\hat{z}_1 = \hat{z}_2 = 1$.

Exercise VII.6

A price increase leaves the household worse off. In order to compensate for this negative event, the household's income has to change by $cv > 0$. Instead of higher prices (which would lead to a lower indifference curve), we can reduce the household's income ($ev < 0$) so that he is as worse off as he would be under a price increase.

Exercise VII.7

Which sum (paid to the consumer) makes him as well off as the price decrease would? We use equation VII.14 to obtain

$$\underbrace{\left(a\frac{m}{p_1^l}\right)^a\left((1-a)\frac{m}{p_2}\right)^{1-a}}_{\text{utility at the new, low price}}$$

$$= \underbrace{\left(a\frac{m+ev\left(p_1^h\to p_1^l\right)}{p_1^h}\right)^a\left((1-a)\frac{m+ev\left(p_1^h\to p_1^l\right)}{p_2}\right)^{1-a}}_{\substack{\text{utility at the old, higher price}\\ \text{and equivalent variation}}}$$

and therefore

$$ev\left(p_1^h\to p_1^l\right)=\left(\left(\frac{p_1^h}{p_1^l}\right)^a-1\right)m>0$$

and finally

$$EV\left(p_1^h\to p_1^l\right)=\left|ev\left(p_1^h\to p_1^l\right)\right|.$$

Exercise VII.8

We obtain the variations from

$$\begin{aligned}
&\ln\frac{p_2}{p_1^h}+\frac{m}{p_2}-1\\
=\;& V\left(p_1^h,m\right)=V\left(p^l,m+cv\right)=\ln\frac{p_2}{p_1^l}+\frac{m+cv}{p_2}-1 \text{ and}\\
&\ln\frac{p_2}{p_1^l}+\frac{m}{p_2}-1\\
=\;& V\left(p^l,m\right)=V\left(p^h,m+ev\right)=\ln\frac{p_2}{p_1^h}+\frac{m+ev}{p_2}-1
\end{aligned}$$

and, solving for cv and ev, respectively, yields

$$\begin{aligned}
cv\left(p_1^h\to p_1^l\right) &= -p_2\left(\ln p_1^h-\ln p_1^l\right)<0 \text{ and}\\
ev\left(p_1^h\to p_1^l\right) &= p_2\left(\ln p_1^h-\ln p_1^l\right)>0.
\end{aligned}$$

8. Further exercises without solutions

PROBLEM VII.1. Determine the expenditure functions and the Hicksian demand function for $U(x_1, x_2) = \min(x_1, x_2)$ and $U(x_1, x_2) = 2x_1 + x_2$. Can you confirm the duality equations

$$\begin{aligned} \chi(p, V(p, m)) &= x(p, m) \text{ and} \\ x(p, e(p, \bar{U})) &= \chi(p, \bar{U})? \end{aligned}$$

PROBLEM VII.2. Derive the Hicksian demand functions and the expenditure functions of the following utility functions:

(1) $U(x_1, x_2) = x_1 \cdot x_2$
(2) $U(x_1, x_2) = \min(a \cdot x_1, b \cdot x_2)$ with $a, b > 0$
(3) $U(x_1, x_2) = a \cdot x_1 + b \cdot x_2$ with $a, b > 0$

PROBLEM VII.3. Verify Roy's identity for the utility function $U(x_1, x_2) = x_1 \cdot x_2$!

PROBLEM VII.4. Draw a figure that shows the equivalent variation following a price increase. *Hint: consult figure VII.10.*

CHAPTER VIII

Production theory

In household theory, preference relations are the primitive concepts from which indifference curves and utility functions are derived. Similarly, production sets form the basis of production theory. Production sets describe all the feasible input-output combinations and allow to define important concepts such as setup or sunk costs and returns to scale. Also, we can derive isoquants and production functions from production sets. A production function is a handy description of the production possibilities – similar to utility functions in preference theory. This chapter borrows from Mas-Colell et al. (1995, chapter 5.B) and Debreu (1959, pp. 37).

1. The production set

1.1. The vector space of goods and inputs. In chapter IV (p. 59) we consider the set of goods bundles

$$\mathbb{R}_+^\ell := \{(z_1, ..., z_\ell) : z_g \in \mathbb{R}_+, g = 1, ..., \ell\}.$$

We now explicitly allow for $z_g < 0$. Goods of a negative amount are called input or factors of production, while goods of a positive amount are called output or produced goods. Thus, we consider $\mathbb{R}^\ell$ rather than $\mathbb{R}_+^\ell$.

The plane $\mathbb{R}^2$ is depicted in figure VIII.1. If z_1 and z_2 are positive, both are produced goods. Of course, one cannot have output without input so these points should be excluded (see the next section). $z_1 < 0$ and $z_2 > 0$ means that good 2 is produced by factor 1 where $-z_1$ units of factor 1 are employed. Finally, consider $z_1 < 0$ and $z_2 < 0$. These points are inefficient. After all, it should be possible to do nothing at all and choose the inaction point $(0, 0)$.

1.2. Definition of a production set. A production set contains feasible input-output combinations. Feasibility means that there is some technological or other process that allows to produce the output (the goods with a positive sign) from the input (the goods with a negative sign):

DEFINITION VIII.1 (PRODUCTION SET). *A production set $Z \subseteq \mathbb{R}^\ell$ is the set of input-output combinations such that*

H. Wiese, *Advanced Microeconomics*,
https://doi.org/10.1007/978-3-658-34959-2_8

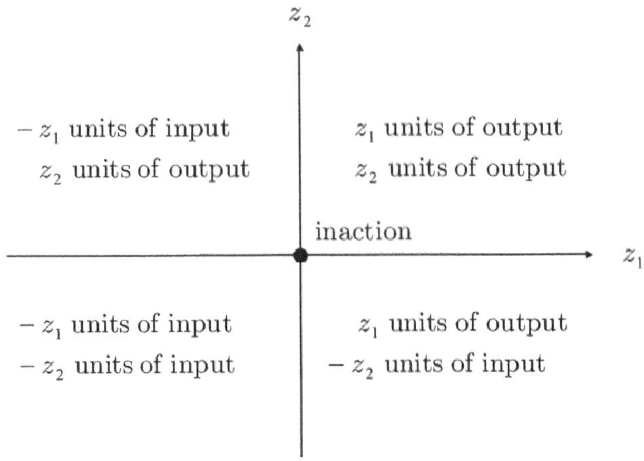

FIGURE VIII.1. Sign convention

- *Z is nonempty,*
- *Z is closed,*
- *for every bundle of inputs* $(z_1, ..., z_m) \in \mathbb{R}^m_-$, *there is a bundle of outputs* $(z_{m+1}, ..., z_\ell) \in \mathbb{R}^{\ell-m}_+$ *such that*

$$\Bigg(\underbrace{z_1, ..., z_m}_{inputs}, \underbrace{z_{m+1}, ..., z_\ell}_{outputs}\Bigg) \in Z$$

  holds,
- $\left\{\Bigg(\underbrace{z_{m+1}, ..., z_\ell}_{outputs}\Bigg) \in \mathbb{R}^{\ell-m}_+ : \Bigg(\underbrace{z_1, ..., z_m}_{inputs}, \underbrace{z_{m+1}, ..., z_\ell}_{outputs}\Bigg) \in Z\right\}$ *is bounded for every input bundle* $\Bigg(\underbrace{z_1, ..., z_m}_{inputs}\Bigg) \in \mathbb{R}^m_-$,
- *Z does not contain any element* $z > 0$ *and*
- $z \in Z$ *implies* $-z \notin Z$.

The elements in Z are called production vectors, production plans, or input-output vectors.

Consider figure VIII.2 and the production set $Z \subseteq \mathbb{R}^2$ (that is the area below the bold curve) passing through $(0, 0)$. For example, point $(\hat{z}_1, \hat{z}_2)$ is an element of Z. By $\hat{z}_1 < 0$ and $\hat{z}_2 > 0$, good 1 is the input (factor) and good 2 the output. We have $\hat{z}_2$ units of output and $-\hat{z}_1$ units of input.

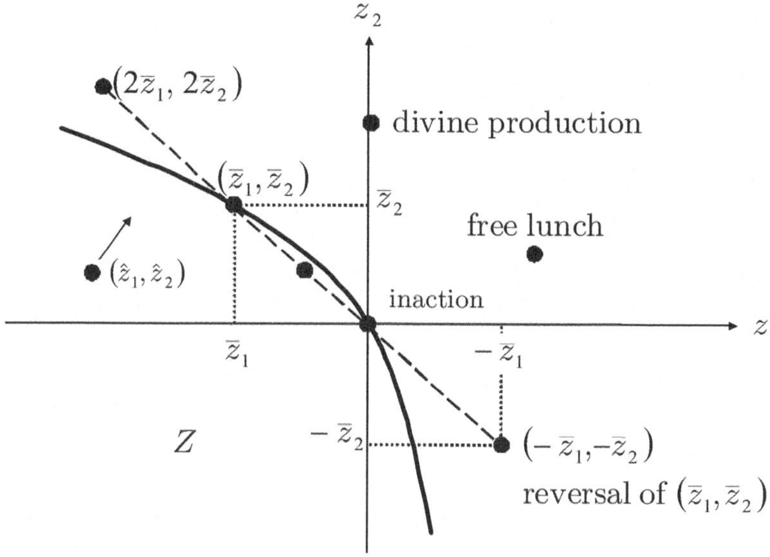

FIGURE VIII.2. Axioms for production sets

The six conditions are readily interpreted. If Z is empty, there is nothing to talk about. The closedness of Z means that if you can realize a sequence of production vectors that converge towards some production vector z, you can also realize z. The third requirement says that there is some output (which can be zero) for every input. These three conditions are more about mathematical convenience than about production.

The fourth requirement means that you cannot have an infinite amount of any good. Turning to the fifth requirement, note that there is no point $z \in Z$ that fulfills $z > 0$ in figure VIII.2. This is the no-free-lunch property or, in Latin, the condition "ex nihilo nihil" (from nothing, nothing). Similarly, "divine production" refers to the verse from a famous Chrismas carol:

> Then let us all with one accord sing praises to our heavenly Lord, who hath made heaven and earth from naught ...

One interesting aspect about the production-set model is that factors of production and goods produced are not determined a priori. For example, one needs electricity as a factor of production to produce coal (to get coal out of the ground). Inversely, we can produce electricity from coal (coal-fired power generation). Now, the sixth property says that inverting does not work 100%. Imagine producing apples juice from apples (easy enough) and reverting the process by forming apples from apple juice (not that easy).

Figure VIII.2 illustrates this property: $(\bar{z}_1, \bar{z}_2)$ is in the production set while its reversal $-(\bar{z}_1, \bar{z}_2)$ is not.

1.3. Further axioms. Apart from the six requirements mentioned in the definition, production sets may also obey or not obey other axioms.

DEFINITION VIII.2 (PRODUCTION SET). *A production set* $Z \subseteq \mathbb{R}^\ell$ *obeys*

- *the possibility of inaction if* $0 \in Z$ *holds,*
- *the property of free disposal if* $z \in Z$ *and* $z' \leq z$ *implies* $z' \in Z$*,*
- *nonincreasing returns to scale if* $z \in Z$ *implies* $kz \in Z$ *for all* $k \in [0, 1]$*,*
- *nondecreasing returns to scale if* $z \in Z$ *implies* $kz \in Z$ *for all* $k \geq 1$*,*
- Z*-convexity if* Z *is convex.*

The production set of figure VIII.2 clearly obeys the possibility of inaction. Indeed, it seems a quite natural requirement. However, the production process may require setup costs. If these costs are sunk, we obtain a production set as in figure VIII.3 (a) where $-z_1$ units of good 1 are a condition sine qua non for producing any positive output. Production set (b) looks very much the same as (a). However, the setup costs are not sunk and the possibility of inaction is still valid. Production set (a), which is taken from Mas-Colell et al. (1995, Figure 5.B.3 (a), p. 131), is problematic because it implies a temporal structure where the sunk costs are already incurred.

Free disposal means that goods can be thrown away and unnecessary factors of production do no harm. With respect to goods, the free-disposal property assumes away any waste problem. After all, goods may be bads.

Figure VIII.2 is an example of a production set fulfilling nonincreasing returns to scale. For example, $\frac{1}{2}(\bar{z}_1, \bar{z}_2)$ (halfway between $(0,0)$ and $(\bar{z}_1, \bar{z}_2)$ is contained in Z. Nonincreasing returns to scale imply the possibility of inaction. Do you see why?

Nonincreasing returns to scale are violated in figure VIII.3 (b) and (c). Nondecreasing returns to scale are not fulfilled by the production set of figure VIII.2. Just compare $(\bar{z}_1, \bar{z}_2) \in Z$ and $2(\bar{z}_1, \bar{z}_2) \notin Z$.

Thus returns to scale are nonincreasing if production can be scaled down and nondecreasing if production can be scaled up. Z-convexity is related to nonincreasing returns to scale. Indeed, Z-convexity and possibility of inaction imply nonincreasing returns to scale. However, there is more to convexity than downscaling. Consider figure VIII.3 (d). We will see that Z-convexity implies convexity (in the sense known from preference theory). As in preference theory, a rough description of convexity is "mixtures are

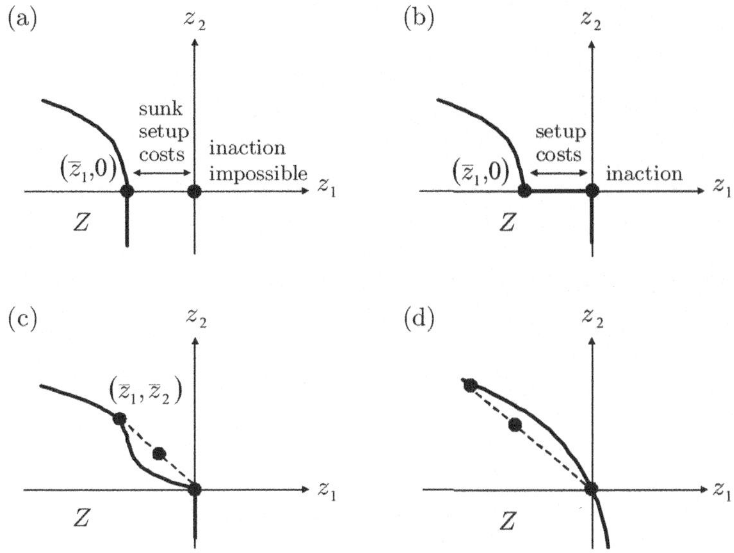

FIGURE VIII.3. Setup costs, inaction and convexity

prefered to extremes". The production vector in the middle uses the average amount of input (where the averaging concerns the two extreme production vectors) and produces the average amount of output. If the production set is strictly convex, the resulting middle point is not efficient and it is possible to produce more than the average with the given input. We return to "mixtures better than extremes" once we have introduced the isoquants.

2. Efficiency

2.1. Input efficiency and output efficiency. If a firm wants to maximize output or minimize input, it can discard all but a few points from the production set. Three different types of inefficiency can be defined. Consider, for example, point $(\hat{z}_1, \hat{z}_2)$ in figure VIII.2. The firm can keep on producing z_2 while reducing $-z_1$ (move to the right) – this is an instance of input inefficiency. $(\hat{z}_1, \hat{z}_2)$ is also output-inefficient, the firm can produce more than z_2 while keeping z_1 constant (move upwards). Finally, we can move to the north-east from $(\hat{z}_1, \hat{z}_2)$.

DEFINITION VIII.3 (INPUT-EFFICIENCY). *Let $Z \subseteq \mathbb{R}^\ell$ be a production set. A point*

$$z = \left(\underbrace{z_1, \ldots, z_m}_{inputs}, \underbrace{z_{m+1}, \ldots, z_\ell}_{outputs} \right)$$

is not input-efficient if another input-output vector

$$\hat{z} = \left(\underbrace{\hat{z}_1, ..., \hat{z}_m}_{inputs}, \underbrace{z_{m+1}, ..., z_\ell}_{outputs} \right)$$

exists such that $(\hat{z}_1, ..., \hat{z}_m) > (z_1, ..., z_m)$. *In that case,* $\hat{z}$ *is called an input improvement over* z.

If z is input-inefficient, we can reduce one input factor without increasing others while still producing the same output.

DEFINITION VIII.4 (OUTPUT-EFFICIENCY). *Let* $Z \subseteq \mathbb{R}^\ell$ *be a production set. A point*

$$z = \left(\underbrace{z_1, ..., z_m}_{inputs}, \underbrace{z_{m+1}, ..., z_\ell}_{outputs} \right)$$

is not output-efficient if another input-output vector

$$\hat{z} = \left(\underbrace{z_1, ..., z_m}_{inputs}, \underbrace{\hat{z}_{m+1}, ..., \hat{z}_\ell}_{outputs} \right)$$

exists such that $(\hat{z}_{m+1}, ..., \hat{z}_\ell) > (z_{m+1}, ..., z_\ell)$. *In that case,* $\hat{z}$ *is called an output improvement over* z.

Output inefficiency means that the given input can produce more of one good without producing less of another one.

DEFINITION VIII.5 (EFFICIENCY). *Let* $Z \subseteq \mathbb{R}^\ell$ *be a production set. A point*

$$z = \left(\underbrace{z_1, ..., z_m}_{inputs}, \underbrace{z_{m+1}, ..., z_\ell}_{outputs} \right)$$

is not efficient if another input-output vector

$$\hat{z} = \left(\underbrace{\hat{z}_1, ..., \hat{z}_\ell}_{inputs\ and\ outputs} \right)$$

exists such that $\hat{z} > z$ *holds. In that case,* $\hat{z}$ *is called an improvement over* z.

Note that the improving input-output vector does not clearly distinguish inputs and outputs. The reasons is that an input becoming an output is one instance of an improvement.

2.2. Definitions: production function and isoquant. We now concentrate on output efficiency. We begin with one output. If we restrict the production set to output-efficient vectors, we are led to the concept of a production function:

DEFINITION VIII.6 (PRODUCTION FUNCTION). *Let $Z \subseteq \mathbb{R}^\ell$ be a production set. Define a function $f : \mathbb{R}_+^{\ell-1} \to \mathbb{R}_+$ by*

$$f(x_1, ..., x_{\ell-1}) = \max \{y \in \mathbb{R}_+ : (-x_1, ..., -x_{\ell-1}, y) \in Z\}.$$

f is called the production function for y.

Note that a production function is well-defined by the requirements laid down in definition VIII.1. In particular, the set

$$\{y \in \mathbb{R}_+ : (-x_1, ..., -x_{\ell-1}, y) \in Z\}$$

is closed, nonempty, and bounded. Obviously,

$$(-x_1, ..., -x_{\ell-1}, f(x_1, ..., x_{\ell-1}))$$

is output-efficient.

If a utility function corresponds to a production function, an indifference curve is preference theory's analogue to an isoquant in production theory:

DEFINITION VIII.7 (BETTER SET, ISOQUANT). *Let f be a production function on $\mathbb{R}_+^{\ell-1}$. Then, we have the better set $B_{\hat{x}}$ of $\hat{x}$:*

$$B_{\hat{x}} := \left\{x \in \mathbb{R}_+^{\ell-1} : f(x) \geq f(\hat{x})\right\}$$

the worse set $W_{\hat{x}}$ of $\hat{x}$:

$$W_{\hat{x}} := \left\{x \in \mathbb{R}_+^{\ell-1} : f(x) \leq f(\hat{x})\right\}$$

and $\hat{x}$'s isoquant $I_{\hat{x}}$:

$$I_{\hat{x}} := B_{\hat{x}} \cap W_{\hat{x}} = \left\{x \in \mathbb{R}_+^{\ell-1} : f(x) = f(\hat{x})\right\}$$

Figure VIII.4 depicts two isoquants in the case of perfect complements (which you are familiar with from utility theory). Obviously, there exist points that are not input-efficient.

The analogy between indifference curves and isoquants should be evident. In spite of the ordinality of preferences, the monotonicity concepts defined for preferences and utility functions carry over to those for a production function:

DEFINITION VIII.8 (MONOTONICITY). *A production function f obeys*

- *weak monotonicity if $x > x'$ implies $f(x) \geq f(x')$,*
- *strict monotonicity if $x > x'$ implies $f(x) > f(x')$, and*

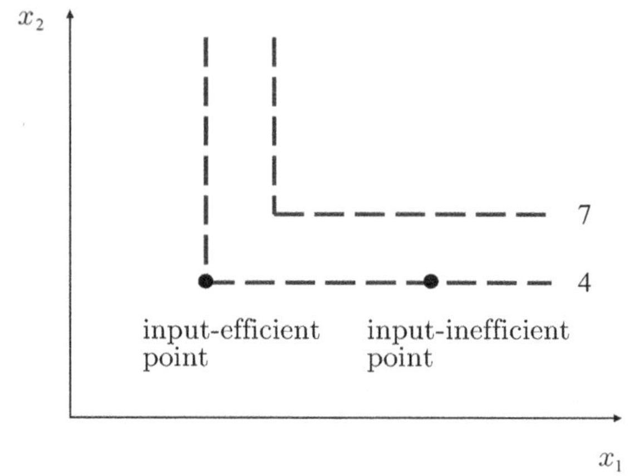

FIGURE VIII.4. Isoquants for perfect complements

- *local non-satiation at x' if, in every ε-ball with center x', a bundle x with $f(x) > f(x')$ can be found .*

2.3. Edgeworth box and transformation curve. Production functions provide an answer to this question: Which points in a production set survive the output-efficiency test in case of one output? There is a neat way to examine the output-efficient elements in a production set in case of two inputs and two outputs if the two output processes are separated. For example, they are not separated in case of wool and milk from sheep.

DEFINITION VIII.9. *Let $Z \subseteq \mathbb{R}^4$ be a production set. Z obeys the separation property if $\left(\underbrace{-x_1, -x_2}_{inputs}, \underbrace{y_A, y_B}_{outputs}\right) \in Z$ is equivalent to the existence of $x_1^A \in [0, x_1]$ and $x_2^A \in [0, x_2]$ such that $x_1 = x_1^A + x_1^B, x_2 = x_2^A + x_2^B$, $\left(\underbrace{-x_1^A, -x_2^A}_{inputs}, \underbrace{y_A, 0}_{outputs}\right) \in Z$ and $\left(\underbrace{-x_1^B, -x_2^B}_{inputs}, \underbrace{0, y_B}_{outputs}\right) \in Z$ hold.*

Separation means that inputs are attributable to specific output. In that case, a production Edgeworth box can be used to explain output efficiency. Consider figure VIII.5. It is called a production Edgeworth box. It consists of two families of isoquants, one for output A and one for output B (turn the book by 180 degrees). The breadth indicates the amount of factor 1, and the height the amount of factor 2. Every point inside that box shows how the inputs 1 and 2 are allocated to produce the outputs A and B. Thus, every point corresponds to a quadruple of two inputs and two ouputs.

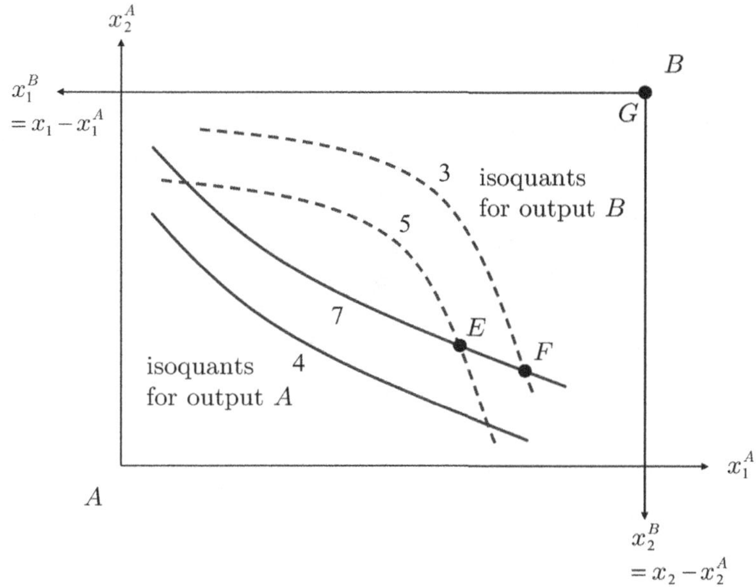

FIGURE VIII.5. A production Edgeworth box

The quantities produced are indicated by the isoquants and the numbers associated with them. Consider, for example, points E and F. They both use the same input tuple (x_1, x_2) (the overall use of both factors), but the output is different, $(7, 5)$ in case of point E and $(7, 3)$ in case of point F.

Point F is not output efficient. It is possible to produce more of output B while leaving output A constant. Figure VIII.6 illustrates what we can do about output inefficiency. If we have two crossing isoquants (and some other conditions), we find a lens between these isoquants. This lens contains the set of all output improvements over the crossing point. Such an improvement is marked by an arrow in our figure.

No further improvements are possible if the lens shrinks down to a point, a point of tangency between two isoquants. The locus of all these points is called the production curve and shown in figure VIII.7.

The production curve can be used to derive the so-called transformation curve (also known as production-possibility frontier). Referring again to figure VIII.7, the output tuples $(9, 5)$ and $(14, 3)$ are efficient. They can be placed in a $y_A - y_B$ diagram and are two points of the transformation curve that is depicted in figure VIII.8.

EXERCISE VIII.1. *Using a transformation curve, discuss output efficiency.*

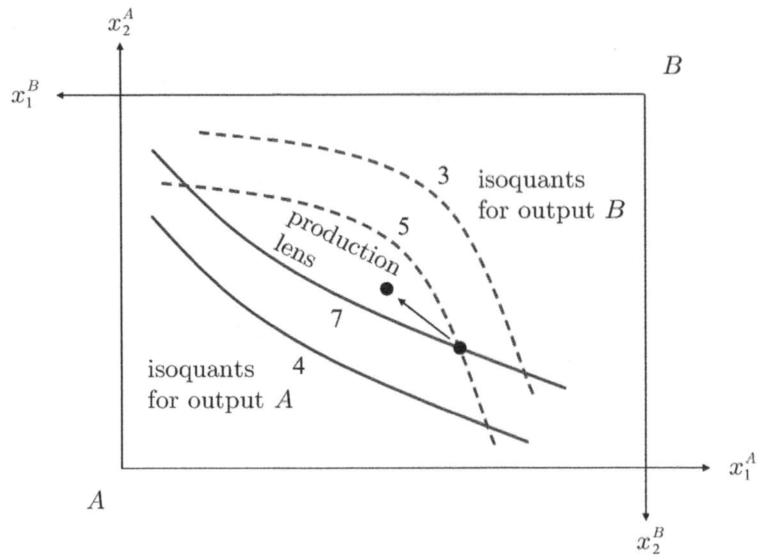

FIGURE VIII.6. An output-inefficient factor combination

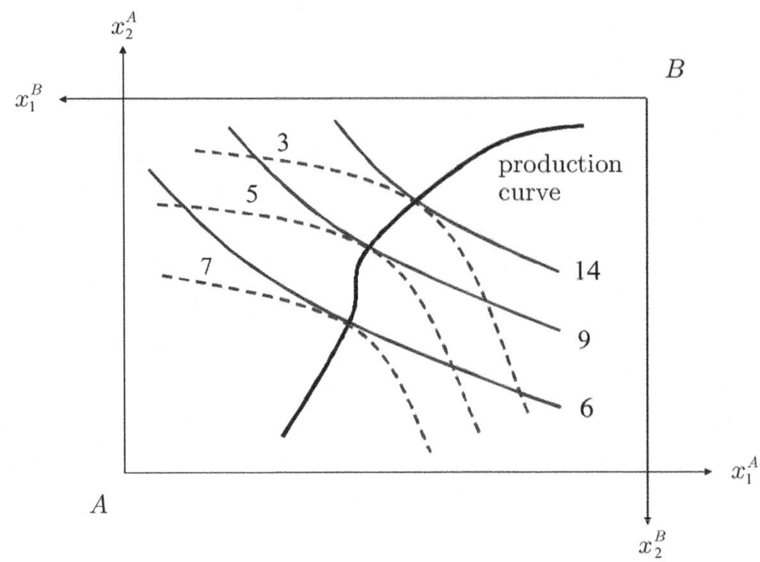

FIGURE VIII.7. The production curve

The slope of the transformation curve tells us by how much output y_B decreases if output y_A is increased by one small unit.

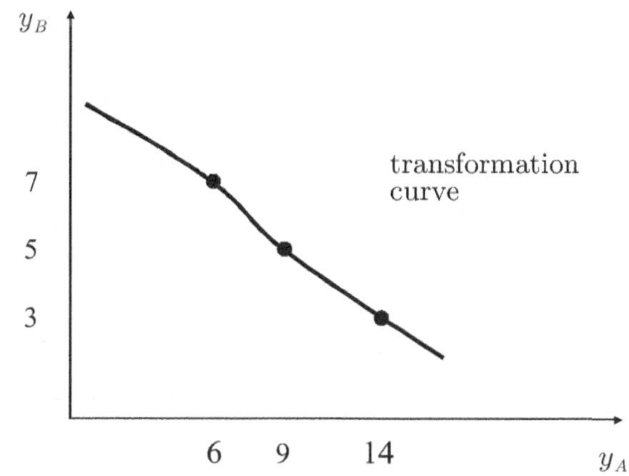

FIGURE VIII.8. The transformation curve

DEFINITION VIII.10 (MARGINAL RATE OF TRANSFORMATION). *Assume that the transformation curve defines a differentiable function* $y_A \mapsto y_B$. *We call*

$$MRT := \left| \frac{dy_B}{dy_A} \right|$$

the marginal rate of transformation between good A and good B.

3. Convex production sets and convave production functions

3.1. Convexity of the production set and concavity of the production function. Above, we have defined a production function $f : \mathbb{R}_+^{\ell-1} \to \mathbb{R}_+$ on the basis of a production set. Inversely, and assuming free disposal, the production set associated with f is

$$Z := \left\{ \left(-\left(x_1, ..., x_{\ell-1}\right), y\right) \in \mathbb{R}_-^{\ell-1} \times \mathbb{R}_+ : f\left(x_1, ..., x_{\ell-1}\right) \geq y \right\}$$

In this section, we want to establish an important equivalence:

LEMMA VIII.1. *Let Z be a production set where the first* $\ell - 1$ *entries are always nonpositive. Let f be the production function associated with Z and let Z obey free disposal. Then, Z is convex if and only if the corresponding production function f is concave.*

PROOF. Concavity of $f : \mathbb{R}_+^{\ell-1} \to \mathbb{R}_+$ means

$$f\left(kx + (1-k)x'\right) \geq kf(x) + (1-k) f\left(x'\right)$$

for all $x, x' \in \mathbb{R}^{\ell-1}$ and for all $k \in [0, 1]$. Assume, now, that Z is convex. We need to show that the concavity of f ensues. For any $x, x' \in \mathbb{R}^{\ell-1}$,

$(-x, f(x))$ and $(-x', f(x'))$ belong to Z. By the convexity of Z,

$$k(-x, f(x)) + (1-k)(-x', f(x'))$$

also belongs to Z for any $k \in [0,1]$. This vector can be rewritten as

$$(k(-x) + (1-k)(-x'), kf(x) + (1-k)f(x')),$$

i.e., $kf(x) + (1-k)f(x')$ is producible from $kx + (1-k)x'$. By the definition of a production function (observe the max operator), we have the above inequality.

For the inverse implication, we assume that f is concave and consider two elements $(-x, y)$ and $(-x', y')$ from Z. By the definition of a production function, we have

$$f(x) \geq y \text{ and } f(x') \geq (y').$$

We now obtain

$$\begin{aligned} f(kx + (1-k)x') &\geq kf(x) + (1-k)f(x') \text{ (concavity of } f\text{)} \\ &\geq ky + (1-k)y' \text{ (the above two inequalities)} \end{aligned}$$

for any $k \in [0,1]$ so that $kx + (1-k)x'$ produces at least $ky + (1-k)y'$ units of output. Free disposal implies

$$\begin{aligned} Z &\ni (k(-x) + (1-k)(-x'), ky + (1-k)y') \\ &= k(-x, y) + (1-k)(-x', y'). \end{aligned}$$

Thus, if $(-x, y)$ and $(-x', y')$ are from Z, so is $k(-x, y) + (1-k)(-x', y')$ for any $k \in [0,1]$. □

3.2. Convex production sets versus convex better sets. It is of some interest to know how Z-convexity (convex production sets) corresponds to convexity in the sense of convex better sets. We know from lemma IV.9 (p. 82) that quasi-concavity is equivalent to convex better sets. However, a function can be quasi-concave without being concave. That is, a production function can have convex better sets without the production set being convex.

Consider, for example, the production function given by

$$f(x, y) = xy.$$

It obeys strict quasi-concavity as we know from chapter IV (p. 86). It is not concave which can be seen from

$$\begin{aligned} f(k(0,0) + (1-k)(1,1)) &= f(1-k, 1-k) = (1-k)^2 < 1-k \\ &= k \cdot 0 + (1-k) \cdot 1 \\ &= kf(0,0) + (1-k)f(1,1). \end{aligned}$$

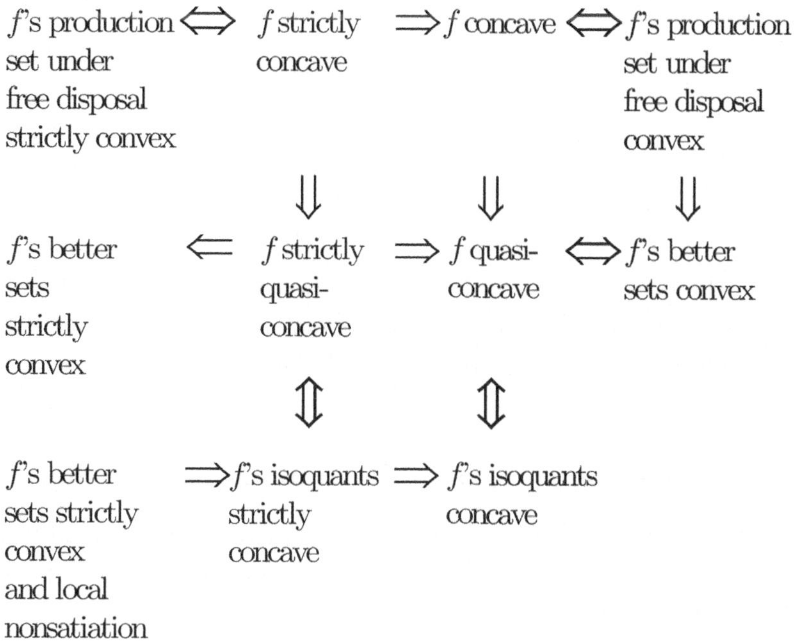

FIGURE VIII.9. Concavity and convexity

for $0 < k < 1$.

We can, however, show the inverse.

EXERCISE VIII.2. *Show that every concave function is quasi-concave.*

The following lemma summarizes the result of the above exercise and important other relationships. The reader is invited to compare figure IV.20 on p. 82.

LEMMA VIII.2. *Let f be a continuous production function on $\mathbb{R}_+^\ell$. Then, the relationships summarized in figure VIII.9 hold.*

3.3. What about concave utility functions? Comparing the above discussion on quasi-concave and concave functions, one may wonder why concave functions did not figure more prominently in preference theory. By exercise VIII.2 above, concave utility functions are quasi-concave and have convex better sets. However, utility functions exist that are not concave but still quasi-concave.

Consider, for example, the utility functions U and V given by $U(x, y) = xy$ and $V(x, y) = x^{\frac{1}{3}}y^{\frac{1}{3}}$. They represent the same preferences because we can apply the strictly increasing function $\tau : \mathbb{R} \to \mathbb{R}$ given by $\tau(U) = U^{\frac{1}{3}}$ and obtain

$$\begin{aligned}(\tau \circ U)(x,y) &= \tau(U(x,y)) \\ &= \tau(xy) \\ &= (xy)^{\frac{1}{3}} \\ &= V(x,y)\end{aligned}$$

Thus, from the point of view of preference theory, U and V are basically the same. However, U is neither convex nor concave (see p. 187) but still quasi-concave, while V is concave (trust me) and, therefore, quasi-concave. Thus, in preference theory, quasi-concavity is more important than concavity.

4. Exploring the production mountain (function)

4.1. Factor variations. When exploring the production mountain, we do not engage in a random walk but follow carefully laid-out paths. These paths can be characterized by factor variations:

- Partial factor variation: We change one factor only and keep the other factors constant (see figure VIII.10).
- Proportional factor variation: We change all the factors while keeping proportions constant (see figure VIII.10).
- Isoquant factor variation: We change the factors so as to keep output constant (see figure VIII.11 on p. 226).
- Isoclinic factor variation: We change the factors so as to keep the marginal rate of technical substitution (to be introduced shortly) constant (see figure VIII.11).

In contrast to the previous notation, we now use ℓ (rather than $\ell - 1$) factors of production, i.e., we have the input vector $x = (x_1, ..., x_\ell)$ and output $y = f(x) = f(x_1, .., x_\ell)$.

4.2. Partial factor variation: marginal productivity, average productivity and production elasticity. In line with utility theory, we can define the marginal productivity of factor i by

$$MP_i := \frac{\partial f(x)}{\partial x_i}.$$

Average productivity is a concept not known from utility theory (why?). It is defined by

$$AP_i := \frac{f(x_i)}{x_i}.$$

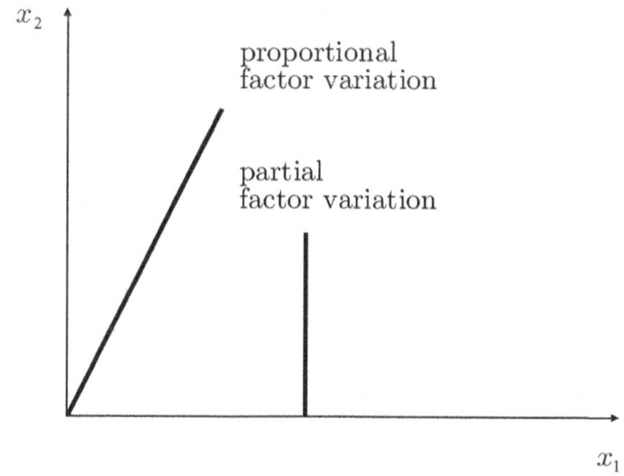

FIGURE VIII.10. Two explorations in the production mountain

EXERCISE VIII.3. *Consult definition VI.8 (p. 147) and suggest a definition of production elasticity. Do you see how the production elasticity depends on the marginal and the average productivity?*

EXERCISE VIII.4. *Calculate factor 1's production elasticity for the Cobb-Douglas production function f given by $f(x_1, x_2) = x_1^a x_2^b$, $a, b \geq 0$.*

4.3. Marginal something equals average something. The marginal productivity sometimes equals the average productivity. An analogy may be helpful:

- If I (Wiese) am the first to enter a lecture hall, the average age (my age divided by one) is equal to the marginal age (the additional age from increasing the number of persons from zero to one).
- If I enter the lecture hall with many students present, the marginal age (the additional age from adding myself) is above (well above, indeed) the average age. Therefore, my entering has the average age increase.

Translating the first point from a discrete formulation to a continuous one, we have the equality of marginal age and average age for 0 persons. Here, we run into the difficulty of obtaining the average age as $\frac{0}{0}$ which is not defined. However, we can apply de l'Hospital's rule:

THEOREM VIII.1. *Let f and g be two functions $\mathbb{R} \to \mathbb{R}$ for which we have $f(x_0) = g(x_0) = 0$ for some $x_0 \in \mathbb{R}$. Then, if f and g are differentiable at*

x_0 and if $g'(x_0) \neq 0$, we have

$$\lim_{\substack{x \to x_0, \\ x \neq x_0}} \frac{f(x)}{g(x)} = \frac{f'(x_0)}{g'(x_0)}.$$

PROOF. It is quite simple to prove de l'Hospital's rule. Since f and g are differentiable at x_0, both $\lim_{x \to x_0} \frac{f(x)-f(x_0)}{x-x_0}$ and $\lim_{x \to x_0} \frac{g(x)-g(x_0)}{x-x_0}$ exist and are denoted by $f'(x_0)$ and $g'(x_0)$, respectively. We find

$$\begin{aligned}
& \frac{f'(x_0)}{g'(x_0)} \ (g'(x_0) \neq 0) \\
= & \frac{\lim_{\substack{x \to x_0, \\ x \neq x_0}} \frac{f(x)-f(x_0)}{x-x_0}}{\lim_{\substack{x \to x_0, \\ x \neq x_0}} \frac{g(x)-g(x_0)}{x-x_0}} \ \text{(definition of derivatives)} \\
= & \lim_{\substack{x \to x_0, \\ x \neq x_0}} \frac{\frac{f(x)-f(x_0)}{x-x_0}}{\frac{g(x)-g(x_0)}{x-x_0}} \ \text{(rules for limits)} \\
= & \lim_{\substack{x \to x_0, \\ x \neq x_0}} \frac{\frac{f(x)}{x-x_0}}{\frac{g(x)}{x-x_0}} \ (f(x_0) = g(x_0) = 0) \\
= & \lim_{\substack{x \to x_0, \\ x \neq x_0}} \frac{f(x)}{g(x)} \ (x - x_0 \neq 0)
\end{aligned}$$

□

With the support of de l'Hospital's rule, we can now formulate the analogon of "me entering first":

LEMMA VIII.3. *Let $f : \mathbb{R} \to \mathbb{R}$ be any differentiable (production) function. We find $\left.\frac{df}{dx}\right|_{x=0} = \left.\frac{f(x)}{x}\right|_{x=0}$ if $f(0) = 0$ holds.*

PROOF. We apply de l'Hospital's rule and define g by $g(x) = x$. For $x_0 := 0$, the conditions laid out in theorem VIII.1 are fulfilled and we obtain

$$\lim_{\substack{x \to 0, \\ x \neq 0}} \frac{f(x)}{x} = \frac{f'(0)}{1} = \left.\frac{df}{dx}\right|_0.$$

□

In contrast, "me entering last" is reflected in the following lemma:

LEMMA VIII.4. *Let $f : \mathbb{R} \to \mathbb{R}$ be any differentiable (production) function. Assume $x > 0$. Then we have*

$$\frac{df}{dx} > (<) \frac{f(x)}{x} \Leftrightarrow \frac{d\frac{f(x)}{x}}{dx} > (<)\, 0.$$

Thus, if the marginal productivity is above the average productivity, the average productivity increases. Also, if the marginal productivity equals the average productivity, the average productivity is constant. Of course, these two lemmata hold for any marginal something and average something:

- marginal revenue and average revenue (see p. 300),
- marginal cost and average cost, and
- marginal profit and average profit.

4.4. Proportional factor variation: returns to scale. Proportional factor variation means multiplying the factors of production x by a scalar t :

$$(x_1, ..., x_\ell) \mapsto t\,(x_1, ..., x_\ell) = (tx_1, ..., tx_\ell)\,.$$

DEFINITION VIII.11 (RETURNS TO SCALE). *A production function* $f : \mathbb{R}_+^\ell \to \mathbb{R}_+$ *is characterized*

- *by constant returns to scale if*
$$f\,(tx) = tf\,(x) \text{ for all } t \geq 0$$
holds for all $x \in \mathbb{R}_+^\ell$,
- *by increasing returns to scale if*
$$f\,(tx) \geq tf\,(x) \text{ for all } t \geq 1$$
holds for all $x \in \mathbb{R}_+^\ell$ *and*
- *by decreasing returns to scale if*
$$f\,(tx) \leq tf\,(x) \text{ for all } t \geq 1$$
holds for all $x \in \mathbb{R}_+^\ell$.

Alternatively, we can define returns to scale by way of the scale elasticity:

DEFINITION VIII.12 (SCALE ELASTICITY). *Let* $f : \mathbb{R}_+^\ell \to \mathbb{R}_+$ *be a production function. The scale elasticity at* $x = (x_1, ..., x_\ell)$ *is defined by*

$$\varepsilon_{y,t} = \left.\frac{\frac{df(tx)}{f(tx)}}{\frac{dt}{t}}\right|_{t=1} = \left.\frac{df\,(tx)}{dt}\frac{t}{f\,(tx)}\right|_{t=1}.$$

LEMMA VIII.5. *We have*

- *increasing returns to scale at* $x \in \mathbb{R}_+^\ell$ *iff* $\varepsilon_{y,t} \geq 1$ *holds,*
- *decreasing returns to scale at* $x \in \mathbb{R}_+^\ell$ *iff* $\varepsilon_{y,t} \leq 1$ *holds, and*
- *constant returns to scale at* $x \in \mathbb{R}_+^\ell$ *iff* $\varepsilon_{y,t} = 1$ *holds.*

EXERCISE VIII.5. *Calculate the scale elasticity for the Cobb-Douglas production function* f *given by* $f\,(x_1, x_2) = x_1^a x_2^b$, $a, b \geq 0$.

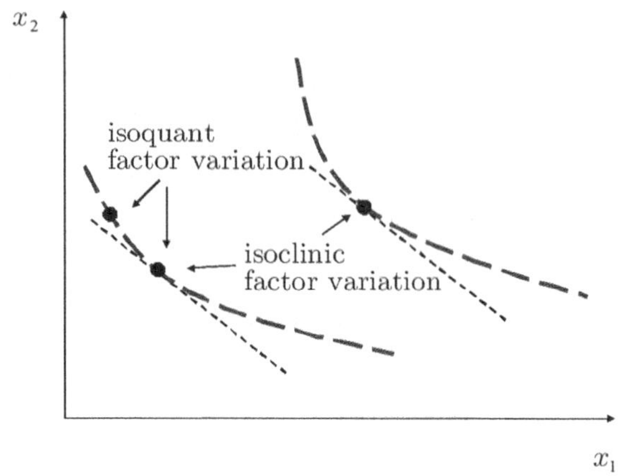

FIGURE VIII.11. Isoquant and isoclinic factor variations

4.5. Isoquant factor variation: Marginal rate of technical substitution. In preference theory, we have come across the marginal rate of substitution. The corresponding concept in production theory is called the marginal rate of technical substitution. If we increase factor 1 by one unit, by how many units do we need to decrease factor 2 in order to hold output constant? In mathematical terms, we have an implicit function

$$I_y : x_1 \mapsto x_2$$

which reflects the amount of factor 2 needed when we want to produce output y with the help of x_1 units of factor 1. Totally analogous to utility theory (see pp. 82) are the following definition and lemma:

DEFINITION VIII.13 (MARGINAL RATE OF TECHNICAL SUBSTITUTION). *If the function I_y is differentiable and if the production function is monotonic, we call*

$$MRTS = \left| \frac{dI_y(x_1)}{dx_1} \right|$$

the marginal rate of technical substitution between factor 1 *and factor* 2 *(or of factor* 2 *for factor* 1*).*

LEMMA VIII.6. *Let f be a differentiable production function, the marginal rate of technical substitution between factor* 1 *and factor* 2 *can be obtained by*

$$MRTS(x_1) = \frac{\frac{\partial f}{\partial x_1}}{\frac{\partial f}{\partial x_2}}.$$

Refer back to pp. 216 where we analyze efficient production in an Edgeworth box. You can see that efficiency requires the equality of the marginal rates of technical substitution. Indeed, let the production of two goods A and B be such that the marginal rates of technical substitution differ:

$$(3 =) \left| \frac{dx_2^A}{dx_1^A} \right| = MRTS^A < MRTS^B = \left| \frac{dx_2^B}{dx_1^B} \right| (= 5) \qquad \text{(VIII.1)}$$

Assume, now, that we decrease factor 1 by one unit in the production of good A. A marginal rate of technical substitution of 3 means that the production of good A can be held at the current level if factor 2 is increased by $MRTS^A = 3$ units.

We now show that the transfer of one unit of factor 1 from the production of A to the production of B allows to produce more of good B without producing less of good A (a so-called Pareto improvement). Indeed, while we need 3 units of factor 2 in order to keep good A's production quantity constant, we can release up to five units of factor 2 in the production of good B. Or, alternatively, we can take the 2 spare units of factor 2 to increase the production of good B.

4.6. Isoclinic factor variation. The importance of isoclinic factor variations becomes obvious in the next chapter where we deal with cost minimization (see pp. 234).

5. Topics

The main topics in this chapter are

- production set
- production function
- no-free-lunch property
- setup costs
- sunk costs
- nonincreasing returns to scale
- nondecreasing returns to scale
- constant returns to scale
- free disposal
- possibility of inaction
- production Edgeworth box
- production curve
- transformation curve
- marginal productivity
- average productivity
- production elasticity
- increasing (constant, decreasing) returns to scale
- scale elasticity
- partial factor variation
- proportional factor variation
- isoquant factor variation
- isoclinic factor variation
- de l'Hospital's rule

6. Solutions

Exercise VIII.1

Every point on the transformation curve is output-efficient, every point inside of it output-inefficient.

Exercise VIII.2

Consider any vectors x, $y \in \mathbb{R}^\ell$ and any $k \in [0,1]$. If f is concave, it is also quasi-concave by

$$\begin{aligned} f(kx+(1-k)y) &\geq kf(x)+(1-k)f(y) \\ &\geq k\min(f(x),f(y))+(1-k)\min(f(x),f(y)) \\ &= \min(f(x),f(y)). \end{aligned}$$

Exercise VIII.3

We define and obtain

$$\varepsilon_{y,x_i} := \underbrace{\frac{\frac{dy}{y}}{\frac{dx_i}{x_i}}}_{\substack{\text{mathematically}\\ \text{doubtful, but}\\ \text{easily interpretable}}} = \frac{\partial y}{\partial x_i}\frac{x_i}{y} = \frac{MP_i}{AP_i}.$$

Exercise VIII.4

We obtain

$$\begin{aligned}
\varepsilon_{y,x_1} &= \frac{\partial\left(x_1^a x_2^b\right)}{\partial x_1}\frac{x_1}{y} \\
&= a x_1^{a-1} x_2^b \frac{x_1}{x_1^a x_2^b} \\
&= a.
\end{aligned}$$

Exercise VIII.5

We find $(tx_1)^a (tx_2)^b = t^{a+b} x_1^a x_2^b$ and obtain the memorable result that, for Cobb-Douglas production functions, the scale elasticity is equal to the sum of the production elasticities:

$$\begin{aligned}
\varepsilon_{y,t} &= \left.\frac{d\left((tx_1)^a (tx_2)^b\right)}{dt}\frac{t}{(tx_1)^a (tx_2)^b}\right|_{t=1} \\
&= \left.(a+b)\, t^{a+b-1} x_1^a x_2^b \frac{t}{(tx_1)^a (tx_2)^b}\right|_{t=1} \\
&= a+b \\
&= \varepsilon_{y,x_1} + \varepsilon_{y,x_2}.
\end{aligned}$$

7. Further exercises without solutions

PROBLEM VIII.1. Sketch a few isoquants that reflect decreasing returns to scale.

PROBLEM VIII.2. Prove lemma VIII.4 (p. 224)! *Hint: Calculate* $\frac{d\frac{f(x)}{x}}{dx}$*!*

PROBLEM VIII.3. Find the production functions for the production set

$$Y = \left\{ (x_1, x_2) \in \mathbb{R}^2 \mid x_2 \leq -(x_1)^2 \text{ if } x_1 \geq 0, \text{ and } x_2 \leq -\tfrac{1}{2}x_1 \text{ if } x_1 < 0 \right\}.$$

PROBLEM VIII.4. Determine the production set for the production function $y = f(x) = \min\{x_1, x_2\}$, $x_1, x_2 \geq 0$.

PROBLEM VIII.5. Let f be a homogeneous function of degree λ (i.e., $f(tx) = t^\lambda \cdot f(x)$). Show

$$\sum_i \frac{\partial f(x)}{\partial x_i} x_i = \lambda t^{\lambda-1} f(x)$$

and, for $\lambda = 1$, Euler's theorem,

$$\sum_i \frac{\partial f(x)}{\partial x_i} x_i = f(x).$$

Hint: Calculate $\frac{\partial f(tx)}{\partial t}$ *and* $\frac{\partial [t^\lambda f(x)]}{\partial t}$.

Comment: Assume an economy where the factors of production are paid their marginal product. Then, the sum of the payments (left-hand side of Euler equation) equals total output (right-hand side). That is, if we have marginal-product payment (as in perfect competition), production and distribution of the social product are equal in the special case of constant returns to scale.

CHAPTER IX

Cost minimization and profit maximization

1. Revisiting the production set

1.1. Definition of profit. The production possibilities have been explored in the previous chapter. We now turn to the two interrelated questions of

- which factors of production does the firm demand and
- which output does it offer.

The production set introduced in the previous chapter allows to address these questions. However, we first need to define the firm's profit. The profit definition is very simple because of the sign convention according to which goods with negative amounts are factors of production, while output goods have positive amounts.

DEFINITION IX.1 (PROFIT (PRODUCTION SET)). *Let $Z \subseteq \mathbb{R}^\ell$ be a production set and $p \in \mathbb{R}^\ell_+$ a price vector. The firm's profit at point $y \in Z$ is given by*

$$\underbrace{p \cdot z}_{profit} := \underbrace{\sum_{\substack{i=1,\\ z_i \geq 0}}^{\ell} p_i z_i}_{revenue} - \underbrace{\sum_{\substack{i=1,\\ z_i < 0}}^{\ell} p_i (-z_i)}_{cost}.$$

For a specific profit level $\bar{\Pi}$,

$$\left\{ z \in \mathbb{R}^\ell : p \cdot z = \bar{\Pi} \right\}$$

is called the isoprofit line.

1.2. Which good is an input and which an output? Consider, now, the production set depicted in figure IX.1 for $\ell = 2$. The firm's problem is to find the highest isoprofit line (with maximal distance from the $(0,0)$ point) that has at least one point in common with Z. We see that for a relatively low price of good 1, the firm produces at (z_1, z_2) so that good 1 (with $z_1 < 0$!) is a factor of production. Thus, in general, we do not need to tell (and cannot tell) which good is an output and which good is an input.

H. Wiese, *Advanced Microeconomics*,
https://doi.org/10.1007/978-3-658-34959-2_9

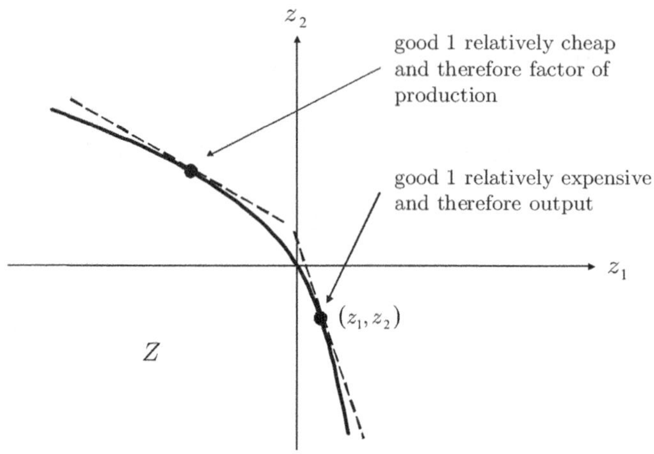

FIGURE IX.1. Factors of production are endogenous

However, from now on, we will assume we can address certain goods as inputs and others as output. We assume ℓ (sometimes $\ell := 2$) input factors.

1.3. Revealed profit maximization. We now employ a technique called revealed profit maximization. The idea is that a firm's choices tell us something about its demand for factors of production and supply of output goods – if we base our analysis on the assumption of profit maximization. We use revealed profit maximzation to show the following lemma.

LEMMA IX.1. *Assume the existence of best elements in the production set Z. Furthermore, assume that price changes do not change the role of input and output goods. Then,*

- *a price increase for a factor of production cannot increase demand for that factor and*
- *a price increase for an output good cannot decrease the supply of that output.*

PROOF. Without loss of generality, we consider two factors of production and one output good, only. Assume two price vectors $\left(p^A, w_1^A, w_2^A\right)$ and $\left(p^B, w_1^B, w_2^B\right)$ and the associated supply-and-demand vectors $\left(y^A, x_1^A, x_2^A\right)$ and $\left(y^B, x_1^B, x_2^B\right)$. Since the A supply-and-demand vector is best at the A prices, we have

$$p^A y^A - w_1^A x_1^A - w_2^A x_2^A \geq p^A y^B - w_1^A x_1^B - w_2^A x_2^B.$$

Inversing the roles, we also find

$$p^B y^B - w_1^B x_1^B - w_2^B x_2^B \geq p^B y^A - w_1^B x_1^A - w_2^B x_2^A$$

and therefore

$$-p^B y^A + w_1^B x_1^A + w_2^B x_2^A \geq -p^B y^B + w_1^B x_1^B + w_2^B x_2^B.$$

We add the first and the third inequality to obtain

$$\begin{aligned} & \left(p^A - p^B\right) y^A - \left(w_1^A - w_1^B\right) x_1^A - \left(w_2^A - w_2^B\right) x_2^A \\ \geq \; & \left(p^A - p^B\right) y^B - \left(w_1^A - w_1^B\right) x_1^B - \left(w_2^A - w_2^B\right) x_2^B \end{aligned}$$

and finally

$$\Delta p \Delta y - \Delta w_1 \Delta x_1 - \Delta w_2 \Delta x_2 \geq 0$$

where $\Delta p := \left(p^A - p^B\right)$, $\Delta x_1 := x_1^A - x_1^B$, etc. □

2. Cost minimization

2.1. The problem. Instead of attacking the maximal-profit problem directly, we prefer a two-step procedure and ask a simpler question first: how can the firm minimize its cost? For the rest of this chapter, we restrict attention to producers of one output, only.

DEFINITION IX.2 (ISOCOST LINE). *For a factor price vector $w \in \mathbb{R}^\ell$, the cost of using the factors of production $x \in \mathbb{R}_+^\ell$ is defined by*

$$w \cdot x.$$

For a specific level of cost $\bar{C}$,

$$\left\{x \in \mathbb{R}_+^\ell : w \cdot x = \bar{C}\right\}$$

is called the isocost line.

EXERCISE IX.1. *Assume two factors of production 1 and 2. Can you tell the slope of the isocost line? Hint: use the household analogy!*

DEFINITION IX.3 (COST-MINIMIZATION PROBLEM). *A firm's cost-minimization problem is a tuple*

$$\Delta = (f, w, y)$$

where f is the production function $\mathbb{R}_+^\ell \to \mathbb{R}_+$, $w \in \mathbb{R}^\ell$ is a vector of factor prices, and $y \in \mathbb{R}_+$ is an element of f's range, the output. The firm's problem is to find the best-response function given by

$$\chi^R(\Delta) := \arg\min_{x \in \mathbb{R}_+^\ell} \{w \cdot x \colon f(x) \geq y\}$$

If $\chi^R(\Delta)$ has one element only, we consider $\chi^R(\Delta)$ an element of $\mathbb{R}_+^\ell$ rather than a subset of $\mathbb{R}_+^\ell$. Depending on the focus, we often write $\chi^R(y)$ or $\chi^R(w, y)$ instead of $\chi^R(f, w, y)$.

Assume that the cost-minimizing problem has a solution. The functions

$$\begin{aligned} C &: \mathbb{R}_+^\ell \times \mathbb{R}_+ \to \mathbb{R}_+, \\ &(w, y) \mapsto C(w, y) = w \cdot \chi^R(w, y) \end{aligned}$$

or

$$\begin{aligned} C &: \mathbb{R}_+ \to \mathbb{R}_+, \\ &y \mapsto C(y) = w \cdot \chi^R(y) \end{aligned}$$

are called cost functions.

The use of the Greek letter χ will become clear in the next subsection.

EXERCISE IX.2. *Fill in the missing term:*

$$\begin{aligned} \chi^R(\Delta) &= \{x \in \mathbb{R}_+^\ell : f(x) \geq y \text{ and,} \\ &\quad \text{for any } x' \in \mathbb{R}_+^\ell,\ f(x') \geq y \Rightarrow w \cdot x' \geq ??\} \end{aligned}$$

EXERCISE IX.3. *Define marginal cost and average cost. Hint: Consult p. 222.*

2.2. A comparison with household theory. The problem of finding a cost-minimizing factor combination turns out to be analogous to the question of expenditure minimization known from household theory (chapter VII). Figure IX.2 juxtaposes expenditure minimization and cost minimization.

EXERCISE IX.4. *Consider the production function f given by $f(x_1, x_2) = x_1 + 2x_2$. Find $C(y)$.*

2.3. Isoclinic factor variation and the graphical derivation of the cost function. At given factor prices, the first-order condition

$$MRTS \overset{!}{=} \frac{w_1}{w_2} \tag{IX.1}$$

makes clear why isoclinic factor variations (change of production factors so as to keep the marginal rate of technical substitution constant) are important. Consider figure IX.3. The upper part of the diagram depicts the cost-minimizing factor combinations in x_1-x_2 space. The lower part shows how to derive the cost curve from this cost-minization curve.

2.4. Cost-minimization and its dual. The cost-minimization problem is depicted in figure IX.4. In order to achieve output level $\overline{y}$, the firm looks out for the isocost line that is closest to the origin. It chooses factor combination B, not the more expensive combination A. Very similar to the development described in chapter VII, we have a dual maximization

	Household theory: expenditure minimization	Theory of the firm: cost minimization
objective function	expenditure $p \cdot x$ (for the consumption of goods x)	expenditure $w \cdot x$ (for the use of factors x)
parameters	prices p, utility $\bar{U}$ (indifference curve)	factor prices w, output y (isoquant)
first-order condition	$MRS \overset{!}{=} \frac{p_1}{p_2}$	$MRTS \overset{!}{=} \frac{w_1}{w_2}$
notation for best bundle(s)	$\chi\left(p, \bar{U}\right)$	$\chi^R\left(w, y\right)$ or $\chi^R\left(y\right)$
name of demand function	Hicksian demand	Hicksian factor demand
minimal value of objective function	$e\left(p, \bar{U}\right)$ $= p \cdot \chi\left(p, \bar{U}\right)$	$C\left(y\right) = C\left(w, y\right)$ $= w \cdot \chi^R\left(w, y\right)$

FIGURE IX.2. Expenditure versus cost

problem: Find the maximal output for a given isocost line. Thus, for expenditure $\bar{C}$, the firm produces at B rather than at C. Take a look at table IX.5 that contrasts household theory and the theory of the firm for the respective maximization problems.

DEFINITION IX.4 (OUTPUT-MAXIMIZATION PROBLEM). *A firm's output-maximization problem is a tuple*

$$\Delta = \left(f, w, \bar{C}\right)$$

where f is the production function $\mathbb{R}^\ell_+ \to \mathbb{R}_+$, $w \in \mathbb{R}^\ell$ is a vector of factor prices, and $\bar{C} \in \mathbb{R}_+$ is expenditure for the factors of production. The firm's best-response function is given by

$$x^R\left(\Delta\right) := \arg\max_{x \in \mathbb{R}^\ell_+} \left\{f\left(x\right) : w \cdot x \leq \bar{C}\right\}.$$

EXERCISE IX.5. *Fill in:*

$$x^R\left(\Delta\right) = \left\{x \in \mathbb{R}^\ell_+ : w \cdot x \leq \bar{C} \text{ and, for any } x' \in \mathbb{R}^\ell_+, w \cdot x' \leq \bar{C} \Rightarrow ...\right\}$$

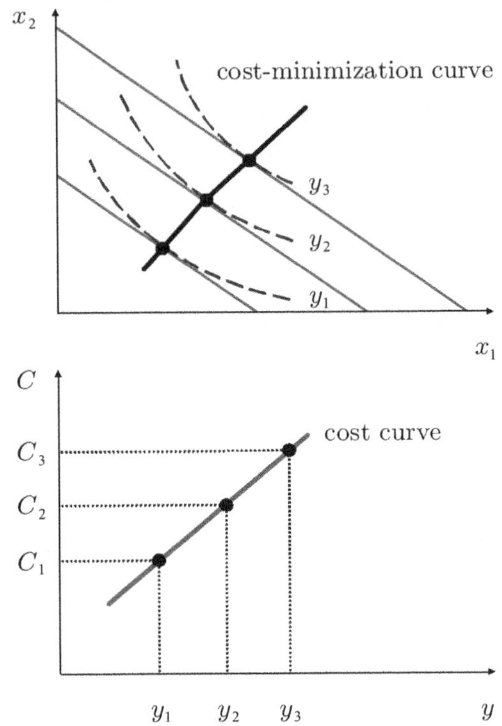

FIGURE IX.3. Deriving the cost curve

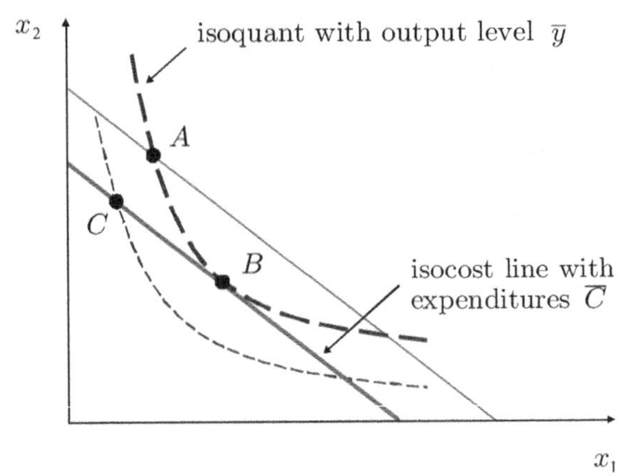

FIGURE IX.4. Cost minimization and output maximization

	Household theory: utility maximization	Theory of the firm: output maximization
objective function	utility function	production function
parameters	prices p, money budget m	factor prices w, cost budget $\bar{C}$
first-order condition	$MRS \overset{!}{=} \frac{p_1}{p_2}$	$MRTS \overset{!}{=} \frac{w_1}{w_2}$
notation for best bundle(s)	$x(p,m)$	$x^R(w,\bar{C})$
name of demand function	Marshallian demand	Marshallian factor demand
maximal value of objective function	$V(p,m)$ $= U(x(p,m))$	$f(x^R(w,\bar{C}))$

FIGURE IX.5. Utility versus output

We can now present the twin theorem of theorem VII.1, p. 177:

THEOREM IX.1. *Let $f:\mathbb{R}^\ell_+ \to \mathbb{R}$ be a continuous production function that obeys local nonsatiation (see p. 215 for the definition) and let $w >> 0$ be a factor-price vector. We then obtain duality in both directions:*

- *If $x^R(\bar{C})$ is the output-maximizing bundle for $\bar{C} > 0$, we have*

$$\chi^R\left(w, f\left(x^R\left(\bar{C}\right)\right)\right) = x^R\left(\bar{C}\right) \tag{IX.2}$$

and

$$C\left(f\left(x^R\left(\bar{C}\right)\right)\right) = \bar{C}. \tag{IX.3}$$

- *If $\chi^R(y)$ is the cost-minimizing bundle for $y > f(0)$, we have*

$$x^R(C(y)) = \chi^R(y) \tag{IX.4}$$

and

$$\max_{x\in\mathbb{R}^\ell_+}\{f(x) : w\cdot x \leq C(y)\} = y. \tag{IX.5}$$

2.5. Main results. In line with chapter VII (pp. 179), we present some important results:

THEOREM IX.2. *Consider a firm with a continuous production function f, cost-minimizing factor combination χ^R and cost function C. We have the following results:*

- *Shephard's lemma: The factor price increase of factor i by one small unit increases the cost by χ_i.*
- *Monotonicity of cost function: In case of local nonsatiation of the production function and strictly positive factor prices, the cost function is monotonic in output.*
- *Concavity of cost function: The cost function is concave in its factor prices.*
- *The Hicksian cross demands are symmetric: $\frac{\partial \chi_i^R(y)}{\partial w_j} = \frac{\partial \chi_j^R(y)}{\partial w_i}$.*
- *Hicksian law of demand: If the price of a factor i increases, the cost-minimizing factor demand χ_i does not increase.*
- *If the production function f is concave, C is a convex function of y.*
- *If the production function f is of constant returns to scale, we have $C(\alpha y) = \alpha C(y)$ for all $\alpha > 0$.*
- *If the production function f is of increasing returns to scale, average costs are a decreasing function of output.*

The first five propositions are basically reformulations of the corresponding household results. The last three propositions link properties of the production function to properties of the cost function. Note that convexity of C implies nondecreasing marginal cost by lemma V.3 (p. 104).

EXERCISE IX.6. *What happens to minimal costs if all factor prices are multiplied by $\alpha > 0$?*

3. Long-run and short-run cost minimization

3.1. Fixed factors and short-run cost function. Some factors of production may be fixed for some time. For example, workers can be released from work (and pay!) only after some time has elapsed. Also, a production hall cannot be downsized easily. We present the definitions and results for two factors of production:

DEFINITION IX.5 (SHORT-RUN COST). *Assume two factors of production 1 and 2 at prices w_1 and w_2. A factor is called fixed at $\bar{x} > 0$ if it cannot be reduced below, or increased above, $\bar{x}$ "in the short run". The long run is the time period after which this factor can be set to any nonnegative value. Assuming that factor 1 is variable (not fixed) and that factor 2 is fixed, the* *short-run cost of using the factor combination* $(x_1, \bar{x}_2)$ *is defined by*

$$w_1 x_1 + w_2 \bar{x}_2.$$

The firm's long-run cost function is the cost function introduced in a previous section. The short-run cost function is denoted by C_s *(where s points to short-run) and given by*

$$C_s(y, \bar{x}_2) := \min_{x_1 \in \mathbb{R}_+} \{w_1 x_1 + w_2 \bar{x}_2 : f(x_1, \bar{x}_2) \geq y\}.$$

This definition is problematic in assuming that exactly $\bar{x}_2$ units of the second factor will be employed. The logic of the above examples would not preclude employment of more than $\bar{x}_2$ units of that factor. The following exercise adheres to the definition of short-run cost.

EXERCISE IX.7. *Consider the production function* f *given by* $f(x_1, x_2) = x_1^{\frac{1}{3}} x_2$ *and the fixed amount of factor* 2, $\bar{x}_2$. *Find the short-run cost function!*

3.2. Fixed and quasi-fixed cost. If factor 2 is fixed at $\bar{x}_2 > 0$, part of the short-run cost is $w_2 \bar{x}_2$. It is called a fixed cost.

DEFINITION IX.6 (FIXED COST). *Let* $C_s : \mathbb{R}_+ \to \mathbb{R}_+$ *be a short-run cost function. We call*

$$F := C_s(0)$$

fixed cost.

$$C_v(y) := C_s(y) - F$$

is called the variable cost of producing y.

Figure IX.6 shows how total cost can be subdivided into fixed and variable cost.

Fixed cost can also be expressed through the production set known from chapter VIII (see figure VIII.3 (a) on p. 213). Expressed in the terminology introduced in that chapter, fixed cost result from the impossibility of inaction. Thus, subfigure (a) stands for the short-run situation where $-\bar{z}_1$ represent sunk cost in terms of good (factor) 1. In contrast, subfigure (b) shows setup costs that are not sunk. This means that inaction is possible (no fixed factors) and $C(0) = 0$ holds. In that case, $w_1(-\bar{z}_1)$ are called quasi-fixed cost.

DEFINITION IX.7 (QUASI-FIXED COST). *Let* $C : \mathbb{R}_+ \to \mathbb{R}_+$ *be a cost function that is not continuous at* 0. *In case of* $C(0) = 0$ *and* $\lim_{y \to 0, y > 0} C(y) > 0$, *we call*

$$F_q := \lim_{\substack{y \to 0, \\ y > 0}} C(y)$$

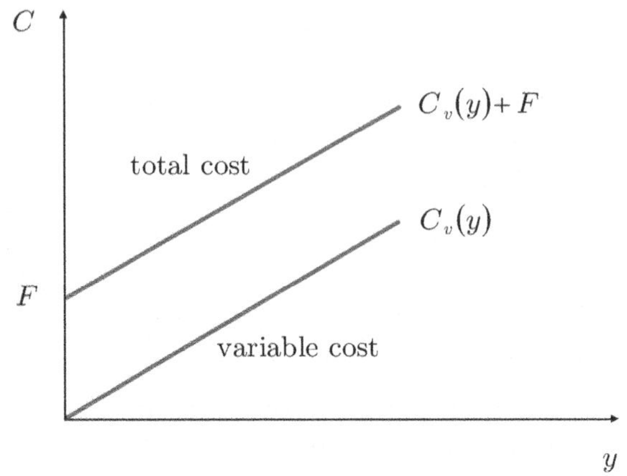

FIGURE IX.6. Fixed and variable cost

quasi-fixed cost.

Thus, quasi-fixed cost can be avoided if nothing is produced. If, however, the firm wants to produce a small quantity, it has to incur the setup cost F_q. Examples are provided by the janitor, who opens the production hall, or by research and development.

4. Profit maximization

4.1. Profit maximization (output space). Once we have found a (short-run or long-run) cost function, we can define profit by

DEFINITION IX.8 (PROFIT (OUTPUT SPACE)). *Let $C : \mathbb{R}_+ \to \mathbb{R}_+$ be a cost function. A firm's profit in output space is defined by*

$$\begin{aligned} \Pi &: \mathbb{R}_+ \to \mathbb{R} \text{ and} \\ \underbrace{\Pi(y)}_{\text{profit}} &:= \underbrace{py}_{\text{revenue}} - \underbrace{C(y)}_{\text{cost}}. \end{aligned}$$

The first-order condition for profit maximization is

$$MC \overset{!}{=} p.$$

DEFINITION IX.9 (SUPPLY FUNCTION). *Let $C : \mathbb{R}_+ \to \mathbb{R}_+$ be a (short-run or long-run) cost function. A firm's supply function is denoted by S and defined by*

$$\begin{aligned} S &: \mathbb{R}_+ \to 2^{\mathbb{R}_+}, \\ & p \mapsto S(p) := \arg\max_{y \in \mathbb{R}_+} \Pi(y). \end{aligned}$$

It is clear from the first-order condition that the inverse supply curve S^{-1} is basically the marginal-cost curve – at least its upward sloping part: By lemma IX.1 on p. 232, the supply curve (or its inverse) is not downward sloping. However, we need to take a closer look.

EXERCISE IX.8. *Consider the short-run cost function C_s given by*

$$C_s(y) := 6y^2 + 15y + 54, y \geq 0$$

and the long-run cost function C defined by

$$C(y) := \begin{cases} 6y^2 + 15y + 54, & y > 0 \\ 0, & y = 0 \end{cases}$$

Determine the short-run supply function S_s and the long-run supply function S. Hint: Use the above first-order condition and compare the profit at the resulting output with the profit at output zero.

Let us now assume that the supply at price p is

- either dictated by the "marginal cost equals price"-rule and thus some output $y^* > 0$
- or equal to zero.

For the short-run supply, we have

$$\begin{aligned} & S_s(p) = y^* \text{ (rather than } S_s(p) = 0) \\ \Leftrightarrow \quad & \Pi_s(y^*) \geq \Pi_s(0) = -F \\ \Leftrightarrow \quad & py^* - C_v(y^*) - F \geq -F \\ \Leftrightarrow \quad & p \geq \frac{C_v(y^*)}{y^*} =: AVC(y^*) \end{aligned}$$

where AVC stands for average variable cost. Turning to long-run supply, we find

$$\begin{aligned} & S(p) = y^* \text{ (rather than } S(p) = 0) \\ \Leftrightarrow \quad & \Pi(y^*) \geq \Pi(0) = 0 \\ \Leftrightarrow \quad & py^* - C(y^*) \geq 0 \\ \Leftrightarrow \quad & p \geq \frac{C(y^*)}{y^*} = AC(y^*). \end{aligned}$$

Thus, we arrive at figure IX.7 which shows the short-run and the long-run supply curves. Note that there is a range of prices at which the firm suffers losses (by $p < AC$) but still prefers to produce a positive quantity because part of the fixed cost is covered (by $p > AVC$).

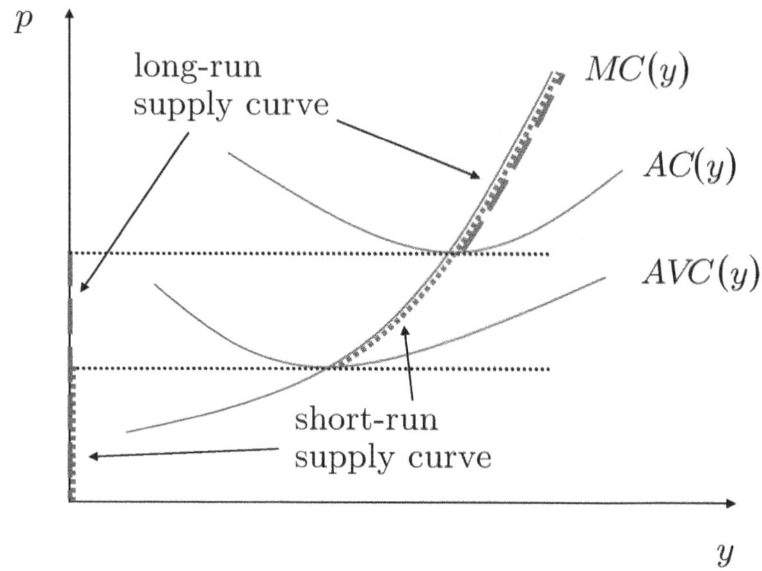

FIGURE IX.7. Short-rund and long-run supply curves

4.2. Producer's rent. We have introduced consumers' rent in chapters VI (pp. 162) and VII (p. 204) and now turn to producers' rent. Let the output price be $\hat{p}$. The producer's rent is the compensating variation $CV(0 \to S(\hat{p}))$ for giving the consumers $S(\hat{p})$ units of quantity at price $\hat{p}$, rather than 0.

DEFINITION IX.10 (PRODUCER'S RENT). *The producer's rent $PR(\hat{p})$ at price $\hat{p}$ is given by*

$$\begin{aligned} PR(\hat{p}) &= CV(0 \to S(\hat{p})) \\ &= \hat{p}S(\hat{p}) - C_v(S(\hat{p})) \\ &= \int_0^{S(\hat{p})} [\hat{p} - MC(X)]\, dX. \end{aligned}$$

Thus, the producer's rent is defined as the difference of revenue and variable cost for quantity $S(\hat{p})$ (see figure IX.8). In the absence of fixed cost, the producer's rent equals profit. However, in the short run, we can have fixed cost. Then, we obtain

$$\begin{aligned} PR(\hat{p}) &= \hat{p}S(\hat{p}) - C_v(S(\hat{p})) \text{ (definition of producer's rent)} \\ &= \hat{p}S(\hat{p}) - [F + C_v(S(\hat{p}))] + F \text{ (adding } 0 = F - F) \\ &= \Pi(S(\hat{p})) + F \text{ (definition of profit).} \end{aligned}$$

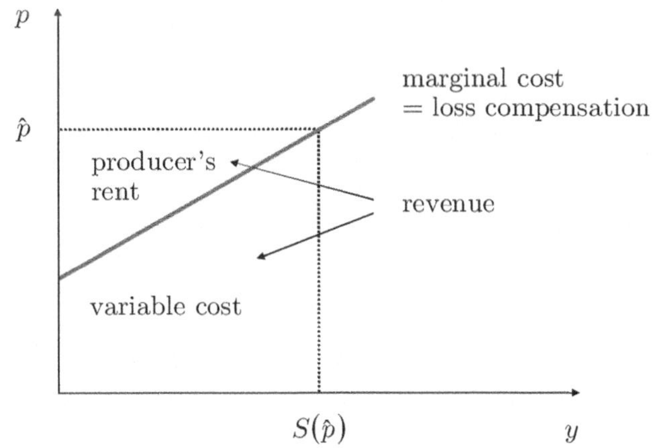

FIGURE IX.8. Producer's rent

4.3. Solution theory, second part. In the chapter on household theory, we have devoted section 4 (pp. 151) to solution theory. We now add some remarks on first-order and second-order conditions that are of general interest and will be applied in the next subsection.

THEOREM IX.3. *Let $f : M \to \mathbb{R}$ be a twice-differentiable real-valued function with open domain $M \subseteq \mathbb{R}^\ell$. Assume some $\hat{x} \in M$ that obeys $f'(\hat{x}) = 0$.*

- *If the Hesse matrix $f''(\hat{x})$ is negative-definite (positive-definite), we have a local maximum (minimum) at $\hat{x}$.*
- *If f is concave (in which case the Hesse matrices $f''(x)$ are negative-semidefinite for all $x \in M$), we have a global maximum at $\hat{x}$.*
- *If f is strictly concave (i.e., the Hesse matrices $f''(x)$ are negative-definite for all $x \in M$), we have a unique global maximum at $\hat{x}$.*

Of course, this theorem holds for $\ell = 1$, too. You obtain propositions about minimization, by considering the second and the third claim above and by substituting concave by convex, negative by positive, and maximum by minimum.

4.4. Profit maximization (input space). Profit is also definable in terms of input factors. Such a definition comes in particularly handy if we do not avail of a cost function.

DEFINITION IX.11 (PROFIT (INPUT SPACE)). *Let* $f : \mathbb{R}_+^\ell \to \mathbb{R}_+$ *be a production function. A firm's profit in input space is defined by*

$$\begin{aligned} \Pi &: \mathbb{R}_+^\ell \to \mathbb{R} \text{ and} \\ \underbrace{\Pi(x)}_{\text{profit}} &: = \underbrace{pf(x)}_{\text{revenue}} - \underbrace{w \cdot x}_{\text{cost}}. \end{aligned}$$

By partial differentiations, we obtain the first-order conditions

$$MVP_i(x) := p\ MP_i = p\frac{\partial f}{\partial x_i} \overset{!}{=} w_i, i = 1, ..., \ell, \tag{IX.6}$$

where MVP_i is factor i's marginal value product. If the firm increases the amount of factor i by one unit, it produces an additional output $\frac{\partial f}{\partial x_i}$ which it can sell at the prevailing market price p. As long as the gain from increasing factor i (MVP_i) exceeds the cost (w_i), the firm is well advised to do so. These are the "marginal value product equals factor price"-rules.

DEFINITION IX.12 (FACTOR DEMAND FUNCTION). *Let* $f : \mathbb{R}_+^\ell \to \mathbb{R}_+$ *be a production function. A firm's factor demand function is denoted by* D *and defined by*

$$\begin{aligned} D &: \mathbb{R}_+^\ell \to \mathbb{R}_+^\ell, \\ w &\mapsto D(w) := \arg\max_{x \in \mathbb{R}_+^\ell} \Pi(x). \end{aligned}$$

We know from lemma IX.1 that demand curves for factors of production are not upward sloping, i.e., production factors are ordinary. The following exercise does not provide a counter example:

EXERCISE IX.9. *Consider the production function* f *given by* $f(x_1, x_2) = x_1^{\frac{1}{3}} x_2$ *and the fixed amount of factor* 2, $\bar{x}_2$. *Find the short-run demand function! How about the long-run demand function? Check the second-order conditions. Show that the firm can achieve any arbitrarily high profit.*

EXERCISE IX.10. *Can you show that profit maximization implies cost minimization? Hint: Divide eq. IX.6 for factor* 1 *by the corresponding equation for factor* 2. *What do you find?*

5. Profit maximization?

5.1. Two problems of our theory of the firm. This chapter is based on the assumption of profit maximization. A not too unfair characterization of a firm is "a production function in search of a profit-maximizing input combination". Indeed, finding the optimal input combination is just an operations-research problem. This is why we can talk about the theory of the firm without mentioning

- principal-agent problems (how can a manager get the workers to work hard?),
- organization (how is the firm subdivided?),
- hierarchy (why are there bosses?, who reports to whom?), or
- delegation (who is allowed to make what kind of decisions?).

Obviously, we have been missing out on some very important aspects of real-world firms. Some of them will be addressed in later chapters.

A second problem with the story told so far is that it takes the existence of firms as given. Economic theory has traditionally been concerned more with markets than with firms. Several high-profile theorems highlight the efficiency of markets. Why, then, do we need firms? A quick answer may

- point to increasing returns to scale realizable by huge organizations,
- stress economies of scope the benefit of which can be reaped by multi-product firms,
- mention "economies of massed reserves" according to which the use of several machines can help to ensure production if, by chance, one of them fails and, finally,
- argue that demand fluctuations are smaller if several markets are served by one (big) firm.

Alas, all these synergies are not a very convincing argument for the existence of firms. After all, the synergies can be realized by contracts between many individual economic units. For example, a janitor firm could offer its service to several firms at the same time. Of course, setting up all these contracts and supervising them may also be a costly procedure. Thus, we may have to compare the costs of different arrangements (which is the subject matter of transaction-cost economics). In any case, we cannot take the existence of firms for granted.

We will not go into a detailed discussion of these problems here. Instead we discuss reasons for which profit maximization is, or is not, a sensible assumption.

5.2. Profit maximization! Arguably, we have a good justification for profit maximization if we can show that it follows from preference theory. Indeed, utility can be expected to increase with a firm's profit if

- the firm is owned by one individual (see subsection 5.3),
- who decides by himself rather than having a manager decide (see chapters XXII and XXIII),
- and without cost to himself (see subsection 5.4),

on how to combine the factors of production

- in a world of certainty
- where prices are not affected by the firm's input-output choice.

If any of these conditions is violated, the assumption of profit maximization becomes dubious. We treat the first and the third point in some detail in the following sections. The second point refers to managers who do not tell the owner all the latter needs to know (asymmetric information) or do not work as hard as the owner would like them to (hidden action). We will take up these problems in later chapters.

The second-to-last item can easily be settled. In case of risk, profit maximization can only mean maximization of expected profit. However, if the individual is not risk neutral, maximization of expected profit is not equivalent to maximization of expected utility (as we know well from chapter V).

Turning to the last point, if the owner consumes goods and offers factors of production, he may be interested in the firm's behavior for reasons other than profit. For example, he may want the firm to produce a lot of the good he plans to buy, or to use factors of production he likes to offer.

5.3. Several owners and risk.

5.3.1. *The consumer-owner economy.* Let us now deal with several owners of a firm.

DEFINITION IX.13 (CONSUMER-OWNER ECONOMY). *A consumer-owner economy is a tuple*

$$\mathcal{E} = \left(N, G, p, y, \left(\theta^i\right)_{i\in N}, \left(\omega^i\right)_{i\in N}, \left(\precsim^i\right)_{i\in N}\right)$$

consisting of

- *the set of consumer-owners* $N = \{1, 2, ..., n\}$,
- *the finite set of goods and factors* $G = \{1, ..., \ell\}$,
- *a price vector* $p \in \mathbb{R}^\ell$,
- *a production plan* $y \in \mathbb{R}^\ell$ *carried out by the firm,*
- *the economy's ownership vector* $\left(\theta^i\right)_{i\in N}$ *where the shares obey* $\theta^i \geq 0$ *for all* $i \in N$ *and* $\sum_{i=1}^n \theta^i = 1$ *holds,*

and for every agent $i \in N$,

- *endowment* $\omega^i \in \mathbb{R}^\ell$, *and*
- *a preference relation* $\precsim^i$.

DEFINITION IX.14 (CONSUMER-OWNER DECISION SITUATION). *Assume a consumer-owner economy* $\mathcal{E}$. *A consumer-owner's decision situation is a tuple*

$$\Delta_i = \left(B\left(p, p\cdot\omega^i + \theta^i p\cdot y\right), \precsim^i\right) \text{ (see p. 136).}$$

Thus, every consumer-owner $i \in N$ with two income sources "consumes" a bundle $x^i \in \mathbb{R}^\ell$ (containing both goods and factors of production) that obeys

$$p \cdot x^i \leq p \cdot \omega^i + \theta^i p \cdot y.$$

We can be sure that all owners want their firm to maximize profit if

- prices are not affected by the firm's input and output choice,
- there is no risk involved (otherwise, different beliefs or risk attitudes may come into play), and
- managers are fully controllable by owners.

We will now elaborate on the second point (no risk). Before presenting a concrete example, we need to introduce the concept of complete financial markets.

5.3.2. *Arrow securities.* In the "Further exercises" section of chapter V, we introduce Arrow securities.

DEFINITION IX.15 (ARROW SECURITY). *Let $W = \{w_1, ..., w_m\}$ be a set of m states of the world. The contingent good $i \in \{1, .., m\}$ that pays one Euro in case of state of the world w_i and nothing in other states is called an Arrow security.*

EXERCISE IX.11. *Consider figure IX.9 and find owner A's willingness to pay for the Arrow security* 1. *Hint: Calculate the expected gain from one unit of Arrow security* 1.

If we are lucky, there are as many Arrow securities as there are states of the world.

DEFINITION IX.16 (COMPLETE FINANCIAL MARKETS). *Let a set of m states of the world $W = \{w_1, ..., w_m\}$ be given. If, for each state of the world w_i, an Arrow security i can be bought and sold for some given price p_i, we say that financial markets are complete.*

We now argue that the prices of all Arrow securities sum to 1 in equilibrium (which we did not, so far, define formally). Assume, to the contrary, that $p_1 + p_2 < 1$ holds for $m = 2$. Then, an agent could buy both Arrow securities and obtain a secure payoff of 1 (either w_1 or w_2 happens) for an investment below 1. In that case, nobody would be prepared to offer Arrow securities. In case of $p_1 + p_2 > 1$, selling Arrow securities is profitable, while buying is not. It should also be clear that an Arrow security cannot have a negative price.

	w_1 (return 80)	w_2 (return 110)	expected value of investment
A's prob. distribution	$\frac{2}{10}$	$\frac{8}{10}$	$\frac{2}{10} \cdot 80 + \frac{8}{10} \cdot 110 - 100 = 4$
B's prob. distribution	$\frac{1}{2}$	$\frac{1}{2}$	$\frac{1}{2} \cdot 80 + \frac{1}{2} \cdot 110 - 100 = -5$

FIGURE IX.9. Different probability distributions for returns on investment

DEFINITION IX.17 (CORRECTLY PRICED ARROW SECURITY). *Let $W = \{w_1, ..., w_m\}$ be a set of m states of the world with complete financial markets. The Arrow securities are said to be priced correctly if*

$$\sum_{i=1}^{m} p_i = 1$$

holds.

Thus, the prices of Arrow securities share the properties of probability distributions.

5.3.3. *An instructive example.* Taking up the topic of differing risk assessments or differing risk attitudes, assume two agents A and B who possess a firm. Following Gravelle & Rees (1992, p. 175), the agents consider an investment with cost 100 that may yield a return of 80 (state of the world w_1) or of 110 (state of the world w_2). The agent's probability distributions differ (see figure IX.9). If both agents are risk neutral, agent A prefers to carry out the investment, while agent B opposes.

It is not difficult to concoct an example where different opinions arise from different risk attitudes rather than different beliefs. Interestingly, both problems can be solved by Arrow securities. Indeed, we can show that the profitability of an investment can be evaluated with the Arrow prices in case of complete financial markets. Beliefs and risk attitudes have no role to play. In terms of the above example, the two owners just check whether

$$p_1 80 + p_2 110 - 100 > 0 \tag{IX.7}$$

holds in which case the investment is to be carried out. If (!) p_1 and p_2 were probabilities, this inequality amounts to checking whether expected profit is positive. We need to show that both owners should agree with this criterion. They consider the investment package:

- spend 100 Euros on the investment,
- buy 20 units of Arrow good 1, and

- sell 10 units of Arrow good 2.

In state of the world 1, the owners obtain

$$\underbrace{80}_{\substack{\text{revenue}\\\text{from}\\\text{investment}}} + \underbrace{20}_{\substack{\text{revenue}\\\text{from}\\\text{Arrow good 1}}} = 100,$$

in state of the world 2, they get

$$\underbrace{110}_{\substack{\text{revenue}\\\text{from}\\\text{investment}}} + \underbrace{-10}_{\substack{\text{revenue}\\\text{from}\\\text{Arrow good 2}}} = 100.$$

Therefore, the two owners are indifferent between the two states. The Arrow securities take all the risk away from them. In state of the world 1, the investment package is profitable iff the following condition holds:

$$\begin{aligned} 0 &< \underbrace{80}_{\substack{\text{revenue}\\\text{from}\\\text{investment}}} - \underbrace{(100 + p_1 20 - p_2 10)}_{\substack{\text{cost of}\\\text{investment}\\\text{package}}} + \underbrace{20}_{\substack{\text{revenue}\\\text{from}\\\text{Arrow good 1}}} \\ &= -p_1 20 + p_2 10 \\ &= -p_1 20 + p_2 10 - 100\underbrace{(1 - p_1 - p_2)}_{0} \\ &= p_1 80 + p_2 110 - 100 \end{aligned}$$

which is nothing but inequality IX.7.

EXERCISE IX.12. *Show that the same inequality is also obtained in state of the world* 2.

Thus, with the help of complete financial markets, beliefs and risk attitudes can be made irrelevant for owners who need to assess the profitability of an investment.

5.4. The cost of managing one's own firm.

5.4.1. *Utility maximization by the owner-manager.* Following Gravelle & Rees (2004, pp. 159), we now assume that the owner manages the firm himself. The owner-manager's effort level and cost of effort is denoted by e, the firm's gross profit is $\Pi(e)$. Thus, the manager's work is not remunerated, while the payment for the other factors of production is taken care of by Π.

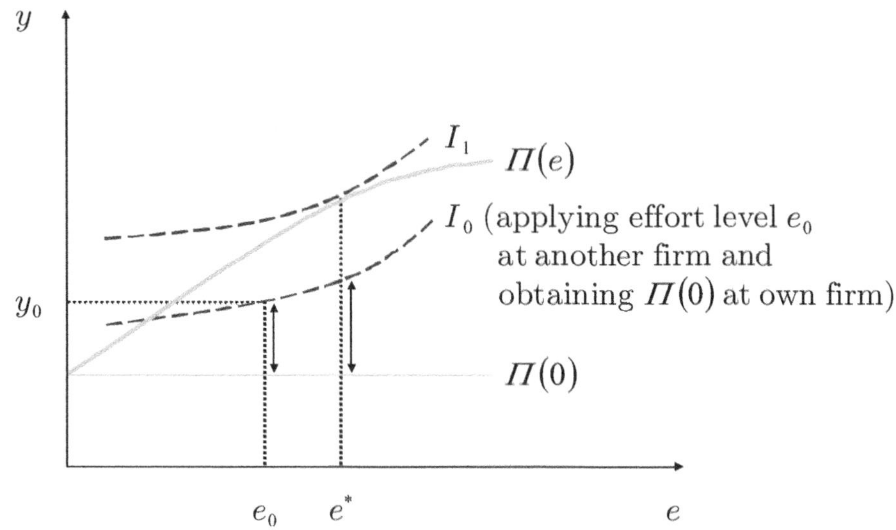

FIGURE IX.10. Utility maximization by the owner-manager

For the manager, profit is a good and effort a bad. Thus, the indifference curves in figure IX.10 (adapted from Gravelle & Rees (2004, p. 159)) are upward-sloping. Furthermore, preferences are convex so that additional units of effort have to be compensating for by ever increasing additional monetary income. Thus, for the time being, we assume that the owner does not pay himself. The owner aims for the highest indifference curve touching the profit curve and chooses effort level e^*.

The owner-manager's preferences are representable by a utility function U so that e^* is given by

$$e^* = \arg\max_{e\geq 0} U\left(e, \Pi\left(e\right)\right).$$

Algebraically, the condition "slope of indifference curve I equals slope of profit curve" is

$$\left.\frac{d\Pi}{de}\right|_{e^*} \overset{!}{=} -\frac{\left.\frac{\partial U}{\partial e}\right|_{(e^*,I(e^*))}}{\left.\frac{\partial U}{\partial y}\right|_{(e^*,I(e^*))}}.$$

Here, I is an indifference curve and also a function of e. For example, we have $I_0\left(e_0\right) = y_0$. The reader is invited to compare pp. 83.

5.4.2. *Does utility maximization imply profit maximization?* We now come back to the original question of whether firms maximize profit. It is obvious from the previous section, that the owner-manager does not maximize profit Π (we have $\left.\frac{d\Pi}{de}\right|_{e^*} > 0$ in figure IX.10). However, it can be argued that the suitable profit concept should be net profit

$$\underbrace{n(e)}_{\text{net profit}} := \underbrace{\Pi(e)}_{\text{gross profit}} - \underbrace{w(e)}_{\text{factor payment}}$$

where w is the wage paid to the manager at effort level e.

Reconsider figure IX.10. We assume that the owner-manager does not work in his own firm, but realizes effort level e_0 in another firm and therefore obtains the salary $w(e_0)$. He also obtains $n(0) = \Pi(0)$ from his own firm. Then, $(e_0, y_0 = w(e^0) + \Pi(0))$ is a realizable point. It lies on the indifference curve denoted by I_0. Thus, if the owner worked in his own firm, his wage at effort level e_0 would have to be at least $y_0 - \Pi(0)$. If the owner worked at some other effort level in his own firm, the opportunity cost is "height of indifference curve I_0 $-$ $\Pi(0)$". This is the amount that the owner-manager should pay himself for working in his own firm. This difference is indicated by the line with two arrowheads above e_0. The wage at e^* is indicated by the longer line with two arrowheads. In this fashion, we define a wage function

$$e \mapsto w(e) := I_0(e) - \Pi(0).$$

The firm's marginal opportunity cost of the manager's effort level is therefore equal to the slope of indifference curve I_0 :

$$\frac{dw(e)}{de} = \frac{dI_0(e)}{de} = -\frac{\left.\frac{\partial U}{\partial e}\right|_{(e,I_0(e))}}{\left.\frac{\partial U}{\partial y}\right|_{(e,I_0(e))}}.$$

We now come back to the above definition of net profit. If the firm wants to maximize net profit $n(e)$, it does so by forming the derivative with respect to e, i.e., by

$$\left.\frac{d\Pi}{de}\right|_{\hat{e}} \overset{!}{=} -\frac{\left.\frac{\partial U}{\partial e}\right|_{(\hat{e},I(\hat{e}))}}{\left.\frac{\partial U}{\partial y}\right|_{(\hat{e},I(\hat{e}))}}$$

Thus, utility maximization means finding the effort level where the slope of the indifference curve I_1 equals the slope of the profit function (at e^*). In contrast, in order to find the profit-maximizing effort level $\hat{e}$, the slope of the indifference curve I_0, rather than I_1, has to be used. Only if the two indifference curves happen to have the same slope at e^*, can we expect $\hat{e} = e^*$. For example, this would be true for the quasi-linear utility function U given by $U(e, y) = -\ln(e) + y$, but not in general. Intuitively, as a business man gets richer, he spends more time playing golf and chess. Such a business man is not a profit maximizer with respect to the firm he owns and manages.

5.4.3. *A market for manager effort.* We now show that a functioning market for management input prevents the problems described in the previous section. Let e be the manager effort offered by our owner-manager and let e_f be the effort level used in the owner's firm. The owner-manager can buy additional effort above the effort supplied by himself ($e < e_f$) or can offer his excess effort on the market ($e > e_f$). In any case, the firm's profit function is

$$n(e_f) = \Pi(e_f) - we_f,$$

where w is the constant wage for managers. The owner-manager's income is

$$y = n(e_f) + we$$

so that the decision about the own effort and the decision about the effort used in the own firm are separated.

The profit-maximizing effort level is

$$e_f^* := \arg\max_{e_f \geq 0} [\Pi(e_f) - we_f]$$

while the utility-maximizing effort level obeys

$$(e^*, \widetilde{e_f}) = \arg\max_{e \geq 0, e_f \geq 0} U(e, \Pi(e_f) - we_f + we).$$

The first-order condition for profit maximization is

$$\left.\frac{d\Pi}{de_f}\right|_{e_f^*} \stackrel{!}{=} w,$$

while utility maximization demands

$$\begin{aligned}
&\left.\frac{\partial U}{\partial e}\right|_{(e^*,e_f)} + \left.\frac{\partial U}{\partial y}\right|_{(e^*,e_f)} w \stackrel{!}{=} 0 \text{ (differentiating w.r.t. } e\text{) and,}\\
&\left.\frac{\partial U}{\partial y}\right|_{(e,\widetilde{e_f})} \left.\frac{d(\Pi(e_f) - we_f + we)}{de_f}\right|_{(e,\widetilde{e_f})}\\
= &\left.\frac{\partial U}{\partial y}\right|_{(e,\widetilde{e_f})} \left(\left.\frac{d\Pi}{de_f}\right|_{\widetilde{e_f}} - w\right) \stackrel{!}{=} 0 \text{ (differentiating w.r.t. } e_f\text{)}
\end{aligned}$$

The last equality shows that utility maximization implies profit maximization because $\left.\frac{d\Pi}{de_f}\right|_{e_f^*} \stackrel{!}{=} w$ (profit maximization) and $\left.\frac{d\Pi}{de_f}\right|_{\widetilde{e_f}} - w \stackrel{!}{=} 0$ (utility maximization) point to the same level of manager efforted exerted in the owner's firm. The owner's income is $\Pi(e_f) - we_f + we$ so that his utility is not maximized if profit $\Pi(e_f) - we_f$ is not maximized.

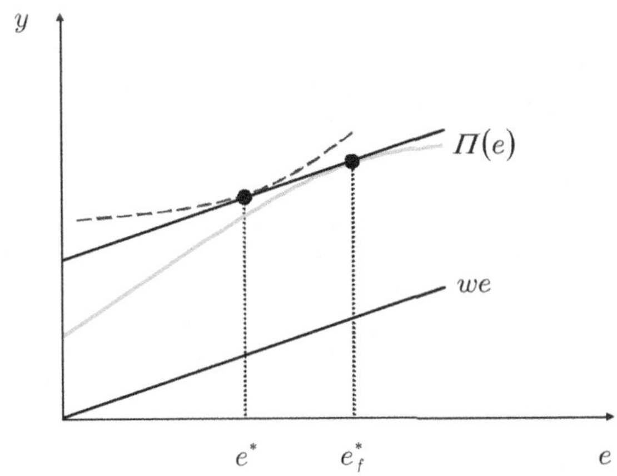

FIGURE IX.11. Separation made possible by a market for manager effort

Figure IX.11 illustrates. The wage for managers equals the slope of the wage line we. Profit maximization determines the effort level $e_f^* = \widetilde{e_f}$ where the slope of the wage line equals the slope of the gross-profit curve. In the figure, the owner-manager works in his own firm and buys additional manager effort $e_f^* - e^*$ on the market. Can you draw a picture for the opposite case?

6. The separation function of markets

We have come to know several examples where markets allow a separation:

- If markets for consumption goods exist, a household with an endowment can consume his endowment but also any other bundle which does not cost more (chapter VI).
- If a market for manager effort is available, an owner-manager can buy additional (on top of his own) manager effort for his firm or supply effort to other firms (this chapter, pp. 249).
- If Arrow securities are correctly priced, the profitability of an investment decision can be assessed separately from the beliefs and risk attitudes of the several owners (this chapter, pp. 246).
- International trade allows an economy to consume a bundle different from the bundle produced.

In all these cases, a higher indifference curve becomes available which shows an important welfare-characteristic of markets.

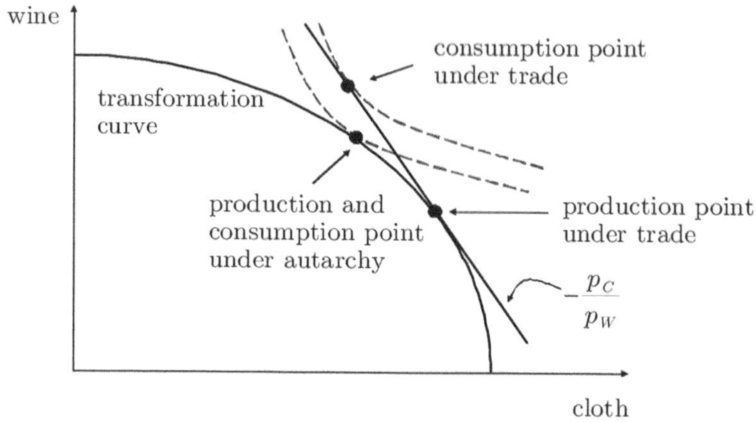

FIGURE IX.12. International trade versus autarky

The example of international trade needs some explanation. Consider the economy-wide transformation curve (for an explanation, see pp. 216) depicted in figure IX.12. The economy produces cloth (C) and wine (W). Under autarchy, the economy produces and consumes the very same bundle. If international trade becomes possible, at prices p_C and p_W, the economy can produce a bundle different from the one it consumes. In our example, the economy exports cloth and imports wine. Indeed, the production decision is separated not only from consumption but also from the community preferences. Optimally, the economy produces where the value of the production

$$p_C C + p_W W$$

is maximal.

7. Topics

The main topics in this chapter are

- cost minimization
- isocost line
- fixed cost
- quasi-fixed cost
- variable cost
- long-run cost minimization
- short-run cost minimization
- supply function
- profit maximization
- input space, output space
- revealed profit
- international trade
- Arrow security
- owner-manager
- effort
- utility versus profit maximization
- producer's rent

8. Solutions

Exercise IX.1

Along an isocost line, we have $\frac{dx_2}{dx_1} = -\frac{w_1}{w_2}$.

Exercise IX.2

A factor combination $x \in \mathbb{R}_+^\ell$ is cost minimizing if any other factor combination x' that also serves to produce y or more is at least as expensive as x:

$$\begin{aligned} \chi^R(\Delta) &= \Big\{x \in \mathbb{R}_+^\ell \colon f(x) \geq y \text{ and,} \\ &\qquad \text{for any } x' \in \mathbb{R}_+^\ell,\ f(x') \geq y \Rightarrow w \cdot x' \geq w \cdot x\Big\} \end{aligned}$$

Exercise IX.3

Marginal cost is defined by $MC(y) = \frac{dC}{dy}$, average cost by $AC(y) = \frac{C(y)}{y}$.

Exercise IX.4

We obtain

$$\chi^R(w,y) = \begin{cases} \left(0, \frac{y}{2}\right), & w_1 > \frac{1}{2}w_2 \\ \left\{\left(x_1, \frac{y-x_1}{2}\right) \in \mathbb{R}_+^2 : x_1 \in [0,y]\right\}, & w_1 = \frac{1}{2}w_2 \\ (y,0), & w_1 < \frac{1}{2}w_2 \end{cases}$$

and

$$\begin{aligned} C(y) &= \min_{x_1+2x_2=y} w_1x_1 + w_2x_2 \\ &= w_1\chi_1 + w_2\chi_2 \\ &= \begin{cases} w_2\frac{y}{2}, & w_1 \geq \frac{1}{2}w_2 \\ w_1 y, & w_1 < \frac{1}{2}w_2 \end{cases} \end{aligned}$$

Exercise IX.5

A factor combination $x \in \mathbb{R}_+^\ell$ is output maximizing if any other factor combination x' that also cost $\bar{C}$ or less produces $f(x)$ or less:

$$\begin{aligned} x^R(\Delta) = &\left\{x \in \mathbb{R}_+^\ell \colon w \cdot x \leq \bar{C} \text{ and,} \right. \\ &\left. \text{for any } x' \in \mathbb{R}_+^\ell, w \cdot x' \leq \bar{C} \Rightarrow f(x') \leq f(x)\right\} \end{aligned}$$

Exercise IX.6

If all factor prices are multiplied by $\alpha > 0$, cost is also multiplied by this factor:

$$\begin{aligned} C(\alpha w, y) &= \min_{x \in \mathbb{R}_+^\ell} \{\alpha w \cdot x \colon f(x) \geq y\} \\ &= \alpha \min_{x \in \mathbb{R}_+^\ell} \{w \cdot x \colon f(x) \geq y\} \\ &= \alpha C(w,y). \end{aligned}$$

Exercise IX.7

We obtain

$$\begin{aligned} C_s(y, \bar{x}_2) &= \min_{x_1 \in \mathbb{R}_+} \left\{w_1x_1 + w_2\bar{x}_2 \colon x_1^{\frac{1}{3}}\bar{x}_2 \geq y\right\} \\ &= \min_{x_1 \in \mathbb{R}_+} \left\{w_1x_1 + w_2\bar{x}_2 \colon x_1 \geq \frac{y^3}{\bar{x}_2^3}\right\} \\ &= w_1\frac{y^3}{\bar{x}_2^3} + w_2\bar{x}_2. \end{aligned}$$

Exercise IX.8

If the firm does not produce ($y = 0$), we find $\Pi_s(0) = 0 - C_s(0) = -54$ in the short run and $\Pi(0) = 0 - C(0) = 0$ in the long run. The first-order condition is

$$MC(y) = 12y + 15 \stackrel{!}{=} p$$

or

$$y \overset{!}{=} \frac{p-15}{12}.$$

In case of $y > 0$, we obtain the profit

$$\begin{aligned}\Pi_s(y) &= \Pi(y) = (12y+15)\,y - \left(6y^2+15y+54\right) \\ &= 6y^2 - 54 \geq \begin{cases} \Pi_s(0) = -54, & y \geq 0 \ (p \geq 15) \\ \Pi(0) = 0, & y \geq 3 \ (p \geq 51) \end{cases}\end{aligned}$$

and hence the short-run supply function

$$S_s(p) = \begin{cases} \frac{p-15}{12}, & p \geq 15 \\ 0, & p < 15 \end{cases}$$

and the long-run supply function

$$S(p) = \begin{cases} \frac{p-15}{12}, & p \geq 51 \\ 0, & p < 51 \end{cases}$$

Exercise IX.9

The first-order conditions are

$$\begin{aligned} p\frac{\partial f}{\partial x_1} &= \frac{1}{3} p x_1^{-\frac{2}{3}} x_2 \overset{!}{=} w_1 \text{ and} \\ p\frac{\partial f}{\partial x_2} &= p x_1^{\frac{1}{3}} \overset{!}{=} w_2 \text{ (if applicable).} \end{aligned}$$

Thus, the short-run demand function is

$$D_s(w) = \left(\left(\frac{1}{3} \frac{p\bar{x}_2}{w_1} \right)^{\frac{3}{2}}, \bar{x}_2 \right)$$

and the long-run demand function is obtained by simultaneously solving the two first-order equations for x_1 and x_2. The second one yields $x_1 \overset{!}{=} \frac{w_2^3}{p^3}$ and, substituting into the first, we obtain

$$D(w) = \left(\frac{w_2^3}{p^3}, 3w_1 \frac{w_2^2}{p^3} \right).$$

However, the comparative-statics results are implausible for the long-run demand function: Factor demand depends positively on factor prices and negatively on the output price, i.e., the comparative-statics results contradict lemma IX.1.

Indeed, a solution to the factor-demand problem need not exist because the second-order condition (Hesse matrix negative-semidefinite) is not fullfilled for the profit function

$$\Pi(x_1, x_2) = p x_1^{\frac{1}{3}} x_2 - w_1 x_1 - w_2 x_2.$$

The Hesse matrix is given by

$$\begin{aligned}
\Pi''(x) &= \begin{pmatrix} \frac{\partial^2 \Pi(x)}{(\partial x_1)^2} & \frac{\partial^2 \Pi(x)}{\partial x_1 \partial x_2} \\ \frac{\partial^2 \Pi(x)}{\partial x_2 \partial x_1} & \frac{\partial^2 \Pi(x)}{(\partial x_2)^2} \end{pmatrix} \\
&= \begin{pmatrix} \left(\frac{1}{3}-1\right)\frac{1}{3} p x_1^{\frac{1}{3}-2} x_2 & \frac{1}{3} p x_1^{\frac{1}{3}-1} \\ \frac{1}{3} p x_1^{\frac{1}{3}-1} & 0 \end{pmatrix} \\
&= \begin{pmatrix} -\frac{2}{9} p \frac{x_2}{x_1^{\frac{5}{3}}} & \frac{1}{3} \frac{p}{x_1^{\frac{2}{3}}} \\ \frac{1}{3} \frac{p}{x_1^{\frac{2}{3}}} & 0 \end{pmatrix}
\end{aligned}$$

and is *not* negative-semidefinite which can be seen from

$$\begin{aligned}
& (z_1, z_2) \begin{pmatrix} -\frac{2}{9} p \frac{x_2}{x_1^{\frac{5}{3}}} & \frac{1}{3} \frac{p}{x_1^{\frac{2}{3}}} \\ \frac{1}{3} \frac{p}{x_1^{\frac{2}{3}}} & 0 \end{pmatrix} \begin{pmatrix} z_1 \\ z_2 \end{pmatrix} \\
= \; & (z_1, z_2) \begin{pmatrix} -\frac{2}{9} p \frac{x_2}{x_1^{\frac{5}{3}}} z_1 + \frac{1}{3} \frac{p}{x_1^{\frac{2}{3}}} z_2 \\ \frac{1}{3} \frac{p}{x_1^{\frac{2}{3}}} z_1 \end{pmatrix} \\
= \; & -\frac{2}{9} p \frac{x_2}{x_1^{\frac{5}{3}}} z_1^2 + \frac{2}{3} \frac{p}{x_1^{\frac{2}{3}}} z_1 z_2
\end{aligned}$$

which is greater than zero for any $z_1 := 1$ and sufficiently large z_2.

Thus, the second-order condition is not fulfilled so that a maximum is not guaranteed. Indeed, we do not have a maximum because

$$\begin{aligned}
\Pi(x_1, x_2) &= p x_1^{\frac{1}{3}} x_2 - w_1 x_1 - w_2 x_2 \\
&= \left(p x_1^{\frac{1}{3}} - w_2\right) x_2 - w_1 x_1
\end{aligned}$$

can be made arbitrarily large by making x_1 so large that $p x_1^{\frac{1}{3}} - w_2$ is positive and by choosing x_2 sufficiently large.

Exercise IX.10

The suggested division yields

$$\frac{p\, MP_1}{p\, MP_2} \overset{!}{=} \frac{w_1}{w_2}$$

and hence the well-known (see p. IX.1) first-order condition for cost minimization

$$MRTS = \frac{MP_1}{MP_2} \overset{!}{=} \frac{w_1}{w_2}.$$

Exercise IX.11

Owner A's expected gain from one unit of Arrow security 1 is

$$\frac{2}{10} \cdot 1 + \frac{8}{10} \cdot 0 = \frac{2}{10}.$$

Therefore, he is indifferent between buying and not buying at price $p_1 = \frac{2}{10}$.

Exercise IX.12

In state of the world 2, the investment package is profitable if

$$\underbrace{110}_{\substack{\text{revenue}\\ \text{from}\\ \text{investment}}} - \underbrace{(100 + p_1 20 - p_2 10)}_{\substack{\text{cost of}\\ \text{investment}\\ \text{package}}} - \underbrace{10}_{\substack{\text{payment}\\ \text{to buyers of}\\ \text{Arrow good 2}}} = -p_1 20 + p_2 10 > 0$$

which is the same inequality as in the main text.

9. Further exercises without solutions

PROBLEM IX.1. A firm has two factories A and B that obey the production functions $f_A(x_1, x_2) = x_1 \cdot x_2$ and $f_B(x_1, x_2) = x_1 + x_2$, respectively. Given the factor-price ratio $w = \frac{w_1}{w_2}$, how to distribute output in order to minimize costs? Hints:

- You are free to assume $w_1 \leq w_2$.
- Find the cost functions for the two factories.
- Are the marginal-cost curves upward-sloping or downward-sloping?

PROBLEM IX.2. Discuss: In the long run, all factors of production are variable and therefore, $C(0) = 0$.

PROBLEM IX.3. Prominent examples for economies of scope are provided by

- sheep that yield both milk and wool or
- corn with the two subproducts flour and straw.

Working with a multi-product cost function, can you suggest a definition of economies of scope?

PROBLEM IX.4. Suppose a profit-maximizing firm has a technology with free disposal. What happens if we have factors/goods with negative prices?

PROBLEM IX.5. Draw the companion figure of figure IX.11 (p. 253) where the owner of a firm works in his own firm and also in another firm.

Part C

Games and industrial organization

The third part of our course in advance microeconomics deals with the basics of monopoly and oligopoly theory – important applications of noncooperative game theory. The Cournot, Bertrand, and Stackelberg models are the workhorse models in IO (industrial organization). We will repeatedly refer to these models when discussing competition and regulatory issues in part F. The reader interested in a broader view of industrial organization and competition theory is invited to throw a glance or two at chapter XXI.

This part consists of four chapters. In chapter X, some game theory is presented: bimatrix games, dominant and mixed strategies, and Nash equilibrium. In chapter XI, we put these concepts to work and analyze price competition (Bertrand) and quantity competition (Cournot) in one-stage models. In order to tackle models of several stages, we deal with games in extensive form in chapter XII where we dwell on the Stackelberg model. In that chapter, we also introduce the important class of multi-stage games, together with a worked-out example of variety competition. A specific subclass of multi-stage games concerns repeated games, the subject matter of chapter XIII. They are relevant for discussing the possibility of cartel formation.

CHAPTER X

Games in strategic form

Game theory is a central piece of microeconomic theory. There are only very few chapters in this book that do not use game theory. In fact, even the decision theory presented in the first part of this book provides preparatory work for game theory. We present strategic games in this chapter by building on chapter II (decisions in strategic form) while chapter III (decisions in extensive form) prepares the ground for extensive-form games.

The aim of game theory is to describe a multiperson decision situation and then to suggest "solutions". Solutions are strategy combinations or sets of strategy combinations for which we find theoretical arguments. These arguments typically point to stability of some sort. In this chapter, we propose solutions by way of dominance arguments and Nash equilibria.

1. Introduction, examples, and definition

1.1. Nobel prices in Game theory. The Sveriges Riksbank Prize in Economic Sciences in Memory of Alfred Nobel was awarded to game theorists several times, most notably in 1994 and in 2005. In 1994, it was awarded to the US economists John C. Harsanyi (University of California, Berkeley), John F. Nash (Princeton University), and to the German Reinhard Selten (Rheinische Friedrich-Wilhelms-Universität, Bonn) "for their pioneering analysis of equilibria in the theory of non-cooperative games." According to the press release by the Royal Swedish Academy of Sciences,

> Game theory emanates from studies of games such as chess or poker. Everyone knows that in these games, players have to think ahead - devise a strategy based on expected countermoves from the other player(s). Such strategic interaction also characterizes many economic situations, and game theory has therefore proved to be very useful in economic analysis.
>
> The foundations for using game theory in economics were introduced in a monumental study by John von Neumann

H. Wiese, *Advanced Microeconomics*,
https://doi.org/10.1007/978-3-658-34959-2_10

> and Oskar Morgenstern entitled Theory of Games and Economic Behavior (1944). Today, 50 years later, game theory has become a dominant tool for analyzing economic issues. In particular, non-cooperative game theory, i.e., the branch of game theory which excludes binding agreements, has had great impact on economic research. The principal aspect of this theory is the concept of equilibrium, which is used to make predictions about the outcome of strategic interaction. John F. Nash, Reinhard Selten and John C. Harsanyi are three researchers who have made eminent contributions to this type of equilibrium analysis.
>
> John F. Nash introduced the distinction between cooperative games, in which binding agreements can be made, and non-cooperative games, where binding agreements are not feasible. Nash developed an equilibrium concept for non-cooperative games that later came to be called Nash equilibrium.
>
> Reinhard Selten was the first to refine the Nash equilibrium concept for analyzing dynamic strategic interaction. He has also applied these refined concepts to analyses of competition with only a few sellers.
>
> John C. Harsanyi showed how games of incomplete information can be analyzed, thereby providing a theoretical foundation for a lively field of research - the economics of information - which focuses on strategic situations where different agents do not know each others' objectives.
>
> [...] As far back as the early nineteenth century, beginning with Auguste Cournot in 1838, economists have developed methods for studying strategic interaction. But these methods focused on specific situations and, for a long time, no overall method existed. The game-theoretic approach now offers a general toolbox for analyzing strategic interaction.

A second time, the prize was awared in 2005, to Robert J. Aumann (Hebrew University of Jerusalem) and Thomas C. Schelling (University of Maryland, USA) "for having enhanced our understanding of conflict and cooperation through game-theory analysis." The press release issued by the Royal Swedish Academy of Sciences mentions both price and atomic wars as an application of game theory:

> [...] Against the backdrop of the nuclear arms race in the late 1950s, Thomas Schelling's book The Strategy of Conflict set forth his vision of game theory as a unifying framework for the social sciences. Schelling showed that a party can strengthen its position by overtly worsening its own options, that the capability to retaliate can be more useful than the ability to resist an attack, and that uncertain retaliation is more credible and more efficient than certain retaliation. These insights have proven to be of great relevance for conflict resolution and efforts to avoid war.
>
> Schelling's work prompted new developments in game theory and accelerated its use and application throughout the social sciences. Notably, his analysis of strategic commitments has explained a wide range of phenomena, from the competitive strategies of firms to the delegation of political decision power.
>
> [...] Robert Aumann was the first to conduct a full-fledged formal analysis of so-called infinitely repeated games. His research identified exactly what outcomes can be upheld over time in long-run relations.
>
> The theory of repeated games enhances our understanding of the prerequisites for cooperation: Why it is more difficult when there are many participants, when they interact infrequently, when interaction is likely to be broken off, when the time horizon is short or when others' actions cannot be clearly observed. Insights into these issues help explain economic conflicts such as price wars and trade wars, as well as why some communities are more successful than others in managing common-pool resources. The repeated-games approach clarifies the raison d'être of many institutions, ranging from merchant guilds and organized crime to wage negotiations and international trade agreements.

All the names mentioned in these press releases are important in this book:

- In this chapter, we will encounter Nash and his existence theorem.
- Selten's contribution is the subject matter of chapter XII on extensive games.
- Harsanyi is rightly famous for his work on Bayesian games that we will explain in chapter XVII.

- Aumann's (and others') work on repeated games is summarized in chapter XIII. Aumann's concept of correlated equilibria is explained in chapter XVII.
- In our view, Cournot has to be credited with founding non-cooperative game theory. We cover his oligopoly theory in the next chapter.

1.2. Some simple bimatrix games. Before formally defining strategic games in subsection 1.3, we have a look at some simple and prominent examples. We first consider the stag hunt:

hunter 1 \ hunter 2	stag	hare
stag	$5,5$	$0,4$
hare	$4,0$	$4,4$

The first number in each field indicates the payoff for player 1 (hunter 1) and the second number is the payoff for player 2 (hunter 2). The two hunters cannot hunt both animals at the same time but have to focus on one of them. Stag hunting requires the communal effort, while the hare can be brought down by a single hunter. Hunter 1 is willing to take part in the stag hunt if hunter 2 also engages in stag hunting. If hunter 2 chases the hare, hunter 1 would just waste his effort trying to catch the stag.

The stag hunt is an instance of a *bimatrix* game. Indeed, we need *two* matrixes to describe the payoffs of the two players – hence the term *bi*-matrix.

Our second example of a bimatrix game is called "matching pennies" or "head or tail". Two players have the choice between head or tail. Player 1 wins one Euro if both choose head or both choose tail. Player 2 wins if their choices differ. We obtain the following game matrix:

player 1 \ player 2	head	tail
head	$1,-1$	$-1,1$
tail	$-1,1$	$1,-1$

The third game is the battle of the sexes. A couple argues about how to spend the evening. Player 1, she, would like to go to the theatre, player

2, he, prefers football. However, both have a preference for spending the evening together. The following matrix captures these preferences:

		he	
		theatre	football
she	theatre	$4,3$	$2,2$
	football	$1,1$	$3,4$

Another famous game is called the game of chicken. Two car drivers head towards each other. Their strategies are "continue" and "swerve". Whoever swerves is a chicken (a coward) and obtains a low payoff. However, continuing is a risky business:

		driver 2	
		continue	swerve
driver 1	continue	$0,0$	$4,2$
	swerve	$2,4$	$3,3$

We finally consider the following game:

		player 2	
		deny	confess
player 1	deny	$4,4$	$0,5$
	confess	$5,0$	$1,1$

It is known as the "prisoners' dilemma". Two prisoners sit in separate cells. They either confess their crime or deny it. If both confess, they will have to go to jail for some moderate time (payoff 1). If one confesses while the other denies, the first will obtain favorable terms (payoff 5), while the second will go to jail for a long time (payoff 0). If both deny, punishment will be small, due to the problem to prove them guilty (payoff 4). Sometimes, the first strategy is called cooperative where the cooperation refers to the other criminal, not the court.

1.3. Definition of a game in strategic form. Bimatrix games (you have seen some examples in the previous subsection) belong to the class of games in strategic form.

DEFINITION X.1 (GAME IN STRATEGIC FORM). *A game in strategic form is a triple*

$$\Gamma = \left(N, (S_i)_{i \in N}, (u_i)_{i \in N}\right) = (N, S, u),$$

where

- $N = \{1, ..., n\}$ *is a nonempty and finite set of* $n := |N|$,
- S_i *is the strategy set for player* $i \in N$,
- $S = \times_{i \in N} S_i$ *is the Cartesian product of all the players' strategy sets with elements* $s = (s_1, s_2, ..., s_n) \in S$, *and*
- $u_i : S \to \mathbb{R}$ *is player i's payoff function.*

Elements of strategy sets are called strategies and elements of S are called strategy combinations.

For example, the "battle of the sexes" has $N = \{she, he\}$, $S_{she} = S_{he} = \{\text{theatre, football}\}$, and the bimatrix reveals that u_{she} is given by

$$\begin{aligned} u_{she}(\text{theatre, theatre}) &= 4, \\ u_{she}(\text{theatre, football}) &= 2, \\ u_{she}(\text{football, theatre}) &= 1, \\ u_{she}(\text{football, football}) &= 3. \end{aligned}$$

A strategy combination $s \in S$ is a n-tuple containing a strategy for every player. Removing player i's strategy leads to a strategy combination s_{-i} of the remaining players from $N \backslash \{i\}$:

$$s_{-i} \in S_{-i} := \times_{\substack{j \in N, \\ j \neq i}} S_j.$$

We can now write player i's payoff as $u_i(s) = u_i(s_i, s_{-i})$ which will come in handy soon.

Note our abuse of notation:

- $S = (S_i)_{i \in N}$ is a tuple of strategy sets that we use to describe the game while
- $S = \times_{i \in N} S_i$ is the set of strategy combinations, the domain for the utility functions.

2. Dominance

2.1. Definition. As in chapter II, dominance means that a strategy is better than the others. Again, we distinguish between (weak) dominance and strict dominance:

DEFINITION X.2 (DOMINANCE). *Let (N, S, u) be a game in strategic form and let i be from N. Strategy $s_i \in S_i$ (weakly) dominates strategy $s_i' \in S_i$ if $u_i(s_i, s_{-i}) \geq u_i(s_i', s_{-i})$ holds for all $s_{-i} \in S_{-i}$ and $u_i(s_i, s_{-i}) > u_i(s_i', s_{-i})$ is true for at least one $s_{-i} \in S_{-i}$. Strategy $s_i \in S_i$ strictly dominates strategy $s_i' \in S_i$ if $u_i(s_i, s_{-i}) > u_i(s_i', s_{-i})$ holds for all $s_{-i} \in S_{-i}$. Then, strategy s_i' is called (weakly) dominated or strictly dominated. A strategy that dominates every other strategy is called dominant (weakly or strictly, whatever the case may be).*

We will spare the reader the corresponding definition for rationalizability. Some games are solvable by dominance (or rationalizability) arguments, applied once or several times. We present some examples.

2.2. The prisoners' dilemma. On p. 267 we present the prisoners' dilemma. It is easy to see that both players have a strictly dominant strategy, "confess". It may seem that the solution is very obvious and the game a boring one. Far from it! The prisoners' dilemma is one of the most widely discussed two-person games in economics, philosophy, and sociology. Although the choice is obvious for each individual player, the overall result is unsatisfactory, viz., Pareto inferior. It is possible to make everybody better off: $u_i(\text{confess}, \text{confess}) = 1 < 4 = u_i(\text{deny}, \text{deny})$ for $i = 1, 2$.

This contradiction between individual rationality (choose the dominant strategy!) and collective rationality (avoid Pareto-inferior payoffs!) is not atypical. For example, tax payers like others to pay for public expenditures while shying away from the burden themselves. However, they (often) prefer the paying of taxes by everybody to nobody paying taxes.

As another example consider two firms with the strategies of charging a high price or a low price:

		firm 2 high	**firm 2** low
firm 1	high	$4, 4$	$0, 5$
	low	$5, 0$	$1, 1$

A high price is the nice, cooperative strategy and corresponds to the strategy of denying in the original game.

In the literature, there are attempts to soften the dilemma. For example, it has been suggested that players promise each other to stick to the cooperative strategy (to choose "deny"). However, such a promise does not help; confess is still a dominant strategy.

The twin argument is also popular and builds on the symmetry of the prisoners' dilemma game. According to the twin argument, the players (should) deliberate as follows: Whatever reason leads one player to choose strategy 1 over strategy 2, is also valid for the other player. Therefore, the players' choice reduces to the diagonal payoffs and they choose the cooperative strategy by $4 > 1$. The problem with this argument is that players choose simultaneously and independently. The choice by one player does not influence the other player's choice of strategy.

EXERCISE X.1. *Is the stag hunt solvable by dominance arguments? How about head or tail, game of chicken, or the battle of the sexes.*

2.3. The second-price auction. William Vickrey is one of the two 1996 Nobel price winners in economics. He is famous for his analysis of auctions. Consider an object for which several players make simultaneous bids. The Vickrey auction is a second-price auction, i.e., the highest bidder obtains the object but pays the second-highest price, only. We will show that every bidder has a dominant strategy.

We restrict attention to two players or bidders, $i = 1, 2$, but the argument basically applies for any finite number of players. We assume that every player $i \in \{1, 2\}$ has a reservation price for the object, r_i. He is indifferent between obtaining the object for a price r_i or not obtaining it. The strategy of player i, s_i, is simply the Euro value bidder i writes on a piece of paper and hands over to the auctioneer. The auctioneer gives the object to the player with the highest bid. If the bidders happen to quote the same value, the auctioneer tosses a fair coin. Thus, bidder 1's payoff function is given by

$$u_1(s_1, s_2) = \begin{cases} 0, & s_1 < s_2, \\ \frac{1}{2}(r_1 - s_2), & s_1 = s_2, \\ r_1 - s_2, & s_1 > s_2 \end{cases}$$

We now show that the strategy $s_1 := r_1$ dominates all the others. We distinguish three cases: $r_1 < s_2$, $r_1 = s_2$, and $r_1 > s_2$.

(1) $r_1 < s_2$

If player 1's reservation price is below player 2's bid, $s_1 = r_1$ leads to payoff 0 for player 1 who does not obtain the object. Player

1 does not receive the object and still obtains payoff zero for any bid $s_1 < s_2$, be the bid below or above r_1. If, however, player 1's bid fulfills $s_1 \geq s_2$, he obtains the object and a negative payoff of $r_1 - s_2 < 0$ or $\frac{1}{2}(r_1 - s_2) < 0$.

(2) $r_1 = s_2$

If player 1's reservation price happens to be equal to player 2's bid, $s_1 = r_1$ leads to the expected payoff of $\frac{1}{2}(r_1 - s_2) = 0$. Understating ($s_1 < r_1$) or overstating ($s_1 > r_1$) affects the chance of getting the object but not the expected payoff which remains at 0.

EXERCISE X.2. *Show that $s_1 = r_1$ is a dominant strategy in case of $r_1 > s_2$.*

Summing up, the auction game due to Vickrey is dominance solvable. Every player has a dominant strategy.

2.4. Take it or leave it. Sometimes, dominance is enough to solve a game. In other cases, we need to apply dominance several times to come to a definite strategy combination. The key idea is to delete dominated strategies. We present two examples, a very simple bargaining game and the Basu game. The most simple bargaining game has player 1 make an offer that player 2 can accept or reject. That's it. Player 2 is not allowed a counteroffer. We will see that this bargaining game gives all the bargaining power to player 1. Indeed, if player 1 knows player 2's reservation price, he can "exploit" player 2.

		player 2 accepts if he is offered at least x_2 coins				player 2 does not accept
		0	1	2	3	
player 1	0	$3, 0$	$0, 0$	$0, 0$	$0, 0$	$0, 0$
offers	1	$2, 1$	$2, 1$	$0, 0$	$0, 0$	$0, 0$
player 2	2	$1, 2$	$1, 2$	$1, 2$	$0, 0$	$0, 0$
x_1 coins	3	$0, 3$	$0, 3$	$0, 3$	$0, 3$	$0, 0$

Assume that the players need to divide three coins of Euros that cannot be subdivided. Player 1's strategy is to quote the number x_1 of coins he offers player 2. Player 2 fixes the minimum number x_2 of coins acceptable to him as seen in the matrix above. If player 2 is rational, he accepts any offer which leads to a payment strictly above 0. He is indifferent between declining or accepting an offer of 0. Thus, player 2 will not choose any of the last three strategies. Deleting these strategies leads to the following payoff matrix:

		player 2 accepts if he is offered at least x_2 coins	
		0	1
player 1	0	$3,0$	$0,0$
offers	1	$2,1$	$2,1$
player 2	2	$1,2$	$1,2$
x_1 coins	3	$0,3$	$0,3$

Now, after this deletion, player 1 finds that the strategies of offering 2 or 3 coins are dominated by the strategy of offering just 1 coin. Let us delete player 1's last two strategies to obtain:

		player 2 accepts if he is offered at least x_2 coins	
		0	1
player 1	0	$3,0$	$0,0$
offers	1	$2,1$	$2,1$

This game is not reducible any more. Player 1 can guarantee two coins for himself by offering one coin to player 2.

2.5. The Basu game (the insurance game). Basu (1994) presents an interesting matrix game which severely questions iterative dominance as a solution concept. Basu concocts this story: Two travelers to a remote island buy an "antique" and "unique" object that is very valuable according to the vendor. The airline smashes both objects that happen to be identical. The travelers turn to the airline company for compensation. The manager of the airline cannot estimate the real worth of the objects and offers the

		Traveler 2 requests so many coins						
		2	3	4	⋯	98	99	100
Trav. 1	2	2, 2	4, 0	4, 0	4, 0	4, 0	4, 0	4, 0
	3	0, 4	3, 3	5, 1	5, 1	5, 1	(5, 1)	5, 1
requests	4	0, 4	1, 5	4, 4	6, 2	6, 2	6, 2	6, 2
	⋮	0, 4	1, 5	2, 6				
so many	98	0, 4	1, 5	2, 6		98, 98	100, 96	100, 96
	99	0, 4	1, 5	2, 6		96, 100	99, 99	101, 97
coins	100	0, 4	1, 5	2, 6		96, 100	97, 101	100, 100

FIGURE X.1. The Basu game

following compensation rule. Both travelers note the worth of the object on a piece of paper. Integer numbers between 2 (two thousand Euro) and 100 (hundred thousand Euro) are allowed. We denote the numbers by s_1 and s_2 for the two travelers 1 and 2, respectively. The manger assumes the lower figure to be more trustworthy. Both obtain the lower figure. However, in order to give the travelers an incentive for honesty, the traveler with the lowest figure obtains this figure +2 while the other obtains the lowest figure -2. Player 1's payoff is

$$u_1(s_1, s_2) = \begin{cases} s_1 + 2, & \text{if} \quad s_1 < s_2, \\ s_1, & \text{if} \quad s_1 = s_2, \\ s_2 - 2, & \text{if} \quad s_1 > s_2; \end{cases}$$

The bimatrix of this game is indicated by the matrix of figure X.1.

EXERCISE X.3. *Find a strategy that dominates another.*

EXERCISE X.4. *Consider the reduced Basu game where players can choose 2 or 3 coins, only. Have you seen it (in general terms) before?*

Deleting the dominated strategies one after the other, 100 for both players first, then 99 for both players etc., the strategies $s_1 = 2$ and $s_2 = 2$ remain, a depressing result from the travelers' points of view. Since many people intuitively feel that the players should be able to do much better, this result has been termed a paradox, the Basu paradox. In chapter XI, we will consider the price-setting behavior of firms and will encounter the so-called Bertrand paradox, a close cousin to the Basu paradox. The Basu paradox poses a serious challenge to the iterative deletion of dominated strategies.

Two other important examples are treated in later chapters. The Clarke-Groves mechanism is dominance solvable and dealt with in chapter XVIII. The Cournot-Dyopol can be solved by iterative rationalizability; it is treated in chapter XI.

3. Best responses and Nash equilibria

3.1. Definition of the Nash equilibrium. For a game in strategic form, a strategy combination is a Nash equilibrium if no player, by himself, can do any better:

DEFINITION X.3 (NASH EQUILIBRIUM). *Let*

$$\Gamma = (N, S, u)$$

be a strategic-form game. The strategy combination

$$s^* = (s_1^*, s_2^*, ..., s_n^*) \in S$$

is a Nash equilibrium (or simply: an equilibrium) if for all i from N

$$u_i\left(s_i^*, s_{-i}^*\right) \geq u_i\left(s_i, s_{-i}^*\right)$$

holds for all s_i from S_i.

At a Nash equilibrium, no player $i \in N$ has an incentive to deviate unilaterally, i.e., to choose another strategy if the other players choose s_{-i}^*.

3.2. Matrix games and best responses. The prisoners' dilemma is dominance solvable. Indeed, any strategy combination consisting of dominant strategies is a Nash equilibrium. If a strategy is best against all strategy combinations of the other players, it is also best against a particular strategy combination of the other players.

EXERCISE X.5. *Determine the equilibria of the following game:*

		player 2	
		left	*right*
player 1	*up*	4, 4	4, 4
	down	0, 0	4, 4

EXERCISE X.6. *Find the equilibrium or the equilibria of the Basu game above.*

We have a simple procedure that helps determine all the equilibria in matrix games. For every row (strategy chosen by player 1) we determine the best strategy or the best strategies for player 2 and mark the corresponding field or fields by a "2". You remember the [R]-procedure introduced in chapter II? For every column, we find the best strategy or the best strategies for player 1 and put a "1" into the corresponding field. Any field with two marks points to an equilibrium strategy combination. No player wins by deviating unilaterally. The stag hunt provides an example of two equilibria:

		hunter 2	
		stag	hare
hunter 1	stag	5, 5 [1][2]	0, 4
	hare	4, 0	4, 4 [1][2]

EXERCISE X.7. *Using the marking technique, determine the Nash equilibria of the following three games:*

		player 2	
		left	*right*
player 1	*up*	1, −1	−1, 1
	down	−1, 1	1, −1

	player 2	
	left	*right*
up	4, 4	0, 5
down	5, 0	1, 1

	player 2	
	left	*right*
up	1, 1	1, 1
down	1, 1	0, 0

The marking technique highlights the best strategies a player can choose.

DEFINITION X.4 (BEST RESPONSE). *Let*

$$\Gamma = (N, S, u)$$

be a strategic-form game and i a player from N. The function $s_i^R : S_{-i} \to 2^{S_i}$ is called a best-response function (a best response, a best answer) for player $i \in N$ if s_i^R is given by

$$s_i^R(s_{-i}) := \arg\max_{s_i \in S_i} u_i(s_i, s_{-i})$$

EXERCISE X.8. *Use best-response functions to characterize $(s_1^*, s_2^*, ..., s_n^*)$ as a Nash equilibrium.*

4. ... for mixed strategies, also

4.1. Introductory remarks. There are games where it is important not to let the other players know your strategy. "Head or tail" is a clear example. In this game, players do not choose "pure" strategies such as head or tail, but probability distributions on the set of these pure strategies. A probability distribution on the set of pure strategies is called a mixed strategy. For example, choosing head with probability $\frac{1}{3}$ and tail with probability $\frac{2}{3}$ is a mixed strategy.

Apart from this literal interpretation of choosing probability distributions, two other interpretations come to mind. According to the first, the player does not actually randomize. His choice depends on information available to him but not observable by the other player. From the point of view of the other player it looks as if a mixed strategy was played when, in fact, it was not. We formalize this idea in chapter XVII (pp. 451). The second interpretation assumes a multitude of players programmed to play a certain pure strategy. Some players are chosen by chance and play against each other. For the individual player it seems as if he is playing against an opponent who mixes his pure strategies according to the shares of the population programmed on these pure strategies.

Games with only pure strategies do not need to have an equilibrium. Extending the strategy sets to include probability distributions on the set of pure strategies removes this problem (if it is one).

4.2. Definitions.

DEFINITION X.5 (MIXED STRATEGY). *Let S_i be a finite strategy set for player $i \in N$. A mixed strategy σ_i for player i is a probability distribution* *on* S_i, *i.e., we have*

$$\sigma_i\left(s_i^j\right) \geq 0 \text{ for all } j = 1, ..., |S_i|$$

and

$$\sum_{j=1}^{|S_i|} \sigma_i\left(s_i^j\right) = 1.$$

The set of player i*'s mixed strategies is denoted by* Σ_i *and we have* $\Sigma := \times_{i \in N} \Sigma_i$ *and* $\Sigma_{-i} := \times_{j \in N, j \neq i} \Sigma_j$ *for the combination of mixed strategies of all players or of all players except player* i*, respectively.* $\sigma_i \in \Sigma_i$ *is called a properly mixed strategy if there is no pure strategy* $s_i \in S_i$ *with* $\sigma_i(s_i) = 1$.

As long as the order of pure strategies is clear, we can also write

$$\sigma_i = \left(\sigma_i\left(s_i^1\right), \sigma_i\left(s_i^2\right), ..., \sigma_i\left(s_i^{|S_i|}\right)\right)$$

The first pure strategy of player i, s_i^1, is identified with the mixed strategy

$$(1, 0, 0, ..., 0).$$

An element from Σ is an n-tuple

$$\sigma = (\sigma_1, \sigma_2, ..., \sigma_n),$$

while an element from Σ_{-i} is an $n-1$-tuple

$$\sigma_{-i} = (\sigma_1, \sigma_2, ..., \sigma_{i-1}, \sigma_{i+1}, ..., \sigma_n).$$

With a little exercise, it is not difficult to calculate the expected payoff.

EXERCISE X.9. *Consider the battle of the sexes (p. 267) and calculate the expected payoff for player 1 if player 1 chooses theatre with probability* $\frac{1}{2}$ *and player 2 chooses theatre with probability* $\frac{1}{3}$*!*

DEFINITION X.6 (PAYOFF FOR MIXED STRATEGY). *The payoff for a mixed strategy is the mean of the payoffs for the pure strategies:*

$$u_i(\sigma_i, \sigma_{-i}) = \sum_{j=1}^{|S_i|} \sigma_i\left(s_i^j\right) u_i\left(s_i^j, \sigma_{-i}\right) \tag{X.1}$$

The definition of the mixed-strategies Nash equilibrium is very familiar:

DEFINITION X.7 (MIXED-STRATEGY NASH EQUILIBRIUM). *Let*

$$\Gamma = (N, S, u)$$

be a strategic-form game. The strategy combination

$$\sigma^* = (\sigma_1^*, \sigma_2^*, ..., \sigma_n^*) \in \Sigma$$

is a Nash equilibrium (or simply: an equilibrium) in mixed strategies if for all i from N

$$u_i\left(\sigma_i^*, \sigma_{-i}^*\right) \geq u_i\left(\sigma_i, \sigma_{-i}^*\right)$$

holds for all σ_i *from* Σ_i.

Alternatively, we could say that $(\sigma_1^*, \sigma_2^*, ..., \sigma_n^*)$ is a Nash equilibrium if $\sigma_i^* \in \sigma_i^R\left(\sigma_{-i}^*\right)$ holds for all $i \in N$.

4.3. Equilibria. "Head or tail" does not have an equilibrium in pure strategies. However, understood as a game in mixed strategies, "head or tail" has an equilibrium. We will see that every matrix game has at least one such equilibrium.

How do we know that "Head or tail" has an equilibrium in mixed strategies? We remind the reader of equation X.1: the payoff for a mixed strategy is the mean of the payoffs for the pure strategies. The reader can also go back to proposition II.1 on p. 21. Assume that an equilibrium in mixed strategies exists. Then, player 1 (or any other player) has no incentive to choose any other strategy. In particular, the payoffs for the pure strategies "head" or "tail" cannot be higher than the payoff for the mixed strategy in question. Therefore, a mixed strategy can only be part of an equilibrium if every pure strategy with a probability above zero has the same payoff as the mixed strategy. As a consequence, every mixed strategy that mixes best pure strategies is a best mixed strategy.

We now turn to "head or tail". If (!) a best properly mixed strategy for player 1 exists, we have $u_1(\text{head}, \sigma_2) = u_1(\text{tail}, \sigma_2)$ which leads to

$$\sigma_2(\text{head}) \cdot 1 + \sigma_2(\text{tail}) \cdot (-1) = \sigma_2(\text{head}) \cdot (-1) + \sigma_2(\text{tail}) \cdot 1$$

By $\sigma_2(\text{head}) + \sigma_2(\text{tail}) = 1$, we obtain

$$\sigma_2(\text{head}) = \frac{1}{2} = \sigma_2(\text{tail})$$

in a Nash equilibrium. This is a peculiar result: player 2 has to mix his strategies in a certain manner so that player 1 is indifferent between his pure strategies. Interchanging players 1 and 2 leads to $\sigma_1(\text{head}) = \frac{1}{2} = \sigma_1(\text{tail})$ so that player 2 is indifferent between his pure strategies. If both players are indifferent between their pure strategies, they are indifferent between any mixed strategies so that a mixed-strategy equilibrium has been found.

There can be no other equilibrium in properly mixed strategies. For example, $\sigma_2(\text{head}) > \frac{1}{2}$ leads to $\sigma_1^R(\sigma_2) = \text{head}$. Indeed, if player 2 chooses "head" with a sufficiently high probability, "head" by player 1 is the unique best response. And, given that player 1 chooses "head" (with probability 1), player 2 should choose "tail" with probability 1 contradicting $\sigma_2(\text{head}) > \frac{1}{2}$.

There exist several manners to find equilibria. My favorite one is this. You find out when head is at least as good as tail for player 1:

$$\begin{aligned} & u_1(\text{head}, \sigma_2) \geq u_1(\text{tail}, \sigma_2) \\ \Leftrightarrow \quad & \sigma_2 \cdot 1 + (1-\sigma_2) \cdot (-1) \geq \sigma_2 \cdot (-1) + (1-\sigma_2) \cdot 1 \\ \Leftrightarrow \quad & 2\sigma_2 - 1 \geq -2\sigma_2 + 1 \\ \Leftrightarrow \quad & \sigma_2 \geq \frac{1}{2} \end{aligned}$$

Apparently, player 1 weakly prefers to choose head if player 2 chooses head with a probability of at least 1/2. If $\sigma_2 = \frac{1}{2}$, head and tail yield the same expected payoff for player 1. Then, any mixture between head and tail is optimal:

$$\sigma_1^R(\sigma_2) = \begin{cases} 1, & \sigma_2 > \frac{1}{2} \\ [0,1], & \sigma_2 = \frac{1}{2} \\ 0, & \sigma_2 < \frac{1}{2} \end{cases}$$

Then, after finding player 2's reaction function in a similar manner, the equilibria are given by the points of intersection.

This procedure can also be applied for other 2×2 games. But, alternatively, one might pursue a more standard method. Consider player 1 in the battle of the sexes. His payoff function is given by

$$u_1(\sigma_1, \sigma_2) = 4\sigma_1\sigma_2 + 2\sigma_1(1-\sigma_2) + (1-\sigma_1)\sigma_2 + 3(1-\sigma_1)(1-\sigma_2)$$

where σ_1 is player 1's (her) probability for theatre and σ_2 is player 2's (his) probability for theatre. We consider the derivative of u_1 with respect to σ_1 to find her best response:

$$\begin{aligned} \frac{\partial u_1}{\partial \sigma_1} &= 4\sigma_2 + 2(1-\sigma_2) - \sigma_2 - 3(1-\sigma_2) \\ &= 4\sigma_2 - 1 \begin{cases} < 0, & \sigma_2 < \frac{1}{4} \\ = 0, & \sigma_2 = \frac{1}{4} \\ > 0, & \sigma_2 > \frac{1}{4} \end{cases} \end{aligned}$$

leads to

$$\sigma_1^R(\sigma_2) = \begin{cases} 0, & \sigma_2 < \frac{1}{4} \\ [0,1], & \sigma_2 = \frac{1}{4} \\ 1, & \sigma_2 > \frac{1}{4} \end{cases}$$

Again, she is indifferent if he mixes in a particular manner. The best-response functions can be drawn in a σ_1-σ_2-diagram. The intersection of best-response functions marks the equilibrium. Can you do it yourself?

EXERCISE X.10. *Find all the equilibria in pure and properly mixed strategies for the four games shown in figure X.2. Please draw the best responses for each game! Hint: a good way to find equilibria is by starting with the inequality* $u_1(up, \sigma_2) \geq u_1(down, \sigma_2)$. *After solving for* σ_2 *(the probability for left), one can find player* 1*'s reaction function.*

		player 2	
		left	right
player 1	up	$5,5$	$0,4$
	down	$4,0$	$4,4$

		player 2	
		left	right
player 1	up	$1,1$	$1,1$
	down	$1,1$	$0,0$

		left	right
player 1	up	$4,3$	$2,2$
	down	$1,1$	$3,4$

		left	right
player 1	up	$4,4$	$0,5$
	down	$5,0$	$1,1$

FIGURE X.2. Four strategic games

4.4. The police game. The players of the police game are the police (tax administration, environmental agency, police department) and an agent (a potential tax evader, a prospective environmental criminal, a car driver) who has to be monitored. This game is from Rasmusen (2001, S. 81f.).

According to the game matrix in figure X.3, the police officer obtains a utility of 4 if she can deter or detect the misdoing. However, she incurs control costs of C with $0 < C < 4$. The potential criminal (you!) has a utility of 1 if he is not caught. Otherwise, he has to suffer a punishment of $F > 1$. (Why is a punishment below 1 ineffective?)

EXERCISE X.11. *Find all the pure-strategy equilibria!*

In order to find the mixed-strategy equilibrium, we assume that the agent commits a crime with probability σ_a. In a properly-mixed equilibrium, σ_a has to be chosen so that the police officer is indifferent between "control" and "no control":

$$\sigma_a (4 - C) + (1 - \sigma_a)(4 - C) \stackrel{!}{=} \sigma_a \cdot 0 + (1 - \sigma_a) 4 \quad \Leftrightarrow \quad \sigma_a \stackrel{!}{=} \frac{1}{4} C.$$

		agent	
		fraud	no fraud
police officer	control	$4-C, 1-F$	$4-C, 0$
	no control	$0, 1$	$4, 0$

FIGURE X.3. The police game

EXERCISE X.12. *Which controlling probability σ_p chosen by the police officer makes the agent indifferent between committing and not committing the crime?*

In equilibrium, the payoffs are

$$u_p = \frac{1}{F}(4-C) + \left(1-\frac{1}{F}\right)\frac{1}{4}C \cdot 0 + \left(1-\frac{1}{F}\right)\left(1-\frac{1}{4}C\right)4 = 4-C$$

for the police and

$$u_a = \frac{1}{4}C\frac{1}{F}(1-F) + \frac{1}{4}C\left(1-\frac{1}{F}\right)1 + \left(1-\frac{1}{4}C\right)\cdot 0 = 0$$

for the prospective criminal. Do you see that we could have found $4-C$ and 0 as equilibrium (!) payoffs also in a different manner?

5. Existence and number of mixed-strategy equilibria

5.1. Number. In this section, we offer two theoretical remarks. The first deals with the number of equilibria we may expect and the second tackles the existence problem. Both theorems work for finite games, only. According to our definition, a game in strategic form (p. 268) has a finite set of players, $|N| < \infty$. A finite strategic game has finite strategy sets, too:

DEFINITION X.8 (FINITE GAME). *A game in strategic form $\Gamma = (N, S, u)$ is called finite if $|S| < \infty$ holds (or $|S_i| < \infty$ for all $i \in N$).*

The number of strategy combinations can be found easily:

$$|S| = |S_1| \cdot |S_2| \cdot \ldots \cdot |S_n|$$

A matrix game $\Gamma = (N, S, u)$ is defined by the payoffs of all players for all strategy combinations. Since for every strategy combination we need the payoff information for every player, a point in $\mathbb{R}^{n|S|}$ defines a finite matrix game $\Gamma = (N, S, u)$. $\mathbb{R}^{n|S|}$ is the set of all vectors with $n \cdot |S|$ real-valued entries.

EXERCISE X.13. *How many payoffs do we need to describe a game for two players with two and three strategies, respectively.*

Wilson (1971) shows that "nearly all" finite strategic games have a finite and odd number of equilibria. Indeed, many of our examples had 1 or 3 equilibria.

THEOREM X.1 (NUMBER OF NASH EQUILIBRIA). *Nearly all finite strategic games have a finite and odd number of equilibria in mixed strategies.*

What does "nearly all" mean? Using the terminology of "distance and balls" (see p. 60), we can make precise what the theorem says:

- If you have a game Γ^* in $\mathbb{R}^{n|S|}$ with a finite and odd number of equilibria,
  - you can jiggle all or some entries of that point (some or all payoffs) by a little, or, alternatively,
  - you can find an ε-ball K with center Γ^* and take any game within that ball

  and you have found another point (another game) with the same number of equilibria.
- If you have a point Γ^* in $\mathbb{R}^{n|S|}$ with an infinite or even number of equilibria,
  - you can find another game very close to this point, or, alternatively,
  - any ε-ball K with center Γ^* contains at least one game

  so that this other game has a finite and odd number of equilibria.

This is a remarkable theorem. As always, the proof of the pudding is in the eating.

EXERCISE X.14. *How many equilibria do the following games have?*

		player 2 *left*	*right*
pl. 1	*up*	1, 1	1, 1
	down	1, 1	0, 0

		player 2 *left*	*right*
pl. 1	*up*	1, 1	0, 0
	down	0, 0	0, 0

Neither game fits the "finite and odd"-bill. From Wilson's theorem, we know that we need to jiggle but a little to obtain games with a finite and odd number of equilibria. We begin with the second game (taken from Fudenberg & Tirole 1991, S. 480) and vary it in the following manner:

		player 2 left	player 2 right
player 1	up	$1,1$	α,α
	down	$0,0$	β,β

where the α and β can be very small. Confirm:

- For $0 < \alpha < 1$ and $\beta \leq 0$, we obtain a game with one equilibrium.
- For $\alpha = 0$ and $\beta < 0$, we also find one equilibrium.
- For $\alpha = 0$ and $\beta > 0$, we have three equilibria.

You can show the effect of a small jiggle for the first game of the above exercise:

EXERCISE X.15. *How many equilibria has the following game for* $\varepsilon > 0$ *and how many for* $\varepsilon < 0$*?*

		player 2 left	player 2 right
player 1	up	$1+\varepsilon, 1+\varepsilon$	$1,1$
	down	$1,1$	$0,0$

5.2. Existence. The existence theorem is due to John Nash (1950, 1951):

THEOREM X.2 (EXISTENCE OF NASH EQUILIBRIA). *Any finite strategic game* $\Gamma = (N, S, u)$ *has a mixed-strategy Nash equilibrium.*

The proof is relegated to chapter XIX (pp. 507) where the formal apparatus needed is ready.

6. Critical reflections on game theory

Equilibria are meant to help us predict how players will play some specific game. Of course, even in games with only one equilibrium, we might be reluctant to accept it as a good prediction, as the Basu game (p. 272) corroborates very nicely.

In case of several equilibria clear-cut solutions are not available. Evolutionary game theory and risk dominance may provide some help. Harsanyi & Selten (1988) offer a very general approach on how to select equilibria in matrix games. Their method is very involved and finally not very convincing. Bowles (2004, p. 53) argues that we should not (always) look for a solution to the multiplicity issue within game theory. Theory, according to his mind, is necessarily insufficient and we need additional information from outside game theory to help to select among several equilibria – history matters. Discuss these issues for the stag hunt, the battle of the sexes, and the chicken game.

7. Topics

The main topics in this chapter are

- games in strategic form
- strategy combination
- Nash equilibrium
- dominance
- iterated dominance
- battle of the sexes
- prisoners' dilemma
- head or tail (matching pennies)
- stag hunt
- game of chicken
- second-price auction
- take it or leave it
- Basu game (insurance game)
- mixed strategies
- police game

8. Solutions

Exercise X.1

There is no dominant strategy in any of these games.

Exercise X.2

If player 1 chooses $s_1 = r_1$, he will obtain the object and pay s_2 so that his payoff is $r_1 - s_2$. This payoff stays the same as long as $s_1 > s_2$. If, however, player 1 chooses $s_1 = s_2$, his payoff is $\frac{1}{2}(r_1 - s_2)$ with $0 < \frac{1}{2}(r_1 - s_2) < r_1 - s_2$ while $s_1 < s_2$ reduces the payoff to $0 < r_1 - s_2$.

Exercise X.3

The game is symmetric, it suffices to consider player 1. Strategy $s_1 = 100$ is dominated by $s_1 = 99$. No strategy other than 100 is dominated.

Exercise X.4

The two-strategies Basu game is the prisoners' dilemma:

		traveller 2	
		2	3
traveler 1	2	$2, 2$	$4, 0$
	3	$0, 4$	$3, 3$

		player 2: left	player 2: right
pl. 1	up	$1,-1$ [1]	$-1,1$ [2]
	down	$-1,1$ [2]	$1,-1$ [1]

	player 2: left	player 2: right
u.	$4,4$	$0,5$ [2]
d.	$5,0$ [1]	$1,1$ [1][2]

		player 2: left	player 2: right
player 1	up	$1,1$ [1][2]	$1,1$ [1][2]
	down	$1,1$ [1][2]	$0,0$

FIGURE X.4. Matrices and markings

where 3 is the cooperative strategy.

Exercise X.5

There are three Nash equilibria, the strategy combinations (up, left), (up, right), and (down, right).

Exercise X.6

In the Basu game, there is exactly one equilibrium, the strategy combination $(2,2)$.

Exercise X.7

After marking, the three matrices look like figure X.4. The first strategic game does not have an equilibrium, the second has exactly one, and the third has three equilibria.

Exercise X.8

$(s_1^*, s_2^*, ..., s_n^*)$ is a Nash equilibrium iff $s_i^* \in s_i^R\left(s_{-i}^*\right)$ for all $i \in N$.

Exercise X.9

Player 1's payoff function is given by

$$u_1(\sigma_1, \sigma_2) = 4\sigma_1\sigma_2 + 2\sigma_1(1-\sigma_2) + (1-\sigma_1)\sigma_2 + 3(1-\sigma_1)(1-\sigma_2)$$

so that $\sigma_1 = \frac{1}{2}$ and $\sigma_2 = \frac{1}{3}$ leads to

$$\begin{aligned} u_1\left(\frac{1}{2}, \frac{1}{3}\right) &= 4 \cdot \frac{1}{2} \cdot \frac{1}{3} + 2 \cdot \frac{1}{2} \cdot \left(1 - \frac{1}{3}\right) + \\ &\quad + \left(1 - \frac{1}{2}\right) \cdot \frac{1}{3} + 3 \cdot \left(1 - \frac{1}{2}\right) \cdot \left(1 - \frac{1}{3}\right) = \frac{5}{2}. \end{aligned}$$

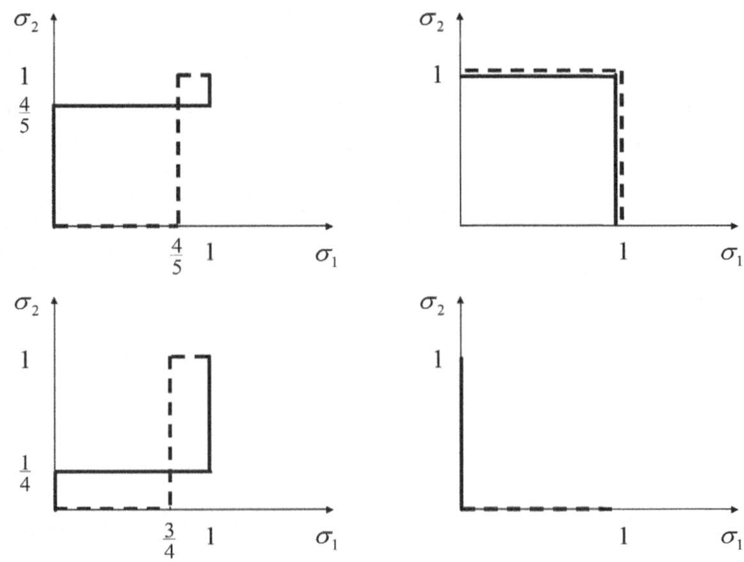

FIGURE X.5. Equilibria

Exercise X.10

The stag hunt is depicted in figure X.5 top left, the battle of the sexes bottom left (did you recognize it?), and the prisoners' dilemma bottom right. The stag hunt and the battle of the sexes admit three equilibria, where one is in purely mixed strategies. The prisoners' dilemma has one equilibrium only, in pure strategies. The top-right game has infinitely many equilibria.

Exercise X.11

There are none. If the police officer controls, the agent will not commit a crime. If the agent is law-abiding, there is no need for control.

Exercise X.12

We calculate:

$$\sigma_p (1 - F) + (1 - \sigma_p) 1 \overset{!}{=} \sigma_p \cdot 0 + (1 - \sigma_p) \cdot 0 \quad \Leftrightarrow \quad \sigma_p \overset{!}{=} \frac{1}{F}.$$

Exercise X.13

We have six strategy combinations and for each strategy combination we have two payoffs. Therefore, we need 12 payoff values.

Exercise X.14

The first matrix game has an infinite number of equilibria, the second has two equilibria in pure strategies, but none in properly mixed strategies.

Exercise X.15

For $\varepsilon > 0$ the game has one equilibrium. For $\varepsilon < 0$, it has two equilibria in pure strategies and one in properly mixed strategies.

9. Further exercises without solutions

PROBLEM X.1. Consider a first-price auction. There are $n = 2$ players $i = 1, 2$ who submit bids $b_i \geq 0$ simultaneously. Player i's willingness to pay for the object is given by w_i. Assume $w_1 > w_2 > 0$. The player with the highest bid obtains the object and has to pay his bid. If both players submit the highest bid, the object is given to player 1. The winning player $i \in \{1, 2\}$ obtains the payoff $w_i - b_i$ and the other the payoff zero. Determine the Nash equilibria in this game!

PROBLEM X.2. (a) Find a game in which a player has a weakly dominant strategy which is not played in one of the Nash equilibria. (b) Is it possible that a player has a strictly dominant strategy that is not played in equilibrium?

PROBLEM X.3. Read the opening scene of Mozart's and Schikaneder's "Magic Flute". Two players $i = 1, 2$ are involved in a dispute over an object. The willingness to pay for the object is w_i, $i = 1, 2$. Assume $w_1 \geq w_2 > 0$. Time is modeled as a continuous variable that starts at 0 and runs indefinitely. Each player i chooses the time $s_i \geq 0$ when to concede the object to the other player. Until the first concession each player loses one unit of payoff per unit of time. Player i's payoff function is given by

$$u_i(s_i, s_j) = \begin{cases} -s_i, & s_i < s_j \\ \frac{w_i}{2} - s_i, & s_i = s_j \\ w_i - s_j, & s_i > s_j. \end{cases}$$

Determine the Nash equilibria in this game!

PROBLEM X.4. Find all (mixed) Nash equilibria of the following game:

		player 2: l	c	r
player 1	o	$(4, 5)$	$(2, 1)$	$(4, 4)$
	m	$(0, 1)$	$(1, 5)$	$(3, 2)$
	u	$(1, 1)$	$(0, 0)$	$(6, 0)$

CHAPTER XI

Price and quantity competition

This chapter presents industrial-organization applications of the previous chapter on strategic games. We concentrate on one-stage price and quantity competition. In contrast to the usual procedure (quantity competition first, then price competition), we begin with price competition.

We prepare the stage by monopoly models. Here, we assume that a single firm is a producer and does not fear entry by other firms. Governmental entry restrictions may be responsible for this state of affairs.

In this chapter, we assume that the products are homogeneous. By this we mean that the consumers consider the products (even if from different suppliers) as equally good or bad. As a consequence, the only difference between the products (if there is any) concerns the price.

1. Monopoly: Pricing policy

1.1. The linear model. Assume the demand function X given by

$$X(p) = d - ep,$$

where d and e are positive constants and p obeys $p \leq \frac{d}{e}$. It is depicted in figure XI.1. The reader is invited to go back to pp. 142 where he can find the definitions for saturation quantity, prohibitive price, and price elasticity of demand.

EXERCISE XI.1. *Find the saturation quantity, the prohibitive price, and the price elasticity of demand of the above demand curve.*

DEFINITION XI.1 (A MONOPOLIST'S PROFIT). *A monopoly's profit in terms of price p is given by*

$$\begin{aligned}\underbrace{\Pi(p)}_{profit} \quad : &= \underbrace{R(p)}_{revenue} - \underbrace{C(p)}_{cost} \\ &= pX(p) - C[X(p)] \qquad \text{(XI.1)}\end{aligned}$$

where X is the demand function. In our linear model, we obtain

$$\Pi(p) = p(d - ep) - c(d - ep), p \leq \frac{d}{e},$$

H. Wiese, *Advanced Microeconomics*,
https://doi.org/10.1007/978-3-658-34959-2_11

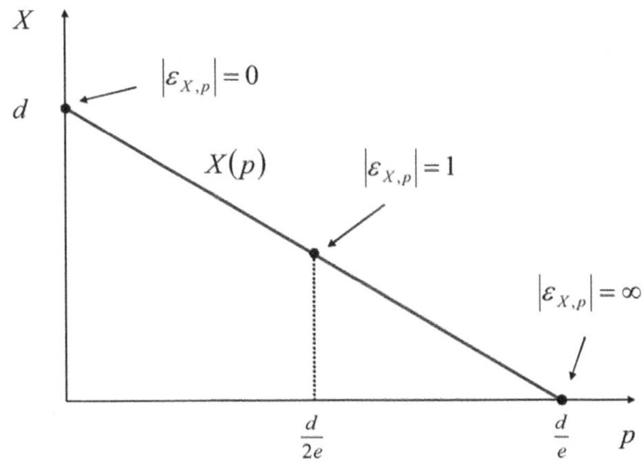

FIGURE XI.1. Price elasticity along a linear demand curve

where d, e, and c are positive parameters.

Note that costs are a function of the quantity produced (as in chapter IX), but that the quantity itself is determined by the price. In figure XI.3, you see that the cost curve is downward-sloping (with the price on the abscissa!). c can be addressed as both marginal and average cost.

DEFINITION XI.2 (A MONOPOLIST'S DECISION SITUATION (PRICE SETTING)). *A monopolist's decision situation with price setting is a tuple*

$$\Delta = (X, C),$$

where

- *X is the demand curve and*
- *C is the cost function.*

The price-setting monopolist's problem is to find the profit-maximizing price given by

$$p^R(\Delta) := \arg\max_{p\in\mathbb{R}} \Pi(p)$$

$p^R(\Delta)$ is also denoted by p^M and referred to as monopoly price.

EXERCISE XI.2. *What is the interpretation of $p^?$ in figure XI.2?*

In order to make meaningful comparisons, units of measurement have to be the same. Prices are measured in $\frac{\text{monetary units}}{\text{quantity units}}$ while revenue (price

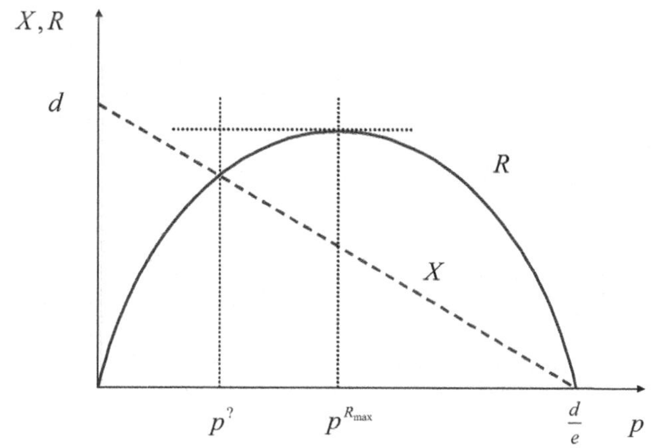

FIGURE XI.2. Find the economic meaning of the question mark!

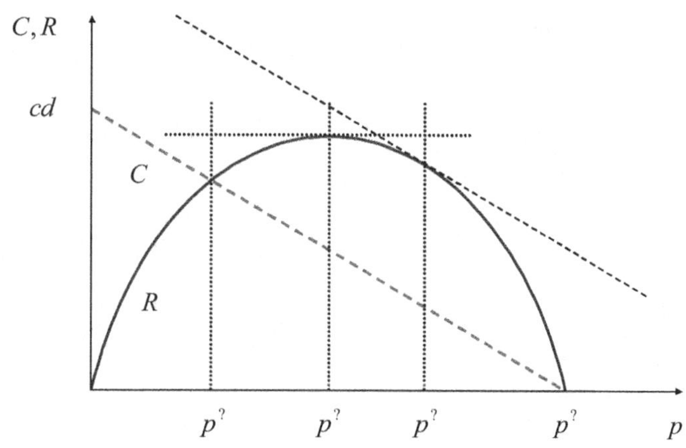

FIGURE XI.3. Same question, different answer

times quantity!) is measured in

$$\begin{aligned} & \frac{\text{monetary units}}{\text{quantity units}} \cdot \text{quantity units} \\ = & \text{ monetary units.} \end{aligned}$$

EXERCISE XI.3. *And how about the interpretation of the $p^?$ in figure XI.3?*

1.2. Marginal revenue and elasticity. Differentiating eq. XI.1 leads to marginal revenue with respect to price and to marginal cost with respect

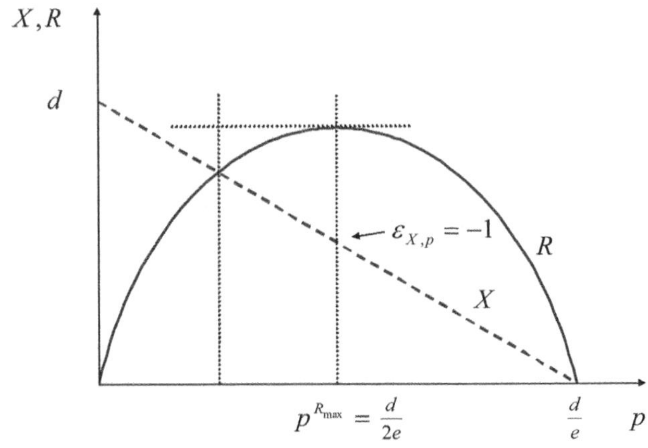

FIGURE XI.4. Revenue and price elasticity

to price. The former is given by

$$\frac{dR(p)}{dp} = \frac{d\left[pX(p)\right]}{dp} = X + p\frac{dX}{dp}$$

and consists of two summands:

- A price increase by one Euro increases revenue by X; for every unit sold the firm obtains an extra Euro.
- A price increase by one Euro affects demand by $\frac{dX}{dp}$ (assumed to be negative) so that sales are reduced by $p\left|\frac{dX}{dp}\right|$.

Using the price elasticity of demand, the marginal revenue with respect to price can also be written as

$$\begin{aligned} \frac{dR(p)}{dp} &= p\frac{dX(p)}{dp} + X(p) \\ &= X(p)\left[1 + \frac{p}{X(p)}\frac{dX(p)}{dp}\right] \\ &= -X(p)\left[\left|\varepsilon_{X,p}\right| - 1\right] \\ &> 0 \quad \text{for} \quad \left|\varepsilon_{X,p}\right| < 1. \end{aligned} \tag{XI.2}$$

Thus, marginal revenue with respect to price is equal to zero (and hence revenue maximal) if the price elasticity of demand is equal to -1. A price increase by one percent reduces demand by one percent and therefore revenue remains constant. Consider figure XI.4, where revenue $R(p) = p(d - ep) = pd - ep^2$ is maximal at $p^{R_{\max}} := \frac{d}{2e}$. Relationships between marginal something and elasticity of something are called Amoroso-Robinson equations. We will encounter several of these.

EXERCISE XI.4. *Comment! A firm can increase profit if it produces at a point where demand is inelastic, i.e., where* $0 > \varepsilon_{X,p} > -1$ *holds.*

Marginal cost with respect to price $\frac{dC}{dp}$ is related to marginal cost (with respect to quantity) $\frac{dC}{dX}$. Indeed, we have

$$\frac{dC}{dp} = \underbrace{\frac{dC}{dX}}_{>0} \underbrace{\frac{dX}{dp}}_{<0} < 0.$$

1.3. Profit maximization. The first-order condition for profit maximization is

$$\frac{dR}{dp} \stackrel{!}{=} \frac{dC}{dp}.$$

As you see on p. 305, the optimal price rule

$$\frac{p - \frac{dC}{dX}}{p} \stackrel{!}{=} \frac{1}{|\varepsilon_{X,p}|}$$

is easily obtained. It shows how far the monopolist can increase the price above marginal cost. We find that the optimal relative price-cost margin is high if demand is inelastic. That makes sense: If demand is inelastic, a relative price increase is met by a relatively low relative quantity decrease. This demand reaction gives the monopolist an incentive to increase the price. In our linear case, we find the profit-maximizing price (also called monopoly price)

$$p^M = \frac{d + ce}{2e} = \frac{1}{2}\left(\frac{d}{e} + c\right)$$

which is the mean of the prohibitive price and the marginal cost.

EXERCISE XI.5. *Confirm the formula for the above monopoly price. Which firm maximizes revenue? How does the monopoly price change if average cost* c *changes?*

1.4. Price differentiation. The monopolist can increase his profit above $\Pi\left(p^M\right)$ by not selling all units at a unique price, i.e., by engaging in price differentiation or price discrimination. In this section, we deal with perfect price differentiation (also called price differentiation of the first degree) and price differentiation of the third degree:

- First-degree price differentiation means that every consumer pays a price equal to his willingness to pay. Using the Marshallian concept of consumers' rent (see pp. 162), the monopolist's revenue is given by the area under the Marshallian demand curve. We return to this case in subsection 3.7 below (pp. 306).

- Third-degree price differentiation means that we have several markets and for each market a (possibly) different price.

EXERCISE XI.6. *A monopolist is active in two markets with demand functions X_1 and X_2 given by*

$$\begin{aligned} X_1(p_1) &= 100 - p_1, \\ X_2(p_2) &= 100 - 2p_2. \end{aligned}$$

He has constant unit cost c of 20.

- *Find the profit-maximizing prices p_1^M and p_2^M.*
- *Assume that price differentiation is not possible. Find the one profit-maximizing price p^M. Hint 1: Find the prohibitve prices for each submarket before summing demand. Hint 2: You arrive at two solutions. Compare profits.*

2. Price competition

2.1. The game. We start with the same market demand as in the monopoly section. Assuming that the products offered by the two firms 1 and 2 are homogeneous, all consumers buy from the firm with the lower price. Finally, we assume that demand is shared equally at identical prices. We then arrive at the demand function for firm 1:

$$x_1(p_1, p_2) = \begin{cases} d - ep_1, & p_1 < p_2 \\ \frac{d-ep_1}{2}, & p_1 = p_2 \\ 0, & p_1 > p_2. \end{cases} \quad \text{(XI.3)}$$

It is shown in figure XI.5.

Assuming unit cost c_1, firm 1's profit is then given by

$$\Pi_1(p_1, p_2) = (p_1 - c_1)x_1(p_1, p_2). \quad \text{(XI.4)}$$

It is a function of firm 1's own price and the competitor's price and depicted in figure XI.6 for several cases.

We can now define the strategic pricing game:

DEFINITION XI.3 (PRICING (BERTRAND) GAME). *A pricing game is the strategic form*

$$\Gamma = \left(N, (S_i)_{i\in N}, (\Pi_i)_{i\in N}\right),$$

where

- *N is the set of firms,*
- *$S_i := \left[0, \frac{d}{e}\right]$ is the set of prices, and*
- *$\Pi_i : S \to \mathbb{R}$ is firm i's profit function analogous to equation XI.4.*

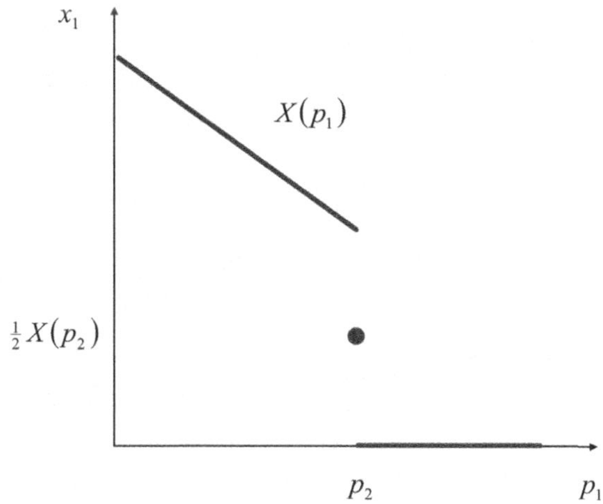

FIGURE XI.5. Firm 1's demand function

Equilibria of this game are called Bertrand equilibria or Bertrand-Nash equilibria.

2.2. Accomodation and Bertrand paradox. We now consider two firms. We are dealing with both actual competition (between firms on the market) and potential competition (by firms as yet outside). In most models, we focus attention on just two firms. With respect to entry, Bain (1956) has taught us to differentiate between

- accomodated entry (both firms are actual competitors),
- blockaded entry (one firm is in the market and can charge the monopoly price while the other firm does not find entry profitable) and
- deterred entry (which is like blockaded entry with the difference that the incumbent firm holds the price sufficiently low so as to deter entry of the other firm).

Throughout this chapter, we will repeatedly refer to these three terms.

We begin with accomodation and assume identical marginal and unit cost below the prohibitive price:

$$c := c_1 = c_2 < \frac{d}{e}.$$

Homogeneity of the products makes undercutting very profitable. Therefore,

$$\left(p_1^B, p_2^B\right) = (c, c)$$

is a good Nash-equilibrium candidate. B stands for "Bertrand" (more in a minute). The strategy combination (p_1^B, p_2^B) leads to the quantities and profits

$$x_1^B = x_2^B = \frac{1}{2}X(c) = \frac{d-ec}{2}, \quad \text{(XI.5)}$$

$$\Pi_1^B = \Pi_2^B = 0. \quad \text{(XI.6)}$$

LEMMA XI.1 (BERTRAND PARADOX). *Assume the game of definition XI.3 with $N = \{1,2\}$ and $c := c_1 = c_2 < \frac{d}{e}$. This game has one and only one equilibrium, namely $(p_1^B, p_2^B) = (c, c)$.*

For a proof, we need to check that (c, c) is an equilibrium, and furthermore that there are no other equilibria. This is not difficult to do.

The result of this lemma is known as Bertrand paradox. In light of the incentives to undercut a rival, the equilibrium may not be really astonishing. However, quantity competition does not lead to the equality of price and marginal cost unless we have perfect price differentiation or perfect competition.

Of course, the Bertrand paradox is somewhat unsatisfactory – from the point of view of the oligopolists and also from theoretical perspectives. Several ways out of the paradox can be pursued:

- Firms typically meet several times under similar conditions. The theory of repeated games presented in chapter XIII explains when and how firms may escape the dire consequences of the Bertrand paradox.
- Different average costs are treated in the next section.
- A price cartel allows firms to share the monopoly profit.
- Products may not be homogeneous but differentiated. We follow up on this possibility in chapter XII.

EXERCISE XI.7. *Assume two firms with identical unit costs of 10 Euro. The strategy sets are $S_1 = S_2 = \{1, 2, ..., \}$. Determine all Bertrand equilibria.*

2.3. Blockaded entry and deterred entry. So far, we have considered the case of identical unit costs and hence accomodation. We now assume $c_1 < c_2$ so that firm 1 produces cheaper than firm 2. Figure XI.6 presents firm 1's profit function depending on both c_1 and p_2.

2.3.1. *Market entry blockaded for both firms.* If the marginal costs of both firms lie above the prohibitive price,

$$c_1 \geq \frac{d}{e}, c_2 \geq \frac{d}{e},$$

market entry is blockaded for both firms.

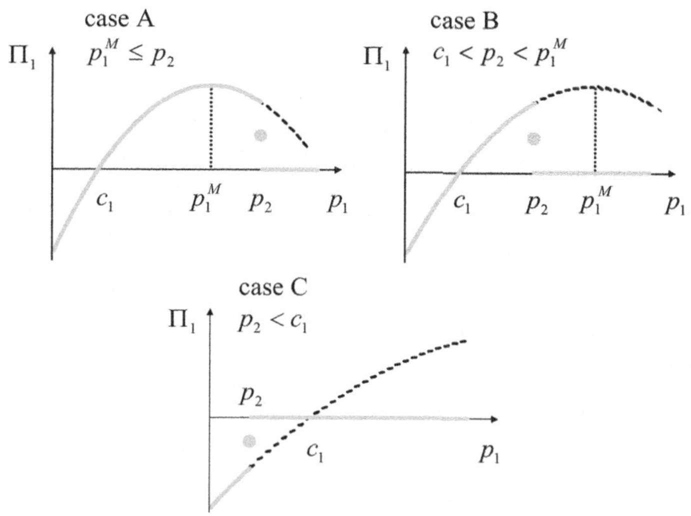

FIGURE XI.6. Firm 1's profit function

2.3.2. *Market entry of firm 2 blockaded.* We now assume

$$c_1 < \frac{d}{e} \text{ and } c_2 \geq p_1^M.$$

Entry is not blockaded for firm 1 and firm 2's average cost lies above firm 1's monopoly price. Firm 2 can avoid losses by setting $p_2 = c_2$. Our assumptions are then reflected in figure XI.6, case A. The strategy combination

$$\left(p_1^B, p_2^B\right) = \left(p_1^M, c_2\right) = \left(\frac{d}{2e} + \frac{c_1}{2}, c_2\right)$$

is then an equilibrium where

$$\begin{aligned} x_1^B &= \frac{d - ec_1}{2}, \quad x_2^B = 0, \\ \Pi_1^B &= \frac{(d - ec_1)^2}{4e}, \quad \Pi_2^B = 0 \end{aligned}$$

hold.

EXERCISE XI.8. *Can you find other equilibria?*

2.3.3. *Market entry of firm 2 deterred.* We now assume that the monopoly price set by firm 1 is not sufficient to keep firm 2 out of the market. This is the case for

$$c_1 < \frac{d}{e} \text{ and } c_2 < p_1^M.$$

The reader is invited to consult figure XI.6, case B. Firm 2's price $p_2 := c_2$ prompts firm 1's best response $c_2 - \varepsilon$ where ε is "very small". Strictly

1. no supply	$c_1 \geq \frac{d}{e}$ and $c_2 \geq \frac{d}{e}$
2. entry of firm 2 blockaded	$0 \leq c_1 < \frac{d}{e}$ and $p_1^M = \frac{d+ec_1}{2e} < c_2$
3. entry of firm 2 deterred	$0 \leq c_1 < \frac{d}{e}$ and $c_1 < c_2 \leq \frac{d+ec_1}{2e} = p_1^M$
4. Bertrand-Paradox	$c_1 = c_2 =: c$ and $0 \leq c < \frac{d}{e}$
5. entry of firm 1 deterred	$0 \leq c_2 < \frac{d}{e}$ and $c_2 < c_1 \leq \frac{d+ec_2}{2e} = p_2^M$
6. entry of firm 1 blockaded	$0 \leq c_2 < \frac{d}{e}$ and $p_2^M = \frac{d+ec_2}{2e} < c_1$

FIGURE XI.7. Solutions to pricing competition for different combinations of unit costs

speaking, and for reasons explained on p. 152, a best response does not exist here. The reader will pardon us to ignore these purely mathematical problems.

In contrast to blockaded entry, firm 1 now strives to prevent entry by firm 2. Thus, firm 1's profit is lower under deterrence than under blockade. The entry-deterring price is also called limit price. It is denoted by

$$p_1^L(c_2) := c_2 - \varepsilon.$$

We have the Bertrand-Nash equilibrium

$$\left(p_1^B, p_2^B\right) = \left(p_1^L(c_2), c_2\right) = (c_2 - \varepsilon, c_2)$$

with the associated quantities and profits

$$\begin{aligned} x_1^B &\approx d - ec_2, \quad x_2^B = 0, \\ \Pi_1^B &\approx (c_2 - c_1)(d - ec_2), \quad \Pi_2^B = 0. \end{aligned}$$

2.3.4. *Summary.* Depending on marginal costs, we find six cases that are listed in figure XI.7 and depicted in figure XI.8. $c_1 = c_2$ yields the Bertrand paradox which is the 45°-line in this figure. The other five cases are reflected in other areas of the c_1-c_2-space. For example, entry is blockaded for firm 2 in case of $c_2 > p_1^M(c_1) = \frac{d}{2e} + \frac{c_1}{2}$. The corresponding demarcation line is the left of the two dashed lines.

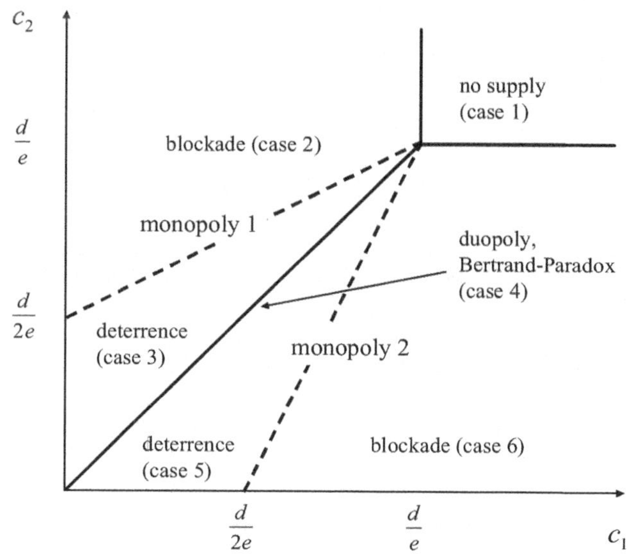

FIGURE XI.8. Deterred and blockaded entry

3. Monopoly: quantity policy

3.1. The linear model. The quantity-setting monopoly presupposes an inverse demand function. Again, we work with a linear specification.

EXERCISE XI.9. *Assume the linear inverse demand function p given by* $p(X) = a - bX$, *with positive constants a and b. Determine*

(1) *the slope of the inverse linear demand function,*
(2) *the slope of its marginal-revenue curve,*
(3) *saturation quantity, and*
(4) *prohibitive price.*

DEFINITION XI.4. *A monopoly's profit in terms of quantity* $X \geq 0$ *is given by*

$$\begin{aligned}\underbrace{\Pi(X)}_{profit} \quad : &= \underbrace{R(X)}_{revenue} - \underbrace{C(X)}_{cost} \\ &= p(X)X - C(X)\end{aligned}$$

where p is the inverse demand function. In our linear model, we obtain

$$\Pi(X) = (a - bX)X - cX, X \leq \frac{a}{b},$$

where a, b, and c are positive parameters.

DEFINITION XI.5 (A MONOPOLIST'S DECISION SITUATION (QUANTITY SETTING)). *A monopolist's decision situation with quantity setting is a tuple*

$$\Delta = (p, C)$$

where

- *p is the inverse demand function and*
- *C is the cost function.*

The quantity-setting monopolist's problem is to find the profit-maximizing quantity given by

$$X^R(\Delta) := \arg\max_{X \in \mathbb{R}_+} \Pi(X)$$

$X^R(\Delta)$ is also denoted by X^M and referred to as monopoly quantity.

3.2. Marginal revenue. In our linear model, marginal cost is simply c. Marginal revenue is more interesting. It is given by

$$MR(X) = p + X\frac{dp}{dX}$$

and consists of two summands:

- If the monopolist increases his quantity by one unit, he obtains the current price for that last unit sold.
- The bad news is that a quantity increase affects the price by $\frac{dp}{dX}$. Without price differentiation, this price decrease applies to all units sold. Thus, in case of a negatively sloped inverse demand curve, revenue is changed by $X\frac{dp}{dX} \leq 0$.

There are three cases where marginal revenue equals price:

- The demand curve is horizontal ($\frac{dp}{dX} = 0$) in which case marginal revenue equals price. Then, the monopoly's output decision has no effect on the market price.
- We have $MR = p$ for the first "small" unit ($X = 0$). Indeed, the price can also be understood as average revenue, $R(X)/X = p$. On pp. 223, we have seen the general argument for any marginal something versus any average something. In our specific case, the firm obtains the prohibitive price (or just below) for the first unit it sells.
- First-degree price differentiation also implies $MR = p$. In a sense, we also have $X = 0$ because the price decrease implied by the quantity increase is not applied to all consumers buying "so far" (the so-called infra-marginal consumers). We will deal with perfect price differentiation in some detail later on, see pp. 306.

Marginal revenue can also be expressed by the price elasticity of demand. In fact, we need its inverse $\frac{1}{\varepsilon_{X,p}}$ which is the relative effect of a one-percent increase in quantity on price. We obtain the Amoroso-Robinson equation

$$\begin{aligned} MR &= p + X\frac{dp}{dX} = p\left[1 + \frac{dp}{dX}\frac{X}{p}\right] \\ &= p\left[1 + \frac{1}{\varepsilon_{X,p}}\right] \\ &= p\left[1 - \frac{1}{|\varepsilon_{X,p}|}\right] > 0 \quad \text{for} \quad |\varepsilon_{X,p}| > 1. \end{aligned}$$

3.3. Monopoly profit. The profit at some quantity $\bar{X}$ is given by

$$\begin{aligned} \Pi\left(\bar{X}\right) &= p(\bar{X})\bar{X} - C\left(\bar{X}\right) \\ &= \left[p(\bar{X}) - AC\left(\bar{X}\right)\right]\bar{X} \text{ (average definition)} \\ &= \int_0^{\bar{X}} \left[MR(X) - MC(X)\right] dX - F \text{ (marginal definition)} \end{aligned}$$

Graphically, this profit can be reflected in two different manners:

- Average viewpoint: For $\bar{X} > 0$, the monopoly profit is equal to average profit $p(\bar{X}) - AC\left(\bar{X}\right)$ times quantity $\bar{X}$. In figure XI.9, the corresponding area is the rectangle EGHF.
- Marginal viewpoint: We add the marginal profit for the first, the second etc. units. Algebraically, we have the above integral, graphically, we address the area between marginal-revenue curve and marginal-cost curve, DCBA.

3.4. Profit maximization. The first-order condition for profit maximization is

$$MC \stackrel{!}{=} MR.$$

EXERCISE XI.10. *Find the profit-maximizing quantity X^M for the inverse demand curve $p(X) = 24 - X$ and constant unit cost $c = 2$!*

EXERCISE XI.11. *Find the profit-maximizing quantity X^M for the inverse demand curve $p(X) = \frac{1}{X}$ and constant unit cost c!*

The profit-maximizing rule "marginal revenue equals marginal cost" determines X^M (see figure XI.10 for the linear case). The consumers have to pay the price $p^M = p\left(X^M\right)$. $M = \left(X^M, p^M\right)$ is sometimes denoted as Cournot point. Antoine Augustin Cournot (1801-1877) was a French philosopher, mathematician, and economist. He is rightly famous for his

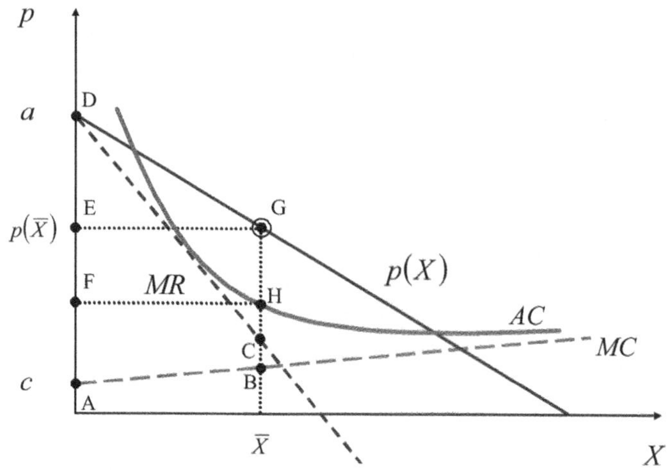

FIGURE XI.9. Average versus marginal profit

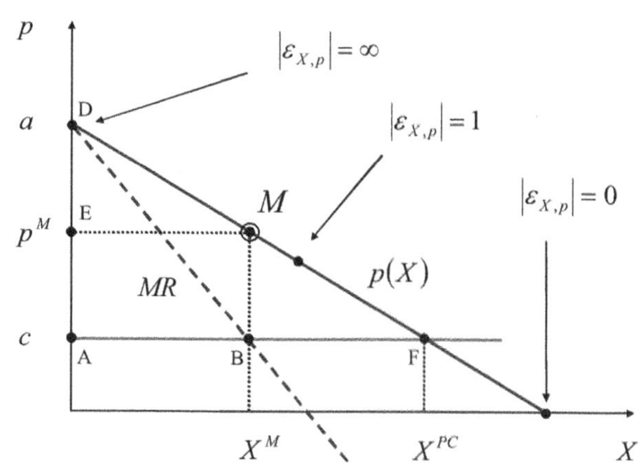

FIGURE XI.10. The Cournot monopoly

1838 treatise “Recherches sur les principes mathématiques de la théorie des richesses”. In chapter 5 of is booklet, Cournot deals with the main elements of monopoly theory and chapter 7 contains oligopoly theory (see section 4 below).

For $c \leq a$, the Cournot point lies in the elastic part of market demand (see exercise XI.4, p. 293). The case of $c > a$ is depicted in figure XI.11.

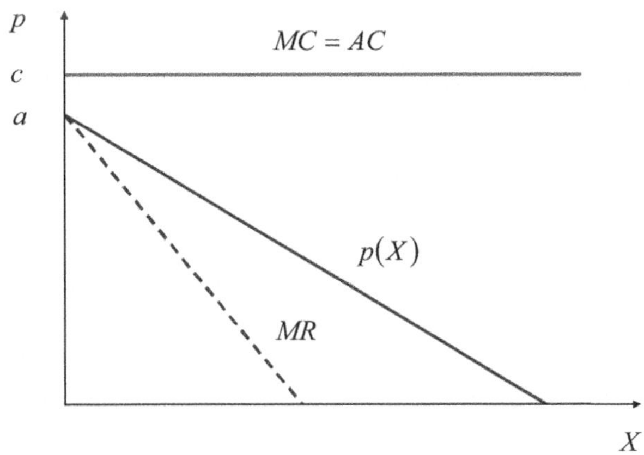

FIGURE XI.11. A blockaded monopolist

The optimal quantiy is summarized by

$$X^M = X^M(c,a,b) = \begin{cases} \frac{1}{2}\frac{a-c}{b}, & c \leq a \\ 0, & c > a. \end{cases} \tag{XI.7}$$

We will assume $c \leq a$ for the rest of this monopoly section.

In our linear case, it is easy to do some comparative statics. For that purpose, we write down the equilibrium variables X^M, p^M, and Π^M together with the partial derivatives with respect to a, b, and c :

$$\begin{aligned} X^M(a,b,c) &= \frac{1}{2}\frac{a-c}{b}, & \text{where } \frac{\partial X^M}{\partial c} < 0; \frac{\partial X^M}{\partial a} > 0; \frac{\partial X^M}{\partial b} < 0, \\ p^M(a,b,c) &= \frac{1}{2}(a+c), & \text{where } \frac{\partial p^M}{\partial c} > 0; \frac{\partial p^M}{\partial a} > 0; \frac{\partial p^M}{\partial b} = 0, \\ \Pi^M(a,b,c) &= \frac{1}{4}\frac{(a-c)^2}{b}, & \text{where } \frac{\partial \Pi^M}{\partial c} < 0; \frac{\partial \Pi^M}{\partial a} > 0; \frac{\partial \Pi^M}{\partial b} < 0. \end{aligned} \tag{XI.8}$$

EXERCISE XI.12. *Consider* $\Pi^M(c) = \frac{1}{4}\frac{(a-c)^2}{b}$ *and calculate* $\frac{d\Pi^M}{dc}$! *Hint: Use the chain rule!*

For a monopolist, it does not matter whether he chooses the profit-maximizing quantity or the profit-maximizing price. The equivalence of these two profit-maximizing rules can be seen from figure XI.12.

3.5. The influence of average cost on maximal profit. We now reconsider the question asked in the last exercise: How does average cost c influence the maximal profit $\Pi^M(c)$? With a glance at

$$\Pi^M = p(X^M)X^M - cX^M,$$

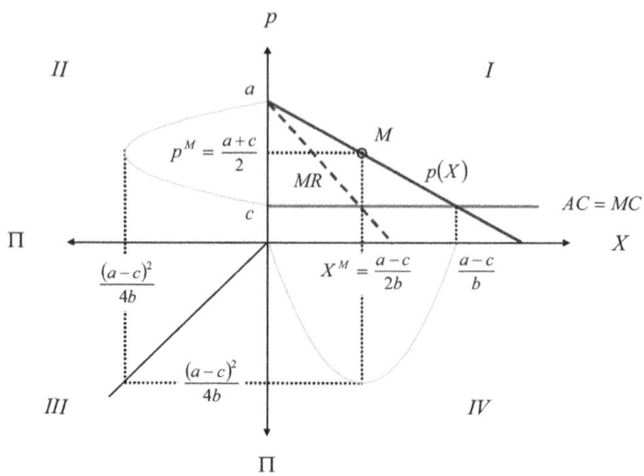

FIGURE XI.12. Monopoly price and quantity

a first guess may be

$$\frac{d\Pi^M(c)}{dc} = -X^M.$$

However, shouldn't a cost increase influence the optimal quantity? Indeed, we have determined

$$X^M(c) = \frac{1}{2}\frac{a-c}{b}.$$

Thus, an increase of c leads to a decrease of the optimal quantity so that we should find $\frac{d\Pi^M(c)}{dc} > -X^M$. If we are dealing with a discrete cost increase, this argument is correct. For example, consider an increase of c from 0 to $\frac{a}{2}$. Holding $X^M(0) = \frac{a}{2b}$ constant, we have $p(X^M(0)) = \frac{1}{2}(a+0) = \frac{a}{2}$ and therefore

$$\Pi^M = \left[p(X^M(0)) - \frac{a}{2}\right] X^M(0) = 0.$$

Now, if we adjust the quantity in an optimal fashion, we obtain $X^M\left(\frac{a}{2}\right) = \frac{1}{2}\frac{a-\frac{a}{2}}{b} = \frac{a}{4b}$ and the price $p(X^M\left(\frac{a}{2}\right)) = \frac{1}{2}(a+\frac{a}{2}) = \frac{3}{4}a$. Then, profit is higher than 0:

$$\begin{aligned} \Pi^M &= \left[p\left(X^M\left(\frac{a}{2}\right)\right) - \frac{a}{2}\right] X^M\left(\frac{a}{2}\right) = \\ &= \left[\frac{3}{4}a - \frac{a}{2}\right]\frac{1}{4}\frac{a}{b} = \frac{1}{16}\frac{a^2}{b} > 0. \end{aligned}$$

Thus, adjusting pays in the descrete case. Interestingly, this argument does not apply to "very small" increases of c. In order to see why, we write the monopoly profit as

$$\Pi^M(c) = \Pi\left(c, X^M(c)\right).$$

The maximal profit depends both directly and indirectly on c. Differentiating both sides of this equation, we find

$$\frac{d\Pi^M(c)}{dc} = \underbrace{\frac{\partial \Pi}{\partial c}}_{\substack{<0 \\ \text{direct effect}}} + \underbrace{\underbrace{\left.\frac{\partial \Pi}{\partial X}\right|_{X=X^M}}_{\substack{=0 \\ \text{first-order condition} \\ \text{for profit maximization}}} \underbrace{\frac{dX^M}{dc}}_{\substack{<0 \\ \text{high marginal cost leads} \\ \text{to a lower optimal quantity}}}}_{\substack{=0 \\ \text{indirect effect}}}.$$

On the right-hand side, we use ∂ because of partial differentiation, with respect to c and with respect to X. Differentiation means considering the effect of very small changes. A very small change in c leads to a very small change in output X. Since $\frac{\partial \Pi}{\partial X}$ is zero at $X = X^M$, it is also zero very close to X^M. This explains why the indirect effect is zero. Indeed, for the left-hand side, the solution to exercise XI.12 shows

$$\frac{d\Pi^M(c)}{dc} = -X^M$$

and the direct effect on the right-hand side leads to the same result:

$$\frac{\partial \Pi\left(c, X^M(c)\right)}{\partial c} = \frac{\partial\left(p\left(X^M\right)X^M - cX^M\right)}{\partial c} = -X^M.$$

Did you realize that this is just another application of the envelope theorem that we introduced in chapter VII (pp. 180)?

EXERCISE XI.13. *In order to apply definition VII.2 and theorem VII.3 (pp. 181), transfer the symbols beginning with* $\Pi := f$.

3.6. Alternative expressions for profit maximization. Using the Amoroso-Robinson equation, the profit-maximization rule can be expressed by

$$MC \overset{!}{=} MR = p\left[1 - \frac{1}{|\varepsilon_{X,p}|}\right], \tag{XI.9}$$

$$p \overset{!}{=} \frac{|\varepsilon_{X,p}|}{|\varepsilon_{X,p}| - 1} MC \text{ or} \tag{XI.10}$$

$$\frac{p - MC}{p} \overset{!}{=} \frac{1}{|\varepsilon_{X,p}|}. \tag{XI.11}$$

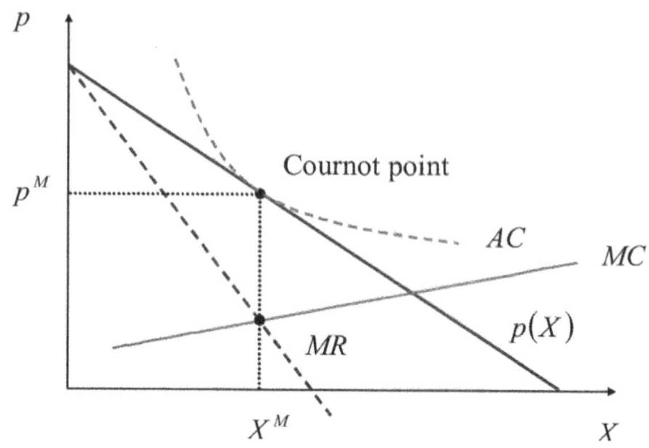

FIGURE XI.13. Monopoly power versus monopoly profit

According to eq. XI.9, the monopolist extends his quantity until marginal cost reach marginal revenue. Eq. XI.10 tells us that the profit-maximizing price can be understood as a multiplicative surcharge on marginal cost. The surcharge itself is a function of the price elasticity of demand.

DEFINITION XI.6 (LERNER INDEX). *Assuming a monopoly, the Lerner index of market power is defined by*

$$\frac{p - MC}{p}.$$

$\frac{p-MC}{p}$ is addressed as the relative price-cost margin, also known as Lerner index. It tells how far the monopolist can increase the price above marginal cost, relative to the price. In perfect competition, we have $p = MC$ so that the Lerner index is a measure for the distance between monopoly and perfect competition. It is sometimes said that perfect competition epitomizes the absence of power. From that perspective, the Lerner index measures market power. According to eq. XI.11, the optimal price-cost margin varies inversely with the absolute value of market elasticity.

Monopoly power and monopoly profit are not the same as figure XI.13 makes clear. The monopolist has monopoly power by $p > MC$ but no profit because of $AC\left(X^M\right) = \frac{C(X^M)}{X^M} = p^M$.

3.7. First-degree price differentiation. First-degree price differentiation means that every consumer pays according to his willingness to pay and that marginal revenue

$$MR = p + X\frac{dp}{dX} = p + 0 \cdot \frac{dp}{dX}$$

equals the price. The price decrease following a quantity increase concerns the marginal, but not the infra-marginal consumers.

Of course, we have derived $MR = p$ in a somewhat rough-and-ready manner. More formally, the price-discriminating monopolist maximizes

$$\begin{aligned} & \text{Marshallian willingness to pay} \; - \; \text{cost} \\ = & \int_0^X p(q)\,dq - C(X) \end{aligned}$$

where Marshallian willingness to pay is defined on p. 163. Differentiating with respect to X now yields

$$MR = p(X) \stackrel{!}{=} \frac{dC}{dX}$$

and thus our "price equals marginal cost" rule.

Consider figure XI.10 on p. 302. The perfectly discriminating monopoly's profit is equal to the triangle AFD.

3.8. Third-degree price differentiation (two markets, one factory) . In this subsection, we deal with a monopolist who produces in one factory, but supplies two markets. In the following subsection, we consider the opposite case: one market, two factories. The two-markets-one-factory case can be used to examine third-degree price differentiation. Let x_1 and x_2 be the output on the two markets 1 and 2. Let p_1 and p_2 be the corresponding inverse demand functions. The monopolist's profit is equal to

$$\Pi(x_1, x_2) = p_1(x_1)\,x_1 + p_2(x_2)\,x_2 - C(x_1 + x_2),$$

with first-order conditions

$$\begin{aligned} \frac{\partial \Pi(x_1, x_2)}{\partial x_1} &= MR_1(x_1) - MC(x_1 + x_2) \stackrel{!}{=} 0, \\ \frac{\partial \Pi(x_1, x_2)}{\partial x_2} &= MR_2(x_2) - MC(x_1 + x_2) \stackrel{!}{=} 0. \end{aligned}$$

We immediately see that the marginal revenues in both markets should coincide. Assume $MR_1 < MR_2$. The monopolist can transfer one unit from market 1 to market 2. Revenue and profit (we have not changed total output $x_1 + x_2$) increases by $MR_2 - MR_1$.

Figure XI.14 provides an illustration. The horizonal line parallel to the abscissa guarantees $MR_1(x_1^*) = MR_2(x_2^*)$. As long as

$$MC(x_1^* + x_2^*) < MR_1(x_1^*) = MR_2(x_2^*)$$

holds, the monopolist should sell more in both markets.

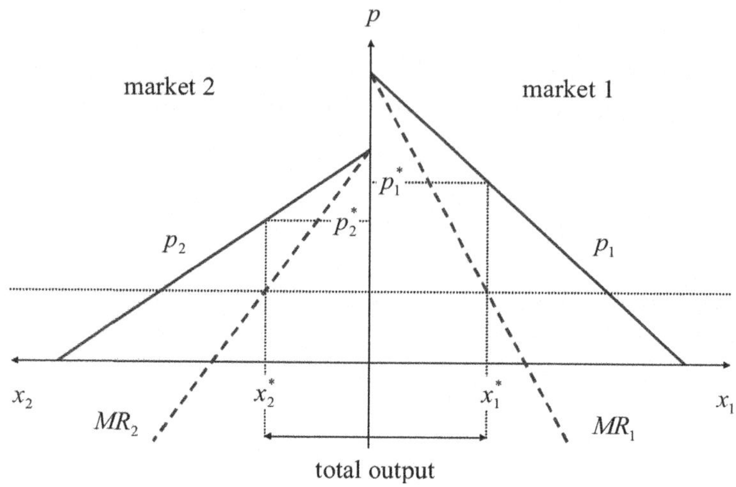

FIGURE XI.14. Equality of marginal revenues

Price differentiation of the third degree obeys the inverse elasticities rule:

$$|\varepsilon_1| > |\varepsilon_2| \Rightarrow p_1^M < p_2^M.$$

Just use the Amoroso-Robinson equation to rewrite the optimization condition $MR_1(x_1^*) = MR_2(x_2^*)$ as follows:

$$p_1^M\left[1 - \frac{1}{|\varepsilon_1|}\right] \stackrel{!}{=} p_2^M\left[1 - \frac{1}{|\varepsilon_2|}\right].$$

EXERCISE XI.14. *A monopolist sells his product in two markets:*

$$\begin{aligned} p_1(x_1) &= 100 - x_1, \\ p_2(x_2) &= 80 - x_2. \end{aligned}$$

(1) *Assume price differentiation of the third degree and the cost function given by $C(X) = X^2$. Determine the profit-maximizing quantities and the profit.*
(2) *Repeat the first part of the exercise with the cost function $C(X) = 10X$.*
(3) *Assume, now, that price differentiation is not possible any more. Using the cost function $C(X) = 10X$, find the profit-maximizing output and price. Hint: You need to distinguish quantities below and above 20.*

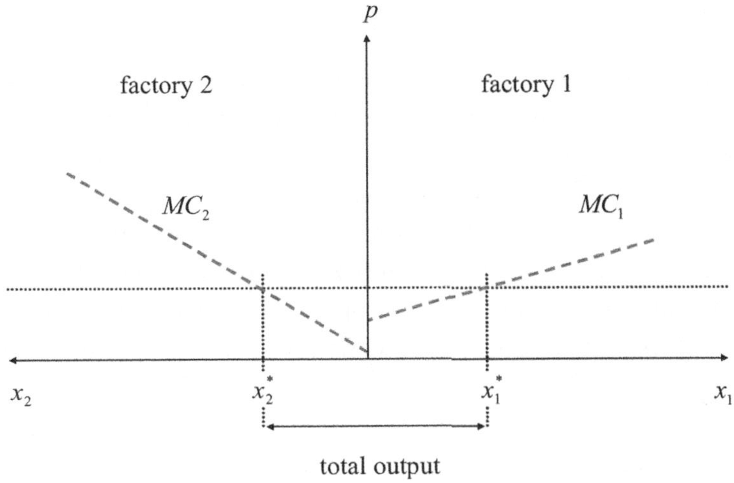

FIGURE XI.15. Equality of marginal costs

3.9. One market, two factories. Mirrowing the above subsection, we now turn to a monopolist who serves one market but operates two factories. With obvious notation, we have

$$\Pi(x_1, x_2) = p(x_1 + x_2)(x_1 + x_2) - C_1(x_1) - C_2(x_2).$$

and

$$\begin{aligned}\frac{\partial \Pi(x_1, x_2)}{\partial x_1} &= MR(x_1 + x_2) - MC_1(x_1) \stackrel{!}{=} 0,\\ \frac{\partial \Pi(x_1, x_2)}{\partial x_2} &= MR(x_1 + x_2) - MC_2(x_2) \stackrel{!}{=} 0.\end{aligned}$$

Thus, the firm produces at $MC_1 \stackrel{!}{=} MC_2$. Assume, to the contrary, $MC_1 < MC_2$. A transfer of one unit of production from the (marginally!) more expensive factory 2 to the cheaper factory 1 decreases cost, and increases profit, by $MC_2 - MC_1$. Figure XI.15 depicts the two marginal-cost curves and the horizontal line where the equality holds.

As a corollary, the equality of the marginal costs also holds for a cartel of two or more firms.

3.10. Welfare-theoretic analysis of monopoly.

3.10.1. *Introduction.* Welfare theory is a branch of normative economics that is concerned with the evaluation of economic policies or, more generally, economic changes of any sort. The most simple analysis just rests on

- Marshallian consumers' rent (see chapter VI, pp. 162),

- producers' rent (chapter IX, pp. 242), and
- taxes.

They measure the monetary advantage accruing to consumers, producers, and the government, respectively. We just sum these three components into an aggregate called welfare. Distributional aspects are disregarded. It is irrelevant which particular group benefits.

Sometimes, we make the assumption that the government tries to maximize welfare. The underlying reason need not be benevolence. Perhaps, the government just maximizes its support (chances of reelection) by benefitting consumers, producers, beneficiaries of publicly provided goods, and tax payers.

3.10.2. *Perfect competition as benchmark.* Perfect competition is a market form with many consumers and producers. In perfect competition (see more in chapter XIX), households and consumers are price takers and the profit-maximizing rule for producers is "price equals marginal cost" (see pp. 240). We talk about it here because perfect competition provides a welfare-theoretical benchmark. This can be seen by two different lines of argumentation.

(1) Consider the demand and supply curves in figure XI.16. At point R, supply and demand meet, this is the equilibrium. The marginal consumer's willingness to pay equals the marginal firm's loss compensation (marginal cost). Therefore, the quantity produced and consumed at R is optimal.
(2) At R, the sum of consumers' and producers' rents is maximal. Just examine what happens if the price is higher or lower than at R.

3.10.3. *Cournot monopoly.* We are now set to analyze the Cournot monopoly from a welfare point of view. The Cournot quantity X^M is lower than the welfare-maximizing quantity X^{PC} produced in perfect competition, as illustrated in figure XI.17. The wefare loss is measured by the shaded area.

EXERCISE XI.15. *Assume that the monopolist cannot price discriminate. His marginal-cost curve is given by* $MC = 2X$ *and he faces the inverse demand curve* $p(X) = 12 - 2X$*. Determine the welfare loss! Hint: Sketch the situation and apply the triangle rule!*

The welfare loss is due to the positive external effect of a quantity increase unrealized in a Cournot monopoly. We have consumers' rent

$$CR(\bar{X}) = \int_0^{\bar{X}} p(X)\,dX - p(\bar{X})\,\bar{X}$$

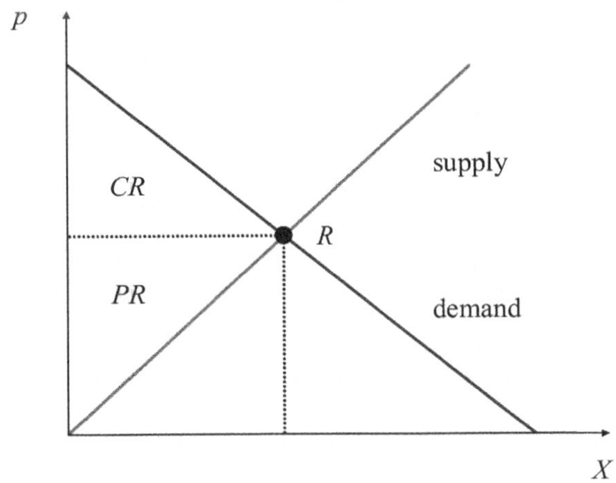

FIGURE XI.16. Welfare is maximal under perfect competition

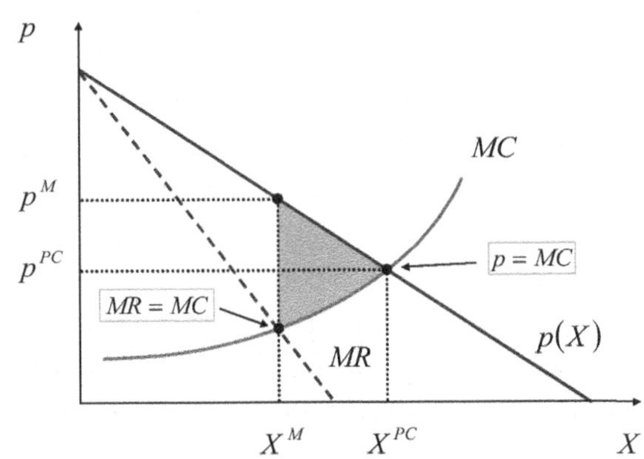

FIGURE XI.17. Wefare loss due to monopoly

and its derivation with respect to $\bar{X}$

$$\begin{aligned}\frac{dCR(\bar{X})}{d\bar{X}} &= \frac{d\int_0^{\bar{X}} p(X)\,dX}{d\bar{X}} - \frac{d\left[p(\bar{X})\,\bar{X}\right]}{d\bar{X}} \\ &= p(\bar{X}) - \left(p(\bar{X}) + \frac{dp}{d\bar{X}}\bar{X}\right) = -\frac{dp}{d\bar{X}}\bar{X} > 0.\end{aligned}$$

The Cournot monopolist offers too small a quantity. If, however, the monopolist were to take account of the positive external effect, he would maximize

$$\left[p(\bar{X})\bar{X} - C(\bar{X})\right] + CR(\bar{X}).$$

Then, the first-order condition would be

$$\left[p(\bar{X}) + \frac{dp}{d\bar{X}}\bar{X} - \frac{dC}{d\bar{X}}\right] - \frac{dp}{d\bar{X}}\bar{X} \overset{!}{=} 0$$

or

$$p(\bar{X}) \overset{!}{=} \frac{dC}{d\bar{X}}$$

and hence the first-order conditions for perfect competition or price differentiation of the first degree.

4. Quantity competition

4.1. Price versus quantity competition. For the development of economic theory, the treatise offered by Cournot in 1838 is of upmost importance. After all, Cournot has to be regarded as the founder of noncooperative game theory. From the point of view of the history of economic thought, we have chosen the wrong sequence of presentation. After all, Cournot's oligopoly theory (for quantity setters) goes back to 1838 while Bertrand's oligopoly theory (for price setters) was published much later, 1883 (compare 1838 versus 1883). The French mathematician Bertrand criticizes Cournot for having firms choose quantities rather than prices.

However, we do not necessarily need to subscribe to Bertrand's criticism. The Cournot model can be defended via two different arguments. First, there are goods where quantity competition seems to be the correct model. Consider, for example, agricultural production or oil production, where some time elapses between sowing (drilling) and the delivery of the product to the consumer. The firms do not have a direct influence on prices. Rather, they produce and can only observe the resulting price some time later. In contrast, the competition of mail-order firms is best modeled by price-setting firms who cannot change prices easily.

Second, and certainly related to the first point, a very interesting vindication of Cournot has been offered by Kreps & Scheinkman (1983), just 100 years after Bertrand's publication. These authors present a two-stage model where the firms decide on capacity at the first stage and on prices at the second. They find that the quantities obtained from the two-stage model equal the Cournot equilibrium quantities. Roughly speaking, they justify the formula

simultaneous capacity construction
\+ Bertrand competition
= Cournot results

4.2. The game. The Cournot model is about simultaneous quantity competition between producers of a homogeneous good:

DEFINITION XI.7 (QUANTITY (COURNOT) GAME). *A quantity game is the strategic form*

$$\Gamma = \left(N, (S_i)_{i\in N}, (\Pi_i)_{i\in N}\right),$$

where

- N *is the set of firms,*
- $S_i := [0, \infty)$ *is the set of quantities and*
- $\Pi_i : S \to \mathbb{R}$ *is firm i's profit function given by*

$$\Pi_i (x_i, X_{-i}) := p(x_i + X_{-i}) x_i - C(x_i).$$

Here, X_{-i} stands for $\sum_{\substack{j=1,\\ j\neq i}}^{n} x_j$. Equilibria of this game are called Cournot equilibria or Cournot-Nash equilibria.

In the previous chapter (pp. 274), we learned that Nash equilibria are definable by the crossing of best-response functions. In case of two firms 1 and 2, the best-response functions (often called reaction functions) are denoted by x_1^R and x_2^R and the Nash equilibrium is given by $\left(x_1^C, x_2^C\right)$ with $x_1^C = x_1^R\left(x_2^C\right)$ and $x_2^C = x_2^R\left(x_1^C\right)$.

4.3. Accomodation.

4.3.1. *Equilibrium.* In the linear case with two firms, we find the first-order condition for firm 1

$$\frac{\partial \Pi_1(x_1,x_2)}{\partial x_1} = MR_1(x_1) - MC_1(x_1) = a - 2bx_1 - bx_2 - c_1 \overset{!}{=} 0$$

and its reaction function

$$x_1^R(x_2) = \frac{a - c_1}{2b} - \frac{1}{2}x_2 = x_1^M - \frac{1}{2}x_2.$$

Thus, in response to the increase of x_2 by one unit, firm 1 wants to decrease x_1 by half a unit. Sometimes, we say that quantities are strategic substitutes. In household theory, butter and margarine are called substitutes because the demand for butter decreases if the demand for margarine increases (due to a price decrease of margarine, see p. 189). See figure XI.18 for an illustration of both reaction functions and their intersection $\left(x_1^C, x_2^C\right)$.

Solving the two best-response functions for x_1 and x_2, we obtain the equilibrium $\left(x_1^C, x_2^C\right)$ with

$$x_1^C = \frac{1}{3b}(a - 2c_1 + c_2), \tag{XI.12}$$

$$x_2^C = \frac{1}{3b}(a - 2c_2 + c_1) \tag{XI.13}$$

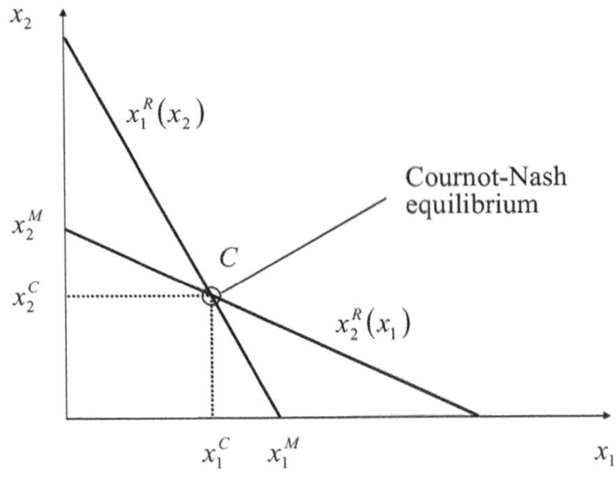

FIGURE XI.18. Reaction functions and Cournot-Nash equilibrium

and hence

$$\begin{aligned}
X^C &= x_1^C + x_2^C = \frac{1}{3b}(2a - c_1 - c_2), \\
p^C &= \frac{1}{3}(a + c_1 + c_2), \\
\Pi_1^C &= \frac{1}{9b}(a - 2c_1 + c_2)^2, \\
\Pi_2^C &= \frac{1}{9b}(a - 2c_2 + c_1)^2, \\
\Pi^C &= \Pi_1^C + \Pi_2^C < \Pi^M.
\end{aligned}$$

4.3.2. *Iterative rationalizability.* The Cournot duopoly can also be solved by rationalizability (see chapter II), applied several times. Examples of iterative dominance can be found in the previous chapter.

With a view to eq. XI.7, p. 303, we see that the reaction functions of figure XI.18 are not complete. Instead, the complete reaction function for firm 2 is given by

$$x_2^R(x_1) = \begin{cases} \frac{a-c_2}{2b} - \frac{x_1}{2}, & x_1 < \frac{a-c_2}{b} \\ 0, & \text{otherwise} \end{cases}$$

and depicted in figure XI.19.

$$x_1^L := \frac{a - c_2}{b}$$

is firm 1's limit quantity. Because of $p\left(x_1^L\right) = c_2$, firm 2 would rather not offer a positive quantity in response to any $x_1 \geq x_1^L$.

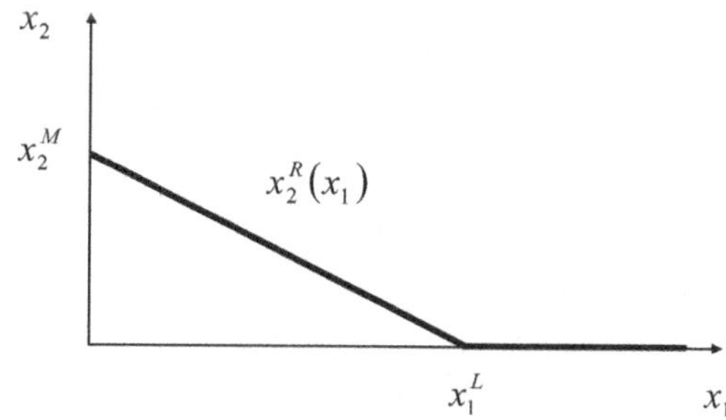

FIGURE XI.19. Firm 2's complete reaction function

Now, our argument rests on figure XI.20. Firm 1 may choose any non-negative quantity. From firm 2's point of view, we can restrict attention to the quantities between 0 and x_1^L. One glance at firm 2's reaction curve makes clear that every quantity x_2 from

$$\begin{aligned} I_1 \quad : &= \left[x_2^R\left(x_1^L\right), x_2^R(0)\right] \\ &= \left[0, x_2^M\right] \end{aligned}$$

can be rationalized, while any quantity $x_2 > x_2^M$ cannot, due to strict dominance by $x_2^M = \frac{1}{2}\frac{a-c}{b}$. Turning to firm 1, the best responses to quantities from I_1 are collected in

$$\begin{aligned} I_2 \quad : &= \left[x_1^R\left(x_2^M\right), x_1^R(0)\right] \\ &= \left[\frac{1}{4}\frac{a-c_1}{b}, x_1^M\right]. \end{aligned}$$

After all, outputs smaller than $x_1 = \frac{1}{4}\frac{a-c}{b}$ are dominated by $\frac{1}{4}\frac{a-c}{b}$. The third step is firm 2's interval $I_3 := \left[\frac{1}{4}\frac{a-c_2}{b}, \frac{3}{8}\frac{a-c_2}{b}\right]$.

The intervals get smaller and smaller and converge towards the intersection point of the two reaction curves. Thus, the Cournot model is solvable by iterative rationalizability.

4.3.3. *Cartel treaty between two duopolists.* We now consider two firms 1 and 2 and their quantities x_1 and x_2, respectively. They form a cartel in order to realize Pareto improvements (from the point of view of the two firms!). We write the cartel profit as

$$\begin{aligned} \Pi_{1,2}(x_1, x_2) &= \Pi_1(x_1, x_2) + \Pi_2(x_1, x_2) \\ &= p(x_1 + x_2) \cdot (x_1 + x_2) - C_1(x_1) - C_2(x_2). \end{aligned}$$

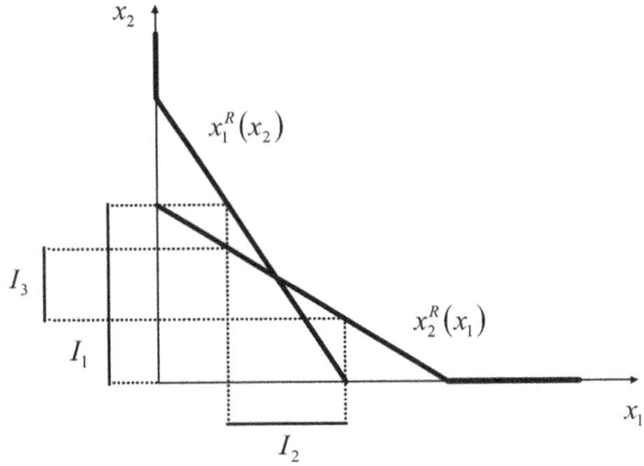

FIGURE XI.20. Iterative dominance in a Cournot duopoly

The first-order conditions are

$$\frac{\partial \Pi_{1,2}}{\partial x_1} = p + \frac{dp}{dX}(x_1 + x_2) - \frac{dC_1}{dx_1} \stackrel{!}{=} 0$$

for the quantity x_1 and

$$\frac{\partial \Pi_{1,2}}{\partial x_2} = p + \frac{dp}{dX}(x_1 + x_2) - \frac{dC_2}{dx_2} \stackrel{!}{=} 0$$

for x_2. As in subsection 3.9 (pp. 309), the marginal costs are the same for both firms (if both of them produce a positive quantity).

More importantly, the Cournot duopolists do not achieve Pareto optimality because they do not take account of the negative externalities that each exerts on the other. As we will explain in chapter XIV (pp. 394), an externality means that consumption or production activities are influenced positively or negatively while no compensation is paid for this influence. The relationship between two suppliers is a case in point. A price decrease caused by an increase in x_1 has a negative effect not only on firm 1's revenue $\left(\frac{dp}{dX}x_1 < 0\right)$ but also on firm 2's revenue $\left(\frac{dp}{dX}x_2 < 0\right)$. Hence, firm 1 exerts a negative externality on firm 2:

$$\frac{\partial \Pi_2}{\partial x_1} < 0.$$

The cartel agreement asks firms not to disregard this negative externality as you can see from the above first-order conditions.

4.3.4. *Comparative statics and cost competition.* Firms may have common interests and conflicting interests. With respect to a and b, firms have common interests. Both profit from an increase in a and a decrease in b.

They may try to enhance overall demand for their products by collaborative advertising campaigns. Lower costs for both firms are also in their common interest. For example, they may lobby for governmental subsidies or take a common stance against union demands.

EXERCISE XI.16. *Two firms sell gasoline with unit costs* $c_1 = 0.2$ *and* $c_2 = 0.5$, *respectively. The inverse demand function is* $p(X) = 5 - 0.5X$.

(1) *Determine the Cournot equilibrium and the resulting market price.*
(2) *The government charges a quantity tax* t *on gasoline. How does the tax affect the price payable by consumers?*

We now use the envelope theorem to analyze how marginal-cost changes affect profit. Cost leadership can be obtained directly (by decreasing one's own cost) or indirectly (by raising the rival's cost):

- Own cost may be decreased by cost-saving measures or by research and development.
- Sabotage or, politically more correct, the support of environmental groups in the competitor's country are examples of the indirect way.

In order to analyze the direct way, we write down firm 1's reduced profit function

$$\Pi_1^C(c_1, c_2) = \Pi_1\left(c_1, c_2, x_1^C(c_1, c_2), x_2^C(c_1, c_2)\right).$$

Differentiating with respect to c_1 yields

$$\frac{\partial \Pi_1^C}{\partial c_1} = \underbrace{\frac{\partial \Pi_1}{\partial c_1}}_{\substack{<0 \\ \text{direct effect}}} + \underbrace{\frac{\partial \Pi_1}{\partial x_1}}_{=0} \frac{\partial x_1^C}{\partial c_1} + \underbrace{\underbrace{\frac{\partial \Pi_1}{\partial x_2}}_{<0} \underbrace{\frac{\partial x_2^C}{\partial c_1}}_{>0}}_{\substack{<0 \\ \text{strategic effect}}} < 0.$$

The first summand is the direct effect. How is profit Π_1 affected if c_1 is increased while x_1^C and x_2^C are not changed? Negative, of course. The second summand vanishes because of the envelope theorem (firm 1's marginal profit is zero in equilibrium). The last summand is the strategic effect. A decrease in c_1 reduces x_2^C (see also figure XI.21) which is an effect welcome to firm 1 because of the associated price increase. Thus, in order to maximize profit, a duopolist invests more in cost reductions than advisable from the direct effect alone.

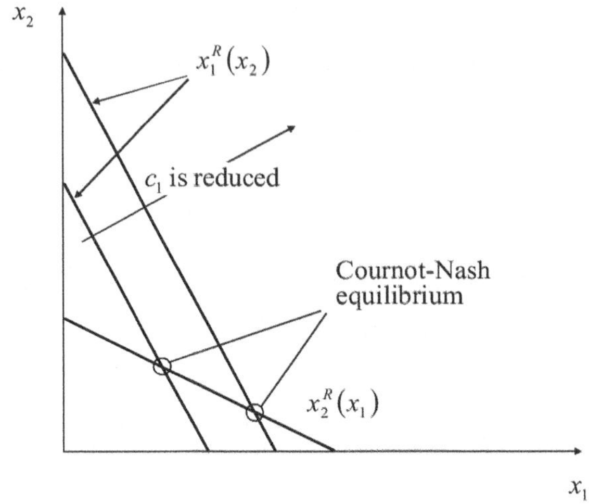

FIGURE XI.21. Cournot-Nash equilibria for decreasing costs

The indirect way can be shown by differentiating Π_1 with respect to c_2. We find

$$\frac{\partial \Pi_1^C}{\partial c_2} = \underbrace{\frac{\partial \Pi_1}{\partial c_2}}_{\substack{=0 \\ \text{direct effect}}} + \underbrace{\frac{\partial \Pi_1}{\partial x_1}}_{=0}\frac{\partial x_1^C}{\partial c_2} + \underbrace{\underbrace{\frac{\partial \Pi_1}{\partial x_2}}_{<0}\underbrace{\frac{\partial x_2^C}{\partial c_2}}_{<0}}_{\substack{>0 \\ \text{strategic effect}}} > 0.$$

The strategic effect is positive. A rival's cost increase increases a firm's profit. Indeed, if a firm 2 produces with higher marginal cost c_2, its equilibrium quantity decreases (see figure XI.22) so that firm 1's profit increases, again due to a price increase.

4.3.5. *Replicating the Cournot model.* Replication means making copies of the agents and their properties. We now want to replicate the Cournot model and hope to arrive at perfect-competition results. We consider an example taken from Bester (2017, pp. 135-136, 244-245).

We assume m identical consumers and n identical firms. Every consumer has the demand function

$$1 - p$$

and every firm j is characterized by the cost function

$$C(x_j) = \frac{1}{2}x_j^2.$$

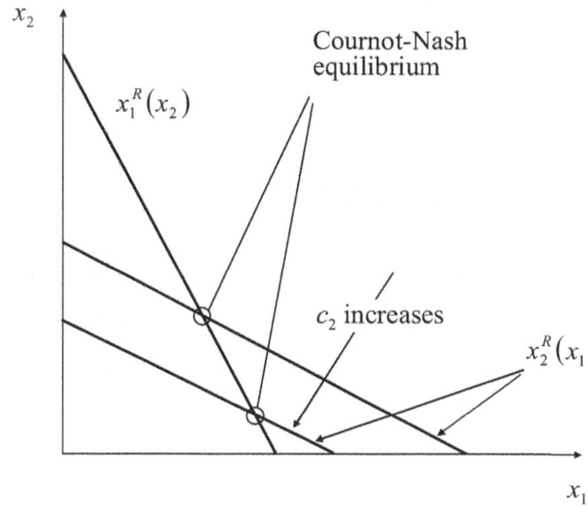

FIGURE XI.22. Cournot-Nash equilibria for increasing rival costs

We now derive the inverse market demand function. The m consumers' demand is $X = m(1-p)$ so that inverse aggregate demand is

$$p(X) = \frac{m - X}{m}.$$

Therefore, firm j's profit function is

$$\begin{aligned} \Pi_j(X) &= p(X)\, x_j - C(x_j) \\ &= \left(1 - \frac{x_j + \sum_{i \neq j} x_i}{m}\right) x_j - \frac{1}{2} x_j^2 \\ &= \left(1 - \frac{x_j + X_{-j}}{m}\right) x_j - \frac{1}{2} x_j^2, \end{aligned}$$

yielding the reaction function x_j^R given by

$$x_j^R(X_{-j}) = \frac{m - X_{-j}}{m+2}.$$

We now restrict attention to symmetric equilibria. Thus, we let $X_{-j} = (n-1)\, x_j$ and find

$$x_j = \frac{m - (n-1)\, x_j}{m+2}$$

and hence (by solving for x_j) the equilibrium quantity

$$x_j^C = \frac{m}{m+1+n}.$$

Aggregate output in equilibrium is

$$X^C = nx_j^C = \frac{nm}{m+1+n}$$

and we obtain the market price

$$p^C = 1 - \frac{n}{m+1+n}.$$

Now, we can set replication into effect. Instead of n firms and m consumers, we consider λn firms and λm consumers. The λ-fold replication yields the market price

$$p^C(\lambda) = 1 - \frac{\lambda n}{\lambda m + 1 + \lambda n}$$

and marginal cost

$$MC_j(\lambda) = \frac{\lambda m}{\lambda m + 1 + \lambda n}.$$

The price-cost margin is

$$\begin{aligned} p^C(\lambda) - MC_j(\lambda) &= 1 - \frac{\lambda n}{\lambda m + 1 + \lambda n} - \frac{\lambda m}{\lambda m + 1 + \lambda n} \\ &= \frac{\lambda m + 1 + \lambda n - \lambda n - \lambda m}{\lambda m + 1 + \lambda n} \\ &= \frac{1}{\lambda m + 1 + \lambda n} \\ &= \frac{1}{\lambda(m+n)+1}. \end{aligned}$$

Letting λ go towards infinity, we find

$$\lim_{\lambda\to\infty} \frac{1}{\lambda(m+n)+1} = 0.$$

Thus, we obtain the same result as under perfect competition: $MC = p$.

4.4. Blockaded entry and deterred entry. It is, of course, no forgone conclusion that both firms offer positive quantities in equilibrium. If the marginal costs are sufficiently different, one firm may choose not to enter. Let us assume $c_1 < c_2$.

4.4.1. *Market entry blockaded for both firms.* Market entry is blockaded for both firm if

$$c_1 \geq a \text{ and } c_2 \geq a$$

holds.

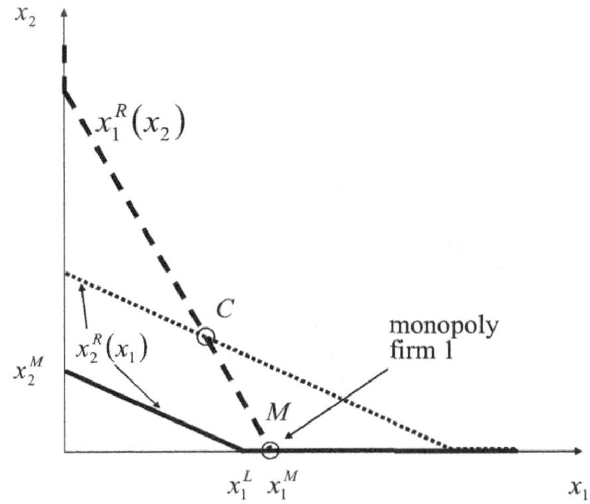

FIGURE XI.23. Cournot equilibrium with blockaded entry

1. no supply	$c_1 \geq a$ and $c_2 \geq a$
2. entry of firm 2 blockaded	$0 \leq c_1 < a$ and $c_2 \geq p^M(c_1) = \frac{1}{2}(a + c_1)$
3. entry of firm 1 blockaded	$0 \leq c_2 < a$ and $c_1 \geq p^M(c_2) = \frac{1}{2}(a + c_2)$
4. accomodation	$c_1 < p^M(c_2)$ and $c_2 < p^M(c_1)$

FIGURE XI.24. Solutions to quantity competition for different combinations of unit costs

4.4.2. *Market entry of firm 2 blockaded.* Market entry for firm 2 is blockaded if its unit cost is above firm 1's monopoly price:

$$c_2 \geq p^M(c_1) = \frac{1}{2}(a + c_1).$$

In terms of the reaction functions, we obtain figure XI.23. You see two reaction functions for firm 2. The lower one represents higher unit cost c_2. This curve intersects firm 1's reaction curve at $(x_1^M, 0)$. In other words, we have $x_1^M > x_1^L$ so that firm 1's monopoly quantity is sufficient to keep firm 2 off the market.

4.4.3. *Summary.* We distinguish four cases that are listed in figure XI.24 and depicted in figure XI.25.

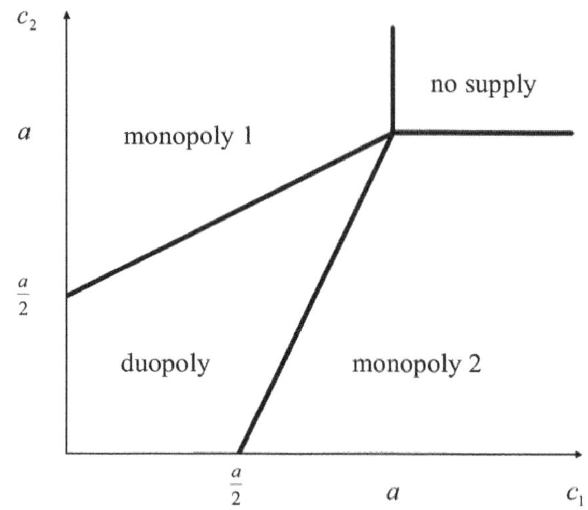

FIGURE XI.25. Cournot competition: accomodation and blockade

5. Topics and literature

The main topics in this chapter are

- monopoly
- oligopoly
- pricing policy
- quantity competition
- potential competition
- rent
- reaction curve
- Cournot point
- marginal revenue
- marginal cost
- price differentiation
- direct and indirect effects
- envelope theorem
- local solution
- global solution

This chapter owes a lot to the German textbook by Pfähler & Wiese (2008).

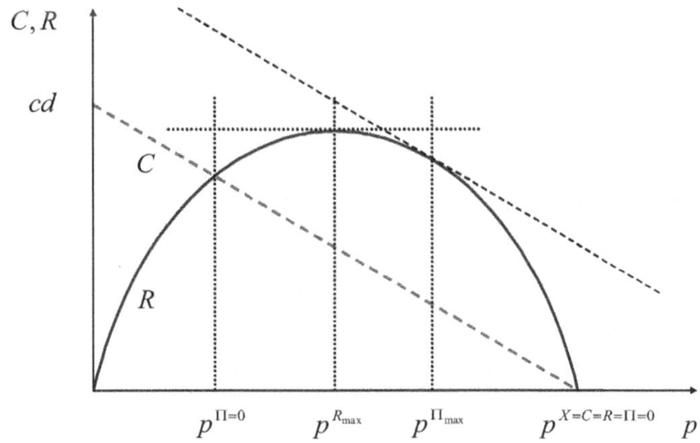

FIGURE XI.26. Solution

6. Solutions

Exercise XI.1

The saturation quantity is $X(0) = d$, the prohibitive price is implicitly defined by $X(p) = 0$ and hence equal to $\frac{d}{e}$. The price elasticity of demand is

$$\varepsilon_{X,p} = \frac{dX}{dp}\frac{p}{X} = (-e)\frac{p}{d - ep}.$$

Exercise XI.2

Sorry, there is none.

Exercise XI.3

The answer is provided in figure XI.26. Note that we have to shift the cost curve upwards in order to find the price at which the difference between revenue and cost is maximal.

Exercise XI.4

If demand is inelastic at the monopoly price ($|\varepsilon_{X,p}| < 1$), a price increase leads to a revenue increase. Thus, revenue can be increased and cost can be decreased (a smaller quantity normally leads to lower cost). Therefore, the claim is correct.

Exercise XI.5

The monopolist's profit function is given by

$$\begin{aligned} \Pi(p) &= X(p)p - cX(p) \\ &= (d - ep)p - c(d - ep) \\ &= dp - ep^2 - cd + cep. \end{aligned}$$

Setting the derivative equal to zero and solving for p yields

$$\begin{aligned}\frac{d\Pi}{dp} &= d - 2ep + ce \stackrel{!}{=} 0 \text{ and, indeed,} \\ p^M &= \frac{d+ce}{2e} = \frac{d}{2e} + \frac{c}{2}.\end{aligned}$$

The revenue-maximizing price is $\frac{d}{2e}$ (just let $c = 0$). We find

$$\frac{\partial p^M}{\partial c} = \frac{1}{2}.$$

Thus, an increase of the unit cost by one unit leads to a price increase by $\frac{1}{2}$.

Exercise XI.6

By the linear cost funcion, if we have price differentiation of the third degree, we just need to solve two isolated profit-maximization problems. We obtain

$$\begin{aligned}p_1^M &= 60, \\ p_2^M &= 35.\end{aligned}$$

The prohibitive prices are 100 and 50, respectively. The aggregate demand is therefore given by

$$X(p) = \begin{cases} 0, & p > 100 \\ 100 - p, & 50 < p \leq 100 \\ 200 - 3p, & 0 \leq p \leq 50. \end{cases}$$

We find two local solutions, at $p = 43\frac{1}{3}$ and $p = 60$. A comparison of profits yields the global maximum at

$$p^M = 43\frac{1}{3}.$$

Exercise XI.7

$(10, 10)$ is still an equilibrium. $(11, 11)$ is a second one.

Exercise XI.8

All strategy combinations $\left(p_1^M, p_2\right)$ fulfilling $p_2 > p_1^M$ are also equilibria.

Exercise XI.9

(1) The slope of the inverse demand curve is $\frac{dp}{dX} = -b$.

(2) Revenue is given by $R(X) = p(X)X = aX - bX^2$ so that we obtain the marginal revenue

$$\frac{dR(X)}{dX} = a - 2bX.$$

The slope of the marginal-revenue curve is $-2b$. Hence this curve is twice as steep as the demand curve itself (see figure XI.27).

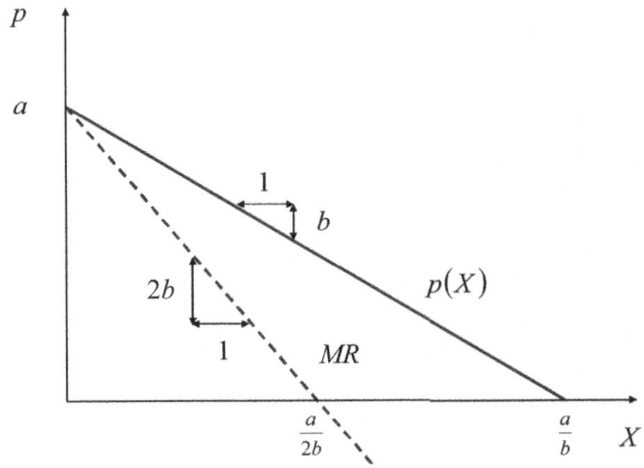

FIGURE XI.27. Demand curve and marginal-revenue curve

(3) The saturation quantity is $\frac{a}{b}$.
(4) a is the prohibitive price.

Exercise XI.10

The monopoly quantity is $X^M = 11$.

Exercise XI.11

Mean question! For any output, revenue is constant, $R(X) = p(X)X = 1$. Differently put, marginal revenue is zero:

$$\begin{aligned} MR &= p + X\frac{dp}{dX} \\ &= \frac{1}{X} + X\left(-\frac{1}{X^2}\right) = 0. \end{aligned}$$

Thus, the monopolist should produce one very small unit, only.

Exercise XI.12

You should have obtained

$$\begin{aligned} \frac{d\Pi^M}{dc} &= \frac{d\left(\frac{1}{4}\frac{(a-c)^2}{b}\right)}{dc} \\ &= \frac{1}{4b}2(a-c)(-1) \\ &= -\frac{a-c}{2b} \end{aligned}$$

Exercise XI.13

The above result is a corollary of theorem VII.3 if we let $\Pi := f, \Pi^M := \hat{f}, c := a$, and $X^M(c) := x^R(a)$.

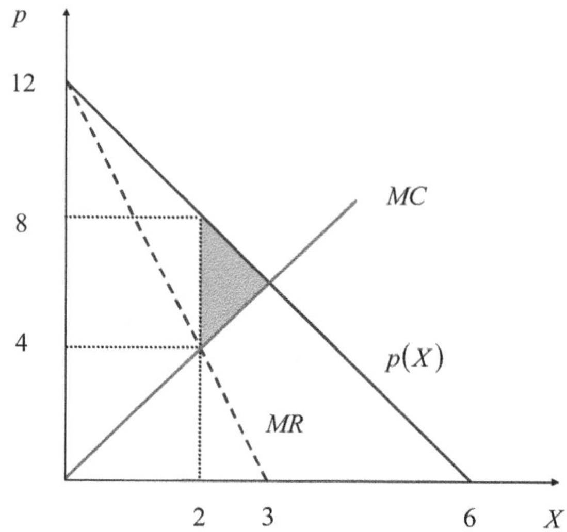

FIGURE XI.28. The welfare loss in a monopoly

Exercise XI.14

Here comes the summary of what you should have found:

(1) The firm's profit function is

$$\begin{aligned}\Pi(x_1, x_2) &= p_1(x_1)x_1 + p_2(x_2)x_2 - C(x_1 + x_2) \\ &= (100 - x_1)x_1 + (80 - x_2)x_2 - (x_1 + x_2)^2.\end{aligned}$$

Partial differentiations yield

$$x_1^M = 20 \text{ and } x_2^M = 10.$$

The monopolist earns $\Pi^M(20, 10) = 1400$.

(2) We find the optimal outputs $x_1^M = 45$ and $x_2^M = 35$; profit is $\Pi^M = 3250$.

(3) The aggregate inverse demand function is

$$p(X) = \begin{cases} 100 - X, & X < 20 \\ 90 - \frac{1}{2}X, & X \geq 20. \end{cases}$$

At $X^M = 80$, the monopolist's profit is 3200<3250.

Exercise XI.15

The situation is depicted in figure XI.28. The welfare loss is equal to

$$\frac{(8-4)(3-2)}{2} = 2.$$

Exercise XI.16

Did you get the following results:

(1) $x_1^C = 3.4$, $x_2^C = 2.8$, and $p^C = 1.9$

(2) The market price is $p^C = 1.9 + \frac{2}{3}t$. Differentiation with respect to t yields $\frac{dp}{dt} = \frac{2}{3}$, i.e., a tax increase by one leads to a price increase by $\frac{2}{3}$.

7. Further exercises without solutions

PROBLEM XI.1. Consider a monopolist with profit function $C(X) = cX$, $c > 0$, and demand function $X(p) = ap^{\varepsilon}$, $\varepsilon < -1$.

(1) Find the price elasticity of demand and the marginal revenue with respect to price!
(2) Express the monopoly price p^M as a function of ε!
(3) Find and interpret $\frac{dp^M}{d|\varepsilon|}$!

PROBLEM XI.2. Assume simultaneous price competition and two firms where firm 2 has capacity constraint cap_2 such that

$$\frac{1}{2}X(c) < cap_2 < X(c).$$

Is (c, c) an equilibrium?

PROBLEM XI.3. Three firms operate on a market. The consumers are uniformly distributed on the unit interval, $[0,1]$. The firms $i = 1, 2, 3$ simultaneously choose their respective location $l_i \in [0,1]$. Each consumer buys one unit from the firm which is closest to her position; if more than one firm is closest to her position, she splits her demand evenly among them. Each firm tries to maximize its demand. Determine the Nash equilibria in this game!

PROBLEM XI.4. Assume a Cournot monopoly. Analyze the welfare effects of a unit tax and a profit tax. Consider the welfare effects of a unit tax in the Cournot oligopoly with $n > 1$ firms, linear demand, and constant average cost. Restrict attention to symmetric Nash equilibria! What happens for $n \to \infty$?

PROBLEM XI.5. Assume a Cournot monopoly. Analyze the quantity effects of a price cap.

CHAPTER XII

Games in extensive form

This chapter on games in extensive form is a convex combination of chapter III (on extensive-form decisions) and chapter X (on games in strategic form). We consider games of (nearly) perfect information without moves by nature. Games with imperfect information due to moves by nature are dealt with in chapter XVII.

With respect to industrial organization, we explain

- the Stackelberg model (as the archetypal extensive-form game),
- product differentiation and the Hotelling model,
- strategic trade policy.

1. Examples

1.1. Non-simultaneous moves in simple bimatrix games. In chapter X, we consider the bimatrix games "stag hunt", "head or tail" and "chicken". We now revisit these games and have player 1 move first and then player 2 follow. We assume that player 2 observes the action undertaken by player 1.

Let us begin with the stag hunt:

		hunter 2	
		stag	hare
hunter 1	stag	$5,5$ [1][2]	$0,4$
	hare	$4,0$	$4,4$ [1][2]

The boxes [1] and [2] indicate the players' best responses (see p. 275). Player 1 chooses "stag" or "hare". He knows that player 2 knows his (player 1's) action. Assuming that player 2 is rational, player 1 predicts that player 2 chooses "stag" in response to "stag" and "hare" in response to "hare". In other words, player 2 acts according to his best response and player 1 knows this. Thus, both players choose the actions "stag".

H. Wiese, *Advanced Microeconomics*,
https://doi.org/10.1007/978-3-658-34959-2_12

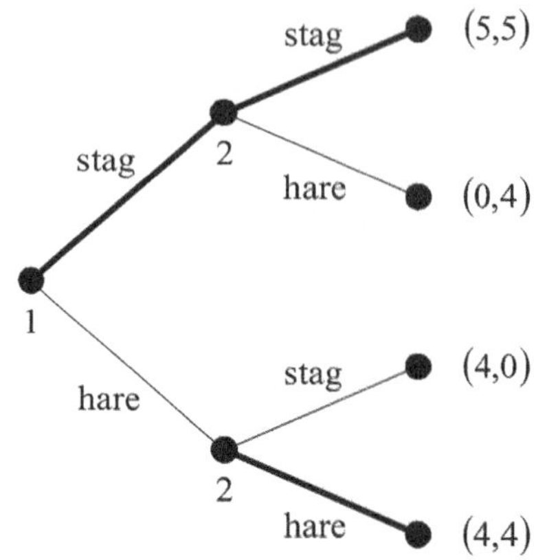

FIGURE XII.1. The stag hunt in extensive form

Of course, the above reasoning is just the backward-induction argument. To visualize backward induction, we draw the game tree of the stag hunt as in figure XII.1. Note that the nonterminal nodes indicate the acting players. The edges are reinforced in basically the same manner as in chapter III. The only difference is that we have two deciders instead of just one.

The backward-induction strategies are $\lfloor$stag, hare$\rfloor$ for player 2 and $\lfloor$stag$\rfloor$ for player 1. Indeed, player 1 has two strategies and player 2 four strategies.

Consider the strategies $\lfloor$stag$\rfloor$ for player 1 and $\lfloor$stag, stag$\rfloor$ for player 2. These strategies are in equilibrium because they lead to the maximal payoff of 5 for both hunters. However, there is something wrong with player 2's strategy. If you do not know, consult p. 36 in a decision-theoretic context.

EXERCISE XII.1. *Find the backward-induction solution for the game of chicken (p. 267).*

1.2. Three Indian fables.

1.2.1. *The tiger and the traveller.* Wiese (2012) argues that backward induction has already been applied in Indian animal tales. We present three examples. The first example is the tale of the tiger and the traveller known from the Hitopadesha collection of fable-based advice (see, for example, Kale & Kale (1967, pp. 7-9) or the comic book by Chandakant & Fowler (1975, pp. 14-18)).

This is the story: A tiger that finds himself on one side of a lake sees a traveller passing by on the opposite side. The tiger attempts to catch and eat the traveller by offering a golden bracelet to him. Since the traveller is suspicious of the tiger's intentions, the tiger argues that he would not

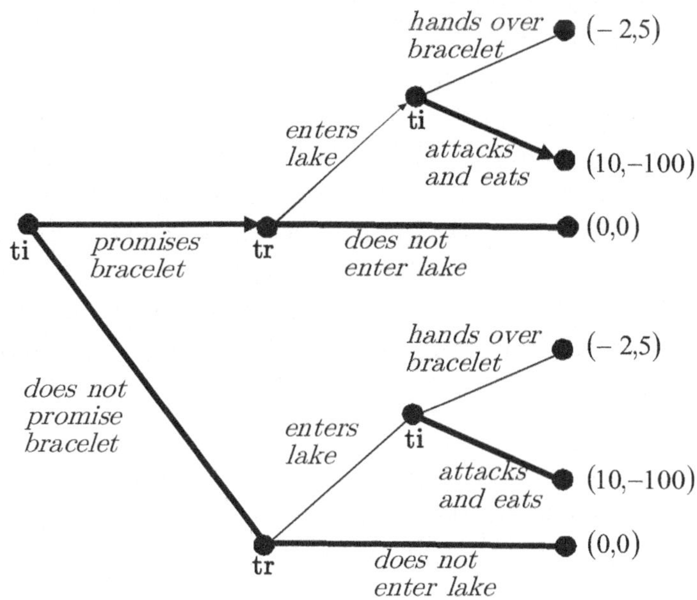

FIGURE XII.2. The tiger and the traveller

(he claims to have profoundly changed his former evil behavior) and could not (he claims to be old and weak) do any harm to the traveller. Finally, the traveller is convinced, gets into the murky waters where he gets stuck. Immediately, the tiger takes advantage of the traveller's misfortune and kills him as planned.

Consider the payoffs in figure XII.2. The first number at the final nodes refers to the tiger, the second one to the traveller. The tiger's payoffs are -2 for giving away the bracelet and not eating the traveller, 10 for keeping the bracelet and enjoying a good meal, and 0 for the status quo of keeping the bracelet but staying hungry. The corresponding traveller's payoffs are 5, -100, and 0.

The tragic sequence of events sketched above is indicated by the arrows. The tiger (ti) moves first by promising the bracelet (upper branch). The traveller (tr) enters the lake (upper branch) and then the tiger kills the traveller (lower branch).

The game tree of this story has three stages. First, the tiger offers the bracelet and talks about his guru who has convinced him to lead a more virtuous life or the tiger refrains from offering the bracelet and/or from talking convincingly. Then, the traveller needs to decide on whether or not to accept the tiger's invitation to join him by crossing the lake. Finally, the tiger fulfills his promise or reneges on it.

One may of course speculate why the traveller is so "stupid". Did "greed cloud the mind" or did he act on some probability assessment about the tiger telling the truth? Indeed, the tiger claims to have studied the Vedas to lend credibility to his peaceful intentions. However, it seems obvious that the fable writer does not think of this example under the heading of "better safe than sorry". Instead he argues that the tiger's preferences being as they are the traveller should have known his fate in advance. Before being killed, the traveller has time for some wise insights to share with the readers (see Kale & Kale 1967, p. 8):

> That he reads the texts of religious law and studies the Vedas, is no reason why confidence should be reposed in a villain: it is the nature that predominates in such a case: just as by nature the milk of cows is sweet.

Knowledge of backward induction would also have led the traveller to avoid the lake. By $10 > -2$, he should have foreseen his being eaten after entering the lake so that keeping clear of the lake is best by $0 > -100$.

Interestingly, the traveller should refrain from entering the lake independent of whether or not the tiger talks about his guru who advised the tiger to pursue a more virtuous life. In game-theory parlance, the tiger's arguments, the first step in our game tree, are just "cheap talk". Both a mischievous and a benevolent tiger could claim their benevolence without any cost. Therefore, this claim is not credible.

Pious appearances are also used by the cat in an animal tale from the Panchatantra (see, for example Olivelle 2006, pp. 393-399). The cat is chosen as a judge in a dispute between a partridge and a hare. Although wary of the danger, the two contestants finally approach the cat who kills them without much ado.

1.2.2. *The lion, the mouse, and the cat.* The second animal tale is also taken from the Hitopadesha (see Kale & Kale 1967, p. 51). A lion that lives in a cave is infuriated by a mouse that also lives in his cave. The mouse regularly gnaws at the sleeping lion's mane. Since the lion does not succeed in catching the mouse, he invites a hungry and desperate cat to live and eat in his cave.

The arrangement between the lion and the cat works out well. The mouse does not dare to show up while the cat is present. Therefore, the lion is happy to share his food with the cat as promised although he does not particularly like the cat's company by itself. One day, the mouse is

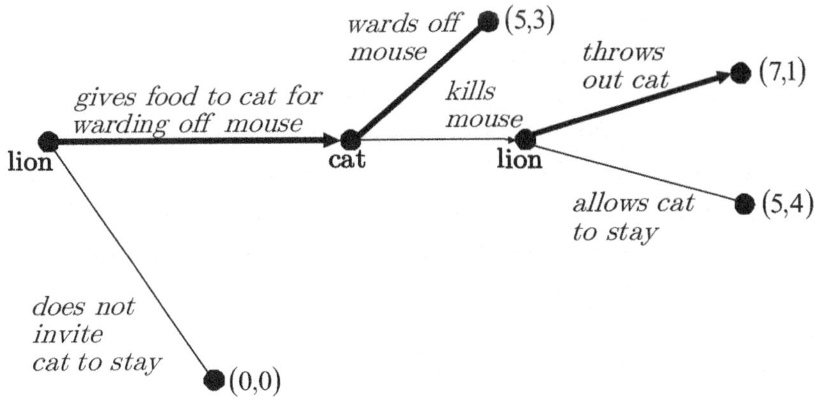

FIGURE XII.3. The lion, the mouse, and the cat

detected by the cat who catches and kills it. The lion does not see any reason to extend his hospitality and makes the cat leave his cave. Soon, the cat returns to its former miserable state.

The moral to be drawn from this fable is obvious: Do your work but see to it that you are also needed in the future.

The reader is invited to have a look at figure XII.3. The first number at the final nodes refers to the lion, the second to the cat. Both players obtain a payoff of 0 if the lion does not invite the cat to stay so that the lion's mouse problem is not solved and the cat cannot eat the food provided by the lion. The lion's payoff is 5 if the mouse does not annoy him and increases up to 7 if, on top, the cat does not stay in the cave. The cat in the cave has a payoff of 3 if it can stay in the cave and an increased payoff of 4 for eating the mouse and staying in the cave.

The arrows indicate the story as told in the Hitopadesha. This is not the backward-induction result which, again, is indicated by the thickened lines. The wise cat would foresee that it is in the best interest of the lion to get rid of it after the mouse is killed ($7 > 5$). Therefore, the cat should have kept on warding off the mouse (payoff 3) rather than killing the mouse and be thrown out of the convenient cave (payoff 1). Working backwards one final step, we see that the lion was right to invite the cat into his cave ($5 > 0$). Indeed, because of the cat's mistake, the lion is even better off obtaining 7 rather than 5.

Again, one may ask the question whether there are defensible reasons for the violation of backward induction. Did the cat think that another mouse would show up promptly so that the lion would need the cat's services again? It seems that the fable's author did not think along these lines, but had the more straight-forward didactic aim of teaching the forward-looking behavior the cat did not master.

A second possibility comes to mind: The cat may have entertained the hope that the lion would show thankfulness to the cat for freeing the lion of the mouse for good. However, we would rather not follow this line of thought. Instead, the fable authors insist on the lesson that friendship and thankfulness is not to be relied upon. The behaviors of humans or animals are dictated by future gains and losses, rather than by friendly acts in the past.

1.2.3. *The cat and the mouse.* In the previous animal tale, the lion profited from the opponent's mistake. Sometimes, however, players hope that opponents react rationally. To show this, we finally present a fable from book 12 of the grand epic Mahabharata (see Fitzgerald 2004, pp. 513-518). A he-cat is caught in a net laid out by a trapper. The mouse is happy to see her enemy in this difficult situation when she realizes that an owl is about to attack from above and a mongoose is sneaking up on her. She offers the cat to destroy the net if the cat gives her shelter. The mouse realizes that her plan needs a good deal of rationality and foresight on the cat's part (p. 514):

> So I will make use of my enemy the cat. I shall contribute to his welfare ... And now may this enemy of mine happen to be smart.

Fortunately, the cat agrees to the bargain. When seeing the mouse under the cat's protection, owl and mongoose look out for other prey. The cat is dismayed to find that the mouse is in no hurry to fulfill her promise. Indeed, the mouse realizes that freeing the cat immediately makes her an easy victim of the cat. In a long dialogue, the logic of the situation is explicitly spelled out. As the mouse remarks (p. 517):

> No one is really an ally to anyone, no one is really a friend to anyone ... When a job has been finished, no one pays any attention to the one who did it; so one should make sure to leave all his tasks with something still to be done. At just the right time, sir, you will be filled with fear of the [trapper] and intent on getting away, and you won't be able to capture me.

Thus, the mouse waits until the trapper approaches. At the very last moment, the mouse liberates the cat that now has better things to do than

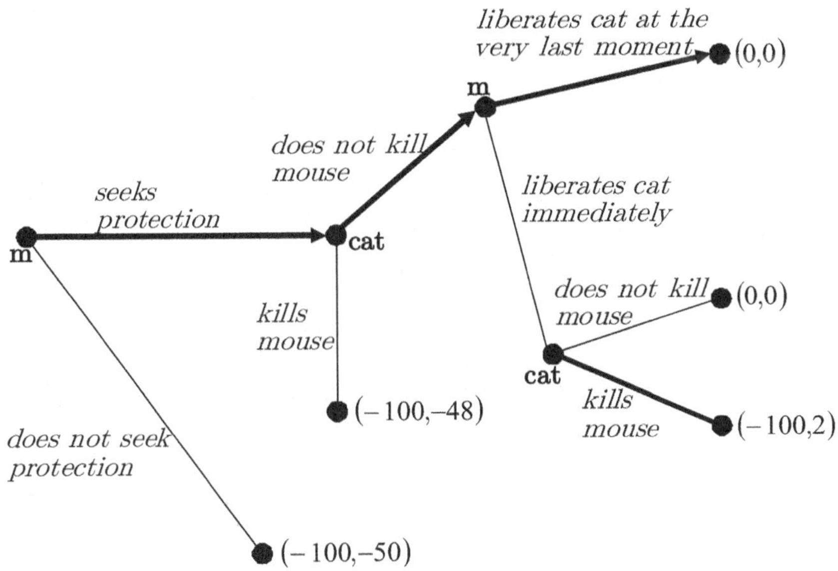

FIGURE XII.4. The cat and the mouse

mouse hunting. Both manage to find a safe place to hide, but certainly not the same.

Figure XII.4 shows the game tree of this animal tale. The first payoff accrues to the mouse (m), the second one to the cat. The mouse obtains 0 for escaping unharmed and suffers the payoff of -100 for being killed by owl, mongoose, or cat. The cat's payoff is zero for escaping unharmed, 2 for escaping and eating the mouse, -50 for being killed by the trapper and -48 for being killed by the trapper after eating the mouse.

Foreseeing that the cat will kill the mouse if liberated well before the trapper arrives ($2 > 0$), the mouse prefers to wait until the trapper approaches ($0 > -100$). The cat is clever enough not to kill the mouse before he is liberated ($0 > -48$). Thus, indeed, the mouse made a clever move to seek the cat's protection ($0 > -100$).

Unlike the first two stories, in this story, the sequence of events is the one predicted by backward induction. Neither the mouse nor the cat makes a mistake.

1.3. Example: the Stackelberg model.

1.3.1. *Recipe: How to solve the Stackelberg model.* The models due to Cournot and Bertrand (see chapter XI) exemplify simultaneous-moves games in IO. The Stackelberg model is the most famous sequential model. We present the basic ideas. We have two quantity-setting firms 1 and 2.

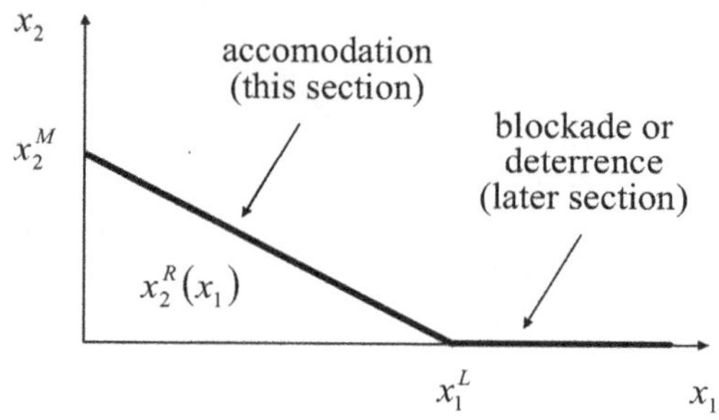

FIGURE XII.5. The follower's reaction curve

Firm 1 is the Stackelberg leader and moves first. Firm 2 is the second mover and called the Stackelberg follower. We assume the profit functions

$$\begin{aligned}\Pi_1(x_1, x_2) &= (a - b(x_1 + x_2))x_1 - c_1 x_1, \\ \Pi_2(x_1, x_2) &= (a - b(x_1 + x_2))x_2 - c_2 x_2.\end{aligned}$$

The leader moves first and chooses some quantity x_1. The follower observes x_1 and chooses his profit-maximizing quantity, i.e.,

$$x_2^R(x_1) = \underset{x_2}{\text{argmax}}\ \Pi_2(x_1, x_2) = \frac{a - c_2}{2b} - \frac{1}{2}x_1.$$

The leader foresees the follower's optimal output choice. Hence, firm 1's *reduced profit function* (a new term for you!) is

$$\Pi_1(x_1) := \Pi_1\left(x_1, x_2^R(x_1)\right) = p\left(x_1 + x_2^R(x_1)\right)x_1 - c_1 x_1.$$

Therefore, firm 1's optimization problem is to find the best point (best for firm 1) on the follower's reaction curve (figure XII.5).

The leader's first-order condition is

$$\begin{aligned}MR_1(x_1) &= a - b\left(x_1 + x_2^R(x_1)\right) + x_1(-b) + x_1(-b)\left(-\frac{1}{2}\right) \\ &= a - b\left(x_1 + \frac{a - c_2}{2b} - \frac{1}{2}x_1\right) + x_1(-b) + x_1(-b)\left(-\frac{1}{2}\right) \\ &= a - bx_1 - \frac{b(a - c_2)}{2b} \overset{!}{=} c_1 = MC_1(x_1)\end{aligned}$$

which yields the profit-maximizing quantity

$$x_1^S = \frac{a - 2c_1 + c_2}{2b}.$$

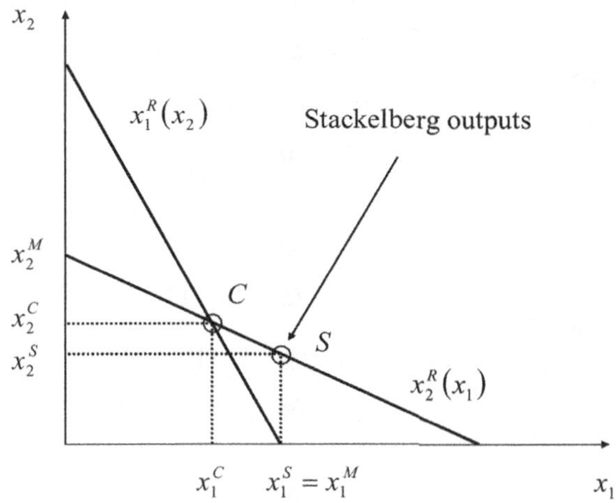

FIGURE XII.6. Stackelberg point for identical unit costs

We also find

$$\begin{aligned} x_2^S &:= x_2^R\left(x_1^S\right) = \frac{a+2c_1-3c_2}{4b}, \\ X^S &:= x_1^S + x_2^S = \frac{3a-2c_1-c_2}{4b}, \\ p\left(X^S\right) &= \frac{1}{4}\left(a+2c_1+c_2\right), \\ \Pi_1^S &= \frac{1}{8}\frac{\left(a+c_2-2c_1\right)^2}{b}, \\ \Pi_2^S &= \frac{1}{16}\frac{\left(a-3c_2+2c_1\right)^2}{b}. \end{aligned}$$

Figure XII.6 shows the Stackelberg point $\left(x_1^S, x_2^S\right)$ on the follower's, not the leader's, reaction curve.

EXERCISE XII.2. *Assume three firms in a homogeneous market with inverse demand curve* $p(X) = 100 - X$. *Average and marginal cost of all firms are equal to zero. Consider the sequence where firm* 1 *moves first and firms* 2 *and* 3 *move second and simultaneously. Solve the model by backward induction. Hint: Solve for a Cournot equilibrium between firms* 2 *and* 3.

So far, the Stackelberg model for Bachelor students.

1.3.2. *Strategies and equilibria.* We sketch the Stackelberg tree in figure XII.7. It is only a sketch because every firm has the action set $[0, \infty)$ in the model, but only some prominent quantities are chosen. Player 1 moves

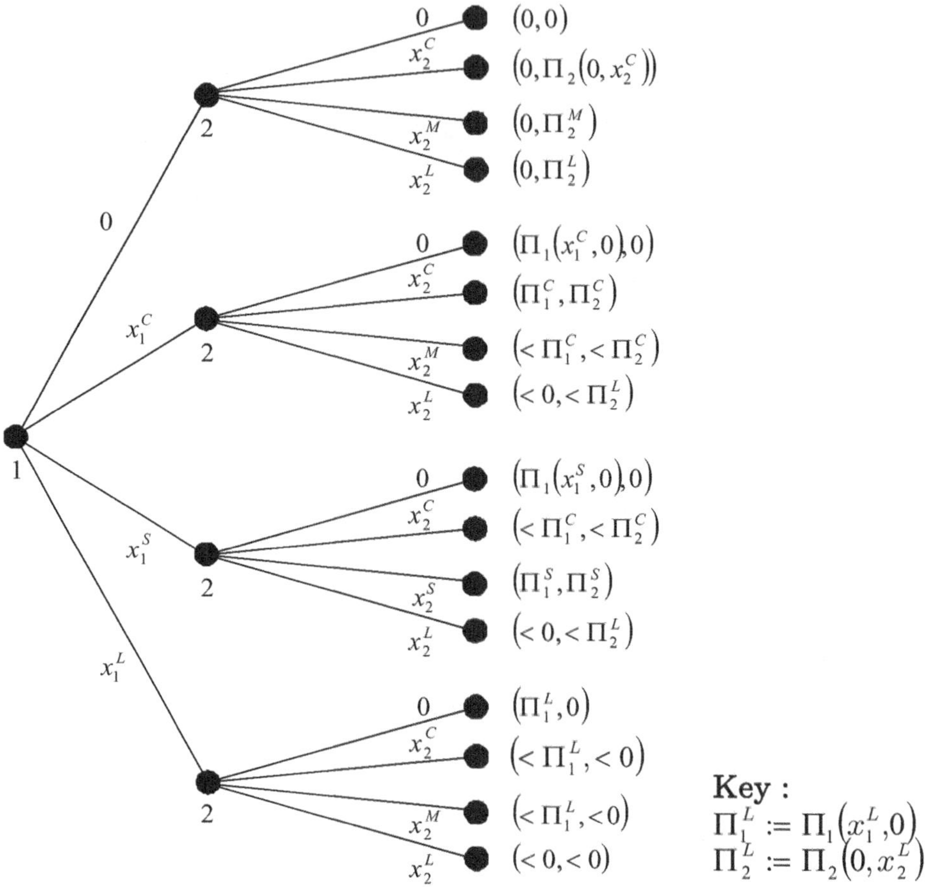

FIGURE XII.7. A sketch of the Stackelberg tree

at the tree's origin. Depending on his action, player 2 moves at the nodes denoted by "2".

Since every 2-node results from a specific quantity chosen by firm 1, we can write firm 2's strategies as functions

$$\begin{aligned} s_2 &: [0,\infty) \to [0,\infty), \\ & \quad x_1 \mapsto s_2(x_1) \end{aligned}$$

that associate an output $x_2 = s_2(x_1)$ with output x_1 chosen by firm 1. A specific example is given by

$$s_2^M : x_1 \mapsto \begin{cases} x_2^M, & x_1 = 0 \\ x_2^L, & x_1 > 0. \end{cases}$$

This example can be put into these words:

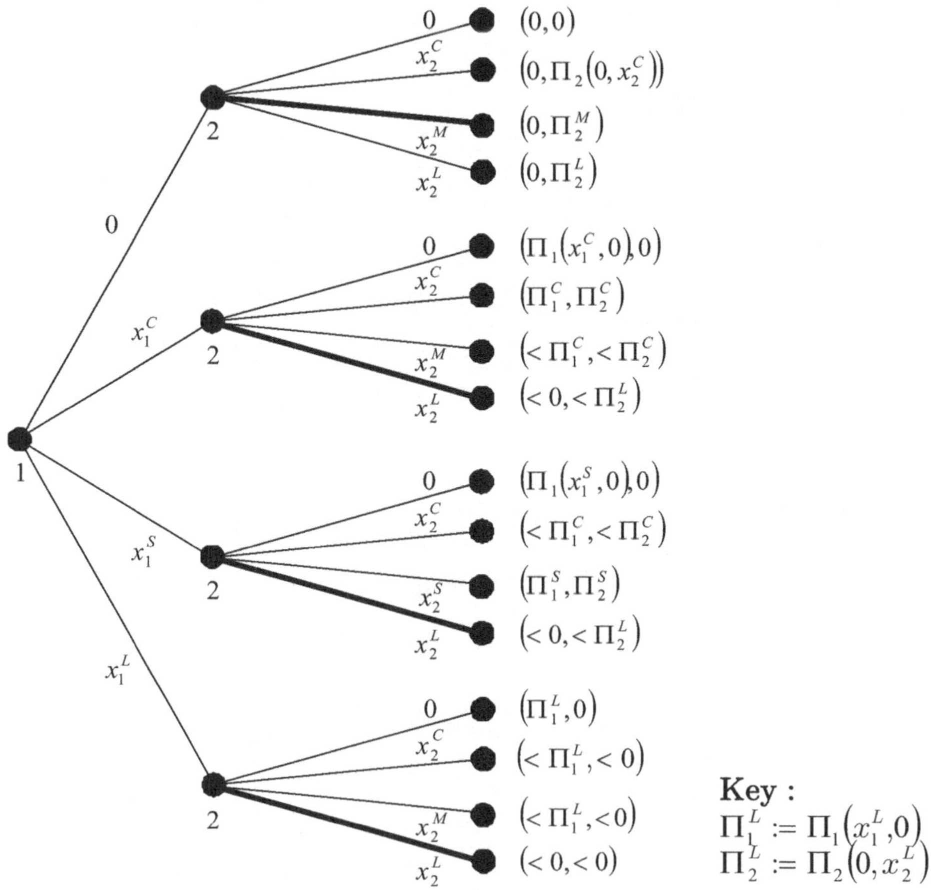

FIGURE XII.8. A threat strategy in the Stackelberg tree

- If firm 1 chooses output 0, firm 2 produces its monopoly output.
- If, however, firm 1 choses any positive quantity, firm 2 floods the market by choosing its limit quantity $x_2^L := \frac{a-c_1}{b}$.

This strategy is a threat. Either firm 1 (the Stackelberg leader) stays out of the market, or it will make a loss.

Figure XII.8 depicts this threat strategy. What is firm 1's best response to that strategy? Since any quantity $x_1 > 0$ leads to a loss, i.e., to profit $\Pi_1\left(x_1, x_2^L\right) < 0$, firm 1's best response to s_2^M is $x_1 = 0$. Therefore,

$$\left(0, s_2^M\right)$$

is a Nash equilibrium.

Of course, the above equilibrium is not the one derived in the previous subsection, which assumes that the follower maximizes his profit after

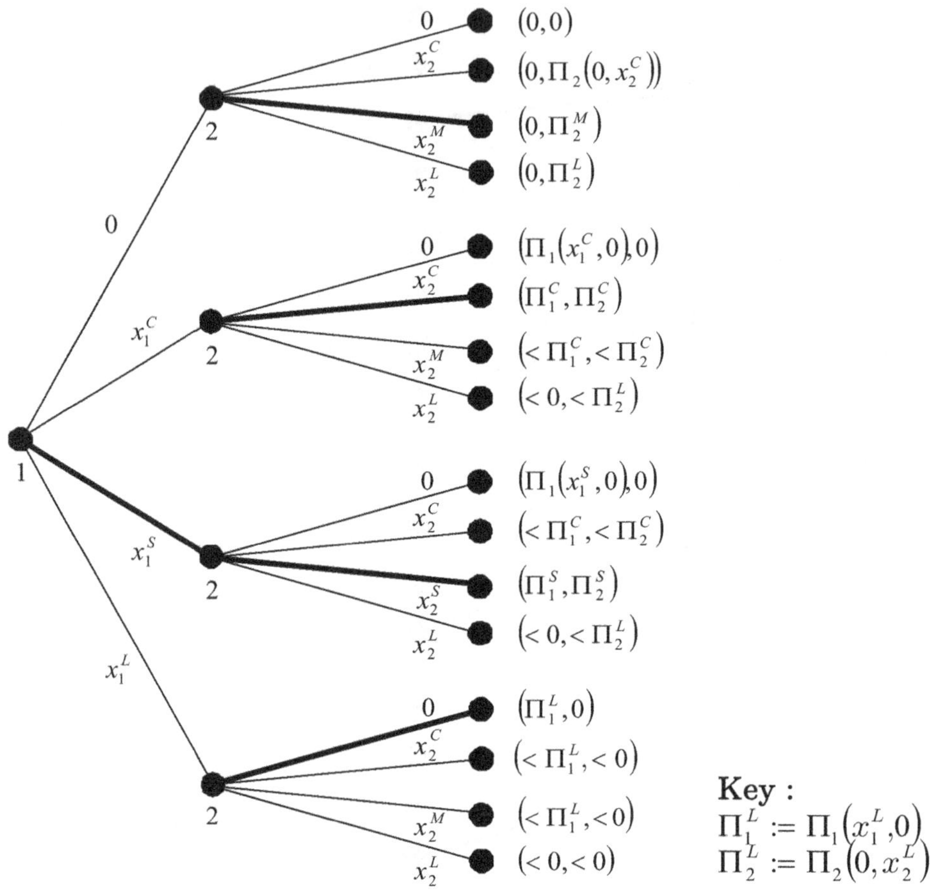

FIGURE XII.9. Backward induction in the Stackelberg tree

observing the leader's choice. Consider figure XII.9 where the backward-induction solution is indicated by bold lines. The corresponding equilibrium is

$$\left(x_1^S, x_2^R\right)$$

where x_1^S is an element from $[0,\infty)$ while x_2^R is the reaction function, i.e., a function $[0,\infty) \to [0,\infty)$.

We elaborate on that point because it is very important. The equilibrium $\left(x_1^S, x_2^R\right)$ has to be distinguished from the Stackelberg quantities

$$\left(x_1^S, x_2^R\left(x_1^S\right)\right).$$

To be very explicit,

- x_2^R is a function $[0,\infty) \to [0,\infty)$, in particular the best-response function of firm 2,

- $x_2^R(x_1^S)$ or $x_2^R(0)$ are values of that function at x_1^S or 0, respectively, i.e., elements from the range $[0,\infty)$.

Elements from $[0,\infty)$ such as 5, x_2^S, or x_2^C can also be considered strategies, namely constant strategies where firm 2 plans to choose quantity 5, x_2^S, or x_2^C, respectively, whatever firm 1's choice. Constant strategies are optimal only by chance.

EXERCISE XII.3. *Which of the following strategy combinations are Nash equilibria of the Stackelberg model?*

(1) $\left(x_1^S, x_2^R\left(x_1^S\right)\right)$
(2) $\left(x_1^S, x_2^R\right)$
(3) $\left(x_1^C, x_2^C\right)$

1.3.3. *Cournot versus Stackelberg.* The difference between the Cournot and the Stackelberg models can be seen from firm 1's marginal revenues in both cases. Let $X = x_1 + x_2^R(x_1)$ and $R_1(x_1) = p(X)x_1$. We then obtain the leader's marginal revenue as

$$\begin{aligned}
MR_1(x_1) &= p(X) + \frac{dp}{dX}\frac{\partial X}{\partial x_1}x_1 \qquad \text{(chain rule)}\\
&= p(X) + \frac{dp}{dX}\frac{d\left(x_1 + x_2^R(x_1)\right)}{dx_1}x_1 \qquad (X = x_1 + x_2^R(x_1))\\
&= p(X) + \frac{dp}{dX}\frac{\partial x_1}{\partial x_1}x_1 + \frac{dp}{dX}\frac{dx_2^R}{dx_1}x_1\\
&= \underbrace{p(X) + x_1\frac{dp(X)}{dX}}_{\text{direct effect}} + \underbrace{x_1\underbrace{\frac{dp(X)}{dX}}_{<0}\underbrace{\frac{dx_2^R(x_1)}{dx_1}}_{<0}}_{\substack{>0\\ \text{follower effect}}} \qquad \left(\frac{dx_1}{dx_1} = 1\right).
\end{aligned}$$

The follower effect is absent in the Cournot model. After all, firms move simultaneously and there is no time for a reaction. In our linear model (and in many other specifications), the follower effect is positive. A quantity increase by the leader firm 1 is partly compensated for by the follower firm so that the price decrease is less pronounced. Therefore, marginal revenue is higher in the Stackelberg than in the Cournot model. This is the reason for $x_1^S > x_1^C$.

2. Defining strategies

After doing all the difficult work in decision theory, life is rather easy now. We copy the definitions from chapter III.

DEFINITION XII.1 (EXTENSIVE FORM FOR PERFECT INFORMATION). *A game in extensive form for perfect information without moves by nature is a tuple*

$$\Gamma = \left(V, N, u, \iota, (A_d)_{d \in D}\right)$$

where

- *V is a tree with the set of non-terminal nodes D and the set of terminal nodes E (see the definition on p. 33),*
- *$N = \{1, ..., n\}$ is the player set,*
- *$u : E \to \mathbb{R}^n$ is a vector $(u_i)_{i \in N}$ of payoff functions $u_i : E \to \mathbb{R}$,*
- *$\iota : D \to N$ is a surjective player-selection function with two consequences:*
  - *player i's set of decision nodes is $D_i := \{d \in D : \iota(d) = i\}$,*
  - *$\{D_1, .., D_n\}$ is a partition of D,*
- *A_d is the set of actions that can be chosen by player $\iota(d)$ at decision node d. Every link at d corresponds to exactly one action. A is the union of all A_d, $d \in D$.*

As in a decision tree, a node is either terminal (and we note the payoff information near that node) or a decision node (and we write the player whose turn it is near that node). Even the player-selection function is not new to us, see definition III.19 on pp. 47.

As an example, we revisit the take-it-or-leave-it game presented on pp. 271. The matrix given in that chapter (on strategic-form games!) is

		player 2 accepts if he is offered at least x_2 coins				player 2 does not accept
		0	1	2	3	
player 1	0	3, 0	0, 0	0, 0	0, 0	0, 0
offers	1	2, 1	2, 1	0, 0	0, 0	0, 0
player 2	2	1, 2	1, 2	1, 2	0, 0	0, 0
x_1 coins	3	0, 3	0, 3	0, 3	0, 3	0, 0

The extensive-form of this game is depicted in figure XII.10.

EXERCISE XII.4. *Consider the bargaining game of figure XII.10. Identify N and D_2 in the game tree. How many strategies does player 2 have?*

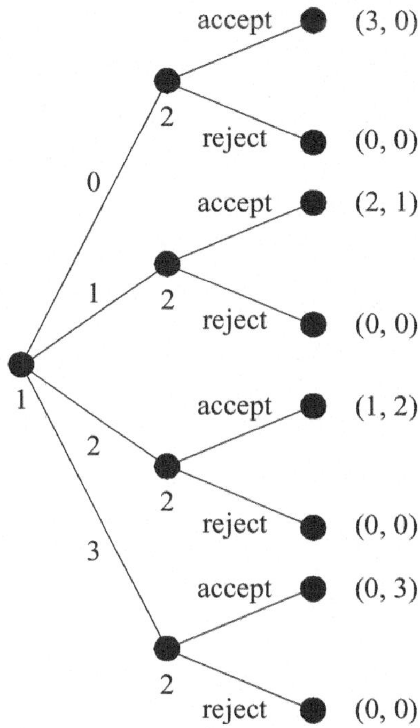

FIGURE XII.10. The take-it-or-leave-it game

Player 1 has four strategies in accordance with the above matrix. However, player 2 has more than two or eight strategies. For example,

$$\lfloor \text{accept}, \text{reject}, \text{accept}, \text{reject} \rfloor$$

is the strategy where, from top to bottom, player 2 accepts when offered 0 or 2 and rejects otherwise.

EXERCISE XII.5. *Write down the strategy (with four actions) that corresponds to*

- *player* 2 *never accepts,*
- *player* 2 *accepts if at least* 2 *coins are offered to him,*
- *player* 2 *accepts if no coin or two coins are offered to him, otherwise he rejects.*

By answering the above exercises, you show that you can transfer the concept of a strategy from the decision context to the game context. We now "copy" the definition of a strategy and some other related definitions:

DEFINITION XII.2. *Let* $\Gamma = \left(V, N, u, \iota, (A_d)_{d \in D}\right)$ *be a game in extensive form for perfect information.*

- *A strategy for player* $i \in N$ *is a function* $s_i : D_i \to A$ *obeying* $s_i(d) \in A_d$.
- *A trail* $T(v_0, v) = \langle v_0, v_1, ..., v_k = v \rangle$ *is provoked or brought about by strategy combination* $s \in S$ *if we arrive at* v *by choosing the actions prescribed by* s. *The terminal node provoked by strategy combination* s *is denoted by* v^s.
- *For every player* $i \in N$, *we define* $u_i : S \to \mathbb{R}$ *by*

$$u_i(s) := u_i(v^s), s \in S,$$

so that best replies and Nash equilibria are defined as in chapter X.

3. Subgame perfection and backward induction

So far, we have shown how to transform a game in extensive form into a strategic-form game. We can now address the problems of Nash equilibria and subgame-perfect Nash equilibria.

EXERCISE XII.6. *Consider the strategy combination*

$$(\lfloor 2 \rfloor, \lfloor \text{reject, reject, accept, accept} \rfloor)$$

in the bargaining game of the previous section. Which terminal node does it provoke? Is it a Nash equilibrium? Answer these questions for the strategy combination $(\lfloor 2 \rfloor, \lfloor \text{accept, reject, accept, reject} \rfloor)$, *too.*

While $(\lfloor 2 \rfloor, \lfloor \text{reject, reject, accept, accept} \rfloor)$ is a Nash equilibrium, it is not subgame perfect. The reason is that player 2 plans to reject an offer of 1 which yields a utility of $1 > 0$. We have encountered this type of problem in the investment-marketing game (p. 36) under the name of subtree imperfection and in the sequential quantity setting model. In any case, it is dealt with by the concept of subtrees (decision theory) or subgames (game theory).

DEFINITION XII.3 (SUBGAME). *Let* $\Gamma = (V, N, u, \iota, (A_d)_{d \in D})$ *be a game and let* $w \in D$. *Let* W *be the set of nodes that comprise* w *and all its successors together with the same links as in* V. *Then,* w*'s subgame of* Γ *is given by*

$$\Gamma^w = \left(W, \iota(D \cap W), u|_{E \cap W}, \iota|_{D \cap W}, (A_d)_{d \in D \cap W}\right)$$

where $u|_{E \cap W}$ *is to be understood as a function* $E \cap W \to \mathbb{R}^{|\iota(D \cap W)|}$. *Let* $A^w = \cup_{d \in D \cap W} A_d$. *For player* $i \in \iota(D \cap W)$, $s_i^w : D_i \cap W \to A^w$ *denotes his substrategy of* $s_i \in S_i$ *in* Γ^w *if* $s_i^w(d) = s_i(d)$ *for all* $d \in D_i \cap W$. *By* S_i^w (S^w) *we denote the set of player* i*'s substrategies (the set of substrategy combinations) in* Γ^w. Γ^w *is called a minimal subtree if* w *is the only decision node in* Γ^w.

We obtain Γ^w from Γ by choosing a $w \in D$ and restricting N, u, ι, and $(A_d)_{d \in D}$ accordingly. Note that $u|_{E \cap W}$ is restricted twofold, with respect to the domain ($E \cap W$ rather than E) and with respect to the number of entries (one entry for each player from $\iota(D \cap W)$ rather than from N).

DEFINITION XII.4 (SUBGAME PERFECTION). *A strategy combination s is subgame perfect if s^w is a Nash equilibrium in every subgame Γ^w.*

ALGORITHM XII.1. *Let $\Gamma = (V, N, u, \iota, (A_d)_{d \in D})$ be a game in extensive form of finite length. Backward-induction proceeds as follows:*

(1) *Consider the minimal subtrees Γ^w and take note of the best strategies in Γ^w. If any of the best-strategy sets is empty (for the reason explained on* *p. 17**), the procedure stops. If several best strategies can be identified, provide appropriate copies. Then, proceed at point (2).*
(2) *Cut the tree by replacing all minimal subtrees Γ^w by a terminal node w carrying the maximal payoff for player $\iota(w)$ received at point (1) and add the payoff information for all the other players from N.*
(3) *If the new trees contain minimal subtrees, turn to point* 1. *Otherwise, the final tree contains (the final trees contain) just one terminal node which is the initial node of the original tree but carries the payoff information for all the players.*

The maximal trails and the strategy combinations generated by the backward-induction procedure are called backward-induction trails and backward-induction strategy combinations, respectively.

EXERCISE XII.7. *Solve the game of figure XII.10 by applying backward induction. How many backward-induction trails and how many backward-induction strategy combinations can you find?*

Without proof, we note the following theorem:

THEOREM XII.1. *Let $\Gamma = (V, N, u, \iota, (A_d)_{d \in D})$ be of finite length. Then,*

- *the set of subgame-perfect strategy combinations and*
- *the set of backward-induction strategy combinations*

coincide.

COROLLARY XII.1. *Every game in extensive form for perfect information $\Gamma = (V, N, u, \iota, (A_d)_{d \in D})$ with $|V| < \infty$ has a subgame-perfect strategy combination.*

4. Multi-stage games

4.1. Definition. The games considered in this chapter and many games in the chapters to come, can be seen as multi-stage games. The idea of multi-stage games is that

- every player chooses an action at every stage and
- the players know all the actions undertaken in previous stages,
- but no action other than one's own in the present stage.

Thus, within each stage, the players choose actions simultaneously so that we are concerned with a mild form of imperfect information. It is cumbersome and not helpful to spell out the many orders in which the players act within a stage. After all, it should not matter to a player whether he moves first or second if, in any case, he does not know the move undertaken by the other player. In a multi-stage game, we still use the notation D, but a node d from D is addressed as a stage node rather than a decision node. At stage nodes, all players simultaneously choose actions. $A_d = \left(A_d^i\right)_{i\in N}$ now stands for tuple of action sets where A_d^i means the set of actions available to player i at stage node d. Every link at d corresponds to exactly one action combination. A strategy s_i for player i picks an action $s_i(d)$ from A_d^i. We now show the usefulness of the concept by way of a few examples. We also introduce a description of extensive-form games without moves by nature, the "very compact form".

4.2. Cournot and Stackelberg games. From the point of view of extensive-form games, the Cournot dyopoly can be depicted as a two-stage game. Indeed, there are two different ways to depict the Cournot tree. Either firm 1 moves first (see the left-hand side of figure XII.11) or firm 2 moves first (the right-hand side of that figure). The important issue is that the second mover does not know the quantity chosen by the first mover.

We propose to use the "very compact form" instead. It is given by figure XII.12. The fact that the two firms move simultaneously is hinted at by placing x_1 and x_2 in a vertical manner.

The Stackelberg dyopoly is also a multi-stage game, albeit a two-stage one. At each stage node, one firm chooses the action "do nothing". Of course, we do not explicitly write down the "do nothing" action. There is also nothing wrong with assuming that one, several, or all players act at a given stage node. In any case, the Stackelberg's very compact form is given in figure XII.13.

EXERCISE XII.8. *Draw the very compact form of the take-it-or-leave-it game!*

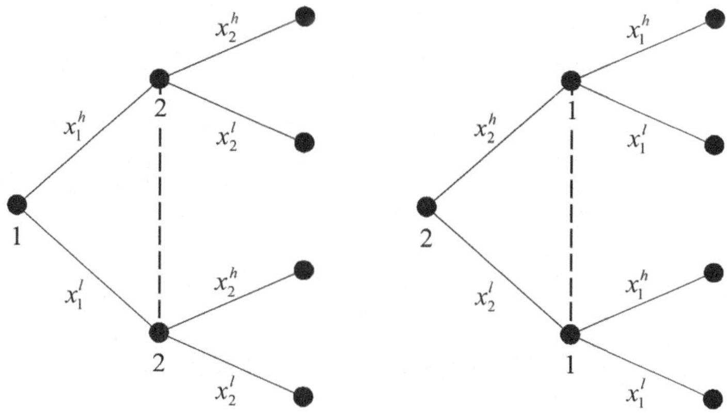

FIGURE XII.11. Two Cournot trees

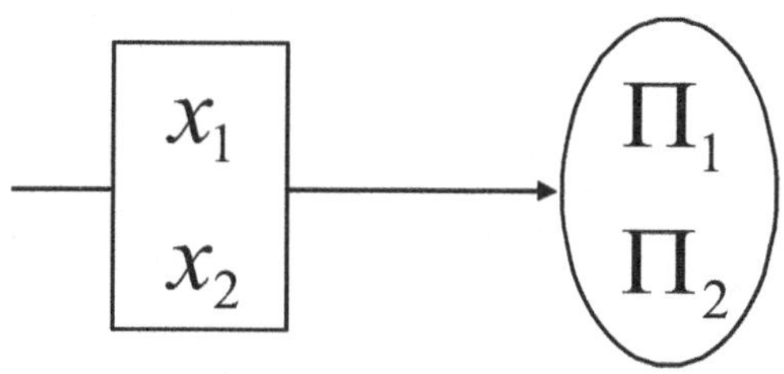

FIGURE XII.12. The very compact form of the Cournot game

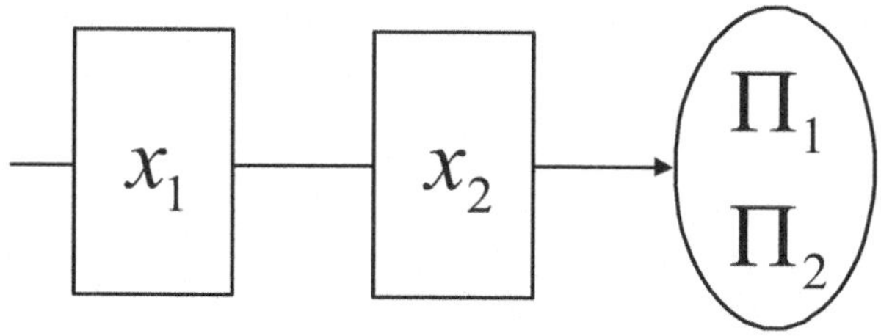

FIGURE XII.13. The very compact form of the Stackelberg game

4.3. Backward induction for multi-stage games. The very compact form is well suited for applying backward induction. You have seen the example of the Stackelberg game. We first concentrate on the last stage and obtain firm 2's reaction function x_2^R. In firm 1's profit function, we substitute x_2 by $x_2^R(x_1)$ so as to obtain firm 1's reduced profit function – a function of x_1, only. We denote the leader's profit-maximizing quantity by x_1^S. We then have

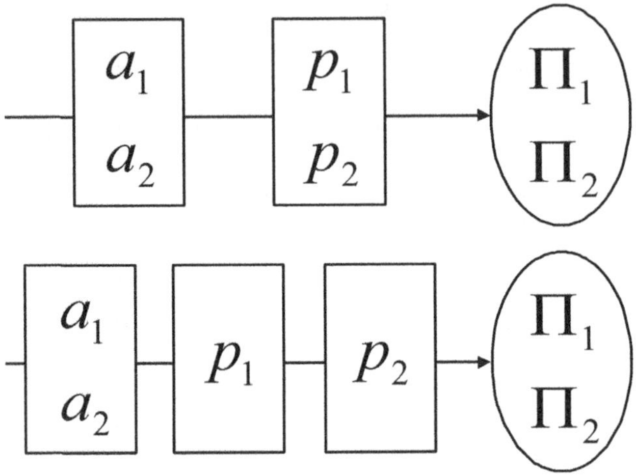

FIGURE XII.14. The positioning and pricing game

- the quantities in equilibrium $\left(x_1^S, x_2^R\left(x_1^S\right)\right)$ and
- the subgame-perfect equilibrium $\left(x_1^S, x_2^R\right)$ (also addressable as backward-induction strategy combination).

This procedure is also applicable to multi-stage games where some or all of the players act simultaneously (without knowing the actions of the other players) within stages. By now, you know how to read the very compact forms of figure XII.14. We have two firms 1 and 2 with actions a_i (chosing a product variety, for example) and p_i (choosing a price), respectively. The upper very compact form depicts a two-stage game where both firms choose product varieties simultaneously and then prices simultaneously. The lower very compact form depicts a three-stage game.

The backward-induction procedure works as follows for the upper-part very compact form.

- At the last stage, the firms choose equilibrium prices which are, in general, a function of first-stage actions: $\left(p_1^B(a_1, a_2), p_2^B(a_1, a_2)\right)$.
- We substitute p_1 and p_2 in the two firms' profit functions by the equilibrium prices $p_1^B(a_1, a_2)$ and $p_2^B(a_1, a_2)$ and obtain the reduced profit functions which depend on a_1 and a_2.
- We can now find the equilibrium varieties $\left(a_1^N, a_2^N\right)$.

The subgame-perfect (backward-induction) equilibrium of the overall game is

$$\left(\left(a_1^N, p_1^B\right), \left(a_2^N, p_2^B\right)\right)$$

where both p_1^B and p_2^B are functions $(a_1, a_2) \mapsto \mathbb{R}$. The prices chosen in equilibrium are $p_1^B\left(a_1^N, a_2^N\right)$ and $p_2^B\left(a_1^N, a_2^N\right)$.

The next section shows how to calculate the prices and varieties in the context of the Hotelling model.

5. Product differentiation

5.1. Hotelling's one-street village. The second to last section of this chapter deals with product differentiation. Consumers pay attention to the prices and also to other characteristics of the goods. For some goods, all consumers may have the same ranking of the goods (when prices are equal), for others, they may differ. The first case – identical ranking – is often referred to as vertical product differentiation or quality differentiation. Most people would agree that a Trabbi is inferior to a Mercedes-Benz C-Class. In contrast, horizontal product differentiation means that some people prefer one good while others prefer another one – think of a Mercedes-Benz C-Class versus an Audi A4.

A simple model for horizontal product differentiation is provided by the Hotelling linear space. It has a normed length of 1 with the corner points 0 and 1 (see figure XII.15). Imagine two firms 1 and 2 that are located at a_1 and a_2, respectively. Without loss of generality, we assume $a_1 \leq a_2$. One may address $[0, a_1]$ as firm 1's "home turf". A position in this space can be interpreted in two different manners:

- If the linear space is geographic space (a one-street village or a beach), you are invited to image firms that sell identical products (ice-cream at the beach, for example) at the locations a_1 and a_2. A consumer who likes to consume at position h has to incur transportation cost in order to buy from 1 or from 2.
- Alternatively, we can think of variants of a good. The ice-cream may be more or less sweet where 0 refers to zero-sugar ice-cream and 1 to a disgustingly sweat version. Consumers differ in their preferences. Some (those near 0) prefer to have a low content of sugar, while others are of the sweet-tooth fraction. A consumer who cannot obtain the very variant he likes best incurs a disutility (the analogue of transportation cost) from having to choose another one.

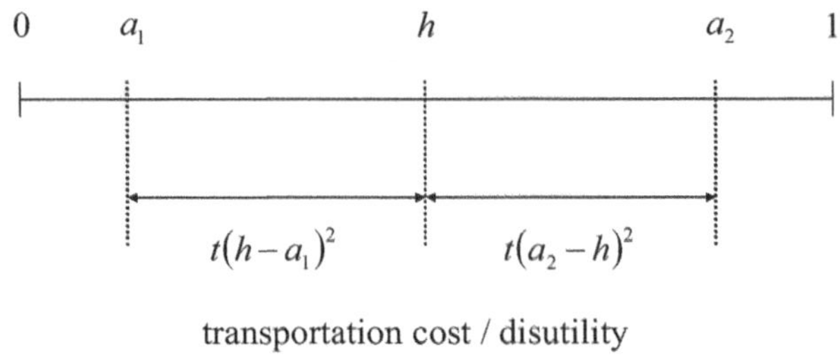

FIGURE XII.15. Hotelling's one-street village

5.2. Demand functions. We assume that consumers are distributed equally along the normed Hotelling space. The consumer at h incurs transportation cost (disutility of not consuming the best-liked variant) of

$$t\,(h-a_1)^2 \text{ or } t\,(a_2-h)^2 \tag{XII.1}$$

depending on the firm he patronizes.

Factor t is the transportation-cost rate (geographic space) or parameter of heterogeneity (variant space). If t is large, the different locations matter a lot. $t=0$ implies homogeneity. Homogeneity also holds for $a_1=a_2$.

DEFINITION XII.5 (HOMOGENEOUS GOODS). *Two products* 1 *and* 2 *are homogeneous if* $p_1 < p_2$ *implies* $x_2(p_1,p_2)=0$ *and if* $p_1>p_2$ *implies* $x_1(p_1,p_2)=0$.

We use a very simple procedure to derive the demand functions. First of all, we assume that every consumer buys one unit of the good from either firm 1 or firm 2. Therefore, we have $x_1+x_2=1$. We also suppose that a consumer at location h buys from firm 1 if

$$p_1+t\,(h-a_1)^2 \leq p_2+t\,(a_2-h)^2 \tag{XII.2}$$

where the sum of price and transportation cost is referred to as effective price. Solving for h, we find that all consumers obeying

$$h \leq \frac{a_2+a_1}{2}+\frac{p_2-p_1}{2t\,(a_2-a_1)} =: h^*. \tag{XII.3}$$

buy good 1. Thus, the demand curve for firm 1 is given by

$$x_1(p_1,p_2,a_1,a_2)=h^*=\underbrace{\overline{a}}_{\substack{\text{demand}\\ \text{for } p_1=p_2}}+\underbrace{\frac{1}{2t\Delta a}}_{\substack{\text{competition}\\ \text{intensity}}}\underbrace{(p_2-p_1)}_{\substack{\text{firm 1's}\\ \text{price advantage}}} \tag{XII.4}$$

with the definitions $\overline{a} := \frac{a_2+a_1}{2}$ and $\Delta a := a_2 - a_1$. Consequently, we have

$$x_2(p_1, p_2, a_1, a_2) = 1 - h^* = (1 - \overline{a}) - \frac{1}{2t\Delta a}(p_2 - p_1). \tag{XII.5}$$

We do not fret about the problem that the quantities may turn out to be negative for one of the firms.

The demand curves are interesting:

- *Product differentiation makes demand inelastic.*
  To make things simple, consider $p_1 = p_2 = p$ and maximal differentiation ($a_1 = 0$, $a_2 = 1$). The price elasticity of demand for firm 1 is

$$\varepsilon_{x_1,p_1}\Big|_{p_1=p_2=p} = \frac{\partial x_1}{\partial p_1}\frac{p_1}{x_1}\Bigg|_{p_1=p_2=p} = \frac{-1}{2t\Delta a}\frac{p_1}{x_1}\Bigg|_{p_1=p_2=p} = -\frac{p}{t}. \tag{XII.6}$$

- *Demand for equal prices.*
  $\overline{a}$ represents the customers who buy from firm 1 if the prices are identical or if product differentiation is very high.
- *Product differentiation lessens competition intensity.*
  The more differentiated the two products, the less important the impact of prices on consumers' decisions and the more monopolistic the firms' behavior in their respective submarket. We interpret $\frac{1}{2t\Delta a} = \left|\frac{\partial x_1}{\partial p_1}\right|$ as a measure of competition intensity. In general, competition intensity is high if small changes in competition variables lead to huge changes of sales (here) or profits.

5.3. The game. We now present the two-stage game where the two firms first decide on positions and then on prices:

DEFINITION XII.6 (PRICING (BERTRAND) GAME). *The two-stage positioning and pricing game is defined by $N = \{1, 2\}$, by the time structure presented in figure XII.16, and by the profit functions given by*

$$\Pi_1(a_1, a_2, p_1, p_2) = (p_1 - c)\, x_1 = (p_1 - c)\left(\overline{a} + \frac{p_2 - p_1}{2t\Delta a}\right),$$

$$\Pi_2(a_1, a_2, p_1, p_2) = (p_2 - c)\, x_2 = (p_2 - c)\left(1 - \overline{a} + \frac{p_1 - p_2}{2t\Delta a}\right).$$

We assume $a_1 \leq a_2$.

Thus, the firms do not incur any costs of positioning but only (identical) unit production costs.

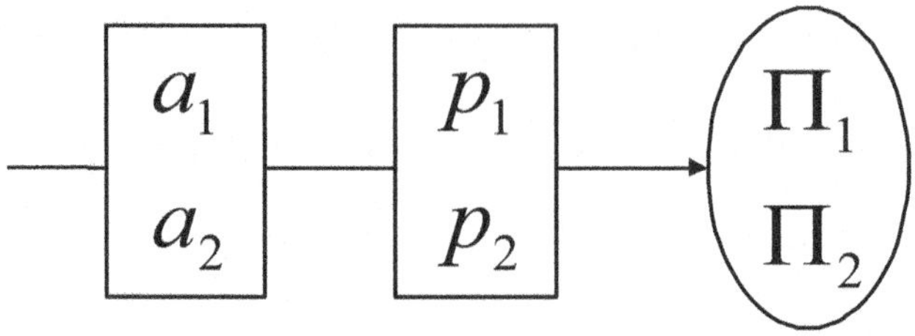

FIGURE XII.16. The positioning and pricing game

EXERCISE XII.9. *Assume that the government regulates prices at* $p_1 = p_2 > c_1 = c_2$ *where* c_1 *and* c_2 *are the average costs of the two firms. The firms* 1 *and* 2 *simultaneously determine their positions* a_1 *and* a_2*, respectively. Can you find an equilibrium?*

5.4. Solving the two-stage game.

5.4.1. *The second stage.* In order to apply backward induction, we first solve for the second stage of our model. Disregarding any corner solutions, we obtain

$$p_1^R(p_2) = \underset{p_1}{\operatorname{argmax}}\ \Pi_1(p_1, p_2) = \frac{p_2 + c + 2t\overline{a}\Delta a}{2}, \qquad \text{(XII.7)}$$

$$p_2^R(p_1) = \underset{p_2}{\operatorname{argmax}}\ \Pi_2(p_1, p_2) = \frac{p_1 + c + 2t(1-\overline{a})\Delta a}{2}. \qquad \text{(XII.8)}$$

Thus, the reaction functions are positively sloped. We also say that the competition variables are strategic complements. You remember from household theory that two goods are called complements if the demand increase for one good (due to a price decrease) leads to a demand increase of the complement (left and right shoe, cinema and popcorn). Similarly, action parameters that are strategic complements decrease both or increase both.

Because of

$$\frac{\partial p_1^R(p_2)}{\partial a_1} = -ta_1 \text{ and } \frac{\partial p_1^R(p_2)}{\partial a_2} = ta_2 \qquad \text{(XII.9)}$$

the optimal price is relatively low if the firms are positioned near each other. This is not astonishing because close positions signal a high competition intensity.

The equilibrium is given by

$$\begin{aligned} p_1^B(a_1, a_2) &= c + \frac{2}{3}t(1+\overline{a})\Delta a, \\ p_2^B(a_1, a_2) &= c + \frac{2}{3}t(2-\overline{a})\Delta a \end{aligned} \qquad \text{(XII.10)}$$

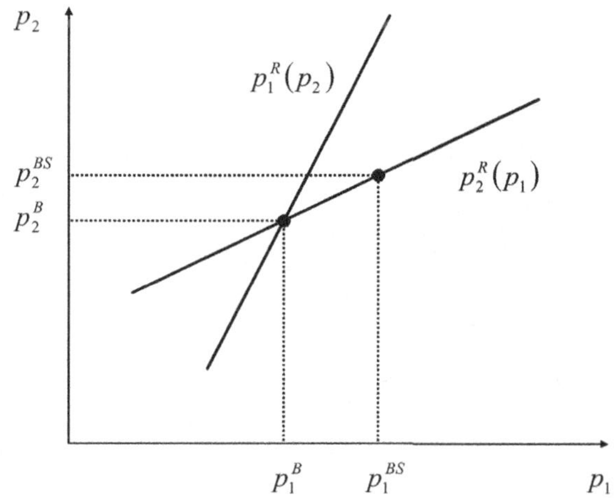

FIGURE XII.17. Pricing equilibria

and depicted in figure XII.17 at the crossing point of the two reaction functions.

We obtain the quantities

$$\begin{aligned} x_1^B(a_1, a_2) &= \frac{1}{3}(1 + \overline{a}) \geq 0, \\ x_2^B(a_1, a_2) &= \frac{1}{3}(2 - \overline{a}) \geq 0 \end{aligned}$$

and the reduced profits

$$\begin{aligned} \Pi_1^B(a_1, a_2) &= \frac{2}{9}t(1 + \overline{a})^2 \Delta a \geq 0, \\ \Pi_2^B(a_1, a_2) &= \frac{2}{9}t(2 - \overline{a})^2 \Delta a \geq 0. \end{aligned} \qquad \text{(XII.11)}$$

In contrast to the Bertrand model presented in chapter XI, product differentiation allows the firms to realize positive profits.

As an exercise, you are invited to consider sequential pricing competition. Firm 1 moves first and firm 2 moves second, just as in the Stackelberg model. You then obtain the equilibrium

$$\left(p_1^{BS}, p_2^R\right),$$

where firm 1 chooses price p_1^{BS} (where B stands for Bertrand and S for sequential) and firm 2 chooses his price optimally, i.e., obeys his reaction function. Reconsider figure XII.17 where you find the prices chosen in the sequential case.

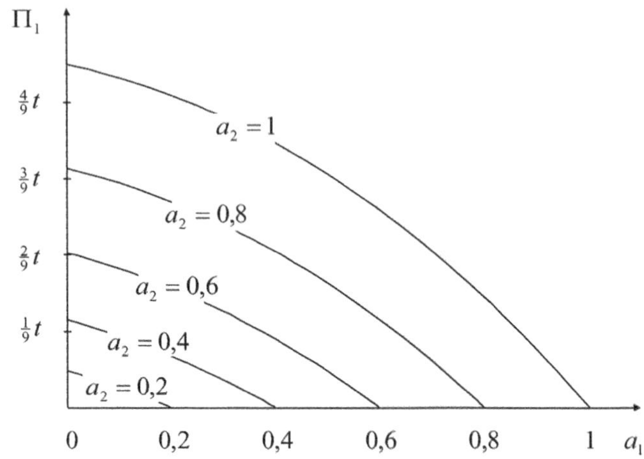

FIGURE XII.18. Firm 1's reduced profit function

In the framework of a specific model, we can find out whether a firm would rather be the first mover (firm 1) or the second mover:

EXERCISE XII.10. *Assume maximal differentiation, i.e.,* $a_1 = 0$ *and* $a_2 = 1$. *Solve the sequential pricing game described above. Show that we have a second-mover advantage. Show also that the leader's profit is higher in the sequential case than in the simultaneous one. Do you see why this is necessarily true?*

5.4.2. *The first stage.* The firms know how the positions affect the pricing game so that they use the reduced (after incorporating the equilibrium prices) profit functions

$$\begin{aligned} \Pi_1^B(a_1, a_2) &= \frac{2}{9}t\,(1+\overline{a})^2\,\Delta a \geq 0, \\ \Pi_2^B(a_1, a_2) &= \frac{2}{9}t\,(2-\overline{a})^2\,\Delta a \geq 0. \end{aligned}$$

Finding the solution to firm 1's maximizing problem,

$$\begin{aligned} a_1^R(a_2) &= \underset{a_1}{\operatorname{argmax}}\ \Pi_1^B(a_1, a_2) = \underset{a_1}{\operatorname{argmax}}\ \frac{2}{9}t(1+\overline{a})^2\Delta a \\ &= \underset{a_1}{\operatorname{argmax}}\ \frac{1}{18}t\,(2+a_1+a_2)^2\,(a_2-a_1)\,, \end{aligned}$$

is not too easy.

Consider figure. XII.18 which shows how firm 1's profit varies as a function of a_1 for some alternative a_2-values. Given $0 \leq a_1 \leq a_2$, we always have a negative marginal profit,

$$\frac{\partial \Pi_1^B}{\partial a_1} = -\frac{t}{18}\left(2 + a_1 + a_2\right)\left(2 + 3a_1 - a_2\right) < 0.$$

Therefore, firm 1's reaction function is given by

$$a_1^R\left(a_2\right) = 0.$$

Again assuming $a_1 \leq a_2$, we also find that firm 2 likes to move away from the other firm as far as possible, $a_2^R\left(a_1\right) = 1$. Obviously, the first-stage equilibrium is

$$\left(a_1^N, a_2^N\right) = (0, 1).$$

We obtain

$$\begin{aligned}
p_1^B &= c + t, & p_2^B &= c + t, \\
x_1^B &= \tfrac{1}{2}, & x_2^B &= \tfrac{1}{2}, \\
\Pi_1^B &= \tfrac{1}{2}t, & \Pi_2^B &= \tfrac{1}{2}t.
\end{aligned}$$

Thus, the firms have positive profits that depend on the heterogeneity parameter (unit transportation cost) t.

5.5. Direct and strategic effects.

5.5.1. *Accomodation.* Similar to the anaylsis found on pp. 316, we evaluate the effect of firm 1 "moving closer" to firm 2 on both firm 1's and firm 2's profits. Firm 1's reduced profit can be written as

$$\Pi_1^B\left(a_1, a_2\right) = \Pi_1\left(a_1, a_2, p_1^B\left(a_1, a_2\right), p_2^B\left(a_1, a_2\right)\right).$$

Forming the derivative with respect to a_1 yields

$$\frac{\partial \Pi_1^B}{\partial a_1} = \underbrace{\underbrace{\frac{\partial \Pi_1}{\partial a_1}}_{>0}}_{\substack{\text{direct or} \\ \text{demand effect}}} + \underbrace{\underbrace{\frac{\partial \Pi_1}{\partial p_1}}_{=0}\frac{\partial p_1^B}{\partial a_1}}_{\substack{=0 \\ \text{first-order condition} \\ \text{at stage 2} \\ \text{(envelope theorem)}}} + \underbrace{\underbrace{\frac{\partial \Pi_1}{\partial p_2}}_{>0}\underbrace{\frac{\partial p_2^B}{\partial a_1}}_{<0}}_{\substack{<0 \\ \text{strategic effect} \\ \text{of positioning}}}.$$

Using $\Pi_1 = \left(p_1 - c\right)x_1$, we obtain

$$\Pi_1^B\left(a_1, a_2\right) = \left(p_1^B\left(a_1, a_2\right) - c\right)x_1\left(a_1, a_2, p_1^B\left(a_1, a_2\right), p_2^B\left(a_1, a_2\right)\right)$$

and

$$\frac{\partial \Pi_1^B}{\partial a_1} = \underbrace{\left(p_1^B(a_1,a_2) - c\right)\frac{\partial x_1}{\partial a_1}}_{\substack{>0 \\ \text{direct or} \\ \text{demand effect}}} + \underbrace{\underbrace{\left(p_1^B(a_1,a_2) - c\right)}_{>0}\underbrace{\frac{\partial x_1}{\partial p_2}}_{>0}\underbrace{\frac{\partial p_2^B}{\partial a_1}}_{<0}}_{\substack{<0 \\ \text{strategic effect} \\ \text{of positioning}}}. \tag{XII.12}$$

The direct effect tends to be positive because moving closer to ones rival increases market share. (In extreme circumstances, the direct effect can also be negative (see Pfähler & Wiese 2008, pp. 269).) The strategic effect is negative – more homogeneity entails more intense competition.

Of course, the decisive question is which of the two effects prevails. In our simple model (with quadratic transportation costs), the strategic effect turns out to outweigh the demand effect.

5.5.2. *Entry deterrence.* We can also examine how the position of firm 1 affects the profit of firm 2. This is an interesting question if firm 1 tries to push firm 2 out of the market or looks for a constellation that makes entry unattractive. Therefore, we consider

$$\Pi_2^B(a_1,a_2) = \Pi_2\left(a_1,a_2,p_1^B(a_1,a_2),p_2^B(a_1,a_2)\right)$$

and

$$\frac{\partial \Pi_2^B}{\partial a_1} = \underbrace{\frac{\partial \Pi_2}{\partial a_1}}_{\substack{<0 \\ \text{direct or} \\ \text{demand effect}}} + \underbrace{\underbrace{\frac{\partial \Pi_2}{\partial p_1}}_{>0}\underbrace{\frac{\partial p_1^B}{\partial a_1}}_{<0}}_{\substack{<0 \\ \text{strategic effect} \\ \text{of positioning}}} + \underbrace{\underbrace{\frac{\partial \Pi_2}{\partial p_2}}_{=0}\frac{\partial p_2^B}{\partial a_1}}_{\substack{=0 \\ \text{first-order condition} \\ \text{at stage 2} \\ \text{(envelope theorem)}}}. \tag{XII.13}$$

Both the demand effect and the strategic effect are negative. Therefore, for the purpose of entry deterrence, it may be a good idea to move closer to the rival and to endure reduced short-term profits.

6. Application: Strategic trade policy

6.1. Free trade or strategic trade policy. Forty years ago, trade theory and policy were analyzed with models of perfect competition. Free trade was a usual implication of these models. Since the beginning of the 1980s, models and recommendations have changed. At first, the researchers used Cournot models. Brander (1981) and Brander & Krugman (1983) show that free trade can lead to the exchange of identical products. From a global perspective, this cannot be an optimal outcome because of transportation costs.

In another strand of the literature, Brander & Spencer (1981, 1983) reason that export subsidies can benefit exporting firms over and above the subsidies. The idea is that the subsidies induce changes in the other firms behavior. This is the subject matter of strategic trade policy.

6.2. The model. We assume two firms that are located in two countries, a domestic firm d in country d (Germany, for example) and a foreign firm f in country f (France). The two firms produce for the market in some third country (Italy). We focus on the domestic welfare effects of an export tax or subsidy levied (or granted) to firm d.

Taking a third country simplifies the analysis because we can disregard the consumers' rent. Thus, our welfare analysis is concerned with profit and taxes, only.

The other assumptions are taken from the linear Cournot model. The inverse demand function in Italy is given by $p(X) = a - bX$ and the two firms have identical marginal and average cost $c := c_d = c_f$ with $c < a$. The German government tries to maximize welfare by choosing an appropriate unit subsidy s benefitting its firm d. Then, welfare is the sum of the firm's profit minus the subsidy payments:

$$W(s) = \Pi_d^C(c - s, c) - s x_d^C(c - s, c).$$

Thus, the German government's subsidy defines a simultaneous game with a specific equilibrium. Assuming accomodation (pp. 313), the domestic firm's quantity is

$$x_d^C(c - s, c) = \frac{1}{3b}(a - 2(c - s) + c) = \frac{1}{3b}(a - c + 2s)$$

and its profit amounts to

$$\Pi_d^C(c - s, c) = \frac{1}{9b}(a - 2(c - s) + c)^2 = \frac{(a - c + 2s)^2}{9b}.$$

Therefore, we find

$$W(s) = \frac{(a - c + 2s)^2}{9b} - s\frac{1}{3b}(a - c + 2s)$$

and, by differentiating,

$$s^* := \arg\max_{s \in \mathbb{R}} W(s) = \frac{a - c}{4} > 0.$$

6.3. Understanding the logic of strategic trade policy. Apparently, strategic policy can pay. The domestic government should fix a strictly positive unit subsidy if its firm engages in quantity competition with a foreign firm on a third market. Why?

The subsidy has a direct effect on welfare and a strategic effect. The direct effect (holding the outputs constant) is zero. From our welfare point of view, it does not matter whether a sum of money ends up in the pockets of the domestic firm or in those of the government. Therefore, we can concentrate on the strategic effect.

The subsidy amounts to a cost decrease for the domestic firm d. On p. 316, we have shown that the strategic effect

$$\underbrace{\underbrace{\frac{\partial \Pi_1}{\partial x_2}}_{<0} \underbrace{\frac{\partial x_2^C}{\partial c_1}}_{>0}}_{<0} < 0$$

is negative. For strategic trade policy, this means

$$\underbrace{\underbrace{\frac{\partial \Pi_d}{\partial x_f}}_{<0} \underbrace{\frac{\partial x_f^C}{\partial s}}_{<0}}_{>0} > 0,$$

where a subsidy increase leads to a reduction of $c_d = c - s$.

Hence, the subsidy for firm d entails an outward shift of d's reaction curve and hence a reduction of firm f's equilibrium quantity x_f^C (see figure XII.19). It is this reduction that implies the sought-after profit increase for firm d – via the associated price increase. Note that the subsidy's overall effect is a price decrease. However, the strategic effect concerns the price increase effected by firm f.

Since the subsidy awarded to firm d has no direct effect on welfare, it should be chosen so as to maximize firm d's profit net of the subsidy. The optimal subsidy picks the profit-maximizing (for firm d!) point on firm f's reaction curve. This is the Stackelberg point! Indeed, for the welfare-maxizing subsidy $s^* = \frac{a-c}{4}$, we find

$$\begin{aligned} x_d^C(c - s^*, c) &= \frac{a - c + 2s^*}{3b} \\ &= \frac{a-c}{2b} \\ &= x_d^S(c, c). \end{aligned}$$

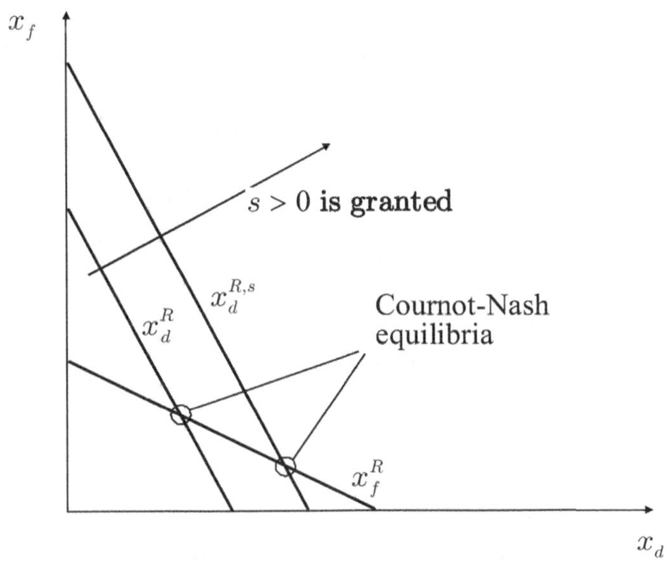

FIGURE XII.19. Effect of an export subsidy

6.4. Strategic trade policy for price competition. As an alternative to quantity competition, we now have a look at pricing competition on the Hotelling linear space. We assume that the two firms d and f offer maximally differentiated products, $\Delta a = 1$. The domestic government would like to support its firm d with a subsidy s because the trade minister had studied Brander & Spencer (1981).

On the basis of equations XII.4 and XII.5 (p. 350), we see that the demand functions are given by

$$\begin{aligned} x_d &= \frac{1}{2} + \frac{p_f - p_d}{2t} \text{ and} \\ x_f &= \frac{1}{2} + \frac{p_d - p_f}{2t}. \end{aligned}$$

Unit profits are $p_d - (c - s)$ for firm d and $p_f - c$ for firm f. Hence, we obtain the equilibrium prices

$$p_d^B = t + c - \frac{2}{3}s$$

for the domestic firm d and

$$p_f^B = t + c - \frac{1}{3}s$$

for the foreign firm f. The equilibrium quantity supplied by firm d is

$$x_d^B = \frac{1}{2} + \frac{p_f^B - p_d^B}{2t} = \frac{1}{2} + \frac{1}{6}\frac{s}{t}.$$

We now obtain

$$\begin{aligned} W(s) &= \Pi_d^B(c-s,c) - s x_d^B(c-s,c) \\ &= \left[p_d^B - (c-s)\right] x_d^B - s x_d^B = \left[p_d^B - c\right] x_d^B \\ &= \left[t + c - \frac{2}{3}s - c\right]\left(\frac{1}{2} + \frac{1}{6}\frac{s}{t}\right) \\ &= \left[t - \frac{2}{3}s\right]\left(\frac{1}{2} + \frac{1}{6}\frac{s}{t}\right) \end{aligned}$$

and hence

$$\frac{dW}{ds} = -\frac{2}{3}\left(\frac{1}{2} + \frac{1}{6}\frac{s}{t}\right) + \left[t - \frac{2}{3}s\right]\frac{1}{6t} = -\frac{1}{18}\left(4\frac{s}{t} + 3\right)$$

Thus, the welfare maximizing "subsidy" is negative at

$$s^* = -\frac{3}{4}t,$$

i.e., the government should tax exports rather than subsidize them.

Again, there is a strategic reason for this intervention. By taxing the domestic firm, the government pushes both firms to an equilibrium with higher prices. The higher foreign price is the reason why this form of strategic trade policy pays: a higher p_f leads to a higher demand for the domestic firm.

Thus, we see that the optimal strategic trade policy depends on whether firms are engaged in price competition (export tax) or in quantity competition (export subsidy). The trade minister should also have noticed Eaton & Grossman (1986) who provide the main ingredients for this section.

6.5. Judging strategic trade policy. We have shown that strategic trade policy can increase a country's welfare. Nevertheless, there are many good reasons not to advocate strategic trade policy. We cite Helpman & Krugman (1989, p. 186): "One can always do better than free trade, but the optimal tariffs or subsidies seem to be small, the potential gains tiny, and there is plenty of room for policy errors that may lead to eventual losses rather than gains. [...] The case for free trade has always rested on an argument that it represents a good rule of thumb given uncertainty about the alternatives, realistic appreciation of the difficulties of managing political intervention, and the need to avoid trade wars."

The remark about trade wars is important. Note that we considered strategic trade policy from the point of view of the domestic government, only. Of course, the other government may have similar ideas. One might envision a two-stage model where both governments simultaneously choose taxes or subsidies s_d and s_f, respectively. ...

7. Topics and literature

The main topics in this chapter are

- sequential moves
- accomodation
- blockade
- deterrence
- multi-stage game
- reduced profit function
- heterogeneous products
- vertical product differentiation
- horizontal product differentiation
- competition advantage
- price advantage
- strategic trade policy
- direct effect
- indirect effect

We cannot but recommend the textbook by Pfähler & Wiese (2008).

8. Solutions

Exercise XII.1

Driver 1 has a first-mover advantage in the game of chicken. He chooses "continue" so that driver 2 is forced to swerve. The game tree and its backward-induction strategies are depicted in figure XII.20.

Exercise XII.2

Firm 2's profit function is

$$\begin{aligned} \Pi_2(x_1, x_2, x_3) &= p(X)\, x_2 - C(x_2) \\ &= (100 - x_1 - x_2 - x_3)\, x_2 - C(x_2). \end{aligned}$$

so that its reaction function is given by

$$x_2^R(x_1, x_3) = \frac{100 - x_1 - x_3}{2}.$$

Firms 2 and 3 are symmetric and we have firm 3's reaction function

$$x_3^R(x_1, x_2) = \frac{100 - x_1 - x_2}{2}.$$

Since these two fims act simultaneously after observing x_1, we obtain the Cournot equilibrium by solving these two equations for x_2 and x_3. We find

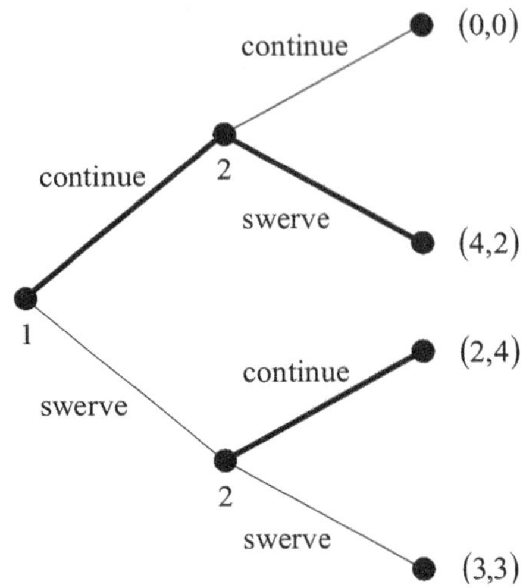

FIGURE XII.20. Backward induction for the game of chicken

$$x_2^C(x_1) = \frac{100 - x_1}{3},$$
$$x_3^C(x_1) = \frac{100 - x_1}{3}.$$

The leader firm 1 plugs these values into its profit function to obtain the reduced profit function

$$\Pi_1\left(x_1, x_2^C(x_1), x_3^C(x_1)\right) = \left(100 - x_1 - x_2^C(x_1) - x_3^C(x_1)\right)x_1 - 0.$$

Finally, we obtain

$$x_1^S = 50 \text{ and } x_2^C(50) = x_3^C(50) = \frac{50}{3}.$$

Exercise XII.3

The first combination is not an equilibrium. Facing $x_2^S = x_2^R\left(x_1^S\right)$, firm 1's optimal choice is $x_1^R\left(x_2^S\right) \neq x_1^S$ (compare figure XII.6, p. 337). The other two combinations are Nash equilibria, but the last one is not subgame perfect (see below).

Exercise XII.4

The answer is indicated in figure XII.21. We have $N = \{1, 2\}$. Player 1 is noted below the initial node and player 2 below the other decision nodes. D_2 is the set of nodes highlighted by the arrows. Player 2 has 16 strategies, among them

$$\lfloor \text{reject, accept, accept, accept} \rfloor \text{ and } \lfloor \text{reject, accept, reject, accept} \rfloor.$$

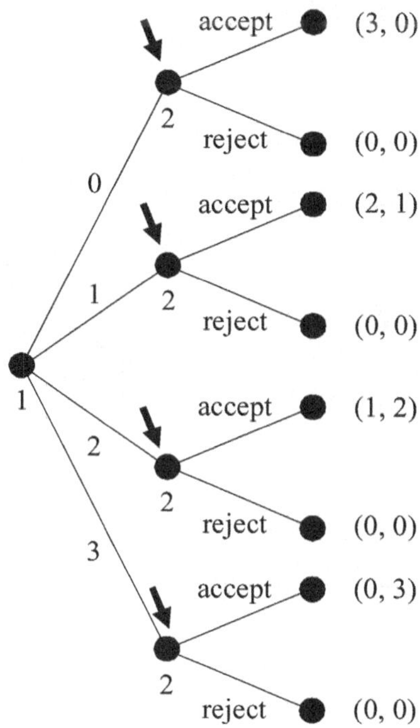

FIGURE XII.21. The take-it-or-leave-it game - the exercise

Exercise XII.5

The first two strategies are reflected in the game matrix, the third is not:

- player 2 never accepts:

$$\lfloor \text{reject, reject, reject, reject} \rfloor$$

- player 2 accepts if at least 2 coins are offered to him:

$$\lfloor \text{reject, reject, accept, accept} \rfloor$$

- player 2 accepts if no coin or two coins are offered to him, otherwise he rejects:

$$\lfloor \text{accept, reject, accept, reject} \rfloor$$

Exercise XII.6

$(\lfloor 2 \rfloor, \lfloor \text{reject, reject, accept, accept} \rfloor)$ is a Nash equilibrium. In contrast, the strategy combination $(\lfloor 2 \rfloor, \lfloor \text{accept, reject, accept, reject} \rfloor)$ is not a Nash equilibrium because player 1 could offer 0 and obtain payoff $3 > 1$.

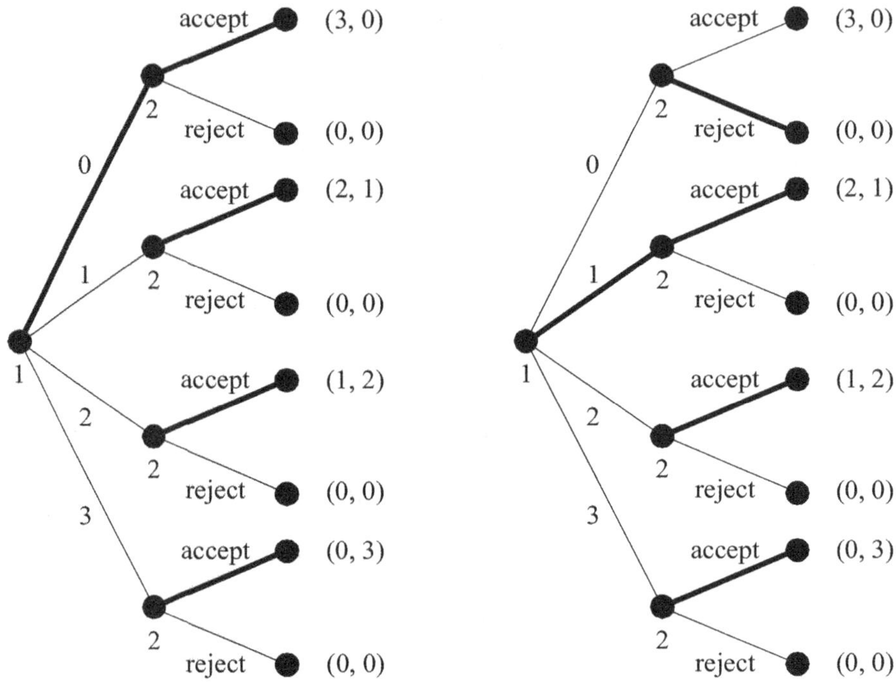

FIGURE XII.22. The backward-induction solution of the bargaining game

Exercise XII.7

The solutions are indicated in figure XII.22. Player 2 is indifferent between accepting and rejecting if offered a payoff of 0. Therefore, we obtain two backward-induction trees, two backward-induction trails and two backward-induction strategy combinations.

Exercise XII.8

The game is a two-stage game and it is depicted in figure XII.23. Player 1 makes an offer x_1 ($x_1 \in \{0,1,2,3\}$) and player 2 gives an answer a_2 ($a_2 \in \{\text{accept, reject}\}$) to that offer.

Exercise XII.9

There is one and only one equilibrium, $(a_1, a_2) = \left(\frac{1}{2}, \frac{1}{2}\right)$. If a firm deviates, its sales are reduced by $\frac{1}{2}$ times the positional change.

Exercise XII.10

You should have found

$$\begin{aligned} p_1^{BS} &= \underset{p_1}{\operatorname{argmax}} \left(\Pi_1(p_1, p_2^R(p_1))\right) \\ &= c + \frac{3}{2}t > c + t \end{aligned}$$

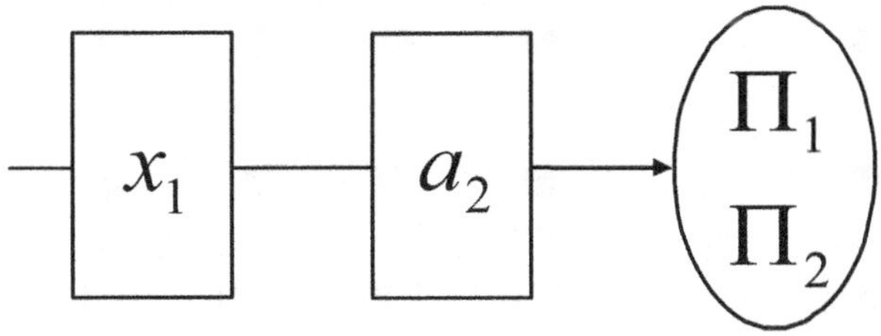

FIGURE XII.23. The very compact form of the take-it-or-leave-it game

and

$$p_2^R\left(p_1^{BS}\right) = c + \frac{5}{4}t < c + \frac{3}{2}t.$$

A second-mover advantage exists because of

$$\begin{aligned}\Pi_1^{BS} &= \left(p_1^{BS} - c\right)\left(\frac{1}{2} + \frac{p_2^R(p_1^{BS}) - p_1^{BS}}{2t}\right) = \frac{18}{32}t \\ &< \frac{25}{32}t \\ &= \left(p_2^R(p_1^{BS}) - c\right)\left(\frac{1}{2} + \frac{p_1^{BS} - p_2^R(p_1^{BS})}{2t}\right) = \Pi_2^{BS}.\end{aligned}$$

However, this does not mean that the price leader would rather be a firm in the simultaneous case. After all, he is free to choose the crossing point of the two reaction functions. He deviates from that point in order to realize a higher profit:

$$\Pi_1^{BS} = \frac{9}{16}t > \frac{8}{16}t = \Pi_1^B.$$

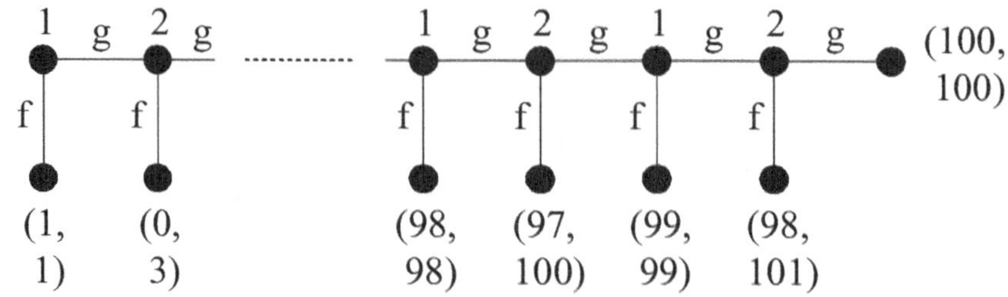

FIGURE XII.24. The centipede game

9. Further exercises without solutions

PROBLEM XII.1. Consider the backward-induction solution for the stag hunt on pp. 329. Do likeweise for head or tail (p. 266).

PROBLEM XII.2. Reconsider the police game (p. 280). Let the police be the first mover and assume that the indifferent agent abstains from committing a crime. Find the optimal control probability!

PROBLEM XII.3. Consider the centipede game depicted in figure XII.24! The players 1 and 2 take turns in choosing between "finish" (action f) or "go on" (action g). For every player, a strategy is a 99-tuple. For example, ⌊g, g, g, g, f, ..., f⌋ is the strategy according to which a player chooses "go on" at his first four decision nodes and "finish" at all the others.

(1) Which strategy would you choose if you were player 1? Does your answer depend on who takes on 2's role?
(2) Can player 1's strategy ⌊g, g, g, g, f, ..., f⌋ be part of a subgame-perfect strategy combination?
(3) Solve the centipede game by backward induction!
(4) Do you want to reconsider your answer to the first question?

PROBLEM XII.4. Work through the innovation chapter in Pfähler & Wiese (2008, pp. 225).

CHAPTER XIII

Repeated games

The last chapter of this part focuses on a specific class of multi-stage games – repeated games. In these games, the actions available at each stage game are the same at each stage. Thus, we have an underlying stage game that gets repeated a finite or an infinite number of times. Arguably, repeated games are relevant for discussing cartel issues. After all, firms typically meet under similar conditions for long stretches of time on one and the same market.

1. Example: Repeating the pricing game

Repeated games are especially simple extensive-form games. A stage game is played repeatedly by the same players. The repetitions are called stages or periods. Since "strategies" refer to a plan for the whole, repeated, game, we use the word "action" for what the players do within the stages. Consider the following pricing game played by two firms:

		firm 2	
		high price	cut price
firm 1	high price	$4,4$	$0,5$
	cut price	$5,0$	$1,1$

If both charge high prices, profits are relatively high for both, with a payoff of 4 (million Euro). However, each firm has an incentive to undercut its rival ($5 > 4$) and to charge a low price if the other firm does ($1 > 0$). Thus, it is likely that both firms charge a low price and obtain low profits 1. The alert reader will have noticed that this game is nothing but the prisoners' dilemma (see p. 267).

One might think and hope (for the firms' sake, not the consumers') that a repetition of this game may help firms to coordinate on a high price. After all, could not a high-price firm punish a low-price firm by charging a

H. Wiese, *Advanced Microeconomics*,
https://doi.org/10.1007/978-3-658-34959-2_13

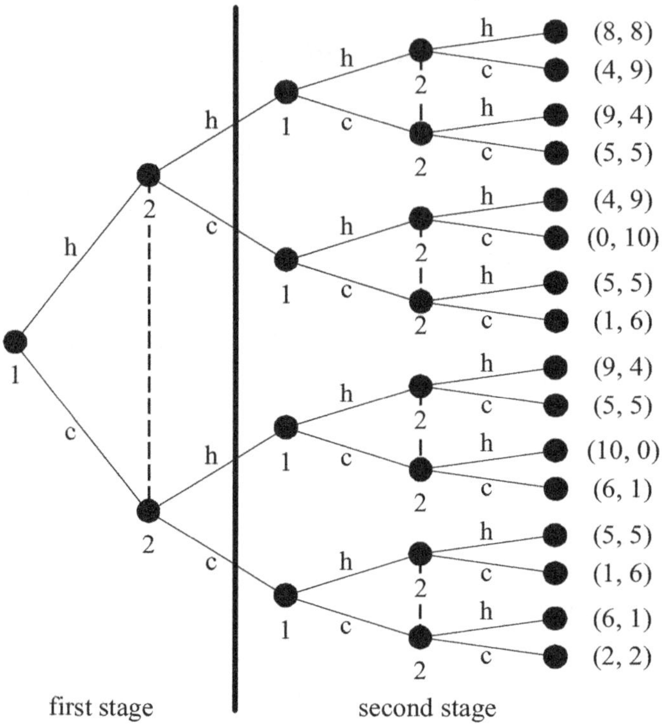

FIGURE XIII.1. The two-stage prisoners' dilemma

low price itself? Let us consider the twofold repetition of the pricing game depicted in figure XIII.1.

Player 1 moves first in each of the two stages. He chooses action h (high price) or c (cut price). These actions correspond to the strategies of the stage game. Player 2 does not learn player 1's choice which is indicated by the dashed line. He also chooses between a high and a low price. Then the first stage is over and both players observe what happened in that stage. Note that the imperfect information of player 2 at the second stage refers not to player 1's action at the first stage but only to player 1's action at the second stage.

EXERCISE XIII.1. *Can you make sense of the payoff pairs* $(4, 9)$ *and* $(10, 0)$*? How many subgames can you find?*

There are two ways to simplify figure XIII.1. First of all, this figure contains information that we do not need. The important issue is that, within each stage, each player makes his decision without knowing the other's move. However, we do need to know which player moves first. For example, the

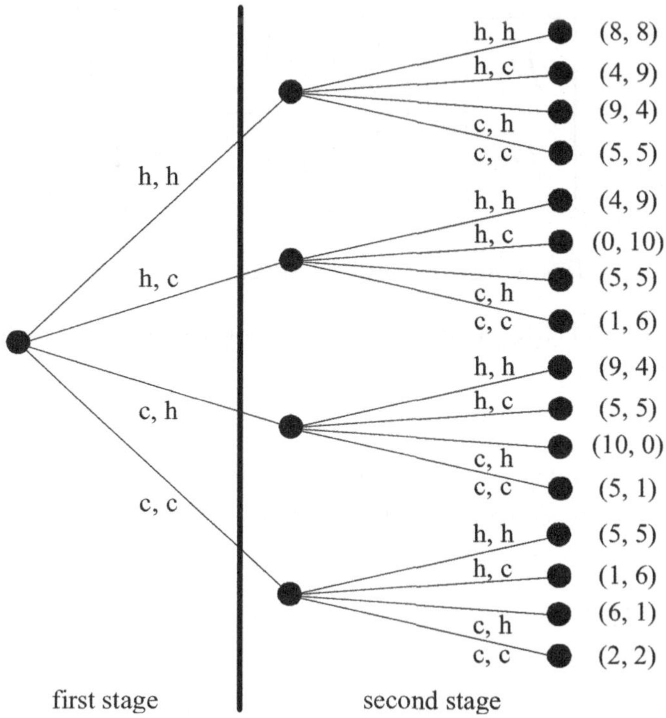

FIGURE XIII.2. The two-stage prisoners' dilemma in a more compact form

Cournot model can be understood as an extensive-form game where firm 1 or firm 2 moves first and the other firm, who does not know the first-mover's action, second. Figure XIII.2 is the simplification for our two-stage prisoners' dilemma game.

If we are mainly interested in the actions that the players can choose and in the information they have, we can employ the "very compact form" introduced in the previous chapter (figure XIII.3). The fact that the prices p_1 and p_2 are chosen simultaneously at stage 1 (or p_1 after p_2 without knowledge of p_2) is reflected in the vertical alignment. At stage 2, both stage-1 prices are known to both players.

Some stage games have equilibria in pure strategies (we do not consider mixed-strategy equilibria in this chapter). For example, there are inefficient price-cutting equilibria in our stage game. If every player chooses "cut price" whenever it is his turn to act, the resulting strategies form an equilibrium. The interesting question (from the point of view of competition policy) is whether we have other equilibria. How about the cooperative tit for tat? Every player begins with the nice action (high price, in our example) and

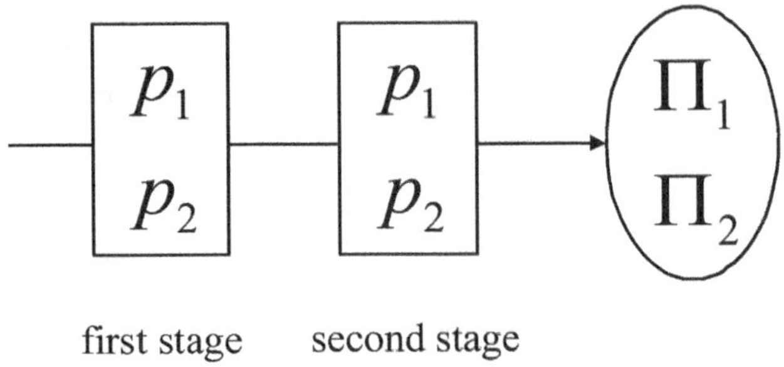

FIGURE XIII.3. The two-stage prisoners' dilemma in compact form

copies the other player's last period action in later stages. Indeed, if both players choose that strategy, each of them obtains payoff 8 rather than just 2.

EXERCISE XIII.2. *Describe player 2's cooperative tit-for-tat strategy as a quintuple of the form*

$$\lfloor \underbrace{c/h}_{\substack{\textit{Action at}\\ \textit{the first stage}}}, \underbrace{c/h}_{\substack{\textit{Action at}\\ \textit{the second stage}\\ \textit{after } h \textit{ by both}}}, \underbrace{c/h}_{\substack{\textit{Action at}\\ \textit{the second stage}\\ \textit{after } (h,\, c)}}, \underbrace{c/h}_{\substack{\textit{Action at}\\ \textit{the second stage}\\ \textit{after } (c,\, h)}}, \underbrace{c/h}_{\substack{\textit{Action at}\\ \textit{the second stage}\\ \textit{after } c \textit{ by both}}} \rfloor !$$

The strategy combination where both players choose cooperative tit for tat is not an equilibrium. For example, player 1 can increase his profits by choosing c at the second stage after both players chose h at the first stage. His payoff incrases from 8 to 9.

EXERCISE XIII.3. *Defining the order of subgames as in the previous exercise, are the strategy combinations*

- $(\lfloor c, c, c, c, c\rfloor, \lfloor c, c, c, c, c\rfloor)$ *or*
- $(\lfloor c, h, h, h, c\rfloor, \lfloor c, h, h, h, c\rfloor)$

Nash equilibria of the twice repeated pricing game?

2. Definitions

Repeated games are a special instance of multi-stage games. They have two peculiarities. First, each player's action set is the same at each stage node. Second, a player's payoff is the (discounted) sum of the payoffs received along a specific trail.

In order to avoid confusion between the symbols used for the stage game and the multi-stage game, we use A_i (rather than S_i) for player i's strategy set in stage game Γ and g (rather than u) for the payoff functions in stage game Γ.

Definition XIII.1 (Finitely repeated game). *Let $\Gamma = \left(N, (A_i)_{i\in N}, g\right)$ be a strategic game and let $\delta \in [0,1]$. The t-fold repetition of Γ with discount factor δ is the multi-stage game $\Gamma^{t,\delta}$ where*

- *at each stage node $d \in D$ every player $i \in N$ chooses an action from A_i,*
- *a strategy for player i is a function $s_i : D \to A_i$ where D is the set of stage nodes that depend on the actions chosen at previous stages,*
- *A is the set of strategy tuples (from the point of view of Γ) or a set of action tuples (within $\Gamma^{t,\delta}$), and*
- *payoffs are given by*

$$u_i(s) = \frac{\sum_{\tau=0}^{t-1} \delta^\tau g_i(s(d_\tau))}{\sum_{\tau=0}^{t-1} \delta^\tau}$$

where

- *d_0 is the initial node,*
- *d_1 the node resulting from the action combination $s(d_0)$*
- *...*
- *d_{t-1} is the node resulting from action combinations $s(d_0)$, $s(d_1)$, ... $s(d_{t-2})$*

While it is quite possible to define the payoff function $u : E \to \mathbb{R}^n$, it is easier to immediately state the payoff as a function of strategies. $s(d_\tau)$ is the action combination chosen at stage node d_τ. The stage game Γ gives the payoff for each stage. However, the payoff is discounted according to a discount factor that lies between 0 and 1. You can think of δ as $\frac{1}{1+r}$ with interest rate r. In case of $\delta = 0$, we have $u_i(s) = g_i(s(d_0))$ for all players $i \in N$ because of $0^0 = 1$. Then, the players are very impatient and the payoff of the first period, only, counts. The other extreme is $\delta = 1$ where payoffs are not discounted at all and we obtain the utility $u_i(s) = \frac{1}{t}\sum_{\tau=0}^{t-1} g_i(s(d_\tau))$. Note that the payoff functions contain the denominator $\sum_{\tau=0}^{t-1} \delta^\tau$. This is a normalization which does not influence best responses or equilibria. It allows a comparison of multi-stage payoffs with payoffs in the stage game.

EXERCISE XIII.4. *Assume that all players use constant strategies, i.e., we have* $s(d) = a \in A$ *for all* $d \in D$. *Find* $u_i(s)$.

So far, we have defined the finite repetition of a stage game Γ. This definition needs to be adapted for infinite repetitions. In particular, in case of infinite repetitions, there are no terminal nodes (we have $V = D$). Therefore, we attach utility information to infinite trails rather than terminal nodes. Also, in order to guarantee the convergence of the infinite sum, we disallow $\delta = 1$:

DEFINITION XIII.2 (INFINITELY REPEATED GAME). *Let* $\Gamma = \left(N, (A_i)_{i\in N}, g\right)$ *be a strategic game and let* $\delta \in [0,1)$. *The infinite repetition of* Γ *with discount factor* δ *is the multi-stage game* $\Gamma^{\infty,\delta}$ *where the first three bullets are those from the definition a finitely repeated game and where payoffs are given by*

$$u_i(s) = \frac{\sum_{\tau=0}^{\infty} \delta^\tau g_i(s(d_\tau))}{\sum_{\tau=0}^{\infty} \delta^\tau}, i \in N.$$

The denominator in the utility functions can be simplified. By $\delta < 1$ and

$$\text{infinite geometric series} = \frac{\text{first term}}{1 - \text{factor}} = \frac{\delta^0}{1-\delta} = \frac{1}{1-\delta},$$

the utility function can be rewritten as

$$u_i(s) = (1-\delta)\sum_{\tau=0}^{\infty} \delta^\tau g_i(s(d_\tau)).$$

3. Equilibria of stage games and of repeated games

In this section, we show how to generate multi-stage equilibria from one-stage equilibria. Indeed, any equilbrium in a stage game Γ gives rise to the obvious equilibrium in the repeated game $\Gamma^{t,\delta}$ or $\Gamma^{\infty,\delta}$:

THEOREM XIII.1. *Let* $\Gamma = \left(N, (A_i)_{i\in N}, g\right)$ *be a strategic game and let* $\delta \in [0,1)$. *Let* $a^* = (a_1^*, a_2^*, ..., a_n^*) \in A$ *be an equilibrium of* Γ. *Then,* s^* *is a subgame-perfect equilibrium of* $\Gamma^{t,\delta}$ *or* $\Gamma^{\infty,\delta}$ *if all the strategies* s_i^* *are constant and equal to* a_i^*, *i.e., if for all* $i \in N$ *we have*

$$s_i^* : D \to A_i, \quad d \mapsto s_i^*(d) = a_i^*.$$

If only one stage equilibrium exists, any subgame-perfect equilibrium of $\Gamma^{t,\delta}$ *is characterized in this manner.*

PROOF. We need to show that a unilateral deviation from s^* does not pay for any player. Let s_i be any strategy for player i from N. For the finite game $\Gamma^{t,\delta}$, we find

$$\begin{aligned} u_i\left(s_i, s_{-i}^*\right) &= \frac{\sum_{\tau=0}^{t-1} \delta^\tau g_i\left(s_i\left(d_\tau\right), s_{-i}^*\left(d_\tau\right)\right)}{\sum_{\tau=0}^{t-1} \delta^\tau} \\ &= \frac{\sum_{\tau=0}^{t-1} \delta^\tau g_i\left(s_i\left(d_\tau\right), a_{-i}^*\right)}{\sum_{\tau=0}^{t-1} \delta^\tau} \\ &\leq \frac{\sum_{\tau=0}^{t-1} \delta^\tau g_i\left(a_i^*, a_{-i}^*\right)}{\sum_{\tau=0}^{t-1} \delta^\tau} \\ &= \frac{\sum_{\tau=0}^{t-1} \delta^\tau g_i\left(s_i^*\left(d_\tau\right), s_{-i}^*\left(d_\tau\right)\right)}{\sum_{\tau=0}^{t-1} \delta^\tau} \\ &= u_i\left(s_i^*, s_{-i}^*\right) \end{aligned}$$

so that s_i^* is a best reply to s_{-i}^*. For the proof of the second claim, apply backward induction. □

Thus, if the stage game admits an equilibrium, we can construct an equilibrium for the multi-stage game in the obvious manner. We can show more: If a stage game has several equilibria, acting accordingly in any pre-specified order, is also an equilibrium. This is the assertion of the following theorem for the case of two one-stage equilibria. It can be extended to any number of one-stage equilibria.

THEOREM XIII.2. *Let* $\Gamma = \left(N, \left(A_i\right)_{i\in N}, g\right)$ *be a strategic game. Let* $a^* = \left(a_1^*, a_2^*, ..., a_n^*\right)$ *and* $b^* = \left(b_1^*, b_2^*, ..., b_n^*\right)$ *be equilibria of* Γ. *Then,* s^* *is a subgame-perfect equilibrium of* $\Gamma^{t,\delta}$ *or* $\Gamma^{\infty,\delta}$ *if there exists a partition* $\{T_{a^*}, T_{b^*}\}$ *of* $\{0, 1, ..., t-1\}$ *or* $\{0, 1, ...\}$, *respectively, such that the strategies* $s_i^* : D \to A_i$, $i \in N$, *obey*

$$d \mapsto s_i^*\left(d\right) = a_i^* \text{ for all } i \in N \text{ and all } d \text{ belonging to stages from } T_{a^*}$$

and

$$d \mapsto s_i^*\left(d\right) = b_i^* \text{ for all } i \in N \text{ and all } d \text{ belonging to stages from } T_{b^*}$$

The theorem claims that we obtain subgame-perfect equilibria of repeated games by sticking to any one stage equilibrium at all stage nodes that belong to the same stage t. The inverse need not hold. See problem XIII.1.

EXERCISE XIII.5. *Apply the above theorem to the battle of the sexes (p. 267) and show that a finitely repeated game with $\delta = 1$ may result in an average equilibrium payoff of $3\frac{3}{4}$ for her.*

Repeated games allow to discuss the punishment one player inflicts on another. For example, one might think that players in a prisoners' dilemma can use some tit-for-tat or similar strategy to threaten the other players into nice, cooperative behavior. Alas, the finitely repeated prisoners' dilemma does not provide this opportunity in a subgame perfect equilibrium.

4. The infinitely repeated prisoners' dilemma

We now focus on the prisoners' dilemma and examine whether players can cooperate if they play an infinite number of times.

4.1. Worst punishment. While finite repetitions of the prisoners' dilemma game admit only one subgame-perfect equilibrium, we obtain a large number of subgame-perfect equilibria in case of infinite repetitions. We will see that infinite repetitions allow for punishment strategies in equilibrium. We begin by defining the worst punishment that players from $N \backslash \{i\}$ can inflict on player i in each stage game:

DEFINITION XIII.3 (WORST PUNISHMENT). *Let $\Gamma = \left(N, (A_i)_{i \in N}, g\right)$ be a strategic game. The worst punishment that can be inflicted on player $i \in N$ is defined by*

$$w_i = \min_{a_{-i}} \max_{a_i} g_i(a_i, a_{-i}).$$

We call

- *$w = (w_1, ..., w_n)$ the worst-punishment point and*
- *$a_{-i}^{pun} := \arg\min_{a_{-i}} \max_{a_i} g_i(a_i, a_{-i})$ the worst-punishment action combination(s) undertaken by players from $N \backslash \{i\}$.*

In order to understand w_i correctly, note

$$\begin{aligned}
g_i(a_i, a_{-i}) &: \quad i\text{'s payoff resulting from } (a_i, a_{-i}), \\
\max_{a_i} g_i(a_i, a_{-i}) &: \quad i\text{'s maximal payoff, given } a_{-i}, \\
\min_{a_{-i}} \max_{a_i} g_i(a_i, a_{-i}) &: \quad i\text{'s worst punishment payoff}, \\
\arg\min_{a_{-i}} \max_{a_i} g_i(a_i, a_{-i}) &: \quad \text{punishing action combination } a_{-i}
\end{aligned}$$

Thus, the players from $N\backslash\{i\}$ choose a punishment action tuple a_{-i}^{pun}. If player i chooses his best response to a_{-i}^{pun}, he obtains

$$w_i = \max_{a_i} g_i\left(a_i, a_{-i}^{pun}\right).$$

Of course, it may well happen that player i obtains less than w_i. After all, the strategies in Γ (which are actions in $\Gamma^{\infty,\delta}$) are chosen simultaneously. w_i is the guaranteed minimum that can be inflicted on player i.

EXERCISE XIII.6. *Consider the following game and determine the worst punishment for player* 1.

		player 2	
		left	*right*
player 1	*up*	$2,1$	$4,0$
	down	$3,0$	$1,1$

The importance of the worst punishment is obvious from the following lemma which we note without proof.

LEMMA XIII.1. *A player's equilibrium payoff in a finitely or infinitely repeated stage game is not smaller than his worst punishment.*

4.2. Folk theorems for equilibria. We now remind the reader of the prisoners' dilemma presented in the introduction:

		firm 2	
		high price	cut price
firm 1	high price	$4,4$	$0,5$
	cut price	$5,0$	$1,1$

Lemma XIII.1 says that the equilibrium payoffs cannot fall below $w_1 = w_2 = 1$. To characterize the equilibrium payoff tuples, we need to define the convex hull of payoff vectors. You are familiar with convex combinations (see p. 70). A convex hull is a generalization from two payoff vectors (or consumption vectors) to any number. In any case, we have nonnegative scalars which sum to 1 and a convex hull is the set of all linear combinations that can be achieved by looking at all those scalar combinations:

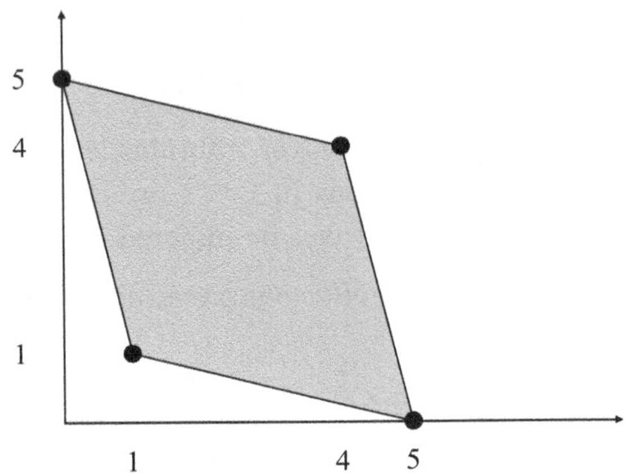

FIGURE XIII.4. The convex hull of the prisoners' dilemma

DEFINITION XIII.4 (CONVEX HULL OF VECTORS). *Let* $M = \{x_1, ..., x_m\} \subseteq \mathbb{R}^n$ *be a set of finite (consumption, payoff, etc.) vectors with* $m = |M|$. *The convex hull of* M *is denoted by* $hull(M)$ *and defined by*

$$\left\{ y \in \mathbb{R}^n : \text{ there exist } \alpha_1, ..., \alpha_m \geq 0 \text{ with } \sum_{j=1}^{m} \alpha_j = 1 \text{ s.t. } y = \sum_{j=1}^{m} \alpha_j x_j \right\}.$$

We can now define the convex hull of a stage game $\Gamma = (N, A, g)$ which is nothing but the convex hull of the payoff vectors.

DEFINITION XIII.5 (CONVEX HULL OF A STAGE GAME). *Let* $\Gamma = (N, A, g)$ *be a stage game. The convex hull of* Γ *is given by*

$$hull(\Gamma) := hull(\{g(a) : a \in A\}).$$

The convex hull of Γ *above* w *is defined by*

$$hull^w(\Gamma) := hull(\Gamma) \cap \{\pi \in \mathbb{R}^n : \pi_i > w_i\}$$

For example, the convex hull of our prisoners' dilemma game is depicted in figure XIII.4. The convex hull above $(1, 1)$ is the subset northeast of the worst-punishment point $(1, 1)$.

THEOREM XIII.3. *Let* $\Gamma = \left(N, (A_i)_{i \in N}, g\right)$ *be a strategic game with worst-punishment point* w. *In the infinite repetition* $\Gamma^{\infty,\delta}$, *every payoff vector from* $hull^w(\Gamma)$ *can be obtained in equilibrium if the discount factor* $\delta < 1$ *is sufficiently large. I.e., for every* $\pi \in hull^w(\Gamma)$, *there is a* $\delta^0 \in (0, 1)$ *such that* $\Gamma^{\infty,\delta}$ *has a Nash equilibrium* s *with* $u(s) = \pi$ *for all* $\delta \in \left(\delta^0, 1\right)$.

The central idea of the proof (which we do not provide) is to punish players who deviate. If the discount factor is large, the punishment can be more severe than the gain obtained by a one-time deviation. In that case, one can construct a subgame-perfect strategy combination for the prisoners' dilemma. Indeed, either the players act in line with this strategy combination. Then, they have nothing to gain by deviating once and being punished forever. Or, the players find themselves in punishment mode. Then, they act according to the equilibrium of the stage game and we have subgame perfection in line with theorem XIII.1.

5. Topics

The main topics in this chapter are

- stage game
- repeated game
- folk theorem
- worse punishment

6. Solutions

Exercise XIII.1

The payoff pair $(4, 9)$ comes about if player 1 chooses "high price" in both periods while player 2 is cooperative at one stage and noncooperative at the other. We obtain the payoff pair $(10, 0)$ if player 1 chooses the cut price in both periods and player 2 the high price in both periods.

Since we have four action combinations at the first stage, the repeated game has five subgames, corresponding to player 1's five decision nodes.

Exercise XIII.2

Player 2's tit-for-tat strategy is

$$\Big\lfloor \underbrace{\text{h}}_{\substack{\text{Action at}\\ \text{the first stage}}}, \underbrace{\text{h}}_{\substack{\text{Action at}\\ \text{the second stage}\\ \text{after h by both}}}, \underbrace{\text{h}}_{\substack{\text{Action at}\\ \text{the second stage}\\ \text{after (h, c)}}}, \underbrace{\text{c}}_{\substack{\text{Action at}\\ \text{the second stage}\\ \text{after (c, h)}}}, \underbrace{\text{c}}_{\substack{\text{Action at}\\ \text{the second stage}\\ \text{after c by both}}} \Big\rfloor!$$

Exercise XIII.3

$(\lfloor \text{c, c, c, c, c} \rfloor, \lfloor \text{c, c, c, c, c} \rfloor)$ is the Nash equilibrium where both players plan to act noncooperatively at both stages. $(\lfloor \text{c, h, h, h, c} \rfloor, \lfloor \text{c, h, h, h, c} \rfloor)$ leads to the same outcome (both choosing c at both stages). If player 2 chooses h at the first stage and c at the second, he can increase his profit from $2 = 1 + 1$ to $5 = 0 + 4$. Therefore, the second strategy combination is not a Nash equilibrium.

Exercise XIII.4

We obtain

$$u_i(s) = \frac{\sum_{\tau=0}^{t-1} \delta^\tau g_i(s(d_\tau))}{\sum_{\tau=0}^{t-1} \delta^\tau} = \frac{\sum_{\tau=0}^{t-1} \delta^\tau g_i(a)}{\sum_{\tau=0}^{t-1} \delta^\tau} = g_i(a) \frac{\sum_{\tau=0}^{t-1} \delta^\tau}{\sum_{\tau=0}^{t-1} \delta^\tau} = g_i(a).$$

Exercise XIII.5

We consider the fourfold repetition of the battle of the sexes. If the players go to the theater three times and to the football match once (the order is irrelevant because of $\delta = 1$), her payoff is

$$\frac{3 \cdot 4 + 3}{4} = \frac{15}{4} = 3\frac{3}{4}.$$

Exercise XIII.6

Player 1's worst punishment is

$$w_1 = \min_{a_2} \max_{a_1} g_1(a_1, a_2) = \min(3, 4) = 3.$$

		player 2		
		a_2^1	a_2^2	a_2^3
player 1	a_1^1	$(3,0)$	$(0,1)$	$(0,1)$
	a_1^2	$(0,0)$	$(2,2)$	$(1,0)$
	a_1^3	$(0,0)$	$(0,0)$	$(0,0)$

FIGURE XIII.5. A stage game with a unique equilibrium

7. Further exercises without solutions

PROBLEM XIII.1. Consider the battle of the sexes:

		he	
		theatre	football
she	theatre	$4,3$	$2,2$
	football	$1,1$	$3,4$

Can you identify all the equilibria of the twice repeated game? Which of these equilibria are subgame perfect? Write the strategies as quintuples

$$\lfloor a, a_{TT}, a_{TF}, a_{FT}, a_{FF} \rfloor,$$

where a is the action (theatre or football) at the first stage and a_{TF} the action if she chose theatre at the first stage and he football.

PROBLEM XIII.2. Consider the stage game of figure XIII.5. How many equilibria can you find? Is there an equilibrium of the twofold repetition where the two players choose the first actions at the first stage and the second actions at the second stage?

PROBLEM XIII.3. Show that a finitely repeated prisoners' dilemma has a unique subgame perfect equilibrium, in which each player chooses the non-cooperative action at every stage node.

PROBLEM XIII.4. Using the two-stage prisoners' dilemma, identify an equilibrium which is not subgame-perfect.

Part D

Bargaining theory and Pareto optimality

The first three parts of our course stay within the confines of non-cooperative game theory. Bargaining theory, our third main topic, is dealt with from the non-cooperative and the cooperative point of view. This part has three chapters. The aim of chapter XIV is to present a wide range of micro models through the lens of the Pareto principle. We then introduce the core and the Pareto principle in the context of cooperative games in some detail in chapter XV. The last chapter deals with the (non-cooperative) Rubinstein bargaining model.

CHAPTER XIV

Pareto optimality in microeconomics

Although the Pareto principle belongs to cooperative game theory, it sheds an interesting light on many different models in microeconomics. We consider bargaining between consumers, producers, countries in international trade, and bargaining in the context of public goods and externalities. We can also subsume profit maximization and household theory under this heading. It turns out that it suffices to consider three different cases with many subcases:

- equality of marginal rates of substitution
- equality of marginal rates of transformation and
- equality of marginal rate of substitution and marginal rate of transformation

Thus, we consider a wide range of microeconomic topics through the lens of Pareto optimality. The reader is invited to consult p. 84 for a definition of the marginal rate of substitution and p. 218 if he is unsure what marginal rate of transformation means.

1. Introduction: Pareto improvements

Economists are somewhat restricted when it comes to judgements on the relative advantages of economic situations. The reason is that ordinal utility does not allow for comparison of the utilities of different people.

However, situations can perhaps be ranked according to their relative Pareto superiority (Vilfredo Pareto, Italian sociologue, 1848-1923). Situation 1 is called Pareto superior to situation 2 if no individual is worse off in the first than in the second while at least one individual is strictly better off. Then, the move from 2 to 1 is called a Pareto improvement. Situations are called Pareto efficient, Pareto optimal, or just efficient if Pareto improvements are not possible. Compare the definition of Pareto inferiority (= absence of Pareto optimality) on p. 269.

EXERCISE XIV.1. *a) Is the redistribution of wealth a Pareto improvement if it reduces inequality?*
b) Can a situation be efficient if one individual possesses everything?

H. Wiese, *Advanced Microeconomics*,
https://doi.org/10.1007/978-3-658-34959-2_14

This chapter rests on the premise that bargaining leads to an efficient outcome under ideal conditions. As long as Pareto improvements are available, there is no reason (so one could argue) not to "cash in" on them.

However, the existence of Pareto improvements does not make their realization a forgone conclusion. This is obvious from the prisoners' dilemma and we revisit this theme later on in the context of bargaining under uncertainty (pp. 464).

2. Identical marginal rates of substitution

2.1. Exchange Edgeworth box.

2.1.1. *Introducing the Edgeworth box for two consumers.* We consider agents or households that consume bundles of goods. A distribution of such bundles among all households is called an allocation. In a two-agent two-good environment, allocations can be visualized via the Edgeworth box. Exchange Edgeworth boxes allow to depict preferences by the use of indifference curves. The reader may remember the production Edgeworth box introduced in chapter VIII, pp. 216.

The analysis of bargaining between consumers in an exchange Edgeworth box is due to Francis Ysidro Edgeworth (1881) who lived 1845-1926. Edgeworth is the author of a book with the beautiful title "Mathematical Psychics". Figure XIV.1 represents the exchange Edgeworth box for goods 1 and 2 and individuals A and B. The exchange Edgeworth box exhibits two points of origin, one for individual A (bottom left corner) and another one for individual B (top right).

Every point in the box denotes an allocation: how much of each good belongs to which individual. One possible allocation is the (initial) endowment. For all allocations $\left(x^A, x^B\right)$ with $x^A = \left(x_1^A, x_2^A\right)$ for individual A and $x^B = \left(x_1^B, x_2^B\right)$ for individual B we have

$$\begin{aligned} x_1^A + x_1^B &= \omega_1^A + \omega_1^B \text{ and} \\ x_2^A + x_2^B &= \omega_2^A + \omega_2^B. \end{aligned}$$

Individual A possesses an endowment $\omega^A = \left(\omega_1^A, \omega_2^A\right)$, i.e., ω_1^A units of good 1 and ω_2^A units of good 2. Similarly, individual B has an endowment $\omega^B = \left(\omega_1^B, \omega_2^B\right)$.

Exercise XIV.2. *Do the two individuals in figure XIV.1 possess the same quantities of good* 1, *i.e., do we have* $\omega_1^A = \omega_1^B$*?*

Exercise XIV.3. *Interpret the length and the breadth of the Edgeworth box!*

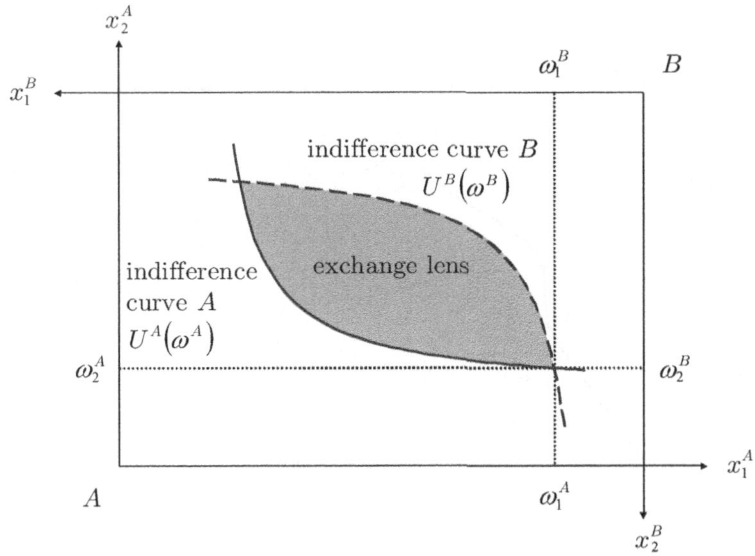

FIGURE XIV.1. The exchange Edgeworth box

2.1.2. *Equality of the marginal rates of substitution.* Seen from the respective points of origin, the Edgeworth box depicts the two individuals' preferences via indifference curves. Refer to figure XIV.1 when you work on the following exercise.

EXERCISE XIV.4. *Which bundles of goods does individual A prefer to his endowment? Which allocations do both individuals prefer to their endowments?*

The two indifference curves in figure XIV.1, crossing at the endowment point, enclose the so-called exchange lens which represents those allocations that are Pareto improvements over the endowment point. A Pareto efficient allocation is achieved if no further improvement is possible. Then, no individual can be made better off without making the other worse off. Oftentimes, we imagine that individuals achieve a Pareto efficient point by a series of exchanges. As long as no Pareto optimum has been reached, they will try to improve their lots.

EXERCISE XIV.5. *Sketch an inequitable Pareto optimum in an exchange Edgeworth box. Is the relation "allocation x is a Pareto improvement over allocation y" complete (see definition IV.13, p. 66)?*

Finally, we turn to the equality of the marginal rates of substitution. Consider an exchange economy with two individuals A and B where the

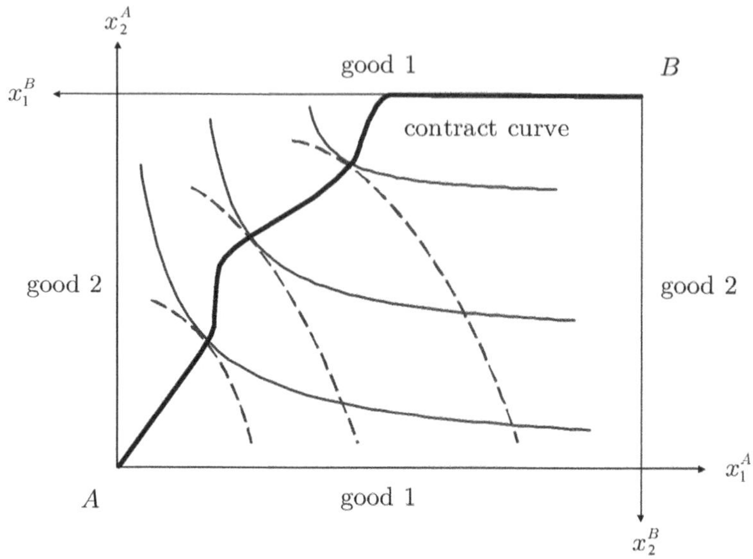

FIGURE XIV.2. The contract curve

marginal rate of substitution of individual A is smaller than that of individual B:

$$(3 =)\left|\frac{dx_2^A}{dx_1^A}\right| = MRS^A < MRS^B = \left|\frac{dx_2^B}{dx_1^B}\right| (= 5)$$

We can show that this situation allows Pareto improvements. Individual A is prepared to give up a small amount of good 1 in exchange for at least MRS^A units (3, for example) of good 2. If individual B obtains a small amount of good 1, he is prepared to give up MRS^B (5, for example) or less units of good 2. Thus, if A gives one unit of good 1 to B, by $MRS^A < MRS^B$ individual B can offer more of good 2 in exchange than individual A would require for compensation. The two agents might agree on $\frac{MRS^A+MRS^B}{2} = 4$ units so that both of them would be better off. Thus, the above inequality signals the possibility of mutually beneficial trade.

Differently put, Pareto optimality requires the equality of the marginal rates of substitution for any two agents A and B and any pair of goods 1 and 2. The locus of all Pareto optima in the Edgeworth box is called the contract curve or exchange curve (see figure XIV.2). In this figure, monotonic and strictly convex preferences are assumed. I generally work with "nice" preferences and technologies in this chapter.

Of special relevance is the intersection of the exchange lens with the contract curve. It is called the core. Such a core allocation has three

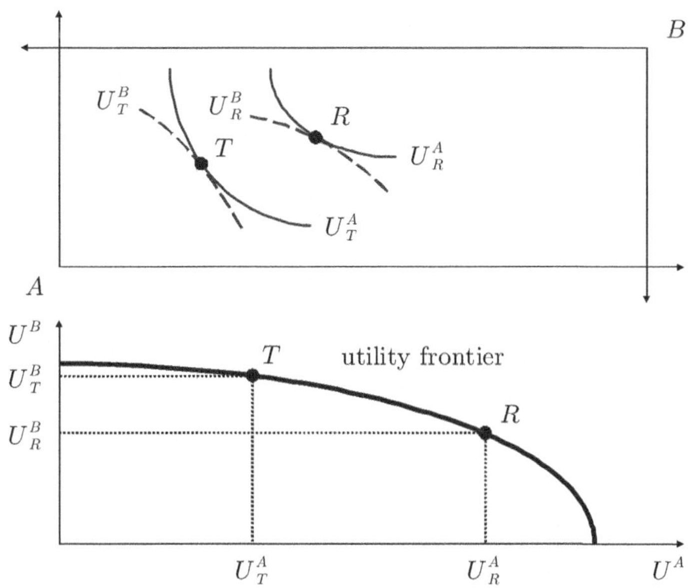

FIGURE XIV.3. Construction of the utility frontier

properties. First, an allocation from the core is feasible because it lies within the Edgeworth box. Second, it is Pareto efficient, i.e., the two agents cannot achieve another allocation that is better for both of them. Third, every individual agent weakly prefers a core allocation to his endowment.

2.1.3. *Deriving the utility frontier.* The contract curve can be transformed into the so-called utility frontier which is nothing but the exchange Edgeworth box' equivalent of the transformation curve (production-possibility frontier) known from the production Edgeworth box. Figure XIV.3 shows how to construct this curve. Take point R in the upper part of figure XIV.3. Here, individual A achieves his utility level U_R^A. Since R is a Pareto-efficient point on the contract curve, it is not possible for individual B to achieve a higher level of utility than U_R^B. The pair of utility levels $\left(U_R^A, U_R^B\right)$ is depicted in the lower part. In a similar fashion, point T (upper part) is transformed into point T (lower part). The resulting curve is called utility frontier. Given some utility level of individual A, this curve represents the maximal utility level possible for individual B.

EXERCISE XIV.6. *Are points S and T in figure XIV.4 Pareto efficient?*

EXERCISE XIV.7. *Two consumers meet on an exchange market with two goods. Both have the utility function $U(x_1, x_2) = x_1 x_2$. Consumer A's endowment is $(10, 90)$, consumer B's is $(90, 10)$.*

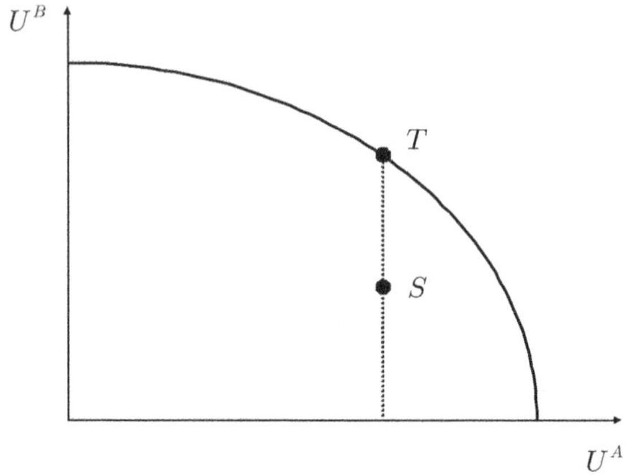

FIGURE XIV.4. Utility-possibility curve

a) Depict the endowments in the Edgeworth box!
b) Find the contract curve and draw it!
c) Find the best bundle that consumer B can achieve through exchange!
d) Draw the Pareto improvements over the endowment point (exchange lens) and draw the Pareto-efficient Pareto improvements!
e) Sketch the utility frontier!

2.1.4. *The generalized Edgeworth box.* We now generalize the two-agents two-goods Edgeworth box to n households, $i = 1, ..., n$, each of which possesses an endowment of ℓ goods. The set of households is denoted by $N := \{1, 2, .., n\}$. ω_g^i stands for household i's endowment of good g. We write ω^i for $\left(\omega_1^i, ..., \omega_\ell^i\right)$ and ω_g for $\left(\omega_g^1, ..., \omega_g^n\right)$. ω is the sum of the endowments of all households $\sum_{i=1}^n \omega^i$. Observe $\sum_{i=1}^n \omega^i \neq \sum_{g=1}^\ell \omega_g$.

EXERCISE XIV.8. *Consider two goods and three households and explain ω^3, ω_1 and ω.*

Households may consume their endowment, but they do not need to if they can exchange goods or buy and sell on a market. Consumption vectors are also called allocations. Here, we will consider vectors with nonnegative entries only.

DEFINITION XIV.1 (ALLOCATION). *Allocations are functions from the set of households N to the goods space $\mathbb{R}_+^\ell$. Differently put, an allocation is a vector $(x^i)_{i=1,...,n}$ or $\left(x^i\right)_{i\in N}$ where x^i is a bundle from $\mathbb{R}_+^\ell$.*

Allocations may or may not be compatible with the endowment of all the households taken together:

DEFINITION XIV.2 (FEASIBLE ALLOCATION). *An allocation is called feasible if*

$$\sum_{i=1}^{n} x^i \leq \sum_{i=1}^{n} \omega^i$$

holds.

Thus, any point in an Edgeworth box is a feasible allocation.

2.2. Production Edgeworth box. With respect to the production Edgeworth box, we can argue in a similar fashion. Remember producer 1's marginal rate of technical substitution $MRTS_1 = \left|\frac{dC_1}{dL_1}\right|$. We now understand this expression as producer 1's marginal willingness to pay for an additional unit of labor in terms of capital units. If two producers 1 and 2 produce goods 1 and 2, respectively, with inputs labor and capital, both can increase their production as long as the marginal rates of technical substitution differ. Thus, Pareto efficiency means

$$\left|\frac{dC_1}{dL_1}\right| = MRTS_1 \stackrel{!}{=} MRTS_2 = \left|\frac{dC_2}{dL_2}\right|$$

so that the marginal willingness to pay for input factors are the same.

2.3. Two markets – one factory. The third subcase under the heading "equality of the marginal willingness to pay" concerns a firm that produces in one factory but supplies two markets 1 and 2. The idea is to consider the marginal revenue $MR = \frac{dR}{dx_i}$ as the monetary marginal willingness to pay for selling one extra unit of good i. How much can a firm pay for the sale of one additional unit?

Thus, the marginal revenue is a marginal rate of substitution $\left|\frac{dR}{dx_i}\right|$. The role of the denominator good is taken over by good 1 or 2, respectively, while the nominator good is "money" (revenue). Now, profit maximization by a firm selling on two markets 1 and 2 implies

$$\left|\frac{dR}{dx_1}\right| = MR_1 \stackrel{!}{=} MR_2 = \left|\frac{dR}{dx_2}\right|$$

as we have seen in chapter XI, p. 307.

2.4. Two firms (cartel). The monetary marginal willingness to pay for producing *and* selling one extra unit of good y is a marginal rate of substitution where the denominator good is good 1 or 2 while the nominator good represents "money" (profit). We know from chapter XI (p. 315) that two cartelists 1 and 2 producing the quantities x_1 and x_2, respectively, maximize their joint profit

$$\Pi_{1,2}(x_1, x_2) = \Pi_1(x_1, x_2) + \Pi_2(x_1, x_2)$$

by obeying the first-order conditions

$$\frac{\partial \Pi_{1,2}}{\partial x_1} \stackrel{!}{=} 0 \stackrel{!}{=} \frac{\partial \Pi_{1,2}}{\partial x_2}$$

so that their marginal rates of substitution are the same if profit is understood as joint profit. If $\frac{\partial \Pi_{1,2}}{\partial x_2}$ were higher than $\frac{\partial \Pi_{1,2}}{\partial x_1}$ the cartel could increase profits by shifting the production of one unit from firm 1 to firm 2.

Note that a similar condition holds for the Cournot equilibrium (see p. 313),

$$\frac{\partial \Pi_1}{\partial x_1} \stackrel{!}{=} 0 \stackrel{!}{=} \frac{\partial \Pi_2}{\partial x_2}.$$

However, this is definitely *not* an example for Pareto optimality (although two marginal rates of substitution coincide). Indeed, both firms could be made better off by producing less and benefitting from a higher price. See subsection 4.4, p. 393.

3. Identical marginal rates of transformation

3.1. Two factories – one market. While the marginal revenue can be understood as the monetary marginal willingness to pay for selling, the marginal cost $MC = \frac{dC}{dy}$ can be seen as the monetary marginal opportunity cost of production. How much money (the second good) must the producer forgo in order to produce an extra unit of y (the first good)? Thus, the marginal cost can be seen as a special case of the marginal rate of transformation, $MRT = \left|\frac{dx_2}{dx_1}\right|^{\text{transformation curve}}$.

According to chapter XI, p. 309, a firm supplying a market from two factories (or a cartel in case of homogeneous goods), obeys the equality

$$MC_1 \stackrel{!}{=} MC_2.$$

The cartel also makes clear that Pareto improvements and Pareto optimality have to be defined relative to a specific group of agents. While the cartel solution (maximizing the sum of profits) is optimal for the producers, it is not, in general, optimal for the economy as a whole because the sum of producers' and consumers' (!) rent may well be below the welfare optimum.

3.2. Bargaining between countries (international trade). David Ricardo (1772–1823) has shown that international trade is profitable as long as the marginal rates of transformation between any two countries are different. Let us consider the classic example of England and Portugal producing wine (W) and cloth (Cl). Suppose that the marginal rates of transformation differ:

$$4 = MRT^P = \left|\frac{dW}{dCl}\right|^P > \left|\frac{dW}{dCl}\right|^E = MRT^E = 2.$$

Both countries can be made better off, irrespective of whether we start from autarchy or not. Indeed, let England produce another unit of cloth Cl that it exports to Portugal. England's production of wine reduces by $MRT^E = 2$ gallons. Portugal, that imports one unit of cloth, reduces the cloth production and can produce additional $MRT^P = 4$ units of wine. Therefore, if England obtains 3 gallons of wine in exchange for the one unit of cloth it gives to Portugal, both countries are better off.

Ricardo's theorem is known under the heading of "comparative cost advantage". So far, it is unclear why this is a good name for his theorem. The answer is provided by the following lemma.

LEMMA XIV.1. *Assume that f is a differentiable transformation function $x_1 \mapsto x_2$. Assume also that the cost function $C(x_1, x_2)$ is differentiable. Then, the marginal rate of transformation between good 1 and good 2 can be obtained by*

$$MRT(x_1) = \left|\frac{df(x_1)}{dx_1}\right| = \frac{MC_1}{MC_2}.$$

PROOF. Reconsider the production Edgeworth box encountered in chapter VIII. We assume a given volume of factor endowments and given factor prices (perfect competition!). Then, the overall cost for the production of goods 1 and 2 is constant and does not change along the transformation curve. Therefore, we can write

$$C(x_1, x_2) = C(x_1, f(x_1)) = \text{constant}.$$

If, now, we produce more of good 1 and less of good 2, the costs do not change:

$$\frac{\partial C}{\partial x_1} + \frac{\partial C}{\partial x_2}\frac{df(x_1)}{dx_1} = 0.$$

Solving for the marginal rate of transformation yields

$$MRT = -\frac{df(x_1)}{dx_1} = \frac{MC_1}{MC_2}.$$

□

Now we have Ricardo's result in the form it is usually presented: As long as the comparative costs (more precise: the ratio of marginal costs) between two goods differ, international trade (more or less of it) is worthwhile for both countries.

Thus, Pareto optimality requires the equality of the marginal opportunity costs between any two goods produced in any two countries. The economists before Ricardo clearly saw that absolute cost advantages make international trade profitable. If England can produce cloth cheaper than Portugal while Portugal can produce wine cheaper than England, we have

$$\begin{aligned} MC_{Cl}^{E} &< MC_{Cl}^{P} \text{ and} \\ MC_{W}^{E} &> MC_{W}^{P} \end{aligned}$$

so that England should produce more cloth and Portugal should produce more wine. Ricardo observed that for the implied division of labor to be profitable, it is sufficient that the ratio of the marginal costs differ:

$$\frac{MC_{Cl}^{E}}{MC_{W}^{E}} < \frac{MC_{Cl}^{P}}{MC_{W}^{P}}.$$

Do you see that this inequality follows from the two inequalities above, but not vice versa?

4. Equality of marginal rates of substitution and transformation

4.1. Base case. Imagine two goods consumed at a marginal rate of substitution MRS and produced at a marginal rate of transformation MRT. We now show that optimality implies $MRS = MRT$. Assume, to the contrary, that the marginal rate of substitution (for a consumer) is lower than the marginal rate of transformation (for a producer):

$$MRS = \left|\frac{dx_2}{dx_1}\right|^{\text{indifference curve}} < \left|\frac{dx_2}{dx_1}\right|^{\text{transformation curve}} = MRT.$$

If the producer reduces the production of good 1 by one unit, he can increase the production of good 2 by MRT units. The consumer has to renounce the one unit of good 1, and he needs at least MRS units of good 2 to make up for this. By $MRT > MRS$ the additional production of good 2 (come about by producing one unit less of good 1) more than suffices to compensate the consumer. Thus, the inequality of marginal rate of substitution and marginal rate of transformation points to a Pareto-inefficient situation.

4.2. Perfect competition. We want to apply the formula

$$MRS \stackrel{!}{=} MRT$$

to the case of perfect competition. For the output space, we have the condition

$$p \stackrel{!}{=} MC.$$

We have derived "price equals marginal cost" as the profit-maximizing condition on p. 240 and have discussed the welfare-theoretic implications on p. 310.

We can link the two formulas by letting good 2 be money with price 1.

- Then, the marginal rate of substitution tells the consumer's monetary marginal willingness to pay for one additional unit of good 1 (see pp. 162). Cum grano salis, the price can be taken to measure this willingness to pay for the marginal consumer (the last consumer prepared to buy the good).
- The marginal rate of transformation is the amount of money one has to forgo for producing one additional unit of good 1, i.e., the marginal cost.

Therefore, we obtain

$$\text{price} = \text{marginal willingness to pay } \stackrel{!}{=} \text{marginal cost.}$$

In a similar fashion, we can argue for inputs. The marginal value product $MVP = p\frac{dy}{dx}$ is the monetary marginal willingness to pay for the factor use, while the factor price w can be understood as the monetary marginal opportunity cost of employing the factor. Thus, we reobtain the optimization condition for a price taker on both the input and the output market introduced on p. 243:

$$\text{marginal value product } \stackrel{!}{=} \text{factor price.}$$

4.3. First-degree price discrimination. In the case of a Cournot monopoly, the "price equals marginal cost" rule is violated. However, first-degree price discrimination fulfills this rule as shown on pp. 306.

4.4. Cournot monopoly. A trivial violation of Pareto optimality ensues if a single agent acts in a non-optimal fashion. Just consider consumer and producer as a single person. For the Cournot monopolist, the $MRS \stackrel{!}{=} MRT$ formula can be rephrased as the equality between

- the monetary marginal willingness to pay for selling – this is the marginal revenue $MR = \frac{dR}{dy}$ (see above p. 389) – and

- the monetary marginal opportunity cost of production, the marginal cost $MC = \frac{dC}{dy}$ (p. 390).

4.5. Household optimum. Let us now turn to a consuming household. He "produces" goods by using his income to buy them, $m = p_1x_1 + p_2x_2$ in case of two goods.

EXERCISE XIV.9. *The prices of two goods* 1 *and* 2 *are* $p_1 = 6$ *and* $p_2 = 2$, *respectively. If the household consumes one additional unit of good* 1, *how many units of good* 2 *does he have to renounce?*

The exercise helps us understand that the marginal rate of transformation is the price ratio,

$$MRT = \frac{p_1}{p_2},$$

that we also know under the heading of "marginal opportunity cost". (Alternatively, consider the transformation function $x_2 = f(x_1) = \frac{m}{p_2} - \frac{p_1}{p_2}x_1$.) Seen this way, $MRS \overset{!}{=} MRT$ is nothing but the famous condition for household optimality dervied on pp. 137.

4.6. External effects and the Coase theorem.

4.6.1. *External effects and bargaining.* The famous Coase theorem can also be interpreted as an instance of $MRS \overset{!}{=} MRT$. We present this example in some detail.

External effects are said to be present if consumption or production activities are influenced positively or negatively while no compensation is paid for this influence. Environmental issues are often discussed in terms of negative externalities. Also, the increase of production exerts a negative influence on other firms that try to sell substitutes. Reciprocal effects exist between beekeepers and apple planters.

Consider a situation where A pollutes the environment doing harm to B. In a very famous and influential paper, Coase (1960) argues that economists have seen environmental and similar problems in a misguided way.

First of all, externalities are a "reciprocal problem". By this Coase means that restraining A from polluting harms A (and benefits B). According to Coase, the question to be decided is whether the harm done to B (who suffers from the pollution) is greater or smaller than the harm done to A (by stopping A's polluting activities).

Second, many problems resulting from externalities stem from missing property rights. While property rights exist for cars and real estate, they normally do not exist for air, water, or quietness. Coase suggests that the agents A and B bargain about the externality. If, for example, A has the right to pollute (i.e., is not liable for the damage caused by him), B can give him some money so that A reduces his harmful (to B) activity. If B has the right not to suffer any pollution (i.e., A is liable), A could approach B and offer some money in order to pursue some of the activity benefitting him. Coase assumes (as we have done in this chapter) that the two parties bargain about the externality so as to obtain a Pareto-efficient outcome.

The Nobel prize winner (of 1991) presents a startling thesis: the equilibrium externality (the pollution etc.) is independent of the initial distribution of property rights. This thesis is also known as the invariance hypothesis.

4.6.2. *Straying cattle.* Coase (1960) discusses the example of a cattle raiser and a crop farmer who possess adjoining land. The cattle regularly destroys part of the farmer's crop. In particular, consider the following table:

number of steers	marginal profit	marginal crop loss
1	4	1
2	3	2
3	2	3
4	1	4

The cattle raiser's marginal profit from steers is a decreasing function of the number of steers while the marginal crop loss increases. Let us begin with the case where the cattle raiser is liable. He can pay the farmer up to 4 (thousand Euros) for allowing him to have one cattle destroy crop. Since the farmer's compensating variation is 1, the two can easily agree on a price of 2 or 3.

The farmer and cattle raiser will also agree to have a second steer roam the fields, for a price of $2\frac{1}{2}$. However, there are no gains from trade to be had for the third steer. The willingness to pay of 2 is below the compensation money of 3.

If the cattle raiser is not liable, the farmer has to pay for reducing the number of steers from 4 to 3. A Pareto improvement can be had for any price between 1 and 4. Also, the farmer will convince the cattle raiser to take the third steer, but not the second one, off the field.

Thus, Coase seems to have a good point – irrespective of the property rights (the liability question), the number of steers and the amount of crop damaged is the same.

The reason for the validity (so far) of the Coase theorem is the fact that forgone profits are losses and forgone losses are profits. Therefore, the numbers used in the comparisons are the same.

It is about time to tell the reader why we talk about the Coase theorem in the $MRS \stackrel{!}{=} MRT$ section. From the cartel example, we are familiar with the idea of finding a Pareto optimum by looking at joint profits. We interpret the cattle raiser's marginal profit as the (hypothetical) joint firm's willingness to pay for another steer and the marginal crop loss incurred by the farmer as the joint firm's marginal opportunity cost for that extra steer.

We close this section by throwing in two caveats:

- If consumers are involved, the distribution of property rights has income effects. Then, Coase's theorem does not hold any more (see Varian 2010, chapter 31).
- More important is the objection raised by Wegehenkel (1980). The distribution of property rights determines who pays whom. Thus, if the property rights were to change from non-liability to liability, cattle raising becomes a less profitable business, while growing crops is more worthwhile as before. In the medium run, agents will move to the profitable occupations with effects on the crop losses (the sign is not clear a priori).

4.7. Public goods. Public goods are defined by non-rivalry in consumption. While an apple can be eaten only once, the consumption of a public good by one individual does not reduce others' consumption possibilities. Often-cited examples include street lamps or national defence.

Consider two individuals A and B who consume a private good x (quantities x^A and x^B, respectively) and a public good G. We define MRS^A and MRS^B as the marginal willingness to pay for the public good in terms of the private good. The optimality condition is

$$\begin{aligned} & MRS^A + MRS^B \\ = & \left| \frac{dx^A}{dG} \right|^{\text{indifference curve}} + \left| \frac{dx^B}{dG} \right|^{\text{indifference curve}} \\ \stackrel{!}{=} & \left| \frac{d\left(x^A + x^B\right)}{dG} \right|^{\text{transformation curve}} = MRT. \end{aligned}$$

Assume that this condition is not fulfilled. For example, let the marginal rate of transformation be smaller than the sum of the marginal rates of substitution. Then, it is a good idea to produce one additional unit of the public good. The two consumers need to forgo MRT units of the private good. However, they are prepared to give up $MRS^A + MRS^B$ units of the private good in exchange for one additional unit of the public good. Thus, they can give up more than they need to. The two consumers are better off than before and the starting point (inequality) does not characterize a Pareto optimum.

Once more, we can assume that good x is the numéraire good (money with price 1). Then, the optimality condition simplifies and Pareto efficiency requires that the sum of the marginal willingness' to pay equals the marginal cost of the public good.

EXERCISE XIV.10. *In a small town, there live 200 people* $i = 1, ..., 200$ *with identical preferences. Person* i*'s utility function is* $U_i(x_i, G) = x_i + \sqrt{G}$*, where* x_i *is the quantity of the private good and* G *the quantity of the public good. The prices are* $p_x = 1$ *and* $p_G = 10$*, respectively. Find the Pareto-optimal quantity of the public good.*

Thus, by the non-rivalry in consumption, we do not quite get a subrule of $MRS \overset{!}{=} MRT$ but something similar.

5. Topics

The main topics in this chapter are

- Pareto efficiency
- Pareto improvement
- exchange Edgeworth box
- contract curve
- exchange lens
- international trade
- external effects
- quantity cartel
- public goods
- first-degree price discrimination

6. Solutions

Exercise XIV.1

a) A redistribution that reduces inequality will harm the rich. Therefore, such a redistribution is not a Pareto improvement.
b) Yes. It is not possible to improve the lot of the have-nots without harming the individual who possesses everything.

Exercise XIV.2

No, ω_1^A is much larger than ω_1^B.

Exercise XIV.3

The length of the exchange Edgeworth box represents the units of good 1 to be divided between the two individuals, i.e., the sum of their endowments of good 1. Similarly, the breadth of the Edgeworth box is $\omega_2^A + \omega_2^B$.

Exercise XIV.4

Individual A prefers all those bundels x_A that lie to the right and above the indifference curve that crosses his endowment point. The allocations preferred by both individuals are those in the hatched part of figure XIV.1.

Exercise XIV.5

For the first question, you should have drawn something like figure XIV.5. Figure XIV.6 makes clear that we can have bundles X and Y where X is no Pareto improvement over Y and Y is no improvement over X. Thus, the relation is not complete.

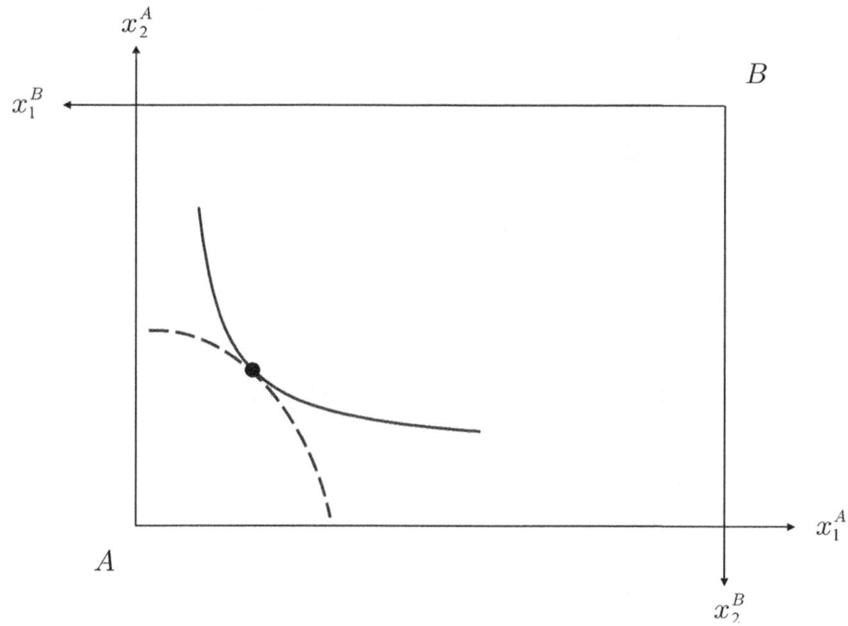

FIGURE XIV.5. Pareto optimality and equality

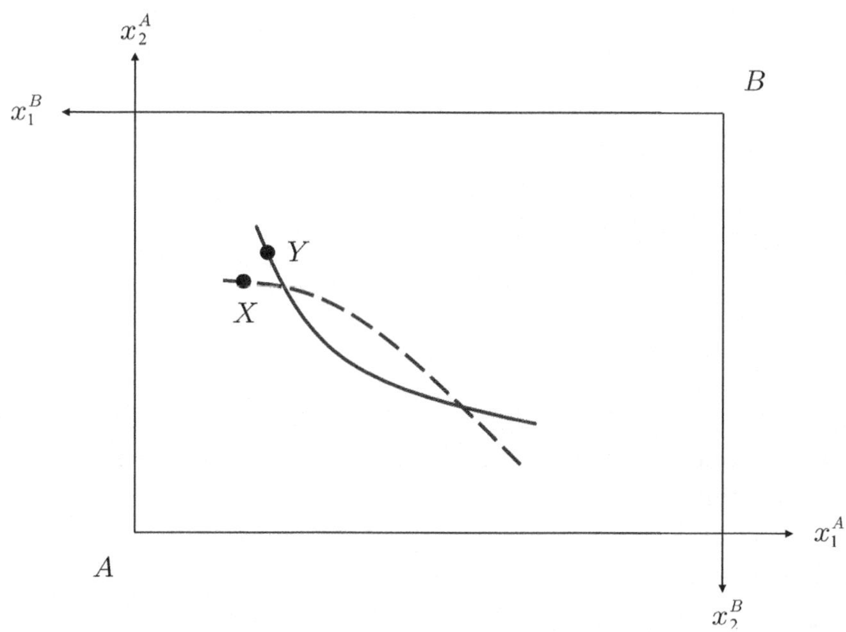

FIGURE XIV.6. Incompleteness

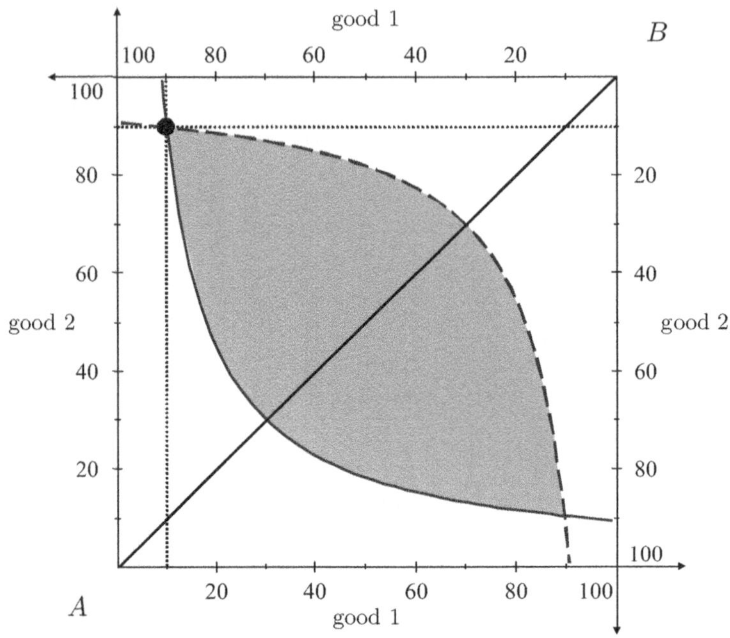

FIGURE XIV.7. The answers to parts a) and d)

Exercise XIV.6

Point S is not Pareto efficient. At T, individual B is better off while A's utility level is the same as at S. From T no Pareto improvement is possible. Therefore, T is Pareto efficient.

Exercise XIV.7

a) See figure XIV.7.

b) $x_2^A = x_1^A$.

c) $(70, 70)$.

d) The exchange lens is dotted in figure XIV.7. The Pareto efficient Pareto improvements are represented by the contract curve within this lens.

e) The utility frontier is downward sloping and given by $U_B(U_A) = (100 - \sqrt{U_A})^2$.

Exercise XIV.8

- $\omega^3 = (\omega_1^3, \omega_2^3)$ is the third household's endowment,
- $\omega_1 = (\omega_1^1, \omega_1^2, \omega_1^3)$ represents the endowment of good 1, distributed among the three households, and
- $\omega = (\omega_1^1 + \omega_1^2 + \omega_1^3, \omega_2^1 + \omega_2^2 + \omega_2^3)$ stands for the endowment present in the whole economy. Thus, ω marks the size of the Edgeworth box.

Exercise XIV.9

If the household consumes one additional unit of good 1, he has to pay Euro 6. Therefore, he has to renounce 3 units of good 2 that also cost Euro 6 = Euro 2 times 3.

Exercise XIV.10

The marginal rate of transformation $\left|\frac{d\left(\sum_{i=1}^{200} x_i\right)}{dG}\right|$ equals $\frac{p_G}{p_x} = \frac{10}{1} = 10$. The marginal rate of substitution for inhabitant i is

$$\left|\frac{dx_i}{dG}\right|^{\text{indifference curve}} = \frac{MU_G}{MU_{x_i}} = \frac{\frac{1}{2\sqrt{G}}}{1} = \frac{1}{2\sqrt{G}}.$$

Applying the optimality condition yields

$$200 \cdot \frac{1}{2\sqrt{G}} \stackrel{!}{=} 10$$

and hence $G = 100$.

7. Further exercises without solutions

PROBLEM XIV.1. Agent A's preferences are represented by $u^A(x_1^A, x_2^A) = x_1^A$. Agent B has preferences, which are represented by the utility function $u^B(x_1^B, x_2^B) = x_2^B$. Agent A starts with $\omega_1^A = \omega_2^A = 5$, and B has the initial endowment $\omega_1^B = 4, \omega_2^B = 6$.

(1) Draw the Edgeworth box, including
   - ω,
   - an indifference curve for each agent through ω!
(2) Is $(x_1^A, x_2^A, x_1^B, x_2^B) = (6, 0, 3, 11)$ a Pareto-improvement compared to the initial allocation?
(3) Find the contract curve!

PROBLEM XIV.2. Consider the player set $N = \{1, ..., n\}$. Player $i \in N$ has 24 hours to spend on leisure or work, $24 = l_i + t_i$ where l_i denotes i's leisure time and t_i the number of hours that i contributes to the production of a good that is equally distributed among the group. In particular, we assume the utility functions $u_i(t_1, ..., t_n) = l_i + \frac{1}{n} \sum_j \lambda t_j, i \in N$. Assume $1 < \lambda$ and $\lambda < n$.

(1) Find the Nash equilibrium.
(2) Is the Nash equilibrium Pareto-efficient?

CHAPTER XV

Cooperative game theory

1. Introduction

The aim of this chapter is to familiarize the reader with cooperative game theory. It is sometimes suggested that non-cooperative game theory is more fundamental than cooperative game theory. Indeed, from an economic or sociological point of view, cooperative game theory seems odd in that it does not model people who "act", "know about things", or "have preferences". In cooperative game theory, people just get payoffs. Cooperative game theory is payoff-centered game theory. Non-cooperative game theory (which deals with strategies and equilibria) could be termed action-centered or strategy-centered. Of course, non-cooperative game theory's strength does not come without cost. The modeller is forced to specify in detail (sequences of) actions, knowledge, and preferences. Cooperative game theory is better at providing a bird's eye view.

Cooperative game theory builds on two pillars. First, the economic (or political or sociological ...) situation is described by a so-called coalition function. We concentrate on the simple transferable-utility case. Second, solution concepts are applied to coalition functions in order to tell the payoffs of all the players. We focus on two central solution concepts from cooperative game theory, the Shapley value and the core.

2. The coalition function

Our discussion uses a specific example, the gloves game. Some players have a left glove and others a right glove. Single gloves have a worth of zero while pairs have a worth of 1 (Euro). The coalition function for the gloves game is given by

$$\begin{aligned} v_{L,R} \quad : \quad & 2^N \to \mathbb{R} \\ & K \mapsto v_{L,R}(K) = \min\left(|K \cap L|, |K \cap R|\right), \end{aligned}$$

where

- N is the set of players (also called the grand coalition),

H. Wiese, *Advanced Microeconomics*,
https://doi.org/10.1007/978-3-658-34959-2_15

- L (set of left-glove holders) and R (set of right-glove holders) form a partition of N,
- $v_{L,R}$ denotes the coalition function for the gloves game,
- 2^N stands for N's power set, i.e., the set of all subsets of N (the domain of $v_{L,R}$), and
- K is a coalition, i.e., $K \subseteq N$ or $K \in 2^N$.

Thus, to each coalition K, the coalition function $v_{L,R}$ attributes that coalition's number of pairs of gloves.

DEFINITION XV.1 (PLAYER SETS AND COALITION FUNCTIONS). *Player sets and coalition functions are specified by the following definitions:*

- *Finite and nonempty player sets are denoted by N. More often than not, we have $N = \{1, ..., n\}$ with $n \in \mathbb{N}$.*
- *$v : 2^N \to \mathbb{R}$ is called a coalition function if v fulfills $v(\emptyset) = 0$. $v(K)$ is called coalition K's worth.*
- *For any given coalition function v, its player set can be addressed by $N(v)$ or, more simply, N.*
- *We denote the set of all games on N by $\mathbb{V}_N$ and the set of all games (for any player set N) by $\mathbb{V}$.*

EXERCISE XV.1. *Assume $N = \{1, 2, 3, 4, 5\}$, $L = \{1, 2\}$ and $R = \{3, 4, 5\}$. Find the worths of the coalitions $K = \{1\}$, $K = \emptyset$, $K = N$, and $K = \{2, 3, 4\}$.*

The above exercise makes clear that $v_{L,R}$ is, indeed, a coalition function. The requirement of $v(\emptyset) = 0$ makes perfect sense: a group of zero agents cannot achieve anything.

EXERCISE XV.2. *Which of the following propositions make sense? Any coalition K and any grand coalition N fulfill*

- *$K \in N$,*
- *$K \subseteq 2^N$,*
- *$K \subseteq N$ and/or*
- *$K \in 2^N$?*

In this chapter, we focus on transferable utility where v attaches real numbers to coalitions. $v(K)$ is the worth or the utility sum created by the members of K. The basic idea is to distribute $v(K)$ or $v(N)$ among the members of K or N. Thus, the utility is "transferable".

Transferability is a severe assumption and does not work well in every model. We need non-transferable utility for the analysis of exchange within an Edgeworth box. Transferable utility is justfied if utility can be measured in terms of money and if the agents are risk neutral.

We can interpret the gloves game as a market game where the left-glove owners form one market side and the right-glove owners the other. We need to distinguish the worth (of a coalition) from the payoff accruing to players. If the owner of a left glove obtains payoff $\frac{7}{10}$, two interpretations come to mind:

- The player sells his left glove for a price of $\frac{7}{10}$.
- The player buys a right glove for the price of $\frac{3}{10}$, assembles a pair of gloves which he uses (or sells for price 1) so that his payoff is $1 - \frac{3}{10} = \frac{7}{10}$.

We can sum payoffs and worths which leads us to the next section.

3. Summing and zeros

Payoffs for players are summarized in payoff vectors:

DEFINITION XV.2 (PAYOFF VECTOR). *For any finite and nonempty player set* $N = \{1, ..., n\}$, *a payoff vector*

$$x = (x_1, ..., x_n) \in \mathbb{R}^n$$

specifies payoffs for all players $i = 1, ..., n$.

It is possible to sum coalition functions and it is possible to sum payoff vectors. Summation of vectors is easy – just sum each entry individually.

EXERCISE XV.3. *Determine the sum of the vectors*

$$\begin{pmatrix} 1 \\ 3 \\ 6 \end{pmatrix} + \begin{pmatrix} 2 \\ 5 \\ 1 \end{pmatrix}!$$

If we have three players, it is obvious that the first entry belongs to player 1, the second to player 2 etc. Note the difference between payoff-vector summation concerning several vectors,

$$x + y = \begin{pmatrix} x_1 \\ x_2 \\ \\ x_n \end{pmatrix} + \begin{pmatrix} y_1 \\ y_2 \\ \\ y_n \end{pmatrix} = \begin{pmatrix} x_1 + y_1 \\ x_2 + y_2 \\ \\ x_n + y_n \end{pmatrix},$$

and payoff summation for one given vector,

$$\sum_{i=1}^{n} x_i.$$

Vector summation is possible for coalition functions, too. For example, we obtain the sum $v_{\{1\},\{2,3\}}+v_{\{1,2\},\{3\}}$ by summing the worths $v_{\{1\},\{2,3\}}(K)+v_{\{1,2\},\{3\}}(K)$ for every coalition K, from the empty set $\emptyset$ down to the grand coalition $\{1,2,3\}$:

$$\begin{pmatrix} \emptyset:0 \\ \{1\}:0 \\ \{2\}:0 \\ \{3\}:0 \\ \{1,2\}:1 \\ \{1,3\}:1 \\ \{2,3\}:0 \\ \{1,2,3\}:1 \end{pmatrix} + \begin{pmatrix} \emptyset:0 \\ \{1\}:0 \\ \{2\}:0 \\ \{3\}:0 \\ \{1,2\}:0 \\ \{1,3\}:1 \\ \{2,3\}:1 \\ \{1,2,3\}:1 \end{pmatrix} = \begin{pmatrix} \emptyset:0 \\ \{1\}:0 \\ \{2\}:0 \\ \{3\}:0 \\ \{1,2\}:1 \\ \{1,3\}:2 \\ \{2,3\}:1 \\ \{1,2,3\}:2 \end{pmatrix}$$

Of course, we need to agree upon a specific order of coalitions.

Mathematically speaking, $\mathbb{R}^n$ and $\mathbb{V}_N$ can be considered as vector spaces. Vector spaces have a zero. The zero from $\mathbb{R}^n$ is

$$\underset{\in\mathbb{R}^n}{0} = \left(\underset{\in\mathbb{R}}{0}, ..., \underset{\in\mathbb{R}}{0}\right)$$

where the zero on the left-hand side is the zero vector while the zeros on the right-hand side are just the zero payoffs for all the individual players. In the vector space of coalition functions, $0 \in \mathbb{V}_N$ is the function that attributes the worth zero to every coalition, i.e.,

$$\underset{\in\mathbb{V}_N}{0}(K) = \underset{\in\mathbb{R}}{0} \text{ for all } K \subseteq N$$

4. Solution correspondences

4.1. Solutions functions and solution correspondences. For the time being, cooperative game theory consists of coalition functions and solution concepts. The task of solution concepts is to define and defend payoffs as a function of coalition functions. That is, we take a coalition function, apply a solution concept, and obtain payoffs for all the players.

Solution concepts may be point-valued (solution function) or set-valued (solution correspondence). In each case, the domain is the set of all games $\mathbb{V}$ for any finite player sets N. A solution function associates each game with exactly one payoff vector, while a correspondence allows for several or no payoff vectors.

DEFINITION XV.3 (SOLUTION FUNCTION, SOLUTION CORRESPONDENCE). *A function σ that attributes, for each coalition function v from $\mathbb{V}$, a payoff* *to each of v's players,*

$$\sigma(v) \in \mathbb{R}^{|N(v)|},$$

is called a solution function (on $\mathbb{V}$*). Player i's payoff is denoted by* $\sigma_i(v)$*. In case of* $N(v) = \{1, ..., n\}$*, we also write* $(\sigma_1(v), ..., \sigma_n(v))$ *for* $\sigma(v)$ *or* $(\sigma_i(v))_{i \in N(v)}$*.*

A correspondence that attributes a set of payoff vectors to every coalition function v,

$$\sigma(v) \subseteq \mathbb{R}^{|N(v)|}$$

is called a solution correspondence (on $\mathbb{V}$*).*

Solution functions and solution correspondences are also called solution concepts (on $\mathbb{V}$*).*

Ideally, solution concepts are described both algorithmically and axiomatically. An algorithm is some kind of mathematical procedure (a more or less simple function) that tells us how to derive payoffs from the coalition functions. Consider, for example, these four solution concepts in algorithmic form:

- player 1 obtains $v(N)$ and the other players obtain 0,
- every player gets 100,
- every player gets $v(N)/n$,
- every player i's payoff set is given by $[v(\{i\}), v(N)]$ (which may be the empty set).

Alternatively, solution concepts can be defined by axioms. For example, axioms might demand that

- all the players obtain the same payoff,
- no more than $v(N)$ is to be distributed among the players,
- player 1 is to get twice the payoff obtained by player 2,
- the names of players are irrelevant,
- every player gets $v(N) - v(N \backslash \{i\})$.

Axioms pin down the players' payoffs, more or less. Axioms may also make contradictory demands. We present prominent axioms in the following sections.

Before looking at some well-known solution concepts, we should ask ourselves which payoffs to expect for the gloves game. How do you feel about these properties:

- Prices of gloves should not be negative.
- The more left-hand gloves we have, the lower the price of left-hand gloves.

- The sum of all the players' payoffs should equal the number of glove pairs.
- Owners of left gloves obtain the same payoff as owners of right gloves.
- All owners of left gloves obtain the same payoff.

4.2. Solution correspondence: Pareto efficiency. Arguably, Pareto efficiency is the single most often applied solution concept in economics – rivaled only by the Nash equilibrium from non-cooperative game theory. For the gloves game, Pareto efficiency is defined by

$$\sum_{i\in N} x_i = v_{L,R}(N).$$

Thus, the sum of all payoffs is equal to the number of glove pairs. It is instructive to write this equality as two inequalities:

$$\begin{aligned}\sum_{i\in N} x_i &\leq v_{L,R}(N) \text{ (feasibility) and}\\ \sum_{i\in N} x_i &\geq v_{L,R}(N) \text{ (the grand coalition cannot block } x).\end{aligned}$$

According to the first inequality, the players cannot distribute more than they (all together) can "produce". This is the requirement of feasibility.

Imagine that the second inequality were violated. Then, $\sum_{i=1}^n x_i < v_{L,R}(N)$ holds and the players would leave "money on the table". All players together could block (or contradict) the payoff vector x. This means they can propose another payoff vector that is both feasible and better for the players. Indeed, the payoff vector $y = (y_1, ..., y_n)$ defined by

$$y_i = x_i + \frac{1}{n}\left(v_{L,R}(N) - \sum_{i=1}^n x_i\right), i \in N,$$

does the trick. y is an improvement upon x.

EXERCISE XV.4. *Show that the payoff vector y is feasible.*

Normally, Pareto efficiency is defined by "it is impossible to improve the lot of one player without making other players worse off". If a sum of money is distributed among the players, we can also define Pareto efficiency by "it is impossible to improve the lot of all players". The additional sum of money that makes one player better off (first definition) can be spread among all the players (second definition).

DEFINITION XV.4 (FEASIBILITY AND EFFICIENCY). *Let $v \in \mathbb{V}_N$ be a coalition function and let $x \in \mathbb{R}^n$ be a payoff vector. x is called*

- *blockable by N in case of*

$$\sum_{i\in N} x_i < v(N),$$

- *feasible in case of*

$$\sum_{i\in N} x_i \leq v(N)$$

- *and efficient or Pareto efficient in case of*

$$\sum_{i\in N} x_i = v(N).$$

Thus, an efficient payoff vector is feasible and cannot be blocked by the grand coalition N. Obviously, Pareto efficiency is a solution correspondence, not a solution function.

EXERCISE XV.5. *Find the Pareto-efficient payoff vectors for the gloves game $v_{\{1\},\{2\}}$!*

For the gloves game, the solution concept "Pareto efficiency" has two important drawbacks:

- We have very many solutions and the predictive power is weak. In particular, a left-hand glove can have any price, positive or negative.
- The payoff for a left-glove owner does not depend on the number of left and right gloves in our simple economy. Thus, the relative scarcity of gloves is not reflected by this solution concept.

We now turn to a solution concept that generalizes the idea of blocking from the grand coalition to all coalitions.

4.3. Solution correspondence: The core. Pareto efficiency demands that the grand coalition should not be in the position to make all players better off. Extending this idea to all coalitions, the core consists of those feasible (!) payoff vectors that cannot be improved upon by any coalition with its own means.

DEFINITION XV.5 (BLOCKABILITY AND CORE). *Let $v \in \mathbb{V}_N$ be a coalition function. A payoff vector $x \in \mathbb{R}^n$ is called blockable by a coalition $K \subseteq N$ if*

$$\sum_{i\in K} x_i < v(K)$$

holds. The core is the set of all those payoff vectors x fulfilling

$$\begin{aligned} \sum_{i\in N} x_i &\leq v(N) \text{ (feasibility) and} \\ \sum_{i\in K} x_i &\geq v(K) \text{ for all } K\subseteq N \text{ (no blockade by any coalition).} \end{aligned}$$

Do you see that every payoff vector from the core is also Pareto efficient? Just take $K := N$.

The core is a stricter concept than Pareto efficiency. It demands that no coalition (not just the grand coalition) can block any of its payoff vectors. Let us consider the gloves game for $L = \{1\}$ and $R = \{2\}$. By Pareto efficiency, we can restrict attention to those payoff vectors $x = (x_1, x_2)$ that fulfill $x_1 + x_2 = 1$. Furthermore, x may not be blocked by one-man coalitions:

$$\begin{aligned} x_1 &\geq v_{L,R}(\{1\}) = 0 \text{ and} \\ x_2 &\geq v_{L,R}(\{2\}) = 0. \end{aligned}$$

Hence, the core is the set of payoff vectors $x = (x_1, x_2)$ obeying

$$x_1 + x_2 = 1, x_1 \geq 0, x_2 \geq 0.$$

Are we not forgetting about $K = \emptyset$? Let us check

$$\sum_{i\in\emptyset} x_i \geq v_{L,R}(\emptyset).$$

Since there is no i from $\emptyset$ (otherwise $\emptyset$ would not be the empty set), the sum $\sum_{i\in\emptyset} x_i$ has no summands and is equal to zero. Since all coalition functions have worth zero for the empty set, we find $\sum_{i\in\emptyset} x_i = 0 = v_{L,R}(\emptyset)$ for the gloves game and also for any coalition function.

EXERCISE XV.6. *Determine the core for the gloves game $v_{L,R}$ with $L = \{1,2\}$ and $R = \{3\}$.*

In case of $|L| = 2 > 1 = |R|$ right gloves are scarcer than left gloves. In such a situation, the owner of a right glove should be better off than the owner of a left glove. The core reflects the relative scarcity in a drastic way. Consider the Pareto-efficient payoff vector

$$y = \left(\frac{1}{10}, \frac{1}{10}, \frac{8}{10}\right)$$

that does not lie in the core. This payoff vector can be blocked by coalition $\{1,3\}$. Its worth is $v(\{1,3\}) = 1$ which can be distributed among its members in a manner that both are better off.

Note that the core is a set-valued solution concept. It can contain one payoff vector (see the above exercise) or very many payoff vectors (in case of $L = \{1\}$ and $R = \{2\}$). Later on, we will see coalition functions with an empty core: every feasible payoff vector is blockable by at least one coalition.

The core for coalition functions has first been defined by Gillies (1959). Shubik (1981, p. 299) mentions that Lloyd Shapley proposed this concept as early as 1953 in unpublished lecture notes. In the framework of an exchange economy, Edgeworth (1881) proposes a very similar concept (see chapter XIV).

5. The Shapley value

5.1. The Shapley formula. In contrast to Pareto efficiency and the core, the Shapley value is a point-valued solution concept, i.e., a solution function. For every coalition function, it spits out exactly one payoff vector. Shapley's (1953) article is famous for pioneering the twofold approach of algorithm and axioms.

We begin with the Shapley formula. It rests on a simple idea. Every player obtains

- an average of
- his marginal contributions.

Beginning with the latter, the marginal contribution of player i with respect to coalition K is

"the value with him" minus "the value without him".

Thus, the marginal contributions reflect a player's productivity.

DEFINITION XV.6 (MARGINAL CONTRIBUTION). *Let $i \in N$ be a player from N and let $v \in \mathbb{V}_N$ be a coalition function on N. Player i's marginal contribution with respect to a coalition K is denoted by $MC_i^K(v)$ and given by*

$$MC_i^K(v) := v(K \cup \{i\}) - v(K \setminus \{i\}).$$

The marginal contribution of a player depends on the coalition function and the coalition. It does not matter whether i is a member of K or not, i.e., we have $MC_i^{K\cup\{i\}}(v) = MC_i^{K\setminus\{i\}}(v)$.

EXERCISE XV.7. *Determine the marginal contributions for $v_{\{1,2,3\},\{4,5\}}$ and*

- $i = 1, K = \{1, 3, 4\}$,
- $i = 1, K = \{3, 4\}$,
- $i = 4, K = \{1, 3, 4\}$,
- $i = 4, K = \{1, 3\}$.

We now need to explain the kind of averaging employed by the Shapley formula. In order to calculate the Shapley value, one considers all rank orders of the n players. $(3,1,2)$ is one rank order of the players 1 to 3. Just imagine that the players 3, 1, and 2 stand outside the door and enter, one after the other. We are interested in the marginal contributions. For rank order $(3,1,2)$, one finds the marginal contributions

$$v(\{3\}) - v(\emptyset), v(\{1,3\}) - v(\{3\}), \text{ and } v(\{1,2,3\}) - v(\{1,3\}).$$

They add up to $v(N) - v(\emptyset) = v(N)$.

Definition XV.7 (rank order). *Let $N = \{1, ..., n\}$ be a player set. Bijective functions $\rho : N \to N$ are called rank orders or permutations on N. The set of all permutations on N is denoted by RO_N. The set of all players "up to and including player i under rank order ρ" is denoted by $K_i(\rho)$ and given by*

$$\rho(j) = i \text{ and } K_i(\rho) = \{\rho(1), .., \rho(j)\}.$$

Player i's marginal contribution with respect to rank order ρ is denoted by $MC_i^\rho(v)$ and given by

$$MC_i^\rho(v) := MC_i^{K_i(\rho)}(v) = v(K_i(\rho)) - v(K_i(\rho) \setminus \{i\}).$$

Exercise XV.8. *Find player 2's marginal contributions for the rank orders $(1,3,2)$ and $(3,1,2)$!*

For every player, his Shapley value is the average of his marginal contributions where each rank order is equally likely. Thus, we can employ the following algorithm:

- We first determine all the possible rank orders.
- We then find the marginal contributions for every rank order.
- For every player, we add his marginal contributions.
- Finally, we divide the sum by the number of rank orders.

Consider the simple example given by $N = \{1,2,3\}$, $L = \{1,2\}$, and $R = \{3\}$. We find the rank orders:

$$\begin{aligned} &(1,2,3), (1,3,2), \\ &(2,1,3), (2,3,1), \\ &(3,1,2), (3,2,1). \end{aligned}$$

For three players, there are $1 \cdot 2 \cdot 3 = 6$ different rank orders. It is not difficult to see, why. For a single player 1, we have just one rank order (1).

The second player 2 can be placed before or after player 1 so that we obtain the $1 \cdot 2$ rank orders

$$(1,2),$$
$$(2,1).$$

For each of these two, the third player 3 can be placed before the two players, in between or after them:

$$(3,1,2), (1,3,2), (1,2,3),$$
$$(3,2,1), (2,3,1), (2,1,3).$$

Therefore, we have $2 \cdot 3 = 6$ rank orders. Generalizing, for n players, we have $1 \cdot 2 \cdot ... \cdot n$ rank orders. We can also use the abbreviation

$$n! := 1 \cdot 2 \cdot ... \cdot n$$

which is to be read "n factorial".

EXERCISE XV.9. *Determine the number of rank oders for* 5 *and for* 6 *players!*

EXERCISE XV.10. *Consider* $N = \{1,2,3\}$, $L = \{1,2\}$, *and* $R = \{3\}$ *and determine player* 1*'s marginal contribution for each rank order.*

DEFINITION XV.8 (SHAPLEY VALUE). *The Shapley value is the solution function* Sh *given by*

$$Sh_i(v) = \frac{1}{n!} \sum_{\rho \in RO_N} MC_i^\rho(v)$$

According to the previous exercise, we have

$$Sh_1\left(v_{\{1,2\},\{3\}}\right) = \frac{1}{6}.$$

The Shapley values of the other two players can be obtained by the same procedure. However, there is a more elegant possibility. The Shapley values of players 1 and 2 are identical because they hold a left glove each and are symmetric (in a sense to be defined shortly). Thus, we have $Sh_2\left(v_{\{1,2\},\{3\}}\right) = \frac{1}{6}$. Also, the Shapley value satisfies Pareto efficiency which means that the sum of the payoffs equals the worth of the grand coalition:

$$\sum_{i=1}^{3} Sh_i\left(v_{\{1,2\},\{3\}}\right) = v(\{1,2,3\}) = 1$$

Thus, we find

$$Sh\left(v_{\{1,2\},\{3\}}\right) = \left(\frac{1}{6}, \frac{1}{6}, \frac{2}{3}\right).$$

The following table reports the Shapley values for an owner of a right glove in a market with r right-glove owners and l left-glove owners:

		number l of left-glove owners				
		0	1	2	3	4
number r	1	0	0,500	**0,667**	0,750	0,800
of	2	0	0,167	0,500	0,650	0,733
right-glove	3	0	0,083	0,233	0,500	0,638
owners	4	0	0,050	0,133	0,271	0,500

This table clearly shows how the payoff increases with the number of players on the other market side. The payoff $Sh_3\left(v_{\{1,2\},\{3\}}\right) = \frac{2}{3}$ is highlighted.

5.2. The Shapley axioms. The Shapley value fulfills four axioms:

- the efficiency axiom: the worth of the grand coalition is to be distributed among all the players,
- the symmetry axiom: players in similar situations obtain the same payoff,
- the null-player axiom: a player with zero marginal contribution to every coalition, obtains zero payoff, and
- additivity axiom: if players are subject to two coalition functions, it does not matter whether we apply the Shapley value to the sum of these two coalition functions or apply the Shapley value to each coalition function separately and sum the payoffs.

A solution function σ may or may not obey the four axioms mentioned above.

DEFINITION XV.9 (EFFICIENCY AXIOM). *A solution function σ is said to obey the efficiency axiom or the Pareto axiom if*

$$\sum_{i \in N} \sigma_i(v) = v(N)$$

holds for all coalition functions $v \in \mathbb{V}$.

In the gloves game, two left-glove owners are symmetric in the sense of the following definition.

DEFINITION XV.10 (SYMMETRY). *Two players i and j are called symmetric (with respect to $v \in \mathbb{V}$) if we have*

$$v(K \cup \{i\}) = v(K \cup \{j\})$$

for every coalition K that contains neither i nor j.

EXERCISE XV.11. *Show that any two left-glove holders are symmetric in a gloves game* $v_{L,R}$.

EXERCISE XV.12. *Show* $MC_i^K = MC_j^K$ *for two symmetric players* i *and* j *fulfilling* $i \notin K$ *and* $j \notin K$.

It may seem an obvious requirement that symmetric players obtain the same payoff.

DEFINITION XV.11 (SYMMETRY AXIOM). *A solution function* σ *is said to obey the symmetry axiom if we have*

$$\sigma_i(v) = \sigma_j(v)$$

for any game $v \in \mathbb{V}$ *and any two symmetric players* i *and* j.

In any gloves game obeying $L \neq \emptyset \neq R$, every player has a non-zero marginal contribution for some coalition K.

DEFINITION XV.12 (NULL PLAYER). *A player* $i \in N$ *is called a null player (with respect to* $v \in \mathbb{V}_N$*) if*

$$v(K \cup \{i\}) = v(K)$$

holds for every coalition K.

Shouldn't a null player obtain nothing?

DEFINITION XV.13 (NULL-PLAYER AXIOM). *A solution function* σ *is said to obey the null-player axiom if we have*

$$\sigma_i(v) = 0$$

for any null player (with respect to $v \in \mathbb{V}$*)* $i \in N$.

EXERCISE XV.13. *Under which condition is a player from* L *a null player in a gloves game* $v_{L,R}$*?*

The last axiom that we consider at present is the additivity axiom. It rests on the possibility to add both payoff vectors and coalition functions (see section 3).

DEFINITION XV.14 (ADDITIVITY AXIOM). *A solution function* σ *is said to obey the additivity axiom if we have*

$$\sigma(v + w) = \sigma(v) + \sigma(w)$$

for any player set N *and any two coalition functions* $v, w \in \mathbb{V}_N$.

Do you see the difference? On the left-hand side, we add the coalition functions first and then apply the solution function. On the right-hand side we apply the solution function to the coalition functions individually and then add the payoff vectors.

Exercise XV.14. *Can you deduce* $\sigma(0) = 0$ *from the additivity axiom? Hint: use* $v = w := 0$.

Now we note a stunning result:

Theorem XV.1 (Shapley axiomatization). *The Shapley formula is the unique solution function that fulfills the symmetry axiom, the efficiency axiom, the null-player axiom, and the additivity axiom.*

The theorem means that the Shapley formula fulfills the four axioms. Consider now a solution function that fulfills the four axioms. According to the theorem, the Shapley formula is the only solution function to do so.

Differently put, the Shapley formula and the four axioms are equivalent – they specify the same payoffs. Cooperative game theorists say that the Shapley formula is "axiomatized" by the set of the four axioms.

Exercise XV.15. *Determine the Shapley value for the gloves game for* $L = \{1\}$ *and* $R = \{2, 3, 4\}$! *Hint: You do not need to write down all* 4! *rank orders. Try to find the probability that player* 1 *does not complete a pair.*

6. Simple games

6.1. Definition. We first define monotonic games and then turn to simple games.

Definition XV.15 (monotonic game). *A coalition function* $v \in \mathbb{V}_N$ *is called monotonic if* $\emptyset \subseteq S \subseteq S'$ *implies* $v(S) \leq v(S')$.

Thus, monotonicity means that the worth of a coalition cannot decrease if other players join. For example, if S' is a superset of S (or S a subset of S'), we cannot have $v(S) = 1$ and $v(S') = 0$.

Definition XV.16 (simple game). *A coalition function* $v \in \mathbb{V}_N$ *is called simple if*

- *we have* $v(K) = 0$ *or* $v(K) = 1$ *for every coalition* $K \subseteq N$,
- *the grand coalition's worth is* 1, *and*
- v *is monotonic.*

Coalitions with $v(K) = 1$ *are called winning coalitions and coalitions with* $v(K) = 0$ *are called losing coalitions. A winning coalition* K *is a minimal winning coalition if every strict subset of* K *is not a winning coalition.*

6.2. Veto players and dictators. Sometimes, certain players are of central importance:

DEFINITION XV.17 (VETO PLAYER, DICTATOR). *Let v be a simple game. A player $i \in N$ is called a veto player if*

$$v(N \backslash \{i\}) = 0$$

holds. i is called a dictator if

$$v(S) = \begin{cases} 1, & i \in S \\ 0, & \text{otherwise} \end{cases}$$

holds for all $S \subseteq N$.

Thus, without a veto player, the worth of a coalition is 0, while a dictator can produce the worth of 1 just by himself.

EXERCISE XV.16. *Can there be a coalition K such that $v(K \backslash \{i\}) = 1$ for a veto player i or a dictator i?*

EXERCISE XV.17. *Is every veto player a dictator or every dictator a veto player?*

6.3. Simple games and voting mechanisms. Simple games can be used to model voting mechanisms. As a matter of consistency, complements of winning coalitions have to be losing coalitions. Otherwise, a coalition K could vote for something and $N \backslash K$ would vote against it, both of them successfully.

DEFINITION XV.18 (CONTRADICTORY, DECIDABLE). *A simple game $v \in \mathbb{V}_N$ is called non-contradictory if $v(K) = 1$ implies $v(N \backslash K) = 0$.*

A simple game $v \in \mathbb{V}_N$ is called decidable if $v(K) = 0$ implies $v(N \backslash K) = 1$.

Thus, a contradictory voting game can lead to opposing decisions – for example, some candidate A is voted president (with the support of some coalition K) and then some other candidate B (with the support of $N \backslash K$) is also voted president. A non-decidable voting game can prevent any decision. Neither A nor B can gain enough support because coalition K blocks candidate B while $N \backslash K$ blocks candidate A.

EXERCISE XV.18. *Show that a simple game with a veto player cannot be contradictory. Also show: A simple game with two veto players cannot be decidable.*

6.4. Unanimity games. Unanimity games are famous games in cooperative game theory. We will use them to prove the Shapley theorem.

DEFINITION XV.19 (UNANIMITY GAME). *For any* $T \neq \emptyset$,

$$u_T(K) = \begin{cases} 1, & K \supseteq T \\ 0, & \text{otherwise} \end{cases}$$

defines a unanimity game.

Thus, a coalition K obtains the payoff of 1 if K contains all the players from T. The players from T are the productive or powerful members of society.

- Every player from T is a veto player and no player from $N\backslash T$ is a veto player.
- In a sense, the players from T exert common dictatorship. For example, each player $i \in T$ possesses part of a treasure map.

While $u_\emptyset$ is also explained, it is not called a unanimity game.

EXERCISE XV.19. *Find the null players in the unanimity game* u_T.

EXERCISE XV.20. *Find the core and the Shapley value for* $N = \{1, 2, 3, 4\}$ *and* $u_{\{1,2\}}$.

6.5. Apex games. The apex game has one important player $i \in N$ who is nearly a veto player and nearly a dictator.

DEFINITION XV.20 (APEX GAME). *For* $i \in N$ *with* $n \geq 2$, *the apex game* h_i *is defined by*

$$h_i(K) = \begin{cases} 1, & i \in K \text{ and } K\backslash\{i\} \neq \emptyset \\ 1, & K = N\backslash\{i\} \\ 0, & \text{otherwise} \end{cases}$$

Player i *is called the main, or apex, player of that game.*

Thus, there are two types of winning coalitions in the apex game:

- i together with at least one other player or
- all the other players taken together.

Mostly, we work with apex games for $n \geq 4$.

EXERCISE XV.21. *Consider* h_1 *for* $n = 2$ *and* $n = 3$. *What do these games look like?*

EXERCISE XV.22. *Is the apex player a veto player or a dictator?*

EXERCISE XV.23. *Show that the apex game is decidable and non-contradictory.*

Let us now find the Shapley value for the apex game. Consider all the rank orders. The apex player $i \in N$ obtains the marginal contribution 1 unless

- he is the first player in a rank order (then his marginal contribution is $v(\{i\}) - v(\emptyset) = 0 - 0 = 0$) or
- he is the last player (with marginal contribution $v(N) - v(N \backslash \{i\}) = 1 - 1 = 0$).

For the following exercise, observe that every position of the apex player in a rank order has the same probability.

EXERCISE XV.24. *Find the Shapley value for the apex game h_1!*

6.6. Weighted voting games.

6.6.1. *Definition.* Weighted voting games form an important subclass of the simple games. We specify weights for every player and a quota. If the sum of weights for a coalition is equal to or above the quota, that coalition is a winning one.

DEFINITION XV.21 (WEIGHTED VOTING GAME). *A voting game v is specified by a quota q and voting weights g_i, $i \in N$, and defined by*

$$v(K) = \begin{cases} 1, & \sum_{i \in K} g_i \geq q \\ 0, & \sum_{i \in K} g_i < q \end{cases}$$

The voting game is denoted by $[q; g_1, ..., g_n]$.

For example,

$$\left[\frac{1}{2}; \frac{1}{n}, ..., \frac{1}{n}\right]$$

is the majority rule, according to which fifty percent of the votes are necessary for a winning coalition. Do you see that $n = 4$ means that this majority rule is contradictory?

The apex game h_1 for n players can be considered a weighted voting game given by

$$\left[n - 1; n - \frac{3}{2}, 1, ..., 1\right].$$

EXERCISE XV.25. *Consider the unanimity game u_T given by $t < n$ and $T = \{1, ..., t\}$. Can you express it as a weighted voting game?*

6.6.2. *UN Security Council.* Let us consider the United Nations' Security Council. It has 5 permanent members and 10 non-permanent ones. The permanent members are China, France, Russian Federation, the United Kingdom, and the United States. The non-permanent members vary. On substantive matters, a motion requires the backing of all five permanent members and the additional support of at least four non-permanent members. Obviously, the UN Security Council has a lot of power and so its voting mechanism deserves analysis. The above rule for "substantive matters" can be translated into the weighted voting game

$$[39; 7, 7, 7, 7, 7, 1, 1, 1, 1, 1, 1, 1, 1, 1, 1]$$

where the weights 7 accrue to the five permanent members and the weights 1 to the non-permanent members.

EXERCISE XV.26. *Using the above voting game, show that every permanent member is a veto player. Show also that the five permanent members need the additional support of four non-permanent ones.*

EXERCISE XV.27. *Is the Security Council's voting rule non-contradictory and decidable?*

It is not easy to calculate the Shapley value for the Security Council. After all, we have

$$15! = 1.307.674.368.000$$

rank orders for the 15 players. Anyway, the Shapley values are

$$0,19627 \text{ for each permanent member}$$
$$0,00186 \text{ for each non-permanent member.}$$

7. Five non-simple games

7.1. Buying a car. Following Morris (1994, S. 162), we consider three agents involved in a car deal. Andreas (A) has a used car he wants to sell, Frank (F) and Tobias (T) are potential buyers with willingness to pay of 700 and 500, respectively. This leads to the coalition function v given by

$$\begin{aligned} v(A) &= v(F) = v(T) = 0, \\ v(A, F) &= 700, \\ v(A, T) &= 500, \\ v(F, T) &= 0 \text{ and} \\ v(A, F, T) &= 700. \end{aligned}$$

One-man coalitions have the worth zero. For Andreas, the car is useless (he believes in cycling rather than driving). Frank and Tobias cannot obtain the car unless Andreas cooperates. In case of a deal, the worth is equal to the (maximal) willingness to pay.

We use the core to find predictions for the car price. The core is the set of those payoff vectors (x_A, x_F, x_T) that fulfill

$$x_A + x_F + x_T = 700$$

and

$$\begin{aligned} x_A &\geq 0, x_F \geq 0, x_T \geq 0, \\ x_A + x_F &\geq 700, \\ x_A + x_T &\geq 500 \text{ and} \\ x_F + x_T &\geq 0. \end{aligned}$$

Tobias obtains

$$\begin{aligned} x_T &= 700 - (x_A + x_F) \text{ (efficiency)} \\ &\leq 700 - 700 \text{ (by } x_A + x_F \geq 700) \\ &= 0 \end{aligned}$$

and hence zero, $x_T = 0$. By $x_A + x_T \geq 500$, the seller Andreas can obtain at least 500.

Summarizing (and checking all the conditions above), we see that the core is the set of vectors (x_A, x_F, x_T) obeying

$$\begin{aligned} 500 &\leq x_A \leq 700, \\ x_F &= 700 - x_A \text{ and} \\ x_T &= 0. \end{aligned}$$

Therefore, the car sells for a price between 500 and 700.

7.2. The Maschler game. Aumann & Myerson (1988) present the Maschler game which is the three-player game given by

$$v(K) = \begin{cases} 0, & |K| = 1 \\ 60, & |K| = 2 \\ 72, & |K| = 3 \end{cases}$$

Obviously, the three players are symmetric. It is easy to see that all players of symmetric games are symmetric.

Definition XV.22 (symmetric game). *A coalition function v is called symmetric if there is a function $f : N \to \mathbb{R}$ such that*

$$v(K) = f(|K|), \ K \subseteq N.$$

Exercise XV.28. *Find the Shapley value for the Maschler game!*

According to the Shapley value, the players 1 and 2 obtain less than their common worth. Therefore, they can block the payoff vector suggested by the Shapley value. Indeed, for any efficient payoff vector, we can find a two-player coalition that can be made better off. Differently put: the core is empty.

This can be seen easily. We are looking for vectors (x_1, x_2, x_3) that fulfill both

$$x_1 + x_2 + x_3 = 72$$

and

$$\begin{aligned} x_1 &\geq 0, x_2 \geq 0, x_3 \geq 0, \\ x_1 + x_2 &\geq 60, \\ x_1 + x_3 &\geq 60 \text{ and} \\ x_2 + x_3 &\geq 60. \end{aligned}$$

Summing the last three inequalities yields

$$2x_1 + 2x_2 + 2x_3 \geq 3 \cdot 60 = 180$$

and hence a contradiction to efficiency.

7.3. The gloves game, once again. Above, we have calculated the core for the gloves game with $L = \{1, 2\}$ and $R = \{3\}$. The core clearly shows the bargaining power of the right-glove owner. We will now consider the core for a case where the scarcity of right gloves is minimal:

$$\begin{aligned} L &= \{1, 2, ..., 100\} \\ R &= \{101, ..., 199\}. \end{aligned}$$

If a payoff vector

$$(x_1, ..., x_{100}, x_{101}, ..., x_{199})$$

is to belong to the core, we have

$$\sum_{i=1}^{199} x_i = 99$$

by the efficiency axiom. We now pick any left-glove holder $j \in \{1, 2, ..., 100\}$. We find

$$v(L \backslash \{j\} \cup R) = 99$$

and hence

$$\begin{aligned} x_j &= 99 - \sum_{\substack{i=1,\\ i\neq j}}^{199} x_i \text{ (efficiency)} \\ &\leq 99 - 99 \text{ (blockade by coalition } L\backslash\{j\}\cup R) \\ &= 0. \end{aligned}$$

Therefore, we have $x_j = 0$ for every $j \in L$.

Every right-glove owner can claim at least 1 because he can point to coalitions where he is joined by at least one left-glove owner. Therefore, every right-glove owner obtains the payoff 1 and every left-glove owner the payoff zero. Inspite of the fact that the scarcity is minimal, the right-glove owners get everything.

If two left-glove owners burned their glove, the other left-glove owners would get a payoff increase from 0 to 1. (Why?)

EXERCISE XV.29. *Consider a generalized gloves game where*

- *player* 1 *has one left glove,*
- *player* 2 *has two left gloves and*
- *players* 3 *and* 4 *have one right glove each.*

Calculate the core. How does the core change if player 2 *burns one of his two gloves?*

The burn-a-glove strategy may make sense if payoffs depend on the scarcity in an extreme fashion as they do for the core.

7.4. The chess game. Chess players enjoy playing. Thus, an even number of players is best.

DEFINITION XV.23 (CHESS GAME). *The chess game v is defined by*

$$v(K) = \begin{cases} \frac{|K|}{2}, & |K| \text{ is even} \\ \frac{|K|-1}{2}, & |K| \text{ is odd} \end{cases}$$

Find the core!

I copied this game from lectures notes by Chris Wallace (Trinity College, Oxford) who calls this game a treasure hunt.

7.5. Cost-division games. We model cost-division games (for doctors sharing a secretarial office or faculties sharing computing facilities) by way of cost functions and cost-savings functions.

DEFINITION XV.24 (COST-DIVISION GAME). *For a player set N, let $c : 2^N \to \mathbb{R}_+$ be a coalition function that is called a cost function. On the basis of c, the cost-savings game is defined by $v : 2^N \to \mathbb{R}$ and*

$$v(K) = \sum_{i \in K} c(\{i\}) - c(K), K \subseteq N.$$

The idea behind this definition is that cost savings can be realized if players pool their resources so that $\sum_{i \in K} c(\{i\})$ is greater than $c(K)$ and $v(K)$ is positive.

Following Young (1994, pp. 1195), we consider a specific example. Two towns A and B plan a water-distribution system. Town A could build such a system for itself at a cost of 11 million Euro and town B would need 7 million Euro for a system tailor-made to its needs. The cost for a common water-distribution system is 15 million Euro. The cost function is given by

$$\begin{aligned} c(\{A\}) &= 11, c(\{B\}) = 7 \text{ and} \\ c(\{A, B\}) &= 15. \end{aligned}$$

The associated cost-savings game is $v : 2^{\{A,B\}} \to \mathbb{R}$ defined by

$$\begin{aligned} v(\{A\}) &= 0, c(\{B\}) = 0 \text{ and} \\ v(\{A, B\}) &= 7 + 11 - 15 = 3. \end{aligned}$$

v's core is obviously given by

$$\left\{(x_A, x_B) \in \mathbb{R}_+^2 : x_1 + x_2 = 3\right\}.$$

The cost savings of $3 = 11 + 7 - 15$ can be allotted to the towns such that no town is worse off compared to going it alone. Turning to cost division, the set of "undominated" cost allocations is

$$\left\{(c_A, c_B) \in \mathbb{R}^2 : c_A + c_B = 15, c_A \leq 11, c_B \leq 7\right\}.$$

8. Topics and literature

The main topics in this chapter are

- coalition
- coalition function
- gloves game
- weighted voting games
- UN Security Council
- unanimity game
- apex game
- simple game
- dictator
- veto player
- Maschler game
- cost-division game
- core
- efficiency
- feasibility
- marginal contribution
- axioms
- symmetry
- null player
- Shapley value

We recommend the textbook by Wiese (2005).

9. Solutions

Exercise XV.1

The values are

$$\begin{aligned} v_{L,R}(\{1\}) &= \min(1,0) = 0, \\ v_{L,R}(\emptyset) &= \min(0,0) = 0, \\ v_{L,R}(N) &= \min(2,3) = 2, \text{ and} \\ v_{L,R}(\{2,3,4\}) &= \min(1,2) = 1. \end{aligned}$$

Exercise XV.2

The first two propositions are nonsensical, the last two propositions are correct.

Exercise XV.3

We obtain the sum of vectors

$$\begin{pmatrix} 1 \\ 3 \\ 6 \end{pmatrix} + \begin{pmatrix} 2 \\ 5 \\ 1 \end{pmatrix} = \begin{pmatrix} 1+2 \\ 3+5 \\ 6+1 \end{pmatrix} = \begin{pmatrix} 3 \\ 8 \\ 7 \end{pmatrix}$$

Exercise XV.4

Feasibility follows from

$$\begin{aligned} \sum_{i=1}^{n} y_i &= \sum_{i=1}^{n} x_i + \sum_{i=1}^{n} \frac{1}{n} \left(v_{L,R}(N) - \sum_{j=1}^{n} x_j \right) \\ &= \sum_{i=1}^{n} x_i + \frac{1}{n} \left(\sum_{i=1}^{n} v_{L,R}(N) - \sum_{i=1}^{n} \sum_{j=1}^{n} x_j \right) \\ &= \sum_{i=1}^{n} x_i + \frac{1}{n} \left(n v_{L,R}(N) - n \sum_{j=1}^{n} x_j \right) \\ &= v_{L,R}(N). \end{aligned}$$

Exercise XV.5

The set of Pareto-efficient payoff vectors (x_1, x_2) is described by $x_1 + x_2 = 1$. In particular, we may well have $x_1 < 0$.

Exercise XV.6

The core obeys the conditions

$$\begin{aligned} x_1 + x_2 + x_3 &= v_{L,R}(N) = 1, \\ x_i &\geq 0, i = 1, 2, 3, \\ x_1 + x_2 &\geq 0, \\ x_1 + x_3 &\geq 1 \text{ and} \\ x_2 + x_3 &\geq 1. \end{aligned}$$

Substituting $x_1 + x_3 \geq 1$ into the efficiency condition yields

$$x_2 = 1 - (x_1 + x_3) \leq 1 - 1 = 0.$$

Hence (because of $x_2 \geq 0$), we have $x_2 = 0$. For reasons of symmetry, we also have $x_1 = 0$. Applying efficiency once again, we obtain $x_3 = 1 - (x_1 + x_2) = 1$. Thus, the only candidate for the core is $x = (0, 0, 1)$. Indeed, this payoff vector fulfills all the conditions noted above. Therefore,

$$(0, 0, 1)$$

is the only element in the core.

Exercise XV.7

The marginal contributions are

$$\begin{aligned}
MC_1^{\{1,3,4\}}\left(v_{\{1,2,3\},\{4,5\}}\right) &= v\left(\{1,3,4\}\cup\{1\}\right)-v\left(\{1,3,4\}\setminus\{1\}\right) \\
&= v\left(\{1,3,4\}\right)-v\left(\{3,4\}\right) \\
&= 1-1=0, \\
MC_1^{\{3,4\}}\left(v_{\{1,2,3\},\{4,5\}}\right) &= v\left(\{3,4\}\cup\{1\}\right)-v\left(\{3,4\}\setminus\{1\}\right) \\
&= v\left(\{1,3,4\}\right)-v\left(\{3,4\}\right) \\
&= 1-1=0, \\
MC_4^{\{1,3,4\}}\left(v_{\{1,2,3\},\{4,5\}}\right) &= v\left(\{1,3,4\}\cup\{4\}\right)-v\left(\{1,3,4\}\setminus\{4\}\right) \\
&= v\left(\{1,3,4\}\right)-v\left(\{1,3\}\right) \\
&= 1-0=1, \\
MC_4^{\{1,3\}}\left(v_{\{1,2,3\},\{4,5\}}\right) &= v\left(\{1,3\}\cup\{4\}\right)-v\left(\{1,3\}\setminus\{4\}\right) \\
&= v\left(\{1,3,4\}\right)-v\left(\{1,3\}\right) \\
&= 1-0=1.
\end{aligned}$$

Exercise XV.8

The marginal contributions are the same: $v\left(\{1,2,3\}\right)-v\left(\{1,3\}\right)$.

Exercise XV.9

We find $5! = 1\cdot 2\cdot 3\cdot 4\cdot 5 = 120$ rank orders for 5 players and $6! = 5!\cdot 6 = 120\cdot 6 = 720$ rank orders for 6 players.

Exercise XV.10

We find the marginal contributions

$$\begin{aligned}
v\left(\{1\}\right)-v\left(\emptyset\right) &= 0-0=0,\text{ rank order } (1,2,3) \\
v\left(\{1\}\right)-v\left(\emptyset\right) &= 0-0=0,\text{ rank order } (1,3,2) \\
v\left(\{1,2\}\right)-v\left(\{2\}\right) &= 0-0=0,\text{ rank order } (2,1,3) \\
v\left(\{1,2,3\}\right)-v\left(\{2,3\}\right) &= 1-1=0,\text{ rank order } (2,3,1) \\
v\left(\{1,3\}\right)-v\left(\{3\}\right) &= 1-0=1,\text{ rank order } (3,1,2) \\
v\left(\{1,2,3\}\right)-v\left(\{2,3\}\right) &= 1-1=0,\text{ rank order } (3,2,1).
\end{aligned}$$

Exercise XV.11

Let i and j be players from L and let K be a coalition that contains neither i nor j. Then $K\cup\{i\}$ contains the same number of left and the same number of right gloves as $K \cup \{j\}$. Therefore,

$$\begin{aligned} v_{L,R}(K \cup \{i\}) &= \min\left(\left|(K \cup \{i\}) \cap L\right|, \left|(K \cup \{i\}) \cap R\right|\right) \\ &= \min\left(\left|(K \cup \{j\}) \cap L\right|, \left|(K \cup \{j\}) \cap R\right|\right) \\ &= v_{L,R}(K \cup \{j\}). \end{aligned}$$

Exercise XV.12

The equality follows from

$$\begin{aligned} MC_i^K &= v(K \cup \{i\}) - v(K \backslash \{i\}) \\ &= v(K \cup \{i\}) - v(K) \\ &= v(K \cup \{j\}) - v(K) \\ &= v(K \cup \{j\}) - v(K \backslash \{j\}) \\ &= MC_j^K. \end{aligned}$$

Exercise XV.13

A player i from L is a null player iff $R = \emptyset$ holds. $R = \emptyset$ implies

$$\begin{aligned} v_{L,\emptyset}(K) &= \min\left(\left|K \cap L\right|, \left|K \cap \emptyset\right|\right) \\ &= \min\left(\left|K \cap L\right|, 0\right) \\ &= 0 \end{aligned}$$

for every coalition K. $R \neq \emptyset$ means that i has a marginal contribution of 1 when he comes second after a right-glove holder.

Exercise XV.14

We first obtain

$$\begin{aligned} \sigma(0) &= \sigma(0+0) \\ &= \sigma(0) + \sigma(0) \text{ (additivity)} \end{aligned}$$

and then the desired result by subtracting $\sigma(0)$ on both sides.

Exercise XV.15

The left-glove holder 1 completes a pair (the only one) whenever he does not come first. The probability for coming first is $\frac{1}{4}$ for player 1 (and any other player). Thus, player 1 obtains $\left(1 - \frac{1}{4}\right) \cdot 1$. The other players share the rest. Therefore, symmetry and efficiency lead to

$$\begin{aligned} Sh_1\left(v_{\{1\},\{2,3,4\}}\right) &= \frac{3}{4}, \\ Sh_2\left(v_{\{1\},\{2,3,4\}}\right) &= Sh_3\left(v_{\{1\},\{2,3,4\}}\right) = Sh_4\left(v_{\{1\},\{2,3,4\}}\right) = \frac{1}{12}. \end{aligned}$$

Exercise XV.16

If i is a veto player, we have $v(K\backslash\{i\}) \leq v(N\backslash\{i\}) = 0$ for every coalition $K \subseteq N$ and hence $v(K\backslash\{i\}) = 0$. Thus, a veto player $i \in N$ cannot fulfill $v(K\backslash\{i\}) = 1$. A dictator i cannot fulfill $v(K\backslash\{i\}) = 1$ because the worth of a coalition is 1 if and only if the dictator belongs to the coalition.

Exercise XV.17

A dictator is always a veto player – without him the coalition cannot win. However, a veto player need not be a dictator. Just consider the simple game v on the player set $N = \{1,2\}$ defined by $v(\{1\}) = v(\{2\}) = 0$, $v(\{1,2\}) = 1$. Players 1 and 2 are veto players but not dictators.

Exercise XV.18

Let v be a simple game with a veto player $i \in N$. Then $v(K) = 1$ implies $i \in K$. By $i \notin N\backslash K$, we obtain $v(N\backslash K) = 0$ – the desired result.

Let v be a simple game with two veto players i and j, $i \neq j$. Then $v(\{i\}) = 0$ (by $j \notin \{i\}$) and $v(K\backslash\{i\}) = 0$ (by $i \notin K\backslash\{i\}$) hold.

Exercise XV.19

For the unanimity game u_T, the null players are the players from $N\backslash T$.

Exercise XV.20

The core is

$$\left\{(x_1, x_2, x_3, x_4) \in \mathbb{R}^4_+ : x_1 + x_2 = 1, x_3 = x_4 = 0\right\}$$

and the Shapley value is given by

$$Sh\left(u_{\{1,2\}}\right) = \left(\frac{1}{2}, \frac{1}{2}, 0, 0\right).$$

Exercise XV.21

For $n = 2$, we have

$$\begin{aligned} h_1(K) &= \begin{cases} 0, & K = \{1\} \text{ or } K = \emptyset \\ 1, & \text{otherwise} \end{cases} \\ &= u_{\{2\}}. \end{aligned}$$

$n = 3$ yields the symmetric game

$$h_1(K) = \begin{cases} 1, & |K| \geq 2 \\ 0, & \text{otherwise} \end{cases}$$

(Symmetry means that the worths depend on the number of the players, only.)

Exercise XV.22

No, the apex player is neither a veto player nor a dictator. If all the other player unite against the apex player, they win:

$$h_i(N\backslash\{i\}) = 1.$$

Exercise XV.23

We first show that h_i is non-contradictory. Assume $h_i(K) = 1$ for any coalition $K \subseteq N$. Then, one of two cases holds. Either we have $K = N\backslash\{i\}$. This implies $h_i(N\backslash K) = h_i(\{i\}) = 0$. Or we have $i \in K$ and $|K| \geq 2$. Then, $h_i(N\backslash K) = 0$. Thus, h_i is non-contradictory.

We now show that h_i is decidable. Take any $K \subseteq N$ with $h_i(K) = 0$. This implies $K = \{i\}$ or $K \subsetneqq N\backslash\{i\}$. The respective complements $N\backslash K = N\backslash\{i\}$ or $N\backslash K \supsetneqq \{i\}$ are winning coalitions.

Exercise XV.24

Since the apex player obtains the marginal contributions for positions 2 through $n-1$, his Shapley payoff is

$$\frac{n-2}{n} \cdot 1.$$

Due to efficiency, the other (symmetric!) players share the rest so that each of them obtains

$$\frac{1}{n-1}\left(1 - \frac{n-2}{n}\right) = \frac{2}{n(n-1)}.$$

Thus, we have

$$Sh(h_1) = \left(\frac{n-2}{n}, \frac{2}{n(n-1)}, \ldots, \frac{2}{n(n-1)}\right).$$

Exercise XV.25

One possible solution is

$$\left[1; \frac{1}{t}, \ldots, \frac{1}{t}, 0, \ldots, 0\right]$$

where $\frac{1}{t}$ is the weight for the powerful T-players, while 0 is the weight for the remaining ones.

Exercise XV.26

Every permanent member is a veto player by $4 \cdot 7 + 10 \cdot 1 = 38 < 39$. Because of $5 \cdot 7 + 4 \cdot 1 = 39$, four non-permanent members are necessary for passing a resolution.

Exercise XV.27

The voting rule is non-contradictory and not decidable. This is just a corollary of exercise XV.18 (p. 417).

Exercise XV.28

By efficiency and symmetry, we have

$$Sh(v) = (24, 24, 24).$$

Exercise XV.29

The core has to fulfill

$$x_1 + x_2 + x_3 + x_4 = 2$$

and also the inequalities

$$\begin{aligned} x_i &\geq 0, i = 1, ..., 4, \\ x_1 + x_3 &\geq 1, \\ x_1 + x_4 &\geq 1, \\ x_2 + x_4 &\geq 1 \text{ and} \\ x_2 + x_3 + x_4 &\geq 2. \end{aligned}$$

We then find

$$x_1 = 2 - (x_2 + x_3 + x_4) \leq 0$$

and hence

$$\begin{aligned} x_1 &= 0 \text{ (because of } x_1 \geq 0), \\ x_3 &\geq 1 \text{ and } x_4 \geq 1. \end{aligned}$$

Using efficiency once more supplies $x_2 = 0$ and

$$(0, 0, 1, 1)$$

is the only candidate for a core. Indeed, this is the core. Just check all the inequalities above and also those omitted. Player 2's payoff is 0 in this situation. If he burns his second glove, we obtain the (non-generalized) gloves game $v_{\{1,2\},\{3,4\}}$ where player 2 may achieve any core payoff between 0 and 1.

CHAPTER XVI

The Rubinstein bargaining model

The aim of this very short chapter is to introduce the Rubinstein (1982) bargaining model. It squarely belongs to non-cooperative game theory and can be seen as the obvious extension of the "take it or leave it" bargaining game in chapter X (pp. 271) and chapter XII (pp. 342) to a multi-period setting. The idea is that players can counter offers by counteroffers and counter-counteroffers.

1. Introduction

The extension of "take it or leave it" bargaining due to Rubinstein allows the players to make offers alternately. The Rubinstein game is infinite. However, the bargainers have an incentive to close the deal fast because the "cake" shrinks. If the agents manage to agree immediately, the cake has size 1, after the first round it shrinks to size $\delta, 0 < \delta < 1$, then to size $\delta^2 < \delta$.

An alternative interpretation points to the utility reduction stemming from consuming the cake later. Then, δ is not a shrinking factor but a discount factor and equal to $\frac{1}{1+r}$ for interest rate r. In any case, δ is a measure for the agents' impatience.

Figure XVI.1 sketches the game. Player 1 makes the first move, an offer of a_1. This means that he wants a_1 for himself and suggests to leave $1 - a_1$ to player 2. If player 2 accepts, the payoffs are a_1 and $1 - a_1$, respectively. If player 2 rejects the offer, he makes a counter offer a_2 at stage 2. This means that player 2 demands $1 - a_2$ for himself and offers a_2 to player 1. However, by the impatience of the players, the payoffs are

$$\delta a_2 \text{ for player 1 and } \delta (1 - a_2) \text{ for player 2}$$

if player 1 accepts the counteroffer.

If player 1 declines the counteroffer, it is up to him to make the counter-counteroffer at stage 3. If bargaining stops at that point, payoffs are

$$\delta^2 a_1 \text{ for player 1 and } \delta^2 (1 - a_1) \text{ for player 2.}$$

Et cetera, et cetera.

H. Wiese, *Advanced Microeconomics*,
https://doi.org/10.1007/978-3-658-34959-2_16

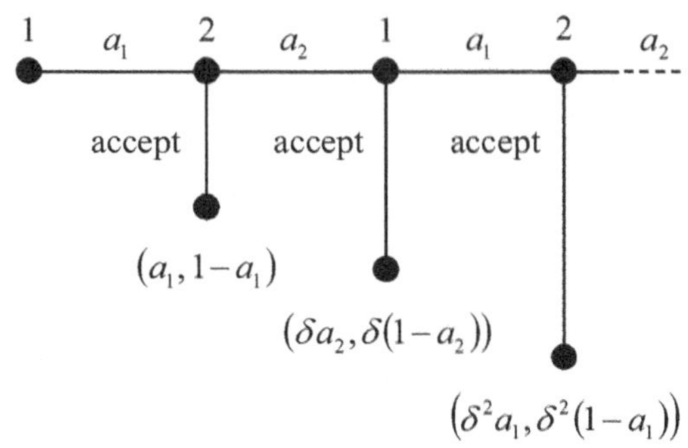

FIGURE XVI.1. The infinite Rubinstein game

It should be clear that the offers a_1 and a_2 may differ from stage to stage.

2. Many equilibria

The Rubinstein game admits many equilibria which is typical for bargaining games. The next exercise helps you to understand the logic behind this multiplicity. Strategies are very complicated objects. They need to specify for every player which offer to decline or accept and what counteroffer to make. These decisions can be a complicated function of the game's history up to the current stage.

EXERCISE XVI.1. *Assume $0 \leq \alpha \leq 1$ and consider the following strategies of two players 1 and 2. Player 1 demands $a_1 = \alpha$ whenever it is his turn to make an offer. He declines any offer obeying $a_2 < \alpha$ and accepts those with $a_2 \geq \alpha$. Player 2 accepts every offer with $a_1 \leq \alpha$ and declines those with $a_1 > \alpha$. Player 2's offer is always $a_2 = \alpha$. Do these strategies form a Nash equilibrium?*

The α-equilibria described in the above exercise suffer from a major drawback. They are not subgame perfect. Note that figure XVI.1 is a sketch of the game tree only. Figure XVI.2 is somewhat more specific, and we will use this figure to discuss the problem of subgame perfection.

Consider player 1 whose offer is not α (for example, 0.6 in figure XVI.2), but slightly higher, at a_1 (for example, 0.7) which obeys $\alpha < a_1 < 1 - \delta(1-\alpha)$ (for $\delta < 0.75$). Acccording to the α-strategies, player 2 should decline the offer although he obtains

$$1 - a_1 \text{ (0.3, to pursue the example)}$$

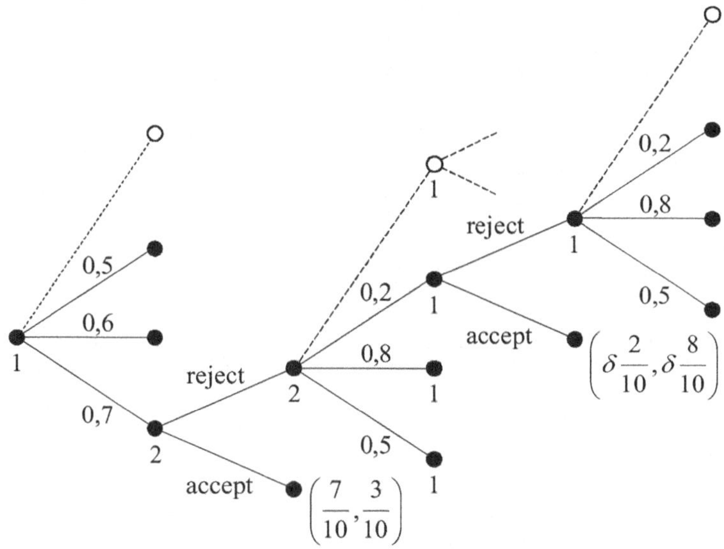

FIGURE XVI.2. Rubinstein's game tree

by accepting while rejection and counteroffer α yield

$$\delta(1-\alpha) < 1 - a_1.$$

Thus, the strategies in the above exercise suggest a behavior that is not optimal, should a deviation from the equilibrium path occur for some reason. Differently put, the α-equilibrium is sustained by empty threats. Player 1 sticks to offer α because he forsees rejection at $a_1 > \alpha$ although the rejection is not in player 2's interest once the worse offer (from player 2's point of view) is on the table. Therefore, the α- equilibria are not subgame perfect.

In order to identify subgame perfect equilibria, backward induction is used. However, the problem with the Rubinstein game is the infinity of the game tree. Despite this handicap, we successfully apply backward induction.

3. Backward induction for a three-stage Rubinstein game

In order to solve the subgame perfection problem, it may be a good idea to describe a player's strategy by two numbers, the first indicating the offer the player makes (a_1 or a_2, respectively), the second reflecting his agreement threshold. We now consider and solve a simplified Rubinstein game, with three stages only. In the next section, we see that the original Rubinstein game is solvable by a clever trick.

The three-stage game is depicted in figure XVI.3. Player 1 makes an offer a_1 that player 2 can counter with a_2. If player 1 rejects the counteroffer,

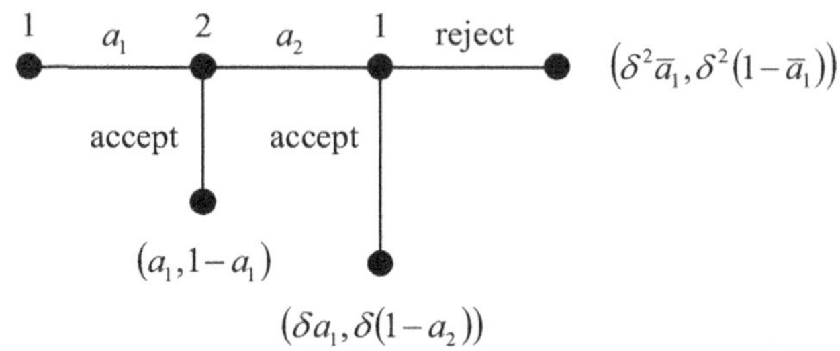

FIGURE XVI.3. The finite Rubinstein game

the game is over and an externally fixed division of the cake is implemented. For example, a mediator gives share $\overline{a}_1$ to player 1 and share $1-\overline{a}_1$ to player 2.

This game is backward solvable. Let us agree on the following principle:

> If a player is indifferent
> between accepting or rejecting an offer,
> he accepts.

This principle is just a technical assumption. The following exercise helps to see why it is needed.

EXERCISE XVI.2. *Consider a "take it or leave it"-game where player* 1 *offers* a_1, $0 \leq a_1 \leq 1$, *and the players get* $(a_1, 1-a_1)$ *if player* 2 *accepts but* $(0,0)$ *if player* 2 *rejects. Assume that player* 2 *rejects when indifferent. What is player* 1*'s optimal offer?*

We now solve the game by backward induction:

- At stage 3, if player 1 rejects player 2's counteroffer, he obtains $\delta^2\overline{a}_1$, otherwise δa_2. Accepting is the best option for player 1 in case of $\delta a_2 \geq \delta^2\overline{a}_1$, i.e., if
$$a_2 \geq \delta\overline{a}_1$$
holds.
- At stage 2, player 2 has to decide whether to accept player 1's offer (made at stage 1). If player 2 rejects player 1's offer, two cases need to be distinguished. Either player 2 offers $a_2 \geq \delta\overline{a}_1$ and player 1 accepts. Or player 2 offers less than $\delta\overline{a}_1$ and player 1 rejects. In the first case, player 2 has no reason to be generous and chooses
$$a_2 = \delta\overline{a}_1$$

which leads to payoff $\delta(1-\delta\overline{a}_1)$ for himself. By

$$\delta(1-\delta\overline{a}_1) > \delta^2(1-\overline{a}_1)$$

he prefers making an acceptable offer to making a non-acceptable one.

We still need to find out, under what circumstances player 2 accepts player 1's offer. Accepting the offer should not make him worse off than the counteroffer, i.e., the condition for accepting is

$$1-a_1 \geq \delta(1-\delta\overline{a}_1) \quad \Leftrightarrow \quad a_1 \leq 1-\delta(1-\delta\overline{a}_1).$$

- At stage 1, player 1 has the option to make the offer

$$a_1 = 1-\delta(1-\delta\overline{a}_1)$$

which is accepted by player 2 or ask for more in which case player 1 accepts player 2's counteroffer $a_2 = \delta\overline{a}_1$ (made at stage 2) at stage 3. Because of

$$\delta^2\overline{a}_1 < 1-\delta(1-\delta\overline{a}_1) \Leftrightarrow 0 < 1-\delta$$

player 1 prefers to make an acceptable offer to player 2.

Payoffs are $1-\delta(1-\delta\overline{a}_1)$ for player 1 and $\delta(1-\delta\overline{a}_1)$ for player 2.

4. Backward induction for the Rubinstein game

Backward induction for the three-stage game leads to the offer made by player 1 at stage 1

$$a_1 = 1-\delta(1-\delta\overline{a}_1) = 1-\delta+\delta^2\overline{a}_1$$

that is accepted by player 2. We now employ a trick to get at the backward-induction solution of the infinite game. At stage 3 of the infinite game, player 1 is basically in the same position as at stage 1. He sees before him an infinite sequence of offers and counteroffers. The only difference is that all payoffs are to be multiplied by δ^2, as a consequence of discounting or shrinking. The shares the players can expect are not affected. Therefore, at stage 3 he can expect the share he offers at stage 1. This more or less convincing reasoning leads to $\overline{a}_1 := a_1$ and therefore $a_1 = 1-\delta(1-\delta a_1)$ which can be written as

$$a_1 = \frac{1-\delta}{1-\delta^2} = \frac{1-\delta}{(1-\delta)(1+\delta)} = \frac{1}{1+\delta}.$$

Maybe you do not find our trick convincing. We therefore offer an alternative argument. The idea is to defer the mediator's decision from the third to the fifth stage. And then to the seventh etc. If the mediator fixes $\overline{a}_1$ for the fifth stage, player 1 can demand

$$[1-\delta(1-\delta\overline{a}_1)]$$

at the third stage and player 2 accepts. Therefore, at the first stage, player 1 makes the acceptable offer

$$1-\delta(1-\delta[1-\delta(1-\delta\overline{a}_1)]) = 1-\delta+\delta^2-\delta^3+\delta^4\overline{a}_1.$$

The general pattern should be clear by now. If we defer the monitor's decision to the seventh stage, player 1 obtains

$$\begin{aligned} & 1-\delta(1-\delta[1-\delta(1-\delta[1-\delta(1-\delta\overline{a}_1)])]) \\ = \; & 1-\delta+\delta^2-\delta^3+\delta^4-\delta^5+\delta^6\overline{a}_1. \end{aligned}$$

The monitor's decision becomes less and less important because of $\delta < 1$. Letting the monitor's stage go to infinity, we obtain player 1's payoff as the limit of the infinite geometric series

$$1-\delta+\delta^2-\delta^3+\delta^4-\delta^5+\delta^6-\delta^7+\delta^8-\ldots$$

Note that one term is the product of the previous term and $(-\delta)$, i.e., we have an infinite geometric series (see p. 97). We find, once again,

$$\text{infinite geometric series} = \frac{\text{first term}}{1-\text{factor}} = \frac{1}{1-(-\delta)} = \frac{1}{1+\delta}.$$

5. Subgame perfect strategies for the Rubinstein game

Note that $a_1 = \frac{1}{1+\delta}$ does not, by itself, tell player 1's strategy. We also need to know when player 1 accepts an offer. Since an offer by player 2 of δa_1 „today“ has the same worth as a_1 „tomorrow“, player 1's strategy is:

Make an offer of $\frac{1}{1+\delta}$ at stages 1, 3, ...,   accept any offer of or above $\delta\frac{1}{1+\delta}$ at stages 2, 4, ..., and   reject any offer below $\delta\frac{1}{1+\delta}$ at stages 2, 4, ...!

Correspondingly, player 2's strategy is

Make an offer of $\delta\frac{1}{1+\delta}$ at stages 2, 4, ...,   accept any offer of or below $\frac{1}{1+\delta}$ at stages 1, 3, ..., and   reject any offer above $\frac{1}{1+\delta}$ at stages 1, 3, ...!

We still need to check these strategies for Nash equilibrium. Given the strategy operated by player 1, player 2 can obtain

$$1-\frac{1}{1+\delta}$$

immediately (by accepting), or making the counteroffer $\delta\frac{1}{1+\delta}$ which (given player 1's strategy) leads to payoff

$$\delta\left(1-\delta\frac{1}{1+\delta}\right).$$

Since these two payoffs are equal, player 2 accepts as his strategy suggests. Therefore, player 2's strategy is a best response to player 1's strategy.

EXERCISE XVI.3. *Confirm that player 1's strategy is a best response to player 2's strategy.*

Can we be sure that the above equilibrium is subgame perfect? Perhaps, this equilibrium rests on empty threats in a similar manner as the α-equilibria? For example, would it be in player 1's interest to accept offers slightly below $\delta\frac{1}{1+\delta}$? No. If, in contrast to the players' strategies (off the equilibrium path), player 2 makes an offer

$$a_2 < \delta\frac{1}{1+\delta},$$

player 1 should reject in accordance with his strategy. Due to shrinkage, his counteroffer yields

$$\delta\frac{1}{1+\delta} > a_2.$$

In a similar fashion, we can confirm that the players' strategies have the mutual best-response property even after we find ourselves (for example by mistake) off the equilibrium path.

6. Patience in bargaining

The time is ripe to reap the benefits of the Rubinstein model. It says that the portion appropriable by player 1 is the higher, the more impatient the players are. If player 2 is very impatient, he is forced to accept a rather bad (in player 2's eyes) offer by player 1. If δ is close to 1 (no impatience), both players obtain half the cake.

The model would be nicer if it allowed for different discount factors, δ_1 for player 1 and δ_2 for player 2. Indeed, this extension is not difficult. You can do it yourself.

EXERCISE XVI.4. *Replace δ by δ_1 and δ_2, respectively, in figure XVI.3 and find player 1's offer for this modified game! Use the same trick as above ($\overline{a}_1 = a_1$) to obtain player 1's offer player 2 cannot reject.*

EXERCISE XVI.5. *Interpret the results from the preceding exercise for*

- $\delta_1 > 0$, $\delta_2 = 0$ *and*
- $\delta_1 = 0$, $\delta_2 > 0$!

7. Topics

The main topics in this chapter are

- alternating offers
- subgame perfection
- bargaining theory

8. Solutions

Exercise XVI.1

These two strategies form an equilibrium. Player 1's offer of $a_1 = \alpha$ is immediately accepted by player 2. If player 1 asks for less than α, player 2 also accepts, but player 1 obtains less than at α. If player 1 asks for more than α, player 2 declines and makes a counteroffer which is worse by the shrinking factor δ. In later periods, player 1 cannot hope for more than a share of α, which amounts to less because the cake has shrunk. In a similar fashion, one ascertains that player 2 cannot find a better strategy.

Exercise XVI.2

Player 1 tries to maximize a_1 under the condition that player 2 accepts. Therefore, player 1 has to make sure that $1 - a_1 > 0$ holds because player 2 rejects otherwise. However, there is no largest number $a_1 < 1$ in the realm of real (or rational) numbers. It is certainly not 0.99 because 0.999 is larger and still smaller than 1.

Exercise XVI.3

Player 1 can make the offer, and obtain, $\frac{1}{1+\delta}$. Given player 2's strategy, asking for less is a bad idea. If he asks for more than $\frac{1}{1+\delta}$, player 2 rejects and makes the counteroffer $\delta\frac{1}{1+\delta}$. This counteroffer (at stage 2) is less attractive to player 1 than obtaining $\frac{1}{1+\delta}$ at stage $1 : \frac{1}{1+\delta} > \delta\left(\delta\frac{1}{1+\delta}\right)$. Rejecting player 2's counteroffer does not help either because then player 1, at stage 3, has to make an offer giving no higher share than the share obtainable at stage 1. Note that shrinking has progressed from stage 1 to stage 3 by δ^2. Therefore, player 1's strategy is, indeed, a best response to player 2's strategy.

Exercise XVI.4

Figure XVI.4 shows how to work with δ_1 and δ_2. Backward solving yields

$$a_1 = 1 - \delta_2 (1 - \delta_1 \overline{a}_1)$$

by the very same arguments as in the main text.

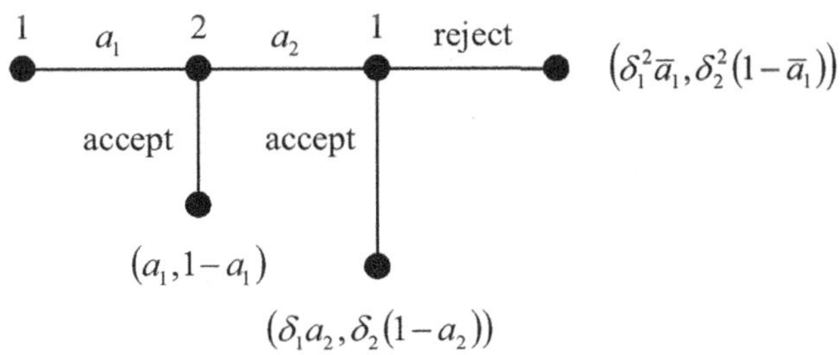

FIGURE XVI.4. The finite Rubinstein game for different discount factors

Letting $a_1 := \overline{a}_1$ leads to

$$\begin{aligned} a_1 &= 1 - \delta_2 (1 - \delta_1 a_1) \\ &= \frac{1 - \delta_2}{1 - \delta_1 \delta_2}. \end{aligned}$$

The result in the main text is obtained for $\delta := \delta_1 = \delta_2$. Obviously, player 1's portion is an increasing function of δ_1. Differentiation with respect to δ_2 shows that player 1 benefits from player 2's impatience.

Exercise XVI.5

If player 2 is so impatient that he cannot enjoy the cake at stage 2, he accepts any offer by player 1. In the extreme case of $\delta_1 > 0$, $\delta_2 = 0$ player 2 gets

$$1 - a_1 = 1 - \frac{1 - \delta_2}{1 - \delta_1 \delta_2} = 1 - \frac{1}{1} = 0.$$

If player 1 is very impatient, he knows that he accepts any counteroffer made by player 2. Therefore, $\delta_1 = 0$, $\delta_2 > 0$ leads to the very modest offer

$$a_1 = \frac{1 - \delta_2}{1 - \delta_1 \delta_2} = 1 - \delta_2.$$

Part E

Bayesian games and mechanism design

This part of the course deals with game theory under uncertainty – Bayesian games and mechanism design. An important example is provided by the first-price auction.

CHAPTER XVII

Static Bayesian games

This chapter has several aims. On the methodological level, we consider static Bayesian games. We use them to shed some light on mixed strategies in strategic games, to introduce correlated equilibria, and to analyze the first-price auction. Finally, we present double auctions and the Myerson-Satterthwaite theorem which argues that we cannot expect Pareto optimality in bargaining if both bargaining parties have imperfect information. Thus, this chapter furnishes the reader with a befitting cautionary note on "bargaining theory and Pareto optimality".

1. Introduction and an example

In this chapter, we deal with imperfect information in extensive-form games where the main source of this imperfection is due to moves by nature. As one of the main applications, we consider the first-price auction. As in the second-price auction (see chapter X, pp. 270), the bidders put down their bids simultaneously and the highest bidder obtains the object. While the successful bidder in the second-price auction pays the highest bid offered by the other bidders, in the first-price auction he has to pay his own bid.

For the second-price auction, we have shown that the strategy to bid according to one's reservation price is a dominant strategy. The first-price auction is more complicated for the bidders. On the one hand, they want to obtain the object if it is obtainable at a price below their willingness to pay. On the other hand, they want to pay as small a price as possible. We assume that the bidders are uncertain about the other bidders' willingness to pay. Therefore, to find an optimal strategy and the equilibrium is not an easy matter.

We assume two bidders 1 and 2. They know their own willingness to pay which is also called their "type". Thus, player 1 is of type t_1 and player 2 of type t_2. The extensive form of the game is rather simple. First, nature decides the players' types, then, the players make their bids. A player can condition his bid on his own type, but not on the other player's type.

H. Wiese, *Advanced Microeconomics*,
https://doi.org/10.1007/978-3-658-34959-2_17

Of course, a player will never bid above his willingness to pay because he may win the auction and obtain the negative payoff "willingness to pay minus bid". Assume a bid below the willingness to pay. An increase of the bid has two effects. First, it increases the bidder's chance of winning the auction. Second, it increases the price to be paid if the bid should win. In order to balance these two effects, a bidder needs to reason about the other player's bid which is possible only on the basis of probabilistic information about that player's type. We spell out this model in section 5.

The first-price auction is a special instance of a static Bayesian game. Bayesian games are characterized by moves by nature that determine the players' types, for example the willingness to pay in the first-price auction. In static Bayesian games, all players act simultaneously after learning their own types. Static Bayesian games are close to strategic-form games. However, each player's action depends on the own type (which is unknown to the other players).

2. Definitions

2.1. Static Bayesian game. We now define static Bayesian games. It is possible to define static Baysian games as extensive-form games where nature moves first and the players second. Consider the extensive form of a static Bayesian game in figure XVII.1. There are two players 1 and 2 with two types each, types t_1^1 and t_1^2 for player 1 and types t_2^1 and t_2^2 for player 2. Thus, we have $T_1 = \{t_1^1, t_1^2\}$ and $T_2 = \{t_2^1, t_2^2\}$. Every player chooses between two actions, $A_1 = \{a, b\}$ and $A_2 = \{c, d\}$.

At the initial node, nature chooses a type combination $t = (t_1, t_2) \in T_1 \times T_2$. The probabilities are not specified in the figure. Then, player 1 chooses between a and b. Since he knows his own type, only, he has two information sets, the upper one for t_1^1, the lower one for t_1^2. Then, player 2 moves. He also has two information sets. He does neither know player 1's type nor player 1's action. The left information set refers to type t_2^1, the right to t_2^2.

While static Bayesian games can be defined as extensive-form games (that is what they are, basically), it is simpler to choose a definition which leans on strategic-form games:

DEFINITION XVII.1 (STATIC BAYESIAN GAME). *A static Bayesian game is a quintuple*

$$\Gamma = \left(N, (A_i)_{i\in N}, (T_i)_{i\in N}, \tau, (u_i)_{i\in N}\right) = (N, A, T, \tau, u),$$

where

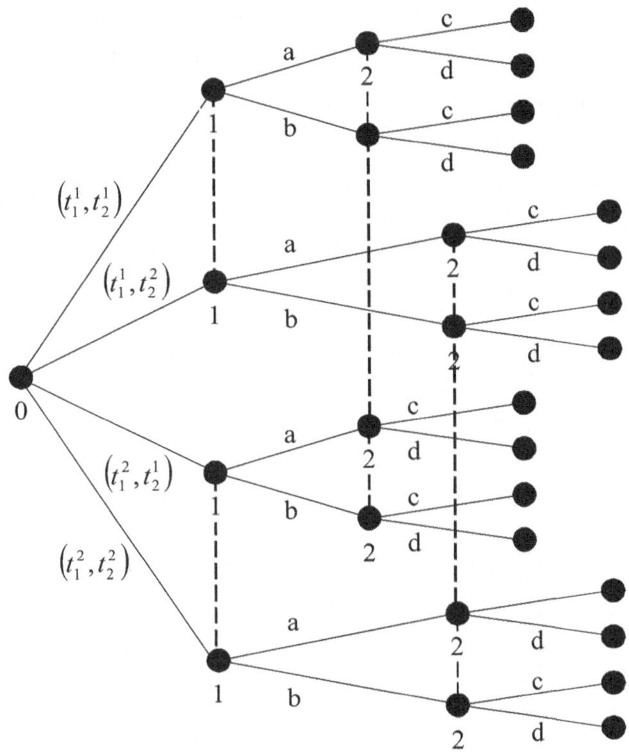

FIGURE XVII.1. A static Bayesian game in extensive form

- $N = \{1, ..., n\}$ *is the player set,*
- A_i *is the action set for player* $i \in N$ *with Cartesian product* $A = \bigtimes_{i \in N} A_i$ *and elements* a_i *and* a*, respectively,*
- $T = \bigtimes_{i \in N} T_i$ *is the tuple of type sets* T_i *for players* $i \in N$*,*
- τ *is the probability distribution on* T*, and*
- $u_i : A \times T_i \to \mathbb{R}$ *is player* i*'s payoff function.*

The probability distribution on T can be understood as a behavioral strategy chosen by nature which we have referred to as β_0 in chapter III (p. 47). For the time being, we assume a finite number of types for every player. For a continuum of types, see pp. 451.

2.2. Beliefs. Ex ante, before the players learn their own types, their beliefs are summarized by τ. The (a priori) probability for type t_i is given by

$$\tau(t_i) := \sum_{t_{-i} \in T_{-i}} \tau(t_{-i}, t_i)$$

After nature reveals the types, the players form (a posteriori) expectations about the other players' types by calculating the conditional probability.

Definition XVII.2 (Belief). *Let* Γ *be a static Bayesian game with probability distribution* τ *on* T. *Player* i*'s ex-post (posterior) belief* τ_i *is the probability distribution on* T_{-i} *given by the conditional probability*

$$\tau_i(t_{-i}) := \tau(t_{-i} | t_i) = \frac{\tau(t_{-i}, t_i)}{\tau(t_i)} = \frac{\tau(t_{-i}, t_i)}{\sum_{t_{-i} \in T_{-i}} \tau(t_{-i}, t_i)}. \quad \text{(XVII.1)}$$

Exercise XVII.1. *Assume two bidders* 1 *and* 2 *with probability distribution* τ *on* T *where* $T_1 = T_2 = \{high,\ low\}$ *refers to the willingness to pay for the good to be auctioned off. Consider*

$$\begin{aligned} \tau(high,\ high) &= \frac{1}{3},\ \tau(high,\ low) = \frac{1}{3}, \\ \tau(low,\ high) &= \frac{1}{9},\ \tau(low,\ low) = \frac{2}{9} \end{aligned}$$

and calculate $\tau(t_2 = high)$ *and also* $\tau_1(high)$ *if player* 1 *has learned that his own willingness to pay is high.*

2.3. Actions, strategies, and equilibria. Static Bayesian games are called static because the players act simultaneously. However, they act on the knowledge of their own type. In fact, we need to differentiate between ex ante and ex post. Ex ante means before the uncertainty about the own type is removed. A strategy is an ex-ante concept: A player i's strategy is a complete plan of how to act for each type t_i from T_i. According to a strategy, actions can be conditioned on the own type, i.e., an action is an ex-post concept. Note that the payoff functions $u_i : A \times T_i \to \mathbb{R}$ are to be understood ex post.

Definition XVII.3. *Let* Γ *be a static Bayesian game.*

- *A strategy for player* $i \in N$ *is a function* $s_i : T_i \to A_i$. *We sometimes write* $s(t)$ *instead of* $(s_1(t_1), ..., s_n(t_n)) \in A$.
- *The tuple of payoffs* $(u_i(a, t_i))_{i \in N}$ *is also denoted by* $u(a, t)$.
- *For each player* $i \in N$, *we define* $u_i : S \to \mathbb{R}$ *by*

$$\underbrace{u_i(s)}_{\text{ex-ante utility}} := \sum_{t \in T} \tau(t) \underbrace{u_i(s(t), t_i)}_{\text{ex-post utility}}, s \in S,$$

Then, best replies and Nash equilibria are defined as in chapter X.

The ex-ante and ex-post dichotomy leads to different (but equivalent!) definitions of equilibria. We start with the ex-ante definition:

DEFINITION XVII.4 (BAYESIAN EQUILIBRIUM (EX ANTE)). *Let Γ be a static Bayesian game. A strategy combination $s^* = (s_1^*, s_2^*, ..., s_n^*)$ is a Bayesian equilibrium (ex ante) if*

$$\begin{aligned} s_i^* &\in \arg\max_{s_i \in S_i} u_i\left(s_i^*, s_{-i}^*\right) \\ &= \arg\max_{s_i \in S_i} \sum_{t_i \in T_i} \sum_{t_{-i} \in T_{-i}} \tau\left(t_i, t_{-i}\right) u_i\left(s_i\left(t_i\right), s_{-i}^*\left(t_{-i}\right), t_i\right) \end{aligned}$$

holds for all $i \in N$.

Thus, in the light of s_{-i}^* and τ, s_i^* maximizes player i's expected payoff. A type t_i that occurs with probability $\tau(t_i) = 0$, can choose any action from A_i. We now turn to the ex-post definition. After learning his type, player i chooses the action that maximizies his expected payoff given his belief τ_i (which is a conditional probability, see above).

DEFINITION XVII.5 (BAYESIAN EQUILIBRIUM (EX POST)). *Let Γ be a static Bayesian game. A strategy combination $s^* = (s_1^*, s_2^*, ..., s_n^*)$ is a Bayesian equilibrium (ex post) if*

$$s_i^*\left(t_i\right) \in \arg\max_{a_i \in A_i} \sum_{t_{-i} \in T_{-i}} \tau_i\left(t_{-i}\right) u_i\left(a_i, s_{-i}^*\left(t_{-i}\right), t_i\right)$$

holds for all $i \in N$ and all $t_i \in T_i$ obeying $\tau(t_i) > 0$.

The qualification $\tau(t_i) > 0$ is necessary because $\tau_i(t_{-i})$ (see eq. XVII.1) is ill-defined otherwise. Again, $\tau(t_i) = 0$ implies "anything goes".

2.4. Example: Cournot duopoly with one-sided uncertainty.

2.4.1. *The model.* We modify the Cournot dyopoly (chapter XI) by introducing one-sided cost uncertainty. Consider two firms 1 and 2 that serve the same market with a homogenous good. The inverse demand function p is given by $p(X) = 80 - X$ where X is the total supply and $p(X)$ the resulting price. Both firms have constant unit costs:

- $c_2 = 20$ for firm 2 and
- $c_1^l = 15$ or $c_1^h = 25$ for firm 1.

The cost c_2 is known to both firms, but the cost $c_1 \in \left\{c_1^l, c_1^h\right\}$ is known to firm 1, only. From the point of view of firm 2, the probability for c_1^l is 1/2. This situation can be described as a static Bayesian game

$$\Gamma = \left(N, \left(A_1, A_2\right), \left(T_1, T_2\right), \tau, \left(u_1, u_2\right)\right)$$

where

- $N = \{1, 2\}$ is the set of the two firms,
- $A_1 = A_2 = [0, \infty)$ are the sets of quantities chosen by the firms,

- $T_1 = \{c_1^l, c_1^h\} = \{15, 25\}$ and $T_2 = \{20\}$ are the type sets,
- τ is the probability distribution on T given by $\tau(15, 20) = \frac{1}{2}$ and $\tau(25, 20) = \frac{1}{2}$ and
- the payoff functions are defined by

$$u_2(x_1, x_2, t_2) = (p(X) - c_2)x_2 = (80 - (x_1 + x_2) - 20)x_2$$

and

$$u_1(x_1, x_2, t_1) = \begin{cases} (80 - (x_1 + x_2) - 15)x_1, & t_1 = c_1^l \\ (80 - (x_1 + x_2) - 25)x_1, & t_1 = c_1^h \end{cases}$$

Since player 2's type set has one element only, a strategy $s_2 : T_2 \to A_2$ is basically just an action from A_2. Player 1's strategy set is the set of functions from T_1 to A_1,

$$S_1 = \left\{s_1 : \left\{c_1^l, c_1^h\right\} \to [0, \infty)\right\}.$$

EXERCISE XVII.2. *Can you show that the a-priori probabilities for the players' types is equal to the belief formed by the players. (In different words: Can you show that the types are independent?*

2.4.2. *The static Bayesian equilibrium.* We employ the ex-post definition of an equilibrium. The strategy combination (s_1^*, s_2^*) is an equilibrium (ex post) if

$$s_i^*(t_i) \in \arg\max_{a_i \in A_i} \sum_{t_{-i} \in T_{-i}} \tau_i(t_{-i})\, u_i\left(a_i, s_{-i}^*(t_{-i}), t_i\right)$$

holds for all $i \in N$ and all $t_i \in T_i$ obeying $\tau(t_i) > 0$. Firm 1's choice depends on its type:

$$\begin{aligned} s_1^R(t_1) &= \begin{cases} \arg\max_{x_1 \in [0,\infty)} \left(80 - (x_1 + x_2) - c_1^l\right)x_1, & t_1 = c_1^l \\ \arg\max_{x_1 \in [0,\infty)} \left(80 - (x_1 + x_2) - c_1^h\right)x_1, & t_1 = c_1^h \end{cases} \\ &= \begin{cases} \frac{65}{2} - \frac{1}{2}x_2, & t_1 = c_1^l \\ \frac{55}{2} - \frac{1}{2}x_2, & t_1 = c_1^h \end{cases} \end{aligned}$$

Firm 2's profit is the expected value

$$\begin{aligned} & \frac{1}{2}\left(80 - \left[x_1^l + x_2\right] - 20\right)x_2 + \frac{1}{2}\left(80 - \left[x_1^h + x_2\right] - 20\right)x_2 \\ = & \left(80 - \left(\frac{1}{2}\left[x_1^l + x_1^h\right] + x_2\right) - 20\right)x_2 \\ = & \left(60 - \frac{1}{2}\left[x_1^l + x_1^h\right]\right)x_2 - x_2^2 \end{aligned}$$

which leads to the reaction function

$$\begin{aligned} x_2^* = s_2^*(20) &= \frac{\left(60 - \frac{1}{2}\left[x_1^l + x_1^h\right]\right)}{2} \\ &= 30 - \frac{1}{4}\left[x_1^l + x_1^h\right]. \end{aligned}$$

We now have three equations in the three unknowns x_2, x_1^l, and x_1^h. They lead to the Nash equilibrium

$$x_2^* = 20, s_1^*\left(c_1^l\right) = \frac{45}{2}, \text{ and } s_1^*\left(c_1^h\right) = \frac{35}{2}.$$

Because of the one-sidedness, the model is easy to solve. Two-sided uncertainty as in the first-price auction is covered below.

3. Revisiting mixed-strategy equilibria

3.1. Continuous types. Before dealing with uncertainty in matrix games, we need to introduce continuous types and probability distributions for these types. We assume two players $i = 1, 2$ with types t_i from $T_i = [0, x]$, $x > 0$. In our Bayesian game (to be defined in the section after next), τ^x is a probability distribution on $T = T_1 \times T_2$. However, for the rest of the chapter, we assume that the events "type t_1 lies between a and b" and "type t_2 lies between c and d" are independent. Then, we can look at the probability for the event "type t_1 lies between a and b" without regard to type 2. We use the symbol, τ^x, to denote this probability, namely as $\tau^x([a, b])$.

Consider, now, the density distribution depicted in figure XVII.2. The density is equal to $1/x$ for types between 0 and x and 0 for other types:

$$\tau^x(a) = \begin{cases} \frac{1}{x}, & a \in [0, x] \\ 0, & a \notin [0, x] \end{cases}$$

For this density distribution, we find

$$\tau^x([a, b]) = \int_a^b \tau^x(t)\, dt = \frac{b-a}{x}, 0 \le a \le b \le x.$$

This formula implies that the probability for a specific type a is zero:

$$\tau^x([a, a]) = \frac{a-a}{x} = 0$$

Thus, while τ^x is a probability distribution on $[0, x] \times [0, x]$ in the first place, it can also be understood as a probability (or density) distribution on $[0, x]$. Since we assume identical densities for players 1 and 2, $\tau^x([a, b])$ can refer to either of them. Please distinguish $\tau^x([a, a])$ from $\tau^x(a)$. $\tau^x([a, a])$ is the

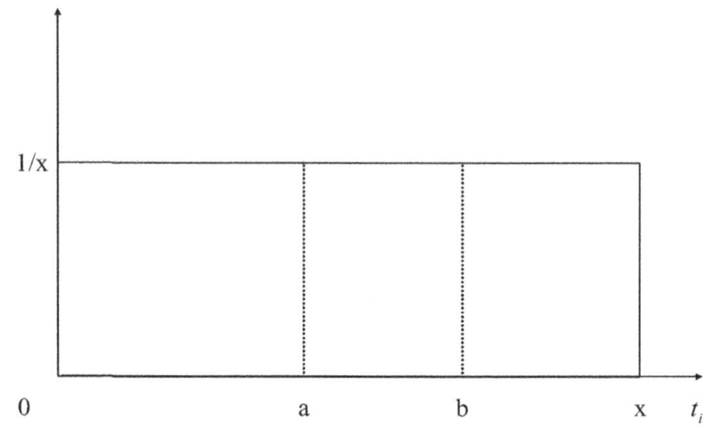

FIGURE XVII.2. The total area is 1.

probability for the specific type a, while $\tau^x(a)$ refers to the density at point a which need not be zero.

3.2. Introducing uncertainty. In chapter X, we consider games in strategic form. John Nash has shown that any finite strategic game $\Gamma = (N, S, u)$ has a mixed-strategy Nash equilibrium. This is a neat result but many people feel uneasy about the very concept of a mixed strategy. In real life, we sometimes wonder whether action a or action b is the better one, but we throw a die in rare cases, only. Static Bayesian games allow a fresh look at mixed equilibria. For a given matrix game with a mixed-strategy equilibrium, we construct a sequence of static Bayesian games that converge towards that game. In a Bayesian game, no player $i \in N$ randomizes. However, from the point of view of the other players from $N\backslash\{i\}$ who do not know the type t_i, it may well seem as if player i is a randomizer. For example, in the first-price auction, player 1 fixes his equilibrium bid according to his strategy s_1 which is given by $s_1(t_1) = \frac{t_1}{2}$. Thus, player 1 with types from the interval $[0, 1]$ does not randomize. But player 2 expects bids between 0 and $\frac{1}{2}$ which are indistinguishable, from 2's perspective, from a probability distribution on this interval.

We illustrate these ideas (due to Harsanyi (1973)) with the battle of the sexes (see p. 267). The concrete numbers are taken from Gibbons (1992, p. 153):

		Peter	
		theatre	football
Cathy	theatre	$2+t_C, 1$	$0, 0$
	football	$0, 0$	$1, 2+t_P$

The types t_C and t_P are called Vollmer parameters and refer to the theatre and football madness entertained by Cathy or Paul, respectively.

EXERCISE XVII.3. *Determine all three equilibria of the above matrix game. Assume* $t_C = t_P = 0$.

We now construct a static Bayesian game. Cathy and Peter know their own types, t_C and t_P, respectively, but not the other player's type. We assume $t_C, t_P \in [0, x]$ for some number $x > 0$ and also a constant density $\frac{1}{x}$. Thus, we have

$$\tau^x([a, b]) = \frac{b-a}{x}. \tag{XVII.2}$$

Formally, the corresponding static Bayesian game $\Gamma^x = (N, A, T, \tau, u)$ is given by

- $N = \{C, P\}$,
- $A_C = A_P = \{\text{theatre, football}\}$,
- $T_C = T_P = [0, x]$,
- the probability distribution τ^x on $T_C \times T_P$ defined in the previous section, and
- the payoff functions $u_C : A_C \times A_P \times T_C \to \mathbb{R}$ and $u_P : A_C \times A_P \times T_P \to \mathbb{R}$ defined by the above matrix.

EXERCISE XVII.4. *What do Peter's strategies look like? Do you see that any strategy chosen by Peter determines a probability distribution on his action set?*

3.3. The equilibria. Let θ_P be Peter's probability for choosing theatre (given the probability distribution for his types and given his strategy). Cathy prefers theatre to football if

$$\underbrace{(2+t_C)\cdot\theta_P + 0\cdot(1-\theta_P)}_{\substack{\text{Cathy's expected payoff}\\ \text{for choosing theatre}}} \geq \underbrace{0\cdot\theta_P + 1\cdot(1-\theta_P)}_{\substack{\text{Cathy's expected payoff}\\ \text{for choosing football}}}$$

holds. Solving for t_C, Cathy's optimal response to θ_P and to Peter's strategy is

$$t_C \mapsto \begin{cases} \text{theatre}, & t_C \geq \frac{1-3\theta_P}{\theta_P}, \\ \text{football}, & t_C < \frac{1-3\theta_P}{\theta_P}. \end{cases}$$

Note that Cathy's optimal response is a threshold strategy. She chooses theatre if her type t_C is equal to or above the threshold $\bar{t}_C := \frac{1-3\theta_P}{\theta_P}$. The same is true for Peter whose optimal response to θ_C (Cathy's probability for theatre) is also a threshold strategy.

EXERCISE XVII.5. *Find Peter's threshold type $\bar{t}_P$ if Cathy's probability for choosing theatre is θ_C!*

Summarizing the results so far,

- Cathy chooses theatre if $t_C \geq \bar{t}_C := \frac{1-3\theta_P}{\theta_P}$ while
- Peter chooses theatre in case of $t_P \leq \bar{t}_P := \frac{3\theta_C-2}{1-\theta_C}$.

Of course, Peter's probability for theatre, θ_P, is related to his threshold type $\bar{t}_P$. By equation XVII.2, we find

$$\theta_P = \tau^x([0, \bar{t}_P]) = \frac{\bar{t}_P - 0}{x}.$$

Similarly, we obtain

$$\theta_C = \tau^x([\bar{t}_C, x]) = \frac{x - \bar{t}_C}{x}.$$

Using the above definitions of $\bar{t}_C$ and $\bar{t}_P$, we get two equations in two unknowns:

$$\begin{aligned} \bar{t}_C &= \frac{1-3\theta_P}{\theta_P} = \frac{1-3\frac{\bar{t}_P}{x}}{\frac{\bar{t}_P-0}{x}} = \frac{x-3\bar{t}_P}{\bar{t}_P} \text{ and} \\ \bar{t}_P &= \frac{3\theta_C-2}{1-\theta_C} = \frac{3\frac{x-\bar{t}_C}{x}-2}{1-\frac{x-\bar{t}_C}{x}} = \frac{x-3\bar{t}_C}{\bar{t}_C}. \end{aligned}$$

EXERCISE XVII.6. *Confirm $\bar{t}_C = \bar{t}_P$! Hint: Consider $\bar{t}_C\bar{t}_P$.*

We now solve the quadratic equation

$$\bar{t}^2 + 3\bar{t} - x = 0$$

to find

$$\bar{t}_{1,2} = -\frac{3}{2} \pm \sqrt{\frac{9}{4} + x}.$$

Since the sought-after threshold types belong to $[0, x]$, we obtain

$$\bar{t}_C^* = \bar{t}_P^* = -\frac{3}{2} + \sqrt{\frac{9}{4} + x}$$

which specify the equilibrium for every $x > 0$:

LEMMA XVII.1. *Let Γ^x be the static Bayesian game defined above. The strategy combination (s_C^*, s_P^*) defined by*

$$\begin{array}{rcl} s_C^* & : & [0,x] \to \{\textit{theatre, football}\} \\ t_C & \mapsto & s_C^*(t_c) = \begin{cases} \textit{theatre}, & t_C \geq -\frac{3}{2} + \sqrt{\frac{9}{4} + x}, \\ \textit{football}, & t_C < -\frac{3}{2} + \sqrt{\frac{9}{4} + x}. \end{cases} \end{array}$$

and

$$\begin{array}{rcl} s_P^* & : & [0,x] \to \{\textit{theatre, football}\} \\ t_P & \mapsto & s_P^*(t_P) = \begin{cases} \textit{theatre}, & t_P \leq -\frac{3}{2} + \sqrt{\frac{9}{4} + x}, \\ \textit{football}, & t_P > -\frac{3}{2} + \sqrt{\frac{9}{4} + x}. \end{cases} \end{array}$$

is an equilibrium.

3.4. Purification. We now turn to the clou of the whole exercise. Above, we have found Cathy's mixed-equilibrium strategy $\left(\frac{2}{3}, \frac{1}{3}\right)$. We can reproduce this strategy by considering $x = 0$. For a given $x > 0$, Cathy's probability for choosing theatre is

$$\begin{aligned} \theta_C &= \frac{x - \bar{t}_C}{x} = \frac{x - \left(-\frac{3}{2} + \sqrt{\frac{9}{4} + x}\right)}{x} \\ &= 1 - \frac{-\frac{3}{2} + \sqrt{\frac{9}{4} + x}}{x}. \end{aligned}$$

Here, we cannot substitute x for 0 because then we have $\frac{0}{0}$ which is not defined. However, we can apply de l'Hospital's rule which is theorem VIII.1 (p. 223):

$$\begin{aligned} \lim_{x\to 0}\left(1 - \frac{-\frac{3}{2} + \sqrt{\frac{9}{4} + x}}{x}\right) &= \lim_{x\to 0} 1 - \lim_{x\to 0} \frac{-\frac{3}{2} + \sqrt{\frac{9}{4} + x}}{x} \\ &= 1 - \lim_{x\to 0} \frac{\frac{1}{2\sqrt{\frac{9}{4}+x}} \cdot 1}{1} \quad \text{(de l'Hospital's rule)} \\ &= 1 - \lim_{x\to 0} \frac{\frac{1}{2\cdot\frac{3}{2}} \cdot 1}{1} \\ &= \frac{2}{3}. \end{aligned}$$

Thus, Cathy's probabilities in the static Bayesian games (which depend on x) converge to the mixed-strategy probability in the matrix game. Since Cathy chooses a pure action (in dependence on her type), substituting a matrix game by a sequence of static Bayesian games and a mixed-strategy equilibrium by a sequence of Bayesian equilibria is called purification.

4. Correlated equilibria

4.1. Telling how to play. In the examples considered so far, the types refer to the players' payoffs. In this section, we consider types that refer to the players' actions. In these games, the players learn the action they are supposed to choose. Of course, they are free to ignore the suggestion. Sometimes, the players will find it beneficial to follow the suggestions. Consider the battle of the sexes and suggestions where both players are told to choose theatre with probability $\frac{1}{2}$ and football with probability $\frac{1}{2}$. Then, it is in Cathy's interest to follow the advice if she thinks that Peter will do as told.

Thus, in these static Bayesian games we have $T_i = A_i$ for each player $i \in N$ so that player i's strategy is a function $s_i : A_i \to A_i$. Obeying the suggestion is tantamount to using the identity function as a strategy, $s_i = id_i : A_i \to A_i$, $a_i \mapsto id_i(a_i) = a_i$. We are particularly interested in knowing when $s = (id_1, id_2, ..., id_n)$ is an equilibrium strategy combination.

Consider, now, the prisoners' dilemma:

		player 2	
		deny	confess
player 1	deny	$4, 4$	$0, 5$
	confess	$5, 0$	$1, 1$

The type sets are $T_1 = T_2 = \{\text{deny, confess}\}$. Assume the probability distribution τ on $T = T_1 \times T_2$ is given by $\tau(\text{deny, deny}) = 1$. This probability distribution implies that both players learn the suggestion "deny" with probability 1. This is a very good suggestion because – if followed – the players obtain 4 rather than 1 in the dominant-strategy equilibrium of the strategic game. Of course, the "if" is essential. Even if player 1 assumes that player 2 will follow the recommendation, player 1 prefers to choose confess.

Thus, we look for suggestions that players like to follow in equilibrium.

4.2. The recommendation game. As in the purification effort, the recommendation game builds on a strategic game. This static Bayesian game has two peculiarities. First, every player's type set is his action set. Second, the payoffs do not depend on the types:

Definition XVII.6 (recommendation game). *Let $\Gamma = (N, A, u)$ be a strategic game. The recommendation game Γ^τ is the static Bayesian game (N, A, T, τ, u) where*

$$T_i = A_i$$

holds for all $i \in N$ and the payoff functions $u_i : A \times T \to \mathbb{R}$ obey

$$\underbrace{u_i(a,t)}_{\substack{\textit{payoff function} \\ \textit{for recommendation game } \Gamma^\tau}} = \underbrace{u_i(a)}_{\substack{\textit{payoff function} \\ \textit{for strategic game } \Gamma}},$$

i.e., the payoffs do not depend on the types. If $s^* = (id_1, id_2, ..., id_n)$ is an equilibrium strategy combination in Γ^τ, τ is called a correlated equilibrium of Γ.

Thus, a correlated equilibrium is a probability distribution on the set of action combinations such that the players are happy to follow the recommendation obtained. If τ puts the whole probability mass on one or several equilibria of the strategic game Γ, τ is a correlated equilibrium. This is the statement of the following theorem.

Theorem XVII.1. *Let $\Gamma = (N, A, u)$ be a strategic game and let Γ^τ be the corresponding recommendation game. Let $a^* = (a_1^*, a_2^*, ..., a_n^*)$ and $b^* = (b_1^*, b_2^*, ..., b_n^*)$ be equilibria of Γ. If $\tau(a^*) + \tau(b^*) = 1$, $s^* = (id_1, id_2, ..., id_n)$ is an equilibrium of Γ^τ.*

Proof. Consider a player $i \in N$ who learns his type $a_i^* \in T_i = A_i$. This player i knows that a_i^* is part of an equilibrium, that the other players learn the strategy combination $a_{-i}^* \in T_{-i}$, and that (a_i^*, a_{-i}^*) is a Nash equilibrium of Γ. Thus, if i believes that the other players $j \in N \backslash \{i\}$ have strategy $s_j^* = id_j : A_j \to A_j$ and will therefore choose $s_j^*\left(a_j^*\right) = a_j^*$, the best that i can do is to choose $a_i^* \in \arg\max_{a_i \in A_i} u_i\left(a_i, a_{-i}^*\right)$ himself. Thus $s_i^* = id_i$ is a best response to $s_{-i}^* = (id_1, ..., id_{i-1}, id_{i+1}, ..., id_n)$. □

By playing a recommendation game, the players can realize any payoff vector that lies in the convex hull (p. 376) of the equilibrium payoff vectors belonging to Γ. However, one can do even better.

4.3. Going beyond the convex hull. Aumann (1974, S. 72) shows that correlated equilibria may lie outside the convex hull of the equilibria – in pure or mixed strategies. Let us consider his example:

This game has two equilibria in pure strategies, $\left(a_1^1, a_2^2\right)$ and $\left(a_1^2, a_2^1\right)$. Because of theorem X.1 (p. 282) it makes sense to look for a third equilibrium.

player 2

player 1		a_2^1	a_2^2
	a_1^1	$6,6$	$2,7$
	a_1^2	$7,2$	$0,0$

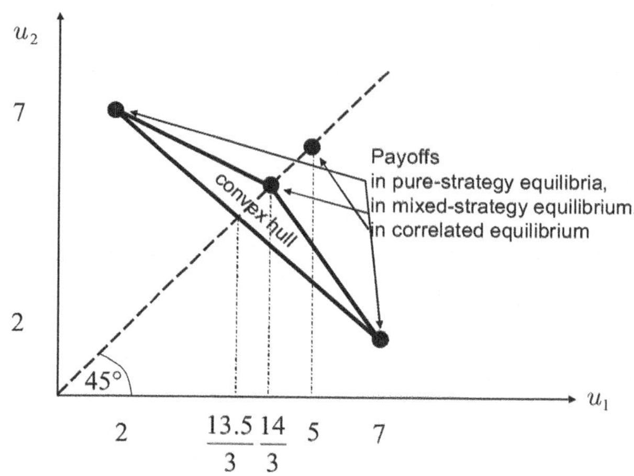

FIGURE XVII.3. The convex hull and beyond

It is $\left(\left(\frac{2}{3},\frac{1}{3}\right),\left(\frac{2}{3},\frac{1}{3}\right)\right)$ with payoffs $\left(\frac{14}{3},\frac{14}{3}\right)$. The convex hull of these equilibria is given in figure XVII.3.

We now look for a correlated equilibrium that is better than the mixed-strategy equilibrium and also better than a "fair" mixture of the two pure-strategy equilibria. Alas, while $\tau\left(a_1^1,a_2^1\right) = 1$ promises the payoff 6, τ is not a correlated equilibrium. If player 2 chooses $s_2 = id_2$ (follows the recommendation), it is in player 1's interest to deviate. Thus, this τ will not do. Consider, however, the probability distribution τ on $T = A$ given by

$$\tau\left(\left(a_1^1,a_2^1\right)\right) = \tau\left(\left(a_1^2,a_2^1\right)\right) = \tau\left(\left(a_1^1,a_2^2\right)\right) = \frac{1}{3}$$

and depicted by

		player 2	
		a_2^1	a_2^2
player 1	a_1^1	$\frac{1}{3}$	$\frac{1}{3}$
	a_1^2	$\frac{1}{3}$	0

The idea behind this probability distribution is to avoid the zero payoff associated with the action combination (a_1^2, a_2^2). It yields the payoff

$$\frac{1}{3} \cdot 7 + \frac{1}{3} \cdot 6 + \frac{1}{3} \cdot 2 = 5$$

for each player (see figure XVII.3). So far, so good. We still need to confirm that $s_1 = id_1$ is a best reply to $s_2 = id_2$ (and vice versa). Assume player 1 observes the recommendation a_1^1. He then believes that player 2 got the recommendation a_2^1 with (conditional) probability

$$\tau_1\left(a_2^1\right) = \tau\left(a_2^1 \left| a_1^1\right.\right) = \frac{\tau\left(a_2^1, a_1^1\right)}{\tau\left(a_1^1\right)} = \frac{\tau\left(a_2^1, a_1^1\right)}{\tau\left(a_1^1, a_2^1\right) + \tau\left(a_1^1, a_2^2\right)} = \frac{\frac{1}{3}}{\frac{2}{3}} = \frac{1}{2}.$$

Thus, player 1 prefers to follow his recommendation and choose a_1^1 if he assumes that player 2 is obedient:

$$\begin{aligned} & \tau_1\left(a_2^1\right) u\left(a_1^1, a_2^1\right) + \tau_1\left(a_2^2\right) u\left(a_1^1, a_2^2\right) \\ > \; & \tau_1\left(a_2^1\right) u\left(a_1^2, a_2^1\right) + \tau_1\left(a_2^2\right) u\left(a_1^2, a_2^2\right) \\ \Leftrightarrow \; & \frac{1}{2} \cdot 6 + \frac{1}{2} \cdot 2 > \frac{1}{2} \cdot 7 + \frac{1}{2} \cdot 0. \end{aligned}$$

Assume, on the other hand, recommendation a_1^2. From observing a_1^2, player 1 knows that player 2 got recommendation a_2^1. Just look at τ or calculate the conditional probability

$$\tau_1\left(a_2^1\right) = \tau\left(a_2^1 \left| a_1^2\right.\right) = \frac{\tau\left(a_2^1, a_1^2\right)}{\tau\left(a_1^2\right)} = \frac{\tau\left(a_2^1, a_1^2\right)}{\tau\left(a_1^2, a_2^1\right)} = \frac{\frac{1}{3}}{\frac{1}{3}} = 1.$$

Given a_2^1, player 1's best response is a_1^2 which is just what he was told.

Thus, we have found a correlated equilibrium that allows to obtain payoffs not reachable by mixing pure-strategy equilibria or by the mixed-strategy equilibrium.

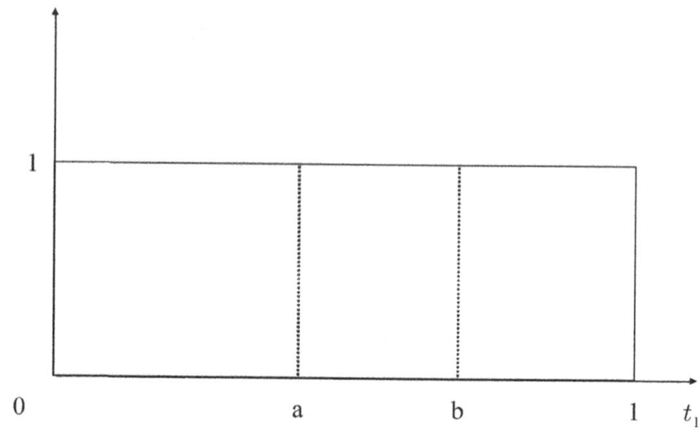

FIGURE XVII.4. The density is 1.

5. The first-price auction

5.1. The model. We now follow up on the first-price auction explained in the introduction to this chapter. We assume two bidders 1 and 2 with independent types $t_1, t_2 \in [0,1]$. A bidder's type stands for his willingness to pay. Consider, now, the density distribution depicted in figure XVII.4 and defined by

$$\tau(a) = \begin{cases} 1, & a \in [0,1] \\ 0, & a \notin [0,1] \end{cases}$$

The probability for types between a and b is given by

$$\tau([a,b]) = \int_a^b \tau(t)\, dt = b - a, 0 \leq a \leq b \leq 1.$$

Formally, the first-price auction is the static Bayesian game

$$\Gamma = (N, (A_1, A_2), (T_1, T_2), \tau, (u_1, u_2))$$

where

- $N = \{1, 2\}$ is the set of the two bidders,
- $A_1 = A_2 = [0, \infty)$ are the sets of bids chosen by the bidders,
- $T_1 = T_2 = [0, 1]$ are the type sets,
- τ is the probability distribution on T given by $\tau([a,b], [c,d]) = (b-a)(d-c)$ where $0 \leq a \leq b \leq 1$ and $0 \leq c \leq d \leq 1$ hold, and

- the payoff functions are defined by

$$u_1(a,t_1) = \begin{cases} t_1 - a_1, & a_1 > a_2, \\ \frac{t_1-a_1}{2}, & a_1 = a_2, \\ 0, & a_1 < a_2, \end{cases} \quad \text{and}$$

$$u_2(a,t_2) = \begin{cases} 0 & a_1 > a_2, \\ \frac{t_2-a_2}{2} & a_1 = a_2, \\ t_2 - a_2 & a_1 < a_2. \end{cases}$$

Thus, the utility is zero if a bidder is outbid by another bidder. If a bidder wins the auction, his payoff is $t_1 - a_1$, i.e., willingness to pay minus the bid. If the bids happen to be identical, a fair coin decides the winner.

5.2. Solution. In order to solve the first-price auction, we use the ex-post equilibrium definition. If player 1 is of type $t_1 \in [0,1]$, his condition for the equilibrium strategy combination (s_1^*, s_2^*) is

$$s_1^*(t_1) \in \arg\max_{a_1 \in A_1} \left((t_1 - a_1) \underbrace{\tau(\{t_2 \in [0,1] : a_1 > s_2^*(t_2)\})}_{\substack{\text{probability that player 1's bid} \\ \text{is higher than player 2's bid}}} + \frac{1}{2}(t_1 - a_1) \underbrace{\tau(\{t_2 \in [0,1] : a_1 = s_2^*(t_2)\})}_{\text{probability for equal bids}} \right).$$

We will obtain a strategy s_2^* that is a strictly increasing function in t_2. Then, the probability for equal bids equals zero. With that assumption, we follow Gibbons (1992, pp. 155) in restricting our search for equilibrium strategies to linear strategies of the forms

$$\begin{aligned} s_1^*(t_1) &= c_1 + d_1 t_1 \ (d_1 > 0), \\ s_2^*(t_2) &= c_2 + d_2 t_2 \ (d_2 > 0). \end{aligned}$$

By

$$\begin{aligned} \tau(\{t_2 \in [0,1] : a_1 > c_2 + d_2 t_2\}) &= \tau\left(\left\{t_2 \in [0,1] : t_2 < \frac{a_1 - c_2}{d_2}\right\}\right) \\ &= \tau\left(\left[0, \frac{a_1 - c_2}{d_2}\right]\right) \\ &= \frac{a_1 - c_2}{d_2}, \end{aligned}$$

player 1's maximization problem is solved by

$$s_1^R(t_1) = \arg\max_{a_1 \in A_1} (t_1 - a_1) \frac{a_1 - c_2}{d_2} = \frac{c_2 + t_1}{2}. \quad \text{(XVII.3)}$$

However, this is not the complete answer.

- A bid below c_2 cannot be better than the bid $a_1 = c_2$ because agent 1 obtains the object with probability 0 for any $a_1 \leq c_2$.
- Any bid $a_1 \geq c_2 + d_2$ means that bidder 1 obtains the object with probability 1. Therefore, a bid above $c_2 + d_2$ increases the price to be paid by bidder 1 without increasing the chance of obtaining the object.

The two restrictions $c_2 \leq a_1 \leq c_2 + d_2$ lead to $\frac{a_1 - c_2}{d_2} \in [0, 1]$ (i.e., the probability that t_2 is smaller than $\frac{a_1 - c_2}{d_2}$). Note also that player 1's best response to c_2 (and hence to player 2's strategy) is a linear strategy with $c_1 = \frac{c_2}{2}$ and $d_1 = \frac{1}{2}$.

Analogously, bidder 2's best response is

$$s_2^R(t_2) = \frac{c_1 + t_2}{2}$$

with $c_2 = \frac{c_1}{2}$ and $d_2 = \frac{1}{2}$. Thus, we cannot have an equilibrium in linear strategies unless $c_1 = \frac{c_2}{2} = \frac{\frac{c_1}{2}}{2} = \frac{c_1}{4}$ and hence $c_1 = 0$ holds. Therefore, the strategy combination

$$s^* = (s_1^*, s_2^*)$$

with

$$s_1^* : [0, 1] \to \mathbb{R}_+, \quad t_1 \mapsto s_1^*(t_1) = \frac{t_1}{2}$$

and

$$s_2^* : [0, 1] \to \mathbb{R}_+, \quad t_2 \mapsto s_2^*(t_2) = \frac{t_2}{2}$$

is a candidate for our equilibrium. These are the "half-bid strategies".

These strategies form an equilibrium because the strategies are best reponses to each other. If player 2 uses the (linear) strategy s_2^*, $s_1^*(t_1) = \frac{t_1}{2}$ is a best response as shown above. Thus, s_1^* turns out to be a linear strategy. Therefore, we have found an equilibrium in linear strategies but cannot exclude the possibility of extra equilibria in non-linear strategies.

5.3. First-price or second-price auction? We now take the auctioneer's perspective and ask the question whether the first-price auction is preferable to the second-price auction. For given bids, the first-price auction is better for the auctioneer. However, in that auction, bidders do not bid in accordance with their types but bid half of the willingness to pay, only. The auctioneer compares the prices

- $\min(t_1, t_2)$ for the second-price auction and
- $\max\left(\frac{1}{2}t_1, \frac{1}{2}t_2\right)$ for the first-price auction.

We assume a risk-neutral auctioneer who maximizes the expected payoff. Let t_1 be a point in the closed interval $[0, 1]$. The integral

$$\int_0^{t_1} 1dt_2 = t_1$$

is the area in figure XVII.4 between 0 and t_1, i.e., the probability for types below t_1. The probability for types above t_1 is given by

$$\int_{t_1}^{1} 1dt_2 = 1 - t_1.$$

We now calculate the auctioneer's expected payoff for the second-price auction and find

$$\begin{aligned}
& \int_{t_1 \in [0,1]} \left(\int_{t_2 \in [0,1]} \min(t_1, t_2)\, dt_2 \right) dt_1 \\
= & \int_{t_1 \in [0,1]} \left(\int_0^{t_1} \min(t_1, t_2)\, dt_2 + \int_{t_1}^{1} \min(t_1, t_2)\, dt_2 \right) dt_1 \\
& \qquad \text{(splitting the } t_2 \text{ integral)} \\
= & \int_{t_1 \in [0,1]} \left(\int_0^{t_1} t_2 dt_2 + t_1 \int_{t_1}^{1} dt_2 \right) dt_1 \\
= & \int_{t_1 \in [0,1]} \left(\frac{1}{2} t_2^2 \Big|_0^{t_1} + t_1 (1 - t_1) \right) dt_1 \\
= & \int_{t_1 \in [0,1]} \left(\frac{1}{2} t_1^2 + t_1 (1 - t_1) \right) dt_1 \\
= & \int_{t_1 \in [0,1]} \left(-\frac{1}{2} t_1^2 + t_1 \right) dt_1 \\
= & \left(-\frac{1}{3}\frac{1}{2} t_1^3 + \frac{1}{2} t_1^2 \right) \Big|_0^1 \\
= & -\frac{1}{6} + \frac{1}{2} - 0 = \frac{1}{3}.
\end{aligned}$$

EXERCISE XVII.7. *Calculate the auctioneer's expected payoff if he chooses the first-price auction and the bidders act non-strategically, i.e., choose actions according to strategies* $s_i(t_i) = t_i$.

According to the above exercise, the first-price auction is better than the second-price auction if (!) the bidders act non-strategically. They cannot be expected to do so. Bidder i will not bid t_i but half his willingness to pay,

$t_1/2$. Therefore, the auctioneer can expect the payoff $\frac{1}{3}$ from the first-price auction. This is the same payoff as for the second-price auction. Therefore, the auctioneer is indifferent between the first-price and the second-price auction!

6. The double auction

6.1. The model. The double auction is a special kind of auction where both the seller and the buyer simultaneously submit a bid r (an announced reservation price) and w (an announced willingness to pay), respectively. If the buyer's willingness to pay w is above the seller's reservation price r, trade occurs at the price just in between,

$$p = \frac{r + w}{2}.$$

EXERCISE XVII.8. *Calculate the seller's and the buyer's rent in case of $r \leq w$ and in case of $r > w$. Assume that the announced reservation price and the announced willingness to pay are equal to the true reservation price and the true willingness to pay, respectively.*

As is obvious from the exercise, Pareto-efficient trade cannot occur unless the true willingness to pay is at least as high as the true reservation price. In that case, the gains from trade (willingness to pay minus reservation price) are positive. Of course, the bids r and w do not necessarily equal the true reservation price and the true willingness to pay, respectively. The true figures (the types) are drawn independently from a constant density distribution.

We model a double auction as the static Bayesian game

$$\Gamma = (\{s, b\}, (A_s, A_b), (T_s, T_b), \tau, (u_s, u_b))$$

where

- $N = \{s, b\}$ is the set of the two players, the seller s and the buyer b,
- $A_s = A_b = [0, \infty)$ are the sets of bids (announced reservation price and announced willingness to pay),
- $T_s = T_b = [0, 1]$ are the type sets (true reservation price and true willingness to pay),
- τ is the the probability distribution on T defined as in the first-price auction, and

- the payoff functions are defined by

$$u_s(a, t_s) = \begin{cases} 0, & a_s > a_b, \\ \frac{a_b+a_s}{2} - t_s, & a_s \leq a_b, \end{cases} \text{ and}$$

$$u_b(a, t_b) = \begin{cases} 0, & a_s > a_b, \\ t_b - \frac{a_b+a_s}{2}, & a_s \leq a_b. \end{cases}$$

  Thus, the utility is zero if the announced reservation price is higher than the announced willingness to pay. Otherwise, the seller obtains the rent "price minus true reservation price" and the buyer the rent "true willingness to pay minus price".

6.2. Equilibrium in linear strategies. As in the analysis of the first-price auction, we look out for linear strategies r and w for the seller and the buyer, respectively, given by

$$\begin{aligned} r &: T_s \to A_s, t_s \mapsto c_s + d_s t_s \ (d_s > 0) \text{ and} \\ w &: T_b \to A_b, t_b \mapsto c_b + d_b t_b \ (d_b > 0). \end{aligned}$$

Given the buyers strategy w, the seller's announced reservation price a_s leads to trade if $a_s < w(t_b)$ holds, i.e., if the reservation price is below the announced willingness to pay of buyer b with true willingness to pay t_b. Since $a_s < w(t_b)$ is equivalent to $a_s < c_b + d_b t_b$ or $t_b > \frac{a_s - c_b}{d_b}$, the seller's announced reservation price determines the probability for trade,

$$\begin{aligned} & \tau(\{t_b \in [0,1] : a_s < w(t_b)\}) \\ = \ & \tau(\{t_b \in [0,1] : a_s < c_b + d_b t_b\}) \\ = \ & \tau\left(\left\{t_b \in [0,1] : t_b > \frac{a_s - c_b}{d_b}\right\}\right) \\ = \ & 1 - \frac{a_s - c_b}{d_b}. \end{aligned}$$

Note

$$\begin{aligned} & 0 \leq 1 - \frac{a_s - c_b}{d_b} \leq 1 \\ \Leftrightarrow \ & c_b \leq a_s \leq c_b + d_b \end{aligned}$$

so that $\tau(\{t_b \in [0,1] : a_s < w(t_b)\})$ is the probability for $c_b \leq a_s \leq c_b + d_b$.

While the average t_b type is $\frac{1+0}{2} = \frac{1}{2}$, the expected t_b type conditional on $t_b > \frac{a_s - c_b}{d_b}$ is larger:

$$E\left[t_b : t_b > \frac{a_s - c_b}{d_b}\right] = \frac{1 + \frac{a_s - c_b}{d_b}}{2} = \frac{1}{2}\frac{a_s - c_b}{d_b} + \frac{1}{2} \geq \frac{1}{2}.$$

We use these equations to derive the seller's best bid (announced reservation price):

$$\begin{aligned}
r^*(t_s) &\in \arg\max_{a_s \in A_s} \left((a_s - t_s) \underbrace{\tau(\{t_b \in [0,1] : a_s = w(t_b)\})}_{\substack{\text{probability that the seller's bid}\\ \text{equals the buyer's bid}}} \right. \\
&\qquad \left. + \left(\frac{E[w(t_b) : a_s < w(t_b)] + a_s}{2} - t_s \right) \underbrace{\tau(\{t_b \in [0,1] : a_s < w(t_b)\})}_{\substack{\text{probability that the seller's bid}\\ \text{is smaller than the buyer's bid}}} \right) \\
&= \arg\max_{a_s \in A_s} \left(\frac{E[w(t_b) : a_s < w(t_b)] + a_s}{2} - t_s \right) \tau(\{t_b \in [0,1] : a_s < w(t_b)\}) \\
&= \arg\max_{a_s \in A_s} \left(\frac{c_b + d_b E\left[t_b : t_b > \frac{a_s - c_b}{d_b}\right] + a_s}{2} - t_s \right) \left(1 - \frac{a_s - c_b}{d_b}\right) \\
&= \arg\max_{a_s \in A_s} \left(\frac{c_b + d_b \frac{1 + \frac{a_s - c_b}{d_b}}{2} + a_s}{2} - t_s \right) \left(1 - \frac{a_s - c_b}{d_b}\right) \\
&= \frac{1}{3}(c_b + d_b) + \frac{2}{3} t_s
\end{aligned}$$

Thus, if we assume that the buyer's best bid is a linear strategy, the seller's best bid turns out to be a linear strategy, too. In a similar fashion, we find

$$\begin{aligned}
&\tau(\{t_s \in [0,1] : r(t_s) < a_b\}) \\
= \; &\tau(\{t_s \in [0,1] : c_s + d_s t_s < a_b\}) \\
= \; &\tau\left(\left\{t_s \in [0,1] : t_s < \frac{a_b - c_s}{d_s}\right\}\right) \\
= \; &\frac{a_b - c_s}{d_s}
\end{aligned}$$

and the buyer's best bid (announced willingness to pay)

$$\begin{aligned}
w^*(t_b) &\in \arg\max_{a_b \in A_b} \left(t_b - \frac{a_b + E[r(t_s) : r(t_s) < a_b]}{2} \right) \tau(\{t_s \in [0,1] : r(t_s) < a_b\}) \\
&= \arg\max_{a_b \in A_b} \left(t_b - \frac{a_b + c_s + d_s E\left[t_s : t_s < \frac{a_b - c_s}{d_s}\right]}{2} \right) \frac{a_b - c_s}{d_s}
\end{aligned}$$

$$
\begin{aligned}
&= \arg\max_{a_b \in A_b} \left(t_b - \frac{a_b + c_s + d_s \frac{a_b - c_s}{2d_s}}{2} \right) \frac{a_b - c_s}{d_s} \\
&= \frac{1}{3}c_s + \frac{2}{3}t_b.
\end{aligned}
$$

Thus, the seller uses the linear strategy

$$
\begin{aligned}
r^*(t_s) &= \frac{1}{3}(c_b + d_b) + \frac{2}{3}t_s \\
&= c_s + d_s t_s
\end{aligned}
$$

while the buyer's linear strategy is

$$
\begin{aligned}
w^*(t_b) &= \frac{1}{3}c_s + \frac{2}{3}t_b \\
&= c_b + d_b t_b.
\end{aligned}
$$

Hence, we have $d_s = d_b = \frac{2}{3}$, $c_s = \frac{1}{3}(c_b + d_b)$, and $c_b = \frac{1}{3}c_s$ whence the linear equilibrium strategies

$$
\begin{aligned}
r^*(t_s) &= \frac{1}{4} + \frac{2}{3}t_s \text{ and} \\
w^*(t_b) &= \frac{1}{12} + \frac{2}{3}t_b
\end{aligned}
$$

result. Non-linear equilibrium strategies also exist.

EXERCISE XVII.9. *Do players ever announce their true reservation price or their true willingness to pay, respectively?*

The seller announces a reservation price above his true reservation price in case of $r^*(t_s) = \frac{1}{4} + \frac{2}{3}t_s > t_s$ which is equivalent to $t_s < \frac{3}{4}$. Sellers with reservation prices above $\frac{3}{4}$ understate their reservation price. Note that this does not contradict optimality. The probability of announcing the true reservation price is zero because this happens for the type $t_s = \frac{3}{4}$, only. The buyer announces his true willingness to pay in case of $t_b = \frac{1}{4}$, only.

6.3. Inefficient trade. It may well happen that the strategies derived in the previous section preclude trade although trade would be mutually beneficial. This unfortunate event is characterized by

$$
\begin{aligned}
t_s &< t_b \text{ (willingness to pay greater than reservation price) and} \\
r^*(t_s) &> w^*(t_b) \text{ (no trade in equilibrium).}
\end{aligned}
$$

Since $\frac{1}{4} + \frac{2}{3}t_s > \frac{1}{12} + \frac{2}{3}t_b$ is equivalent to $t_s + \frac{1}{4} > t_b$, Pareto inefficiency results in case of

$$
t_s < t_b < t_s + \frac{1}{4}
$$

and with probability

$$\begin{aligned}
\int_0^1 \left(\int_{t_s}^{\min\{t_s+\frac{1}{4},1\}} dt_b \right) dt_s &= \int_0^{\frac{3}{4}} \left(\int_{t_s}^{t_s+\frac{1}{4}} dt_b \right) dt_s + \int_{\frac{3}{4}}^1 \left(\int_{t_s}^1 dt_b \right) dt_s \\
&= \int_0^{\frac{3}{4}} \frac{1}{4} dt_s + \int_{\frac{3}{4}}^1 (1 - t_s)\, dt_s \\
&= \frac{1}{4}\left(\frac{3}{4} - 0\right) + \left(t_s - \frac{1}{2}t_s^2\right)\Big|_{\frac{3}{4}}^1 \\
&= \frac{3}{12} + \left(1 - \frac{1}{2}\right) - \left(\frac{3}{4} - \frac{1}{2}\frac{9}{16}\right) \\
&= \frac{9}{32}
\end{aligned}$$

Myerson & Satterthwaite (1983) show that this negative result is not only exemplary but hints at a very general result. In case of two-sided uncertainty we cannot always expect trade to occur even if trade would in principle be efficient. Thus, the link between bargaining theory and the Pareto principle is not as close as the title of the previous part of the book might suggest.

EXERCISE XVII.10. *Show that in our model, we cannot have "too much" trade, i.e., it is not possible that trade occurs while the true reservation price is above the true willingness to pay.*

7. Topics

The main topics in this chapter are

- Bayesian games
- cost uncertainty
- mixed strategies and purification
- correlated equilibrium
- first-price auction
- double auction

8. Solutions

Exercise XVII.1

$\tau(t_2 = \text{high})$ is the a priori probability for player 2 entertaining a high willingness to pay. It is equal to $\frac{1}{3} + \frac{1}{9} = \frac{4}{9}$. Player 1's belief for this event is

$$\tau_1(t_2 = \text{high}) = \tau(t_2 = \text{high}\,|t_1 = \text{high}) = \frac{\tau(\text{high, high})}{\tau(t_1 = \text{high})} = \frac{\frac{1}{3}}{\frac{2}{3}} = \frac{1}{2}.$$

Exercise XVII.2

The ex-ante probabilities for $c_2 = 20$, $c_1 = 15$, and $c_1 = 25$ are

$$\begin{aligned}\tau(20) &= \tau(15,20) + \tau(25,20) = \frac{1}{2} + \frac{1}{2} = 1, \\ \tau(15) &= \tau(15,20) = \frac{1}{2}, \text{ and} \\ \tau(25) &= \tau(25,20) = \frac{1}{2}.\end{aligned}$$

The ex-post probability for $c_2 = 20$, i.e., player 1's belief, is

$$\tau_1(20) = \frac{\tau(t_1,20)}{\tau(t_1)} = \frac{\frac{1}{2}}{\frac{1}{2}} = 1 = \tau(20).$$

We also find the equality of $\tau_2(15)$ and $\tau(15)$:

$$\tau_2(15) = \frac{\tau(15,20)}{\tau(20)} = \frac{\frac{1}{2}}{1} = \frac{1}{2} = \tau(15).$$

Exercise XVII.3

The game has three equilibria, two of which are in pure strategies: (theatre, theatre) and (football, football). The mixed-strategy equilibrium is $\left(\left(\frac{2}{3}, \frac{1}{3}\right), \left(\frac{1}{3}, \frac{2}{3}\right)\right)$; she chooses theatre with probability $\frac{2}{3}$ and he with probability $\frac{1}{3}$.

Exercise XVII.4

Peter's strategies are functions $s_P : [0, x] \to \{\text{theatre, football}\}$. Thus, his probability for choosing theatre (from Cathy's point of view) is

$$\tau^x \left(\{ t_P \in [0, x] : s_P (t_P) = \text{theatre} \} \right).$$

Exercise XVII.5

By

$$\underbrace{1 \cdot \theta_C + 0 \cdot (1 - \theta_C)}_{\substack{\text{Peter's expected payoff} \\ \text{for choosing theatre}}} \geq \underbrace{0 \cdot \theta_C + (2 + t_P) \cdot (1 - \theta_C)}_{\substack{\text{Peter's expected payoff} \\ \text{for choosing football}}}$$

we obtain Peter's strategy s_P given by

$$t_P \mapsto \begin{cases} \text{theatre,} & t_P \leq \frac{3\theta_C - 2}{1 - \theta_C}, \\ \text{football,} & t_P > \frac{3\theta_C - 2}{1 - \theta_C}. \end{cases}$$

Exercise XVII.6

We obtain $x - 3\bar{t}_P = \bar{t}_C \bar{t}_P = x - 3\bar{t}_C$ and hence the desired equality.

Exercise XVII.7

In case of non-strategic bidding, the first-price auction yields the price $\max(t_1, t_2)$ for given t_1 and t_2 and hence the expected payoff

$$\begin{aligned}
& \int_{t_1 \in [0,1]} \left(\int_{t_2 \in [0,1]} \max(t_1, t_2) \, dt_2 \right) dt_1 \\
= & \int_{t_1 \in [0,1]} \left(\int_0^{t_1} \max(t_1, t_2) \, dt_2 + \int_{t_1}^1 \max(t_1, t_2) \, dt_2 \right) dt_1 \\
& \qquad \text{(splitting the } t_2 \text{ integral)} \\
= & \int_{t_1 \in [0,1]} \left(t_1 \int_0^{t_1} dt_2 + \int_{t_1}^1 t_2 dt_2 \right) dt_1 \\
= & \int_{t_1 \in [0,1]} \left(t_1^2 + \frac{1}{2} t_2^2 \Big|_{t_1}^1 \right) dt_1 \\
= & \int_{t_1 \in [0,1]} \left(t_1^2 + \frac{1}{2} - \frac{1}{2} t_1^2 \right) dt_1 \\
= & \int_{t_1 \in [0,1]} \left(\frac{1}{2} t_1^2 + \frac{1}{2} \right) dt_1 \\
= & \left(\frac{1}{3} \frac{1}{2} t_1^3 + \frac{1}{2} t_1 \right) \Big|_0^1 \\
= & \frac{1}{6} + \frac{1}{2} - 0 = \frac{2}{3}
\end{aligned}$$

Exercise XVII.8

The seller's rent is

$$p - r = \frac{r+w}{2} - r = \frac{w-r}{2} \begin{cases} \geq 0, & r \leq w \\ < 0, & r > w \end{cases}$$

while the buyer's rent is

$$w - p = w - \frac{r+w}{2} = p - r.$$

Thus, the rents are equalized at a price in the middle of reservation price and willingness to pay.

Exercise XVII.9

Yes, but only for the special cases of $t_s = \frac{3}{4}$ and $t_b = \frac{1}{4}$, respectively.

Exercise XVII.10

Too much trade means

$$\begin{aligned} t_b &< t_s \text{ (willingness to pay smaller than reservation price) and} \\ r^*(t_s) &< w^*(t_b) \text{ (trade in equilibrium).} \end{aligned}$$

Since $\frac{1}{4} + \frac{2}{3}t_s = r^*(t_s) < w^*(t_b) = \frac{1}{12} + \frac{2}{3}t_b$ is equivalent to $t_s + \frac{1}{4} < t_b$, we obtain the contradiction

$$t_s + \frac{1}{4} < t_b < t_s.$$

9. Further exercises without solutions

PROBLEM XVII.1. There are two players 1 and 2. They play one of two games. In both games, player 1 chooses from his action set $A_1 = \{up, down\}$ and player 2 chooses from his action set $A_2 = \{left, right\}$. Assume $L > M > 1$. They play the game G_A with probability $p > \frac{1}{2}$ and the game G_B with probability $1 - p$.

G_A	left	right
up	M , M	1 , $-L$
down	$-L$, 1	0 , 0

G_B	left	right
up	0 , 0	1 , $-L$
down	$-L$, 1	M , M

(1) Assume that both players are informed about which game they play before they choose their actions, respectively. Formulate this game as a static Bayesian game!
(2) Assume that player 1 learns whether they play G_A or G_B, while player 2 does not. Formulate this game as a static Bayesian game and determine all of its equilibria!

CHAPTER XVIII

The revelation principle and mechanism design

This chapter is a follow-up on the previous one. There, we were concerned with static Bayesian games. In these games, players learn their own types and they choose actions in dependence of their types. The main focus in this chapter is on mechanism design. By mechanism design, we understand the problem of setting up a game so as to benefit the principal. Normally, the principal is not as well informed as the agents who possess some private information. For example, in the previous chapter, we consider the question of which auction, first-price or second-price, yields the highest expected payoff. In that example, the auctioneer is the principal. In other examples, one tries to find a Pareto-efficient solution to some problem; then, a benevolent dictator is the principal. The problem of finding the best mechanism is made complicated by the fact that players do not, in general, "tell the truth". For example, the bidders in the first-price auction or in the double auction do not announce their true reservation price or their true willingness to pay. Here, the revelation principle allows a considerable simplification. According to this principle, we do not lose anything by restricting attention to mechanisms where players "tell the truth" or "reveal their own type". Thus, the revelation principle helps to find the best mechanism by restricting the set of candidate mechanisms.

1. Introduction

In 2007, the Sveriges Riksbank Prize in Economic Sciences in Memory of Alfred Nobel was awarded to the US economists Leonid Hurwicz (University of Minnesota), Eric S. Maskin (Institute for Advanced Study, Princeton), and Roger B. Myerson (University of Chicago) "for having laid the foundations of mechanism design theory." According to the press release by the Royal Swedish Academy of Sciences,

> Adam Smith's classical metaphor of the invisible hand refers to how the market, under ideal conditions, ensures an efficient allocation of scarce resources. But in practice conditions are usually not ideal; for example, competition is not

H. Wiese, *Advanced Microeconomics*,
https://doi.org/10.1007/978-3-658-34959-2_18

> completely free, consumers are not perfectly informed and privately desirable production and consumption may generate social costs and benefits. Furthermore, many transactions do not take place in open markets but within firms, in bargaining between individuals or interest groups and under a host of other institutional arrangements. How well do different such institutions, or allocation mechanisms, perform? What is the optimal mechanism to reach a certain goal, such as social welfare or private profit? Is government regulation called for, and if so, how is it best designed?
>
> These questions are difficult, particularly since information about individual preferences and available production technologies is usually dispersed among many actors who may use their private information to further their own interests. Mechanism design theory, initiated by Leonid Hurwicz and further developed by Eric Maskin and Roger Myerson, has greatly enhanced our understanding of the properties of optimal allocation mechanisms in such situations, accounting for individuals' incentives and private information. The theory allows us to distinguish situations in which markets work well from those in which they do not. It has helped economists identify efficient trading mechanisms, regulation schemes and voting procedures. Today, mechanism design theory plays a central role in many areas of economics and parts of political science.

A cute example of a mechanism can be seen at the house Homburgstr. 55 in 37619 Bodenwerder (Germany) where the author found the following sentence:

> Wünsch mir ein jeder, was er will – Gott gebe ihm dreimal so viel. Anno 1726

Thus, everybody will wish the house owner well, if only to obtain the threefold amount.

2. Revisiting the first-price auction

As an example, we revisit the first-price auction considered in chapter XVII (pp. 460). The first-price auction is a mechanism which allots the object to the highest bidder who has to pay according to his announced willingness to pay. Therefore, every bidder $i \in \{1, 2\}$ acts strategically because his bid does not only influence the chance of getting the object but also the price he has to pay in case of winning the auction. On learning his true willingness to pay t_1, player 1 finds it profitable to announce $t_1/2$ if he assumes that player 2 will also announce half his true willingness to pay.

Thus, we found the equilibrium strategy combination

$$s^* = (s_1^*, s_2^*)$$

defined by

$$s_1^* : [0,1] \to \mathbb{R}_+, \quad t_1 \mapsto s_1^*(t_1) = \frac{t_1}{2}$$

and

$$s_2^* : [0,1] \to \mathbb{R}_+, \quad t_2 \mapsto s_2^*(t_2) = \frac{t_2}{2}.$$

In that combination, every player uses the half-bid strategy.

Now consider another mechanism for the problem of allocating the object. In this new mechanism, the object also goes to the highest bidder but the payment is set at half the announced willingness to pay. Let us call this mechanism the half-price auction. This auction is constructed from the first-price auction and the half-bid strategies in the obvious manner:

- The half-bid strategies have every bidder halve his type, $a_i = \frac{t_i}{2}$. Under the first-price auction, the successful bidder pays $a_i = \frac{t_i}{2}$.
- The half-price auction tries to achieve the same outcome for players who tell the truth. Thus, the factor $\frac{1}{2}$ (stemming from the players' strategic behavior) is brought into the half-price auction by requiring that the successful bidder pays $\frac{a_i}{2}$, only.

Assume, now, that both players truthfully announce their respective willingness to pay. Expressed formally, player i's strategy is

$$\begin{aligned} id_i \ : \ & [0,1] \to [0,1] \\ & t_i \mapsto id_i(t_i) = t_i. \end{aligned}$$

Then, the first-price auction and the half-price auction lead to the same outcome:

- Both give the object to the bidder with the highest willingness to pay. This follows from

$$\underbrace{t_1 \leq t_2}_{\substack{\text{comparison of the player's} \\ \text{truthful announcements} \\ \text{under the half-price auction}}} \quad \Leftrightarrow \quad \underbrace{\frac{t_1}{2} \leq \frac{t_2}{2}}_{\substack{\text{comparison of the player's} \\ \text{strategic announcements} \\ \text{under the first-price auction}}}.$$

- The successful bidder i pays $t_i/2$. In the first-price auction, he understates his willingness to pay and pays this announced willingness to pay, $t_i/2$. In the half-price auction, he truthfully announces his willingness to pay, but the price set by the mechanism is half this value, again $t_i/2$.

We now want to show that (id_1, id_2) is a Bayesian equilibrium of the half-price auction. Of course, the very fact that the strategy combination id leads to the same outcome in the half-price auction as the equilibrium (!) strategy combination s^* in the first-price auction, is not sufficient proof. This is clear from price versus quantity competition. For example, if firms 1 and 2 choose the quantities that result from Bertrand competition, these quantities do, in general, not constitute a Cournot Nash equilibrium.

We now proceed to the proof. The main input is the fact that s^* (which requires each bidder to halve his willingness to pay) is an equilibrium of the first-price auction. If $s_1 = id_1$ were not the best response to $s_2 = id_2$, we would have a type t_1 and an action (announced type) $a_1 \neq t_1$ such that player 1's payoff under the untruthful bid $a_1 \neq t_1$ is higher than his payoff under the truthful bid:

$$\underbrace{\left(t_1 - \frac{a_1}{2}\right) \underbrace{\tau\left(\{t_2 \in [0,1] : a_1 > t_2\}\right)}_{\substack{\text{probability that player 1's untruthful bid} \\ \text{is higher than player 2's truthful bid}}}}_{\substack{\text{half-price auction:} \\ \text{expected payoff for type } t_1 \\ \text{using action } a_1 \neq t_1}}$$

$$> \underbrace{\left(t_1 - \frac{t_1}{2}\right) \underbrace{\tau\left(\{t_2 \in [0,1] : t_1 > t_2\}\right)}_{\substack{\text{probability that player 1's truthful bid} \\ \text{is higher than player 2's truthful bid}}}}_{\substack{\text{half-price auction:} \\ \text{expected payoff for type } t_1 \\ \text{using truthful announcement } a_1 = t_1}}$$

We now rewrite this inequality by just dividing the inequalities in the τ terms by 2. We then obtain an inequality for the first-price rather than the half-price auction:

$$\underbrace{\left(t_1 - \frac{a_1}{2}\right) \quad \underbrace{\tau\left(\left\{t_2 \in [0,1] : \frac{a_1}{2} > \frac{t_2}{2}\right\}\right)}_{\substack{\text{probability that player 1's bid } a_1/2 \neq t_1/2 \\ \text{is higher than player 2's half-bid}}}}_{\substack{\text{first-price auction:} \\ \text{expected payoff for type } t_1 \\ \text{using action } \frac{a_1}{2} \neq \frac{t_1}{2}}}$$

$$> \underbrace{\left(t_1 - \frac{t_1}{2}\right) \quad \underbrace{\tau\left(\left\{t_2 \in [0,1] : \frac{t_1}{2} > \frac{t_2}{2}\right\}\right)}_{\substack{\text{probability that player 1's half-bid} \\ \text{is higher than player 2's half-bid}}}}_{\substack{\text{first-price auction:} \\ \text{expected payoff for type } t_1 \\ \text{using action } \frac{a_1}{2} = \frac{t_1}{2}}}$$

This inequality is a contradiction to s_1^* being a best response to s_2^*. Therefore, our initial assumption ($s_1 = id_1$ is not the best response to $s_2 = id_2$) is false.

3. Social choice problems and mechanisms

The set-up for mechanism design is somewhat complicated. Before going into the formal definitions, we would like the reader to know where we are heading. According to figure XVIII.1,

- (first,) the principal designs a mechanism which defines a Bayesian game for the players who
- (second,) learn their types and
- (third,) choose a message according to an equilibrium of that Bayesian game whereupon
- (fourth,) the principal effects an outcome that
- (fifth,) leads to certain payoffs for the players.

This is the overview, now the nitty-gritty. We first introduce the concept of a social choice problem. This is a set of feasible outcomes Z for a player set N together with type and utility information. In the example of an auction, Z may contain the elements

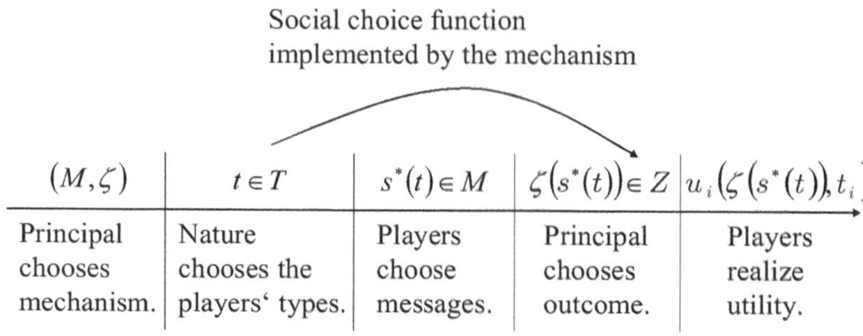

FIGURE XVIII.1. Mechanism design and the social choice function

- the object goes to bidder 1 and bidder 2 pays 3 Euros to bidder 3 and 2 Euros to the principal or
- the object goes to bidder 1 who pays 3 Euros to the principal, or
- the object stays with the principal.

These outcomes cannot be judged without knowing the bidders' types. If bidder 2 has the highest willingness to pay among the bidders and if this willingness to pay is higher than the seller's (the principal's) reservation price, neither of the three outcomes is Pareto efficient. Thus, the description of a social choice problem also includes types and utility functions. We let player i's payoff depend on an outcome z from Z and on his type t_i from T_i. Thus, we have this definition:

DEFINITION XVIII.1 (SOCIAL CHOICE PROBLEM). *A tuple* (N, T, τ, Z, u) *is called a social choice problem where*

- N, T, *and* τ *are defined as in the previous chapter,*
- Z *is a (feasible) outcome set, and*
- $u = (u_i)_{i \in N}$ *is the tuple of utility functions* $u_i : Z \times T_i \to \mathbb{R}$, $i \in N$.

Knowing the social choice problem, one can specify a "best" outcome for each type combination $t \in T$. For example, "best" may refer to Pareto efficiency.

DEFINITION XVIII.2 (SOCIAL CHOICE FUNCTION). *Let* (N, T, τ, Z, u) *be a social choice problem. A function*

$$f : T \to Z$$

is called a solution to this problem or a social choice function. f *is called ex-post efficient if it is impossible to improve upon* f, *i.e., if no* $(z, t) \in Z \times T$

exists such that

$$\begin{aligned} u_i(z, t_i) &\geq u_i(f(t), t_i) \text{ for all } i \in N \text{ and} \\ u_i(z, t_i) &> u_i(f(t), t_i) \text{ for some } i \in N. \end{aligned}$$

We envision a principal who is not identical to one of the players from N. This principal may be the auctioneer who tries to maximize his expected payoff or a benevolent dictator on the look-out for a Pareto-efficient social choice function. The principal is confronted with the social-choice problem and tries to design a best mechanism. He asks the n players for a message (which may, or may not, be related to the players' types) and executes an outcome $z \in Z$ as a function ζ of these messages. ζ is the Greek letter for z.

DEFINITION XVIII.3 (MECHANISM). *Let* (N, T, τ, Z, u) *be a social choice problem. A mechanism for* (N, T, τ, Z, u) *is a tuple* (M, ζ) *where*

- $M = (M_i)_{i \in N}$ *is the tuple of message spaces and*
- ζ *is the mechanism function*

$$\zeta : M \to Z.$$

Note the similarity between the social choice function f and the mechanism function ζ. Both pick an outcome. The social choice function's argument is a type combination (the true characteristics of the players) and the argument feeding into a mechanism function is a message combination. Thus, the outcome chosen by the mechanism function does not directly depend on the players' types which are unknown to the principal.

The principal's mechanism leads to a game for the players:

DEFINITION XVIII.4 (MESSAGE GAME). *Let* (M, ζ) *be a mechanism for* (N, T, τ, Z, u). *The static Bayesian game*

$$\Gamma = \left(N, M, T, \tau, (u_i)_{i \in N}\right)$$

is called the message game induced by (M, ζ) *(and* (N, T, τ, Z, u)*). In that game, the message spaces* M_i *take over the role of the action spaces* A_i *known from chapter XVII and* $u_i : M \times T_i \to \mathbb{R}$ *is defined by* $u_i(m, t_i) := u_i(\zeta(m), t_i)$.

We (and the principal!) are interested in finding a mechanism that yields a desired social choice function. The principal tries to achieve certain outcomes for certain type combinations. Since he does not know the players' types, he asks them for a message. He then takes these messages and effects an outcome according to the mechanism function. If he is lucky, the players with type combination t choose messages $s(t)$ such that the outcome

$\zeta(s(t))$, determined by the mechanism function, equals the outcome $f(t)$ that the social choice function f would pick. In fact, we prefer to depend on equilibrium strategies rather than on luck.

DEFINITION XVIII.5 (IMPLEMENTATION OF A SOCIAL-CHOICE FUNCTION). *Let* (M, ζ) *be a mechanism for* (N, T, τ, Z, u) *and* $\Gamma = \left(N, M, T, \tau, (u_i)_{i \in N}\right)$ *the corresponding message game. We say that* (M, ζ) *implements the social choice function* f

- *through a dominant-strategy equilibrium if a dominant-strategy equilibrium* s^* *exists such that* $f(t) = \zeta(s^*(t))$ *for all* $t \in T$ *or*
- *through a Bayesian equilibrium if a Bayesian equilibrium* s^* *exists such that* $f(t) = \zeta(s^*(t))$ *for all* $t \in T$.

In a sense, the mechanism defined by the principal is the first step of a two-stage game. Note, however, that we did not specify a strategy set for the principal. Such a set would contain all the message tuples from which the principal could choose plus all the functions from all these message tuples to the outcome set Z. Things get worse for the agents. In this two-stage game, player i's strategy is a very complicated object. For every message set M_i plus mechanism function chosen by the principal, player i's strategy specifies the function $T_i \to M_i$. Thus, we better stay with the informal description of the two-stage setup.

4. The revelation principle

As we have seen in the previous section, mechanisms can use any message sets. Thanks to the revelation principle, we can restrict attention to message sets $M_i = T_i$. Furthermore, we can even restrict attention to cases where players "tell the truth".

DEFINITION XVIII.6 (DIRECT MECHANISM). *A mechanism* (M, ζ) *is called direct if* $M = T$ *holds. A social choice function* f *is truthfully implementable by a direct mechanism* (T, ζ) *if* $s^* = id$ *is a dominant-strategy equilibrium or a Bayesian equilibrium and* $f(t) = \zeta(id(t)) = \zeta(t)$ *holds for all* $t \in T$.

From the previous chapter, we know the recommendation game (p. 457). In that game, we also have $A_i = T_i$ for all the players. However, the recommendation game is a very different game from the message game induced by a direct mechanism:

- A recommendation game builds on a strategic game with action sets A_i, $i \in N$. By $T_i := A_i$, the types are the recommended actions.

- In a message game, a principal asks for information about the players' types so that the action sets equal the type sets, $(M_i =) A_i := T_i$.

We now consider any mechanism (M, ζ) and a strategy combination s^* for the corresponding message game. s^* can (but need not) be an equilibrium. With the help of s^* we can transform (M, ζ) into a direct mechanism.

DEFINITION XVIII.7 (DERIVED DIRECT MECHANISM). *Let (M, ζ) be a mechanism for (N, T, τ, Z, u) and let s^* be a strategy combination of the corresponding message game $\Gamma = \left(N, M, T, \tau, (u_i)_{i\in N}\right)$. Then, $\left(T, \zeta^{s^*}\right)$ defined by*

$$\begin{aligned} \zeta^{s^*} &= \zeta \circ s^* : \\ T &\to Z, \\ t &\mapsto \zeta^{s^*}(t) = \zeta\left(s^*(t)\right) \end{aligned}$$

is the direct mechanism derived from (M, ζ) by s^*.

In the introductory section, we present a suitable example. There, the mechanism

$$(T, \text{ first-price auction})$$

is used to derive the mechanism

$$(T, \text{ half-price auction})$$

by the half-bid strategy combination s^*. Figure XVIII.2 depicts ζ^{s^*} as the composition $\zeta \circ s^*$ of functions ζ and s^* where

- first, s^* is applied to a type combination $t \in T$ to yield the message combination $s^*(t) \in M$ and
- second, ζ is applied to that message combination so that we obtain an outcome $\zeta\left(s^*(t)\right) \in Z$.

We now turn to the celebrated revelation principle. We give a proof for implementation via dominant strategies (theorem XVIII.1) and via Bayesian equilibria (theorem XVIII.2).

THEOREM XVIII.1 (REVELATION PRINCIPLE, DOMINANT-STRATEGY EQUILIBRIUM). *Let (M, ζ) be a mechanism for (N, T, τ, Z, u) and s^* be a dominant-strategy equilibrium of $\Gamma = \left(N, M, T, \tau, (u_i)_{i\in N}\right)$, the corresponding message game. Let ζ^{s^*}, defined by $\zeta^{s^*}(t) := \zeta\left(s^*(t)\right)$, be the social choice function implemented by (M, ζ) and s^*. Then, the direct mechanism $\left(T, \zeta^{s^*}\right)$ (for $\left(N, T, T, \tau, (u_i)_{i\in N}\right)$) derived from (M, ζ) by s^* has the dominant-strategy*

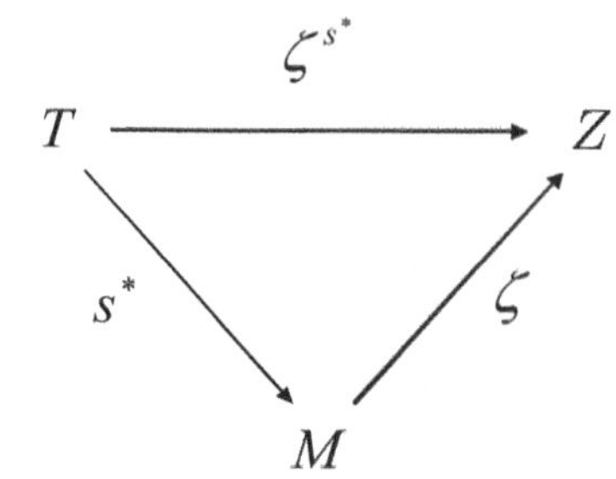

$$\zeta^{s^*} = \zeta \circ s^* : T \to Z$$
$$t \mapsto \zeta^{s^*}(t) = \zeta(s^*(t))$$

FIGURE XVIII.2. Deriving a direct mechanism

equilibrium $id = (id_1, ..., id_n)$ *and* ζ^{s^*} *is implemented by* $\left(T, \zeta^{s^*}\right)$ *as well as by* (M, ζ).

PROOF. We need to show that $id_i : T_i \to T_i$ is a best strategy for player $i \in N$ whatever strategies the other players choose. Assume player $i \in N$ and his true type $t_i \in T_i$. Let m_i be any announced type from T_i and m_{-i} be any announced type combination from T_{-i}. Under the direct mechanism $\left(T, \zeta^{s^*}\right)$, we have

$$\begin{aligned}
& u_i\left(\zeta^{s^*}\left(\underbrace{m_i}_{\in T_i}, \underbrace{m_{-i}}_{\in T_{-i}}\right), t_i\right) \\
= \; & u_i\left(\zeta\left(s^*\left(m_i, m_{-i}\right)\right), t_i\right) \text{ (definition of } \zeta^{s^*}) \\
= \; & u_i\left(\zeta\left(\underbrace{s_i^*\left(m_i\right)}_{\in M_i}, \underbrace{s_{-i}^*\left(m_{-i}\right)}_{\in M_{-i}}\right), t_i\right) \\
\leq \; & u_i\left(\zeta\left(s_i^*\left(t_i\right), s_{-i}^*\left(m_{-i}\right)\right), t_i\right) \; (s_i^* \text{ is a dominant strategy in } \Gamma) \\
= \; & u_i\left(\zeta\left(s^*\left(t_i, m_{-i}\right)\right), t_i\right) \\
= \; & u_i\left(\zeta^{s^*}\left(t_i, m_{-i}\right), t_i\right) \text{ (definition of } \zeta^{s^*}) \\
= \; & u_i\left(\zeta^{s^*}\left(\underbrace{id_i\left(t_i\right)}_{\in T_i}, \underbrace{m_{-i}}_{\in T_{-i}}\right), t_i\right)
\end{aligned}$$

for all $t_{-i} \in T_{-i}$ and hence

$$\begin{aligned} & \sum_{t_{-i}\in T_{-i}} \tau_i(t_{-i})\, u_i\left(\zeta^{s^*}\left(\underbrace{m_i}_{\in T_i}, \underbrace{m_{-i}}_{\in T_{-i}}\right), t_i\right) \\ \leq & \sum_{t_{-i}\in T_{-i}} \tau_i(t_{-i})\, u_i\left(\zeta^{s^*}\left(\underbrace{id_i(t_i)}_{\in T_i}, \underbrace{m_{-i}}_{\in T_{-i}}\right), t_i\right). \end{aligned}$$

□

THEOREM XVIII.2 (REVELATION PRINCIPLE, BAYESIAN EQUILIBRIUM). *Let (M,ζ) be a mechanism for (N,T,τ,Z,u) and let s^* be a Bayesian equilibrium of the corresponding message game $\Gamma = \left(N,M,T,\tau,(u_i)_{i\in N}\right)$. Let ζ^{s^*}, defined by $\zeta^{s^*} := \zeta(s^*(t))$, be the social choice function implemented by (M,ζ) and s^*. Then, the direct mechanism $\left(T,\zeta^{s^*}\right)$ (for $\left(N,T,T,\tau,(u_i)_{i\in N}\right)$) derived from (M,ζ) by s^* has the equilibrium strategy combination $id = (id_1,...,id_n)$ and ζ^{s^*} is implemented by $\left(T,\zeta^{s^*}\right)$ as well as by (M,ζ).*

PROOF. We need to show that $id_i : T_i \to T_i$ is a best reply to $id_{-i} \in S_{-i}$ in $\left(N,T,T,\tau,(u_i)_{i\in N}\right)$ for every player $i \in N$. We use the ex-post Bayesian equilibrium. Assume player $i \in N$ and his true type $t_i \in T_i$. Let m_i be any announced type from T_i. Under the mechanism $\left(T,\zeta^{s^*}\right)$, we have

$$\begin{aligned} & u_i\left(\zeta^{s^*}\left(\underbrace{m_i}_{\in T_i}, \underbrace{(id_j(t_j))_{j\in N\backslash\{i\}}}_{\in T_{-i}}\right), t_i\right) \\ = & u_i\left(\zeta^{s^*}(m_i,t_{-i}), t_i\right) \text{ (definition of } id_j : T_j \to T_j, j \neq i) \\ = & u_i\left(\zeta\left(s^*(m_i,t_{-i})\right), t_i\right) \text{ (definition of } \zeta^{s^*}) \\ = & u_i\left(\zeta\left(\underbrace{s_i^*(m_i)}_{\in M_i}, \underbrace{s_{-i}^*(t_{-i})}_{\in M_{-i}}\right), t_i\right) \\ \leq & u_i\left(\zeta\left(s_i^*(t_i), s_{-i}^*(t_{-i})\right), t_i\right) \text{ (} s^* \text{ is an equilibrium in } \Gamma) \\ = & u_i\left(\zeta\left(s^*(t_i,t_{-i})\right), t_i\right) \end{aligned}$$

$$= u_i\left(\zeta^{s^*}(t_i, t_{-i}), t_i\right) \text{ (definition of } \zeta^{s^*}\text{)}$$

$$= u_i\left(\zeta^{s^*}\left(\underbrace{id_i(t_i)}_{\in T_i}, \underbrace{(id_j(t_j))_{j\in N\backslash\{i\}}}_{\in T_{-i}}\right), t_i\right)$$

for all $t_{-i} \in T_{-i}$ and hence

$$\sum_{t_{-i}\in T_{-i}} \tau_i(t_{-i})\, u_i\left(\zeta^{s^*}\left(m_i, (id_j(t_j))_{j\in N\backslash\{i\}}\right), t_i\right)$$

$$\leq \sum_{t_{-i}\in T_{-i}} \tau_i(t_{-i})\, u_i\left(\zeta^{s^*}\left(id_i(t_i), (id_j(t_j))_{j\in N\backslash\{i\}}\right), t_i\right).$$

□

Thus, in terms of implementable social choice functions, it is sufficient to look for direct mechanisms and for truthful equilibria $s^* = id$ of the corresponding message game.

Let us try to explain these results in a more informal fashion. Imagine a mechanism (M, ζ) and an equilibrium (Bayesian equilibrium or dominant-strategy equilibriuim) s^* for the corresponding Bayesian game. Now, we consider the derived mechanism $\left(T, \zeta^{s^*}\right) = (T, \zeta \circ s^*)$. In a sense, it relieves the players from the burden to act strategically. In fact, if a player $i \in N$ truthfully reveals his type t_i, the derived mechanism transforms this type into the message $s_i^*(t_i) \in M_i$, i.e., the very message the player intends to send under the original mechanism. If, now, the other players truthfully reveal their types (use the strategy combination id_{-i}), the derived mechanism generates the message combination $s_{-i}^*(t_{-i}) \in M_{-i}$. For the original mechanism (M, ζ), player i's best reply to this message combination is $s_i^*(t_i) \in M_i$. As noted above, this message is generated by $\left(T, \zeta^{s^*}\right)$ if i tells the truth, too.

5. The Clarke-Groves mechanism

5.1. Public-goods problems and functions. Imagine the provision of a public good, i.e., a good with non-rivalry in consumption. We assume players $i \in N$ with types $t_i \in \mathbb{R}$ which denote the willingness to pay for the public good in question. (We allow for $t_i < 0$.) The cost of this good is denoted by C and we assume that every player contributes $\frac{C}{n}$ to the cost. The public-good problem is concerned with whether the public good in question should be provided $(b = 1)$ or not $(b = 0)$. The outcome comprises b and also a tax $z_i \in \mathbb{R}$ to be paid by every consumer. This tax is called the Clarke-Groves tax and it is used to induce player i to reveal his true willingness to pay, $m_i = t_i$.

DEFINITION XVIII.8 (PUBLIC-GOOD PROBLEM). *The social choice problem* (N, T, τ, Z, u) *is called a public-good problem where*

- N *is the set of consumers of the public good,*
- T *is defined by* $T_i = \mathbb{R}$ *for all* $i \in N$,
- τ *is a probability distribution on* T,
- $Z = \{0, 1\} \times \mathbb{R}^n$ *is the outcome set, and*
- $u = (u_i)_{i \in N}$ *is the tuple of utility functions* $u_i : Z \times T_i \to \mathbb{R}$, $i \in N$, *given by* $(b, z_1, ..., z_n) \mapsto \begin{cases} t_i - \frac{C}{n} - z_i, & b = 1 \\ -z_i, & b = 0 \end{cases}$

The public good should be provided if and only if the aggregate willingness to pay is not smaller than the cost of providing the good. Thus, efficiency has nothing to do with who contributes how much to the cost C.

DEFINITION XVIII.9. *Let* (N, T, τ, Z, u) *be a public-good problem.*

$$f : T \to Z$$

is called a public-good (social choice) function if

$$\sum_{i \in N} t_i \geq C \Leftrightarrow b = 1$$

holds.

5.2. The definition of the Clarke-Groves mechanism. We now present the Clarke-Groves mechanism. It is a direct mechanism that induces truth-telling. The general idea is this: A player whose willingness to pay changes the decision of the other players creates an externality for which he has to pay a tax. A player changes the decision of the others if

- the other players would have opted for the public good (which is a translation for $\sum_{j \in N \setminus \{i\}} m_j \geq \frac{n-1}{n} C$) but the good is not provided because of player i ($\sum_{j \in N} m_j < C$) or
- the other players are against the public good ($\sum_{j \in N \setminus \{i\}} m_j < \frac{n-1}{n} C$) but the public good is provided due to player i ($\sum_{j \in N} m_j \geq C$).

In both cases, player i is to "blame" for his announced willingness to pay. He has to pay for the negative externality that his announcement creates. Note, however, that the above inequalities use $\frac{n-1}{n} C$ if one player is absent. Of course, without player i, the other players still have to pay C.

DEFINITION XVIII.10 (CLARKE-GROVES MECHANISM). *Let T and Z be defined as above. The direct mechanism (T,ζ) with tuple of message sets $M := T$ is called the Clarke-Groves mechanism if $\zeta : M \to Z$ is defined by*

$$\zeta(m) = (b(m), \zeta_1(m), ..., \zeta_n(m))$$

with

$$b(m) = \begin{cases} 1, & \sum_{i\in N} m_i \geq C \\ 0, & \sum_{i\in N} m_i < C \end{cases}$$

and, for every player $i \in N$,

$$\zeta_i(m) = \begin{cases} \frac{n-1}{n}C - \sum_{j\in N\backslash\{i\}} m_j, & \sum_{j\in N} m_j \geq C \text{ and } \sum_{j\in N\backslash\{i\}} m_j < \frac{n-1}{n}C \\ \sum_{j\in N\backslash\{i\}} m_j - \frac{n-1}{n}C, & \sum_{j\in N} m_j < C \text{ and } \sum_{j\in N\backslash\{i\}} m_j \geq \frac{n-1}{n}C \\ 0, & \text{otherwise} \end{cases}$$

Note that the additional payment as defined by ζ_i is always non-negative. If the other players prefer not to have the public good (and pay for it), the damage they suffer is $\frac{n-1}{n}C - \sum_{j\in N\backslash\{i\}} m_j$, their share of the cost burden minus their (announced) willingness to pay.

5.3. Lemma and proof. The Clarke-Groves mechanism leads to a static Bayesian game for the consumers of the public good. By definition, it is clear that a public-good function is realized. It can be shown that every player $i \in N$ can do no better than choose $m_i = t_i$ in this Bayesian game, irrespective of the messages m_{-i} sent by the other players.

LEMMA XVIII.1. *Let (T,ζ) be the Clarke-Groves mechanism. $s^* = id$ is a dominant-strategy equilibrium in the message game induced by that mechanism.*

PROOF. Assume the message tuple $m_{-i} \in T_{-i}$. We need to consider four cases:

(1) $\sum_{j\in N\backslash\{i\}} m_j + t_i \geq C$ and $\sum_{j\in N\backslash\{i\}} m_j \geq \frac{n-1}{n}C$ (a truthful player i does not change the other players' decision in favor of the public good)
(2) $\sum_{j\in N\backslash\{i\}} m_j + t_i \geq C$ and $\sum_{j\in N\backslash\{i\}} m_j < \frac{n-1}{n}C$ (a truthful player i is responsible for the provision)
(3) $\sum_{j\in N\backslash\{i\}} m_j + t_i < C$ and $\sum_{j\in N\backslash\{i\}} m_j \geq \frac{n-1}{n}C$ (the public good is not provided and a truthful player i is to blame)

(4) $\sum_{j\in N\backslash\{i\}} m_j + t_i < C$ and $\sum_{j\in N\backslash\{i\}} m_j < \frac{n-1}{n}C$ (a truthful player i does not change the other players' decision against the public good)

Let us begin with the first case and consider player 1. The truthful announcement of his type leads to the provision of the public good and $z_1 = 0$. His utility is

$$u_1(1,0,...) = t_1 - \frac{C}{n}.$$

Player 1 has no incentive to overstate his willingness to pay (that will not change anything). Indeed, a utility change occurs only if player 1 understates his willingness to pay so that the public good is not provided. Then, player 1 has to pay $\sum_{j\in N\backslash\{1\}} m_j - \frac{n-1}{n}C$ (third case). Player 1 is harmed by this understatement:

$$\begin{aligned} u_1(1,0,...) &= t_1 - \frac{C}{n} \\ &\geq \left(C - \sum_{j\in N\backslash\{1\}} m_j\right) - \frac{C}{n} \text{ (case 1)} \\ &= -\left(\sum_{j\in N\backslash\{1\}} m_j - \frac{n-1}{n}C\right) \\ &= u_1\left(0, \sum_{j\in N\backslash\{1\}} m_j - \frac{n-1}{n}C, ...\right) \end{aligned}$$

We now turn to the second case where player 1's enthusiasm leads to the provision of the public good and externality payments $\frac{n-1}{n}C - \sum_{j\in N\backslash\{i\}} m_j$. Overstating the willingness to pay brings no change. Understatement may abolish the externality payment but will reduce his utility:

$$\begin{aligned} & u_1\left(1, \frac{n-1}{n}C - \sum_{j\in N\backslash\{1\}} m_j, ...\right) \\ = \; & t_1 - \frac{C}{n} - \left(\frac{n-1}{n}C - \sum_{j\in N\backslash\{1\}} m_j\right) \\ \geq \; & \left(C - \sum_{j\in N\backslash\{1\}} m_j\right) - \frac{C}{n} - \left(\frac{n-1}{n}C - \sum_{j\in N\backslash\{1\}} m_j\right) \text{ (case 2)} \end{aligned}$$

$$= 0$$
$$= u_1(0, 0, ...)$$

The cases 3 and 4 can be treated similarly. □

5.4. Discussion. The Clarke-Groves mechanism suffers from the problem that it cannot guarantee a balanced budget. In the way we presented the lemma, there will be a budget surplus. If this surplus is divided among the players or used for causes that the players take an interest in, the lemma does not hold any more.

The reader will note a similarity to the second-price auction. The announced willingness to pay determines

- who obtains the object in the second-price auction and
- whether the public good will be provided according to the Clarke-Groves mechanism

but has no effect

- on how much the successful bidder pays for the object (second-price auction) and
- on the size of the externality payment (Clarke-Groves mechanism).

6. Topics and literature

The main topics in this chapter are

- mechanism design
- revelation principle
- Clarke-Groves mechanism
- truth telling
- social choice problem
- social choice function

We recommend the textbook by Baron (1989) who looks at regulatory mechanisms, i.e., he provides an application of mechanism design theory to industrial organization.

7. Further exercises without solutions

PROBLEM XVIII.1. Consider the third case in the proof of lemma XVIII.1 where the inequalities $\sum_{j\in N\backslash\{1\}} m_j + t_1 < C$ and $\sum_{j\in N\backslash\{1\}} m_j \geq \frac{n-1}{n}C$ hold (the public good is not provided and a truthful player 1 is to blame). Show that player 1 cannot do any better than tell the truth.

Part F

Perfect competition and competition policy

This part has three chapters. The first one presents general equilibrium theory as a highly stylized model of Walras' ideas. We then present some applications of GET and shed critical light on this model. In particular, we introduce the jungle economy, a Pareto efficient but definitely unpleasant alternative to GET. Furthermore, we report on the views of Austrian and other economists as they critically relate to markets and to GET. The last chapter in this part introduces competition theory and policy.

CHAPTER XIX

General equilibrium theory

1. Introduction to General Equilibrium Theory

1.1. Introductory remarks. Allocation of goods takes place in two different modes, the first of which being person-to-person. For example, voluntary exchange in a peaceful economy is person-to-person, making all agents better off (see pp. 385). Person-to-person does not necessarily imply face-to-face. Indeed, Abu Abdullah Muhammad Ibn Battuta, a fourteenth century prolific traveler and geographer, describes the silent trade that took place along the Volga (see Bowles 2004, p. 233):

> Each traveler ... leaves the goods he has brought ... and they retire to their camping ground. Next day they go back to ... their goods and find opposite them skins of sable, miniver, and ermine. If the merchant is satisfied with the exchange he takes them, but if not he leaves them. The inhabitants then add more skins, but sometimes they take away their goods and leave the merchant's. This is their method of commerce. Those who go there do not know whom they are trading with ...

The second mode is impersonal trading (not just anonymous, silent trading), expounded by General Equilibrium Theory (GET). Here, agents observe prices and choose their bundles accordingly. GET envisions a market system with perfect competition. This means that all agents (households and firms) are price takers. The aim is to find prices such that

- all actors behave in a utility, or profit, maximizing way and
- the demand and supply schedules can be fulfilled simultaneously.

In that case, we have found a Walras equilibrium. Note that the price-finding process is not addressed in GET. Walras suggests that an auctioneer might try to inch towards an equilibrium price vector. This is the so-called tâtonnement. A careful introduction into General Equilibrium Theory is presented by Hildenbrandt & Kirman (1988).

H. Wiese, *Advanced Microeconomics*,
https://doi.org/10.1007/978-3-658-34959-2_19

Finding equilibrium prices for the whole economy is an ambitious undertaking. We need an elaborated mathematical mechanism and a list of restrictive assumptions:

- The goods are private and there are no external effects.
- The individuals interact via market transactions only.
- The individuals take prices as given.
- There are no transaction costs.
- The goods are homogeneous but there can be many goods.
- The preferences are monotonic and convex.

1.2. Nobel prizes. The Sveriges Riksbank Prize in Economic Sciences in Memory of Alfred Nobel was awarded for work on General Equilibrium Theory three times:

- in 1972, to the British economist John Hicks (Oxford University) and to the US economist Kenneth Arrow (Harvard University) "for their pioneering contributions to general economic equilibrium theory and welfare theory",
- in 1982, to the French-born US economist Gerard Debreu (University of California, Berkeley) "for having incorporated new analytical methods into economic theory and for his rigorous reformulation of the theory of general equilibrium", and
- in 1988, to the French economist Maurice Allais (Ecole Nationale Supérieure des Mines de Paris, France) "for his pioneering contributions to the theory of markets and efficient utilization of resources".

With respect to Debreu's prize, the press release by the Royal Swedish Academy of Sciences states:

> This year's prize is awarded for penetrating basic research work in one of the most central fields of economic science, the theory of general equilibrium.
>
> In a decentralized market system, individual consumers and firms make decisions on the purchase and sale of goods and services solely on the basis of self-interest. Adam Smith had already raised the question of how these decisions, apparently independent of one another, are coordinated, and result in a situation whereby sellers usually find outlets for their planned production, while consumers realize their planned consumption. Smith's answer was that, given price and wage flexibility, price systems automatically bring about the desired coordination of individual plans. Towards the end of

the 19th century, Léon Walras formulated this idea in mathematical terms as a system of equations to represent consumers' demand for goods and services, producers' supply of these same goods and services and their demand for factors of production, and equality between supply and demand, i.e., equilibrium in each market. But it was not until long afterward that this system of equations was scrutinized to ascertain whether it had an economically meaningful solution, i.e., whether this theoretical structure of vital importance for understanding the market system was logically consistent.

Gerard Debreu's major achievement is his work in proving the existence of equilibrium-creating prices. His first fundamental contribution came in the early 1950s in collaboration with Professor Kenneth Arrow. Arrow received the 1972 Prize in Economic Sciences in Memory of Alfred Nobel for his work in this and other adjacent fields.

Arrow and Debreu designed a mathematical model of a market economy where different producers planned their output of goods and services and thus also their demand for factors of production in such a way that their profit was maximized. Thus, connections were generated within the model between the supply of goods and demand for factors of production on the one hand, and all prices, on the other. By making additional assumptions about consumer behaviour, Arrow and Debreu were able to generate demand functions or "correspondences", i.e., relations between prices and supplied and demanded quantities. In this model, Arrow and Debreu managed to prove the existence of equilibrium prices, i.e., they confirmed the internal logical consistency of Smith's and Walras' model of the market economy.

Subsequent to these pioneering efforts, there has been considerable development and extensions of such proofs with Gerard Debreu at the forefront. His book, *Theory of Value*, from the late 1950s has already become a classic both for its universality and for its elegant analytical approach. The theory developed in this study lends itself to many far-reaching interpretations and applications. The concept of "goods", for instance, is defined so broadly that the theory may be used in pure static equilibrium analysis, the analysis of the spatial distribution of production and consumption activities, intertemporal analysis and the analysis of uncertainty. Thus, within the same model, Debreu's general equilibrium theory integrates the theory of location, the theory of capital, and the theory of economic behaviour under uncertainty.

We will explain in some detail the simple case of an exchange economy, the positive theory (existence and uniqueness of an equilibrium) in section 2 and the normative theory (the welfare theorems) in section 4. The more general case of an exchange and production economy is briefly mentioned in section 3. The next chapter presents alternative views, criticism, and applications of General Equilibrium Theory.

Following Hildenbrandt & Kirman (1988), it is helpful to differentiate between

- the implications of Pareto efficiency on the one hand (this is the Edgeworthian theme of cooperation) and
- the implications of individual utility and profit maximization for markets (the Walrasian theme of decentralization).

2. Exchange economy: positive theory

2.1. Exchange Edgeworth box: prices and equilibria. Before delving into the General Equilibrium Theory, we will give you a short preview of where we are heading. General Equilibrium Theory has two grand themes. The first is Pareto improvements through exchanges (see pp. 385). The second topic is decentralization through prices.

It is possible to add price information into Edgeworth boxes. If household A buys a bundle $\left(x_1^A, x_2^A\right)$ with the same worth as his endowment, we have

$$p_1 x_1^A + p_2 x_2^A = p_1 \omega_1^A + p_2 \omega_2^A.$$

Starting from an endowment point, positive prices p_1 and p_2 lead to negatively sloped budget lines for both individuals. In figure XIX.1, two price lines with prices $p_1^l < p_1^h$ are depicted. The indifference curves indicate which bundles the households prefer.

EXERCISE XIX.1. *Why do the two price lines in figure XIX.1 cross at the endowment point ω?*

Of course, we would like to know whether these prices are compatible in the sense of allowing both agents to demand the preferred bundle. If that is the case, the prices and the bundles at these prices constitute a Walras equilibrium.

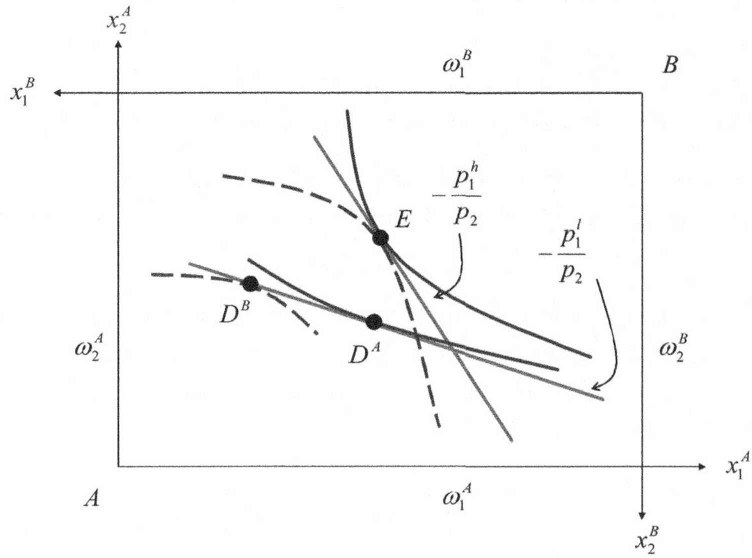

FIGURE XIX.1. Walras equilibrium

EXERCISE XIX.2. *The low price* p_1^l *is not possible in a Walras equilibrium, because there is excess demand for good* 1 *at this price:*

$$x_1^A + x_1^B > \omega_1^A + \omega_1^B.$$

Do you see that? How about good 2*?*

2.2. Definition of an exchange economy. We now proceed to the formal definition of an exchange economy.

DEFINITION XIX.1 (EXCHANGE ECONOMY). *An exchange economy is a tuple*

$$\mathcal{E} = \left(N, G, \left(\omega^i\right)_{i\in N}, \left(\precsim^i\right)_{i\in N}\right)$$

consisting of

- *the set of agents* $N = \{1, 2, ..., n\}$,
- *the finite set of goods* $G = \{1, ..., \ell\}$,

and for every agent $i \in N$

- *an endowment* $\omega^i = \left(\omega_1^i, ..., \omega_\ell^i\right) \in \mathbb{R}_+^\ell$, *and*
- *a preference relation* $\precsim^i$.

Thus, every agent has property rights on endowments. The total endowment of an exchange economy is given by $\omega = \sum_{i\in N} \omega^i$. A household's consumption possibilities are described by the budget. We refer the reader to chapter VI.

Definition XIX.2 (feasible allocation). *Consider an exchange economy $\mathcal{E}$.*

- *A bundle $(y^i)_{i\in N} \in \mathbb{R}_+^{\ell\cdot n}$ is an allocation.*
- *An allocation $(y^i)_{i\in N}$ is called feasible if $\sum_{i\in N} y^i \leq \sum_{i\in N} \omega^i$ holds.*

2.3. Excess Demand and Market Clearance. In this section we deal with the question whether the demand for one good is greater than the supply for this good.

Definition XIX.3 (excess demand). *Assume an exchange economy $\mathcal{E}$, a good $g \in G$ and a price vector $p \in \mathbb{R}^\ell$. If every household $i \in N$ has a unique household optimum $x^i(p, \omega^i)$, good g's excess demand is denoted by $z_g(p)$ and defined by*

$$z_g(p) := \sum_{i=1}^n x_g^i(p, \omega^i) - \sum_{i=1}^n \omega_g^i.$$

The corresponding excess demand for all goods $g = 1, ..., \ell$ is the vector

$$z(p) := (z_g(p))_{g=1,...,\ell}\,.$$

The excess demand is a quantity of goods and a vector of quantities of goods respectively. In contrast, the value of the excess demand, which is given by

$$p \cdot z(p),$$

is a scalar amount of money. We remind the reader of Walras' law (p. 141) which immediately implies the following version:

Lemma XIX.1. *Every consumer demands a bundle of goods obeying $p \cdot x^i \leq p \cdot \omega^i$ where local nonsatiation implies equality. For all consumers together, we have*

$$p \cdot z(p) = \sum_{i=1}^n p \cdot (x^i - \omega^i) \leq 0$$

and, assuming local-nonsatiation, $p \cdot z(p) = 0$.

Walras' law is of great importance for General Equilibrium Theory. We will later look at the conditions under which excess demand is zero. Then, the problem is to get from $z(p) \cdot p = 0 \in \mathbb{R}$ to $z(p) = 0 \in \mathbb{R}^\ell$.

Definition XIX.4 (market clearance). *A market g is called cleared if excess demand $z_g(p)$ on that market is equal to zero.*

The following two exercises are adapted from Leach (2004, pp. 54) .

EXERCISE XIX.3. *Consider a market where the excess demand of three individuals* 1, 2, *and* 3 *is given by*

$$z_1(p) = \frac{8}{p} - 4, z_2(p) = \frac{4}{p} - 2, z_3(p) = \frac{12}{p} - 2.$$

Find the market-clearing price. Is individual 3 *a buyer or a seller?*

EXERCISE XIX.4. *Abba (A) and Bertha (B) consider buying two goods* 1 *and* 2, *and face the price* p *for good* 1 *in terms of good* 2. *Think of good* 2 *as the numéraire good with price* 1. *Abba's and Bertha's utility functions* u_A *and* u_B, *respectively, are given by* $u_A(x_1^A, x_2^A) = \sqrt{x_1^A + x_2^A}$ *and* $u_B(x_1^B, x_2^B) = \sqrt{x_1^B + x_2^B}$. *Endowments are* $\omega^A = (18, 0)$ *and* $\omega^B = (0, 10)$. *Find the bundles demanded by these two agents. Then find the price* p *that fulfills* $\omega_1^A + \omega_1^B = x_1^A + x_1^B$ *and* $\omega_2^A + \omega_2^B = x_2^A + x_2^B$.

In the above exercise, what if only market 1 is cleared? The following lemma shows that local nonsatiation excludes this possibility.

LEMMA XIX.2. *In case of local nonsatiation,*

(1) *if all markets but one are cleared, the last one also clears or its price is zero,*
(2) *if at prices* $p \gg 0$ *all markets but one are cleared, all markets clear.*

PROOF. If $\ell - 1$ markets are cleared, the excess demand on these markets is 0. Without loss of generality, markets $g = 1, ..., \ell - 1$ are cleared. Applying Walras's law we get

$$\begin{aligned} 0 &= p \cdot z(p) \\ &= p_\ell z_\ell(p), \end{aligned}$$

and hence both claims. □

2.4. Walras equilibrium.

2.4.1. *Definition.* Are there prices for all ℓ goods for which all individual demands are possible at the same time? Differently put, is there a price vector $\widehat{p}$, such that the demand for all ℓ goods does not exceed the initial endowment?

DEFINITION XIX.5 (WALRAS EQUILIBRIUM). *A price vector* $\widehat{p}$ *and the corresponding demand system* $\left(\widehat{x}^i\right)_{i=1,...,n} = (x^i(\widehat{p}, \omega^i))_{i=1,...,n}$ *is called a Walras equilibrium if*

$$\sum_{i=1}^n \widehat{x}^i \leq \sum_{i=1}^n \omega^i$$

or

$$z(\widehat{p}) \leq 0$$

holds.

The equilibrium condition requires that

(1) all households choose an optimal bundle, i.e., every household i chooses the bundle of goods $x^i(\widehat{p}, \omega^i)$ (or a bundle from $x^i(\widehat{p}, \omega^i)$) at given prices $\widehat{p}$,
(2) the resulting allocation is feasible, or, differently put, for every good, the quantity demanded is not larger than the available quantity.

EXERCISE XIX.5. *Rewrite the equilibrium condition*

$$\sum_{i=1}^{n} x^i(\widehat{p}, \omega^i) \leq \sum_{i=1}^{n} \omega^i$$

so that it is clear that the inequalities must hold for each good.

The equilibrium condition excludes the possibility that the demand for one good is greater than the supply for this good. The reader might find this definition of the equilibrium confusing at first glance. Why do we not define equilibrium through the equality of supply and demand? Our requirement in the definition is weaker, but we will show that under certain conditions a nonpositive excess demand implies an excess demand of zero.

2.4.2. *Market clearing in the Walras equilibrium.* We now present the conditions for which a market in equilibrium has an excess demand of zero, i.e. the market is cleared. Consider the following definitions and lemmata.

DEFINITION XIX.6 (FREE GOOD). *A good is called free if its price is equal to zero.*

LEMMA XIX.3. *Assume local nonsatiation and weak monotonicity for all households. If* $\left[\widehat{p}, \left(\widehat{x}^i\right)_{i=1,\dots,n}\right]$ *is a Walras equilibrium and the excess demand for a good is negative, this good must be free.*

PROOF. By lemma VI.4 (p. 141), prices cannot be negative in a household optimum under the preferences stated in the current lemma. Therefore, we can assume non-negative prices for all goods. Assume a good g with $p_g > 0$ and a negative excess demand for that good. We obtain a contradiction to Walras law for local nonsatiation:

$$\begin{aligned} p\cdot z(p) &= \underbrace{p_g z_g(p)}_{<0} + \sum_{\substack{g'=1,\\ g'\neq g}}^{\ell} p_{g'} z_{g'}(p) \quad (z_g(p)<0) \\ &< \sum_{\substack{g'=1,\\ g'\neq g}}^{\ell} \underbrace{p_{g'}}_{\geq 0} \underbrace{z_{g'}(p)}_{\leq 0} \text{ (lemma VI.4, p. 141)} \\ &\leq 0. \end{aligned}$$

□

Finally (for now), we need to define the desiredness of goods:

DEFINITION XIX.7 (DESIRED GOOD). *A good is desired if the excess demand at price zero is positive.*

LEMMA XIX.4. *If all goods are desired and if local nonsatiation and weak monotonicity hold and if $\widehat{p}$ is a Walras equilibrium, then $z(\widehat{p}) = 0$.*

PROOF. Suppose that there is a good g with $z_g(\widehat{p}) < 0$. Then g must be a free good according to lemma XIX.3 and have a positive excess demand by the definition of desiredness, $z_g(\widehat{p}) > 0$. □

2.4.3. *Example: The Cobb-Douglas exchange economy with two agents.* We remember from chapter VI that income m and Cobb-Douglas utility function

$$u(x_1, x_2) = x_1^a x_2^{1-a}$$

implies the household optimum

$$\begin{aligned} x_1 &= a\frac{m}{p_1}, \\ x_2 &= (1-a)\frac{m}{p_2}. \end{aligned}$$

Consider, now, individual 1 with Cobb-Douglas utility function u^1 and parameters a_1 (for good 1) and $1-a_1$ (for good 2). The initial endowment of individual 1 equals $\omega^1 = (1,0)$. Individual 2 possesses a Cobb-Douglas utility function u^2 with parameters a_2 (for good 1) and $1-a_2$ (for good 2). His initial endowment is $\omega^2 = (0,1)$. Parameters a_1 and a_2 obey the following conditions: $0 < a_1 < 1$ and $0 < a_2 < 1$. Both goods are desired and the preferences obey local nonsatiation and weak monotonicity. According to lemma XIX.4, the market is in equilibrium only if it is cleared. Substituting the value of the endowment for income, we get individual 1's demand for good 1

$$\begin{aligned} & x_1^1 \left(p_1, p_2, \omega^1 \cdot p\right) \\ & = a_1 \frac{\omega^1 \cdot p}{p_1} \\ & = a_1 \end{aligned}$$

and individual 2's demand for good 1

$$\begin{aligned} & x_1^2 \left(p_1, p_2, \omega^2 \cdot p\right) \\ & = a_2 \frac{\omega^2 \cdot p}{p_1} \\ & = a_2 \frac{p_2}{p_1}. \end{aligned}$$

Assuming positive prices, lemma XIX.2 (p. 499) says that both markets are cleared if one is cleared. Market 1 is cleared if demand equals supply, i.e., if

$$a_1 + a_2 \frac{p_2}{p_1} = 1,$$

which is equivalent to

$$\frac{p_2}{p_1} = \frac{1 - a_1}{a_2}.$$

All prices, which satisfy these equations, are equilibrium prices. Obviously, only relative prices are determined.

Figure XIX.2 sketches the equilibrium in the two-goods case.

2.5. Existence of the Walras equilibrium.

2.5.1. *Proposition.* So far we have not questioned the existence of the Walras equilibrium. Fortunately, the following theorem holds:

THEOREM XIX.1 (EXISTENCE OF THE WALRAS EQUILIBRIUM). *If the following conditions hold:*

- *aggregate excess demand is a continuous function (in prices),*
- *the value of the excess demand is zero for all price vectors p, and*
- *the preferences are strictly monotonic,*

there exists a price vector $\widehat{p}$ such that $z(\widehat{p}) \leq 0$.

The proof of this theorem uses Brouwer's fixed-point theorem. Therefore, we first introduce this theorem and then present the proof of the proposition.

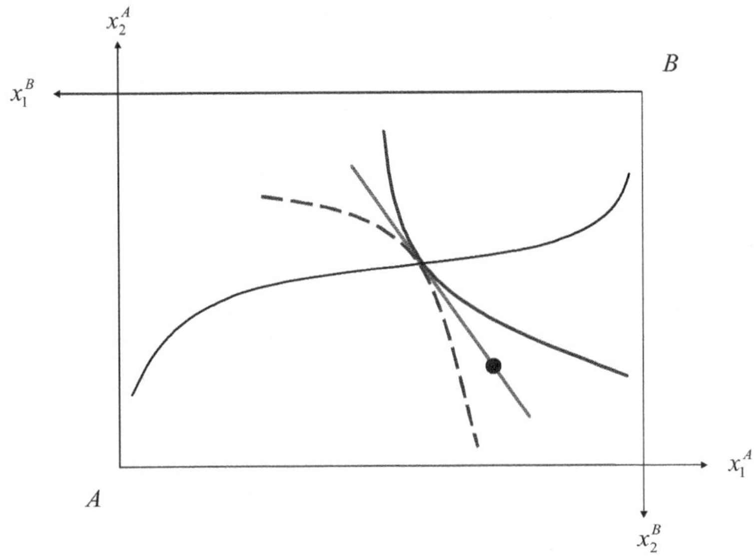

FIGURE XIX.2. The General Equilibrium in the exchange Edgeworth box

2.5.2. *Brouwer's fixed-point theorem.*

THEOREM XIX.2. *Suppose $f : M \to M$ is a function on the nonempty, compact, and convex set $M \subseteq \mathbb{R}^\ell$. If f is continuous, there exists at least one $x \in M$ such that $f(x) = x$. x is called a fixed point.*

Note that the range of f is M. Consider a continuous function on the closed unit interval $[0,1]$. If either $f(0) = 0$ or $f(1) = 1$ holds, a fixed point is found. Otherwise, figure XIX.3 shows a continuous function fulfilling $f(0) > 0$ and $f(1) < 1$. The graph of f necessarily cuts the $45°$-line. The projection of this intersection point onto the x- or the y-axis is the sought-after fixed point.

Take a second example. Put a handkerchief on the square $[0,1] \times [0,1]$ from $\mathbb{R}^2$. This subset is nonempty, compact, and convex. A continuous function

$$f : [0,1] \times [0,1] \to [0,1] \times [0,1]$$

corresponds to the following process:

- rumple the handkerchief,
- put the rumpled handkerchief again on the square and
- press it flat.

The handkerchief must not be torn because tearing corresponds to a non-continuous f. Brouwer's fixed-point theorem now claims that there is at

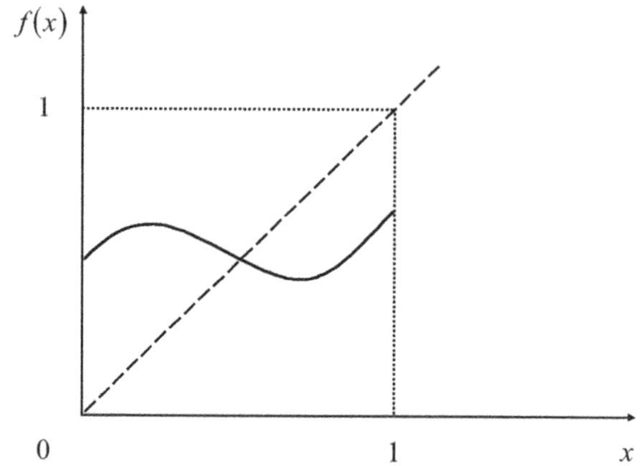

FIGURE XIX.3. Brouwer fixed-point theorem

least one spot on the handkerchief which, before and after rumpling, lies on the same place of the square. Alternatively, one can imagine stirring cake dough with a wooden spoon so that the dough does not lose its coherence. At least one participle of the dough does not change its place despite the stirring movements. But, of course, no Gugelhupf or kougelhopf.

We do not prove the theorem but ask you to try the following exercise.

EXERCISE XIX.6. *Assume that one of the requirements for the fixed-point theorem does not hold. Then, a fixed point need not exist. Specifically, assume that*
a) M is not compact,
b) M is not convex, or
c) f is not continuous.

German-speaking people may learn Brouwer's fixed-point theorem by memorizing the poem due to Hans-Jürgen Podszuweit (found in Homo Oeconomicus, XIV (1997), p. 537):

Das Nilpferd hört perplex:
Sein Bauch, der sei konvex.
Und steht es vor uns nackt,
sieht man: Er ist kompakt.
Nimmt man 'ne stetige Funktion
von Bauch
in Bauch

– Sie ahnen schon –,
dann nämlich folgt aus dem
Brouwer'schen Theorem:
Ein Fixpunkt muß da sein.
Dasselbe gilt beim Schwein
q.e.d.

2.5.3. *Proof of the existence theorem XIX.1.* In order to apply Brouwer's fixed-point theorem to theorem XIX.1, we first construct a convex and compact set. The prices of the ℓ goods are normed such that the sum of the nonnegative (!, we have strict monotonicity) prices equals 1. Just divide all prices by the sum of the prices. We can restrict our search for equilibrium prices to the $\ell - 1$- dimensional unit simplex:

$$S^{\ell-1} = \left\{ p \in \mathbb{R}_+^\ell : \sum_{g=1}^{\ell} p_g = 1 \right\}.$$

$S^{\ell-1}$ is nonempty, compact (closed and bounded as a subset of $\mathbb{R}^\ell$), and convex.

EXERCISE XIX.7. *Draw* $S^1 = S^{2-1}$.

The idea of the proof is as follows: First, we define a continuous function $f : S^{\ell-1} \to S^{\ell-1}$. Brouwer's theorem says that there is at least one fixed point of this function. Second, we show that such a fixed point fulfills the conditions of the Walras equilibrium.

The continuous function is defined by

$$f = \begin{pmatrix} f_1 \\ f_2 \\ \cdot \\ \cdot \\ \cdot \\ f_\ell \end{pmatrix} : S^{\ell-1} \to S^{\ell-1}$$

and

$$f_g(p) = \frac{p_g + \max(0, z_g(p))}{1 + \sum_{g'=1}^{\ell} \max(0, z_{g'}(p))}, g = 1, ..., \ell.$$

f is continuous because every f_g, $g = 1, ..., \ell$, is continuous. The latter is continuous because z (according to our assumption) and max are continuous functions. Finally, we can confirm that f is well defined, i.e., that $f(p)$ lies in $S^{\ell-1}$ for all p from $S^{\ell-1}$:

$$\begin{aligned}\sum_{g=1}^{\ell} f_g(p) &= \sum_{g=1}^{\ell} \frac{p_g + \max(0, z_g(p))}{1 + \sum_{g'=1}^{\ell} \max(0, z_{g'}(p))} \\ &= \frac{1}{1 + \sum_{g'=1}^{\ell} \max(0, z_{g'}(p))} \sum_{g=1}^{\ell} (p_g + \max(0, z_g(p))) \\ &= \frac{1}{1 + \sum_{g'=1}^{\ell} \max(0, z_{g'}(p))} \left(1 + \sum_{g=1}^{\ell} \max(0, z_g(p))\right) \\ &= 1.\end{aligned}$$

The function f increases the price of a good g in case of $f_g(p) > p_g$, only, i.e., if

$$\frac{p_g + \max(0, z_g(p))}{1 + \sum_{g'=1}^{\ell} \max(0, z_{g'}(p))} > p_g$$

or, equivalently,

$$\max(0, z_g(p)) > p_g \sum_{g'=1}^{\ell} \max(0, z_{g'}(p))$$

or, equivalently,

$$\frac{\max(0, z_g(p))}{\sum_{g'=1}^{\ell} \max(0, z_{g'}(p))} > \frac{p_g}{\sum_{g'=1}^{\ell} p_{g'}}$$

holds.

The last formula has a nice interpretation: when the relative excess demand for a good is greater than the relative price for the same good, the function f increases the price. Behind f, we imagine the workings of the Walras auctioneer, who changes prices upon observing excess demands. This so-called tâtonnement may (or may not) converge towards an equilibrium price vector.

We now complete the proof: according to Brouwer's fixed-point theorem there exists at least one $\widehat{p}$ such that

$$\widehat{p} = f(\widehat{p}),$$

from which we have

$$\widehat{p}_g = \frac{\widehat{p}_g + \max(0, z_g(\widehat{p}))}{1 + \sum_{g'=1}^{\ell} \max(0, z_{g'}(\widehat{p}))}$$

and finally

$$\widehat{p}_g \sum_{g'=1}^{\ell} \max\left(0, z_{g'}\left(\widehat{p}\right)\right) = \max\left(0, z_g\left(\widehat{p}\right)\right)$$

for all $g = 1, ..., \ell$.

Next we multiply both sides for all goods $g = 1, ..., \ell$ by $z_g(\widehat{p})$:

$$z_g(\widehat{p})\widehat{p}_g \sum_{g'=1}^{\ell} \max\left(0, z_{g'}\left(\widehat{p}\right)\right) = z_g(\widehat{p}) \max\left(0, z_g\left(\widehat{p}\right)\right).$$

Summing up over all g yields

$$\sum_{g=1}^{\ell} z_g(\widehat{p})\widehat{p}_g \sum_{g'=1}^{\ell} \max\left(0, z_{g'}\left(\widehat{p}\right)\right) = \sum_{g=1}^{\ell} z_g(\widehat{p}) \max\left(0, z_g\left(\widehat{p}\right)\right).$$

Since the value of the excess demand is assumed to be zero, the left-hand term is equal to zero. The right-hand term consists of a sum of expressions, which either equal zero (in case of $z_g\left(\widehat{p}\right) \leq 0$) or $\left(z_g\left(\widehat{p}\right)\right)^2$ (in case of $z_g\left(\widehat{p}\right) > 0$). Therefore, $z_g\left(\widehat{p}\right) \leq 0$ for all $g = 1, ..., \ell$. This is what we wanted to show.

2.6. Existence of the Nash equilibrium. In chapter X we note the Nash theorem:

THEOREM XIX.3 (EXISTENCE OF NASH EQUILIBRIA). *Any finite strategic game $\Gamma = (N, S, u)$ (i.e., $|N| < \infty$ and $|S| < \infty$) has a Nash equilibrium.*

The proof follows Nash's (1951) second proof which rests upon Brouwer's fixed-point theorem and is somewhat similar to the proof of the Walras equilibrium. That is the reason why we present it here rather than in chapter X.

PROOF. We construct a continuous function $f : \Sigma \to \Sigma$ whose fixed point is the sought-after Nash equilibrium. Note that the set of mixed strategy combinations Σ is compact, convex, and non-empty. Now, fix a player $i \in N$ and a strategy $s_i \in S_i$. The function $\phi_{s_i} : \Sigma \to \mathbb{R}$ is defined by

$$\phi_{s_i}(\sigma) := \max(0, u_i(s_i, \sigma_{-i}) - u_i(\sigma)) \geq 0, \sigma \in \Sigma. \tag{XIX.1}$$

$\phi_{s_i}(\sigma)$ is strictly positive iff the pure strategy s_i yields a higher payoff than the mixed strategy σ_i, given that the other players choose σ_{-i}. Remember that the payoff $u_i(\sigma)$ is the mean of the payoffs for pure strategies:

$$u_i\left(\sigma_i, \sigma_{-i}\right) = \sum_{j=1}^{|S_i|} \sigma_i\left(s_i^j\right) u_i\left(s_i^j, \sigma_{-i}\right)$$

Consider the strategies s_i^j with $\sigma_i\left(s_i^j\right) > 0$. At least one of these pure strategies yields a payoff $u_i\left(s_i^j, \sigma_{-i}\right)$ that is equal to or smaller than $u_i(\sigma_i, \sigma_{-i})$. For this strategy $\phi_{s_i}(\sigma)$ is zero:

LEMMA XIX.5. *For every $i \in N$ and every $\sigma \in \Sigma$, there is some $\hat{s}_i \in S_i$ such that $\sigma_i(\hat{s}_i) > 0$ and $\phi_{\hat{s}_i}(\sigma) = 0$.*

We now consider the function

$$\begin{aligned} f_i &: \Sigma \to \Sigma_i \\ &\sigma \mapsto f_i(\sigma) \text{ defined by} \\ &f_i(\sigma)(s_i) := \frac{\sigma_i(s_i) + \phi_{s_i}(\sigma)}{1 + \sum_{s_i' \in S_i} \phi_{s_i'}(\sigma)} \end{aligned}$$

Obviously, we have $f_i(\sigma)(s_i) \geq 0$. The reader will note the parallel construction of f_g in the previous subsection above. Confirm $\sum_{j \in S_i} f_i(\sigma)\left(s_i^j\right) = 1$ for yourself. Therefore, the functions f_i, $i \in N$, are well-defined.

By

$$\begin{aligned} & f_i(\sigma)(s_i) > \sigma_i(s_i) \\ \Leftrightarrow\ & \frac{\sigma_i(s_i) + \phi_{s_i}(\sigma)}{1 + \sum_{s_i' \in S_i} \phi_{s_i'}(\sigma)} > \sigma_i(s_i) \\ \Leftrightarrow\ & \phi_{s_i}(\sigma) > \sigma_i(s_i) \sum_{s_i' \in S_i} \phi_{s_i'}(\sigma) \\ \Leftrightarrow\ & \frac{\phi_{s_i}(\sigma)}{\sum_{s_i' \in S_i} \phi_{s_i'}(\sigma)} > \frac{\sigma_i(s_i)}{\sum_{s_i' \in S_i} \sigma_i(s_i')} \end{aligned}$$

the function f_i makes player i increase the probability attached to pure strategy s_i if the relative utility surplus of s_i is above the relative probability. The utility functions u_i are continuous and so are the functions ϕ_{s_i} and f_i. Therefore,

$$f = \begin{pmatrix} f_1 \\ f_2 \\ \cdot \\ \cdot \\ \cdot \\ f_\ell \end{pmatrix} : \Sigma \to \Sigma$$

is also continuous.

The fixed point σ^* which we know to exist due to Brouwer's theorem, obeys

$$f(\sigma^*) \stackrel{!}{=} \sigma^*,$$
$$f_i(\sigma^*) \stackrel{!}{=} \sigma_i^* \text{ for all } i \in N, \text{ and hence}$$
$$f_i(\sigma^*)(s_i) = \frac{\sigma_i^*(s_i) + \phi_{s_i}(\sigma^*)}{1 + \sum_{s_i' \in S_i} \phi_{s_i'}(\sigma^*)} \stackrel{!}{=} \sigma_i^*(s_i) \text{ for all } i \in N \text{ and all } s_i \in S_i$$

Now, fix a player $i \in N$. By lemma XIX.5, we can be sure of some $\hat{s}_i \in S_i$ such that $\sigma_i^*(\hat{s}_i) > 0$ and $\phi_{\hat{s}_i}(\sigma^*) = 0$. For such a strategy $\hat{s}_i$, we obtain

$$\frac{\sigma_i^*(\hat{s}_i)}{1 + \sum_{s_i' \in S_i} \phi_{s_i'}(\sigma^*)} = \frac{\sigma_i^*(\hat{s}_i) + \phi_{\hat{s}_i}(\sigma^*)}{1 + \sum_{s_i' \in S_i} \phi_{s_i'}(\sigma^*)} \stackrel{!}{=} \sigma_i^*(\hat{s}_i)$$

and hence $\sum_{s_i' \in S_i} \phi_{s_i'}(\sigma^*) = 0$. This implies $\phi_{s_i'}(\sigma^*) = \max(0, u_i(s_i', \sigma_{-i}^*) - u_i(\sigma^*)) = 0$ for each $s_i' \in S_i$. Thus, there is no pure strategy $s_i' \in S_i$ that yields a higher payoff for player i than his mixed strategy σ_i^*, given σ_{-i}^*. This means that σ_i^* is a best response against σ_{-i}^*. Applying the same reasoning to all the players from N establishes that σ^* is a Nash equilibrium.

3. Exchange and production economy: positive theory

GET can deal with production and exchange at the same time. The necessary production and profit theory has been covered in chapters VIII and IX. We do not delve seriously into this more complicated theory but just show that the necessary conditions can be supplied. We remind the reader that he has seen an ownership structure before in the simple case of just one firm, on p. 246.

DEFINITION XIX.8 (PRODUCTION AND EXCHANGE ECONOMY). *A tuple* $\mathcal{E} = \left(N, M, G, (\omega^i)_{i\in N}, (\precsim^i)_{i\in N}, (Z^j)_{j\in M}, (\theta_j^i)_{\substack{i\in N,\\ j\in M}}\right)$ *is a production and exchange economy which consists of*

- *the set of households* $N = \{1, 2, ..., n\}$,
- *the set of firms* $M = \{1, 2, ..., m\}$,
- *the set of goods* $G = \{1, ..., \ell\}$,
- *for every household* $i \in N$
  - *an endowment* $\omega^i \in \mathbb{R}_+^\ell$ *and*
  - *a preference relation* $\precsim^i$,
- *for every firm* $j \in M$ *a production set* $Z^j \subseteq \mathbb{R}^\ell$, *and*

- *the economy's ownership structure* $(\theta_j^i)_{\substack{i \in N, \\ j \in M}}$ *where* $\theta_j^i \geq 0$ *for all* $i \in N$, $j \in M$ *and* $\sum_{i=1}^n \theta_j^i = 1$ *for all* $j \in M$ *hold.*

We now turn to the feasibility of such an economy. Do the production and consumption plans match?

Definition XIX.9 (feasibility). *Let* $\mathcal{E}$ *be a production and exchange economy. The production plans* z^j, $j \in M$, *and the consumption plans* x^i, $i \in N$, *are called feasible if they fulfill*

- $z^j \in Z^j$ *for all* $j \in M$ *and*
- $\sum_{j \in M} z_g^j \geq \sum_{i \in N} (x_g^i - \omega_g^i)$ *for all* $g \in G$.

Finally, we are set to define the Walras equilibrium:

Definition XIX.10 (Walras equilibrium). *A price vector* $\widehat{p} \in \mathbb{R}^\ell$, *together with the corresponding production plans* $(\widehat{z}^j)_{j \in M}$ *and consumption plans* $(\widehat{x}^i)_{i \in N}$, *is called a Walras equilibrium of a production and exchange economy* $\mathcal{E}$ *if*

- *the production and consumption plans are feasible,*
- *for every household* $i \in N$, $\widehat{x}^i$ *is a best bundle for consumer* i *from the budget set*

$$B\left(\widehat{p}, \omega^i, (\theta_j^i)_{j \in M}\right) := \left\{ x^i \in \mathbb{R}_+^\ell : \widehat{p} \cdot x^i \leq \widehat{p} \cdot \omega^i + \sum_{j \in M} \theta_j^i \widehat{p} \cdot \widehat{z}^j \right\},$$

and

- *for every firm* $j \in M$, $\widehat{z}^j$ *is from*

$$\arg \max_{z^j \in Z^j} \widehat{p} \cdot z^j.$$

4. Normative theory

4.1. The first welfare theorem from the point of view of partial analysis.

4.1.1. *Marginal this and that.* In previous chapters, we have come to know some handy formulae for the marginal rate of substitution (p. 85), the marginal rate of technical substitution (p. 226), and the marginal rate of transformation (p. 391). Figure XIX.4 provides a summary.

4.1.2. *Three steps, twice.* A theoretical reason for the confidence of many economists in the efficiency of the market mechanism lies in the first theorem of welfare economics. This theorem states that a system of perfectly competitive markets is Pareto efficient.

slope of	holding constant	algebraic expression
indifference curve	utility $U(x_1, x_2)$	$MRS = \frac{\frac{\partial U}{\partial x_1}}{\frac{\partial U}{\partial x_2}}$
isoquant	output $f(x_1, x_2)$	$MRTS = \frac{\frac{\partial f}{\partial x_1}}{\frac{\partial f}{\partial x_2}}$
transformation curve	cost $C(x_1, x_2)$	$MRT = \frac{\frac{\partial C}{\partial x_1}}{\frac{\partial C}{\partial x_2}}$

FIGURE XIX.4. All the same

We attack this claim by way of partial analysis first and turn to total analysis later on. Our partial analysis (we concentrate on one or two markets leaving the repercussions on and from other markets aside) concerns

- exchange optimality (is it possible to make a consumer better off without making another one worse off?),
- production optimality (is it possible to produce more of one good without producing less of any other good?), and
- the optimal product mix (is it better to produce more of one good and less of another one?).

In the next three subsections, we consider each type of optimality separately. Within each type, we go through three steps:

(1) We first characterize Pareto optimality (concerning exchange, production, and product mix). Clearly, for this specification prices are unimportant.
(2) We then consider how households and firms adapt to given good and factor prices.
(3) We finally show that the adaption according to the second point (price-taking assumption) is done in a way compatible with the first (Pareto optimality).

4.1.3. *Exchange optimality.* We know the exchange Edgeworth box from chapter XIV (pp. 385). Assume two households A and B and two goods 1 and 2. Along the contract curve or exchange curve,

$$\left|\frac{dx_2^A}{dx_1^A}\right| = MRS^A \stackrel{!}{=} MRS^B = \left|\frac{dx_2^B}{dx_1^B}\right|$$

holds. This is the first step.

The second step consists of determining the decisions taken by the households at the given prices. Since at each of the two household optima, the marginal rate of substitution equals the price ratio, we find

$$MRS^A \overset{!}{=} \frac{p_1}{p_2} \overset{!}{=} MRS^B.$$

Thus, the Walras equilibrium implies exchange optimality.

4.1.4. *Production optimality.* In chapter VIII (p. 227), we have come to know the efficiency condition for the use of inputs. Assuming two goods 1 and 2 produced by factors of production C (capital) and L (labor), Pareto efficiency implies

$$\left|\frac{dC_1}{dL_1}\right| = MRTS_1 \overset{!}{=} MRTS_2 = \left|\frac{dC_2}{dL_2}\right|.$$

The two cost-minimizing firms adjust to factor prices by letting

$$MRTS_1 \overset{!}{=} \frac{w}{r} \overset{!}{=} MRTS_2$$

where w is the wage rate and r the interest rate. Thus, production optimality holds in the Walras equilibrium.

4.1.5. *Optimal product mix.* Every point on the production-possibility frontier defines the extent of an exchange Edgeworth box as you can see in figure XIX.5. Exchange optimality demands that the marginal rates of substitution inside the exchange Edgeworth box are the same for consumers A and B. Production optimality is reflected by the very fact that a point on (rather than below) the production-possibility frontier is chosen. The optimal product mix now refers to the question of whether consumers can be made better off if more of good 1 and less of good 2 is produced (along the transformation curve, of course).

Mixing products optimally requires the equality of the marginal rates of substitution for each individual with the marginal rate of transformation as we have shown on p. 392. Profit-maximizing firms obey the "marginal cost equals price" condition so that the equality implied by Pareto efficiency holds in a Walras equilibrium:

$$\begin{aligned} MRS &= \left|\frac{dx_2}{dx_1}\right|^{\text{indifference curve}} \underbrace{\overset{!}{=}}_{\substack{\text{household}\\ \text{optimum}}} \frac{p_1}{p_2} \underbrace{\overset{!}{=}}_{\substack{\text{profit}\\ \text{maximization}}} \frac{MC_1}{MC_2} \\ &= MRT = \left|\frac{dx_2}{dx_1}\right|^{\text{transformation curve}} \end{aligned}$$

Consider figure XIX.5. Point B lies on the transformation curve so that production optimality is fulfilled. This point defines an exchange Edgeworth box for two individuals, A and B. Within the Edgeworth box, the marginal

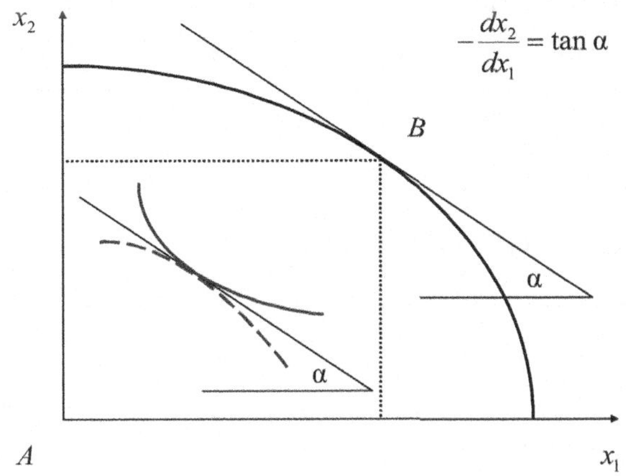

FIGURE XIX.5. Exchange Edgeworth box and production-possibility frontier

rates of substitutions are equal (exchange optimality). Finally, the equality of the marginal rate of transformation and the marginal rates of substitution ensures the optimal product mix.

Note that figure XIX.5 implies $\omega_1^A = \omega_2^A = 0$. How would you adapt this figure for $\omega_1^A > 0$ and $\omega_2^A > 0$? There are two possibilities.

The following exercise is adapted from Leach (2004, pp. 78).

EXERCISE XIX.8. *Consider Abba (A) and Bertha (B), whose utility functions are given by* $u_A(x_1^A, x_2^A) = \sqrt{x_1^A}\sqrt{x_2^A}$ *and* $u_B(x_1^B, x_2^B) = x_1^B + 2x_2^B$*, respectively. The production possibility frontier obeys* $x_2 = 2\sqrt{10 - x_1}$*. Calculate the marginal rates of substitution and the marginal rate of transformation. Find the Pareto-optimal allocation that leads to a utility level of* 8 *for Bertha.*

4.1.6. *Summary.* Figure XIX.6 summarizes the proof. The conditions for Pareto optimality are shown on the left-hand side of the diagram, the equilibrium equations for price takers on the right.

The first welfare theorem is a very remarkable result. Every household and every producer follows the incentives given by prices. Still, all behave in such a way,

(1) that it is impossible to improve the situation of one household without making another household worse off,
(2) that it is impossible to reallocate the given input factors such that we produce more of one good without producing less of another good, and

Pareto optimality requires	**in case of perfect competition**
$MRS^A \overset{!}{=} MRS^B$ optimal exchange	$MRS^A \overset{!}{=} \frac{p_1}{p_2} \overset{!}{=} MRS^B$
$MRTS_1 \overset{!}{=} MRTS_2$ optimal production	$MRTS_1 \overset{!}{=} \frac{w}{r} \overset{!}{=} MRTS_2$
$MRS \overset{!}{=} MRT$ optimal product mix	$MRS \overset{!}{=} \frac{p_1}{p_2} \overset{!}{=} \frac{MC_1}{MC_2} = MRT$

FIGURE XIX.6. Pareto optimality in three steps

(3) that producing more of one good and producing less of another good cannot improve the situation of any household without worsening the situation of another one.

Thus, in a system with perfect competition, the selfish behavior of the individuals yields a "good" result.

Free markets are wonderful.

4.2. The first welfare theorem from the point of view of general equilibrium analysis. We now turn to general equilibrium analysis and consider the total system of markets simultaneously. For an exchange economy, we will be able to show more than just Pareto efficiency (compare chapter XIV, pp. 385). We will show that every Walras allocation lies in the core in case of weak monotonicity and local non-satiation. The core presented in this section is related to the core introduced in chapter XV. While in that chapter, the core is defined within the framework of coalition functions, we present a definition for allocations in the present section. In both cases, a core is defined by Pareto efficiency and the impossibility to block. As in chapter XV, we call a subset $S \subseteq N$ a coalition.

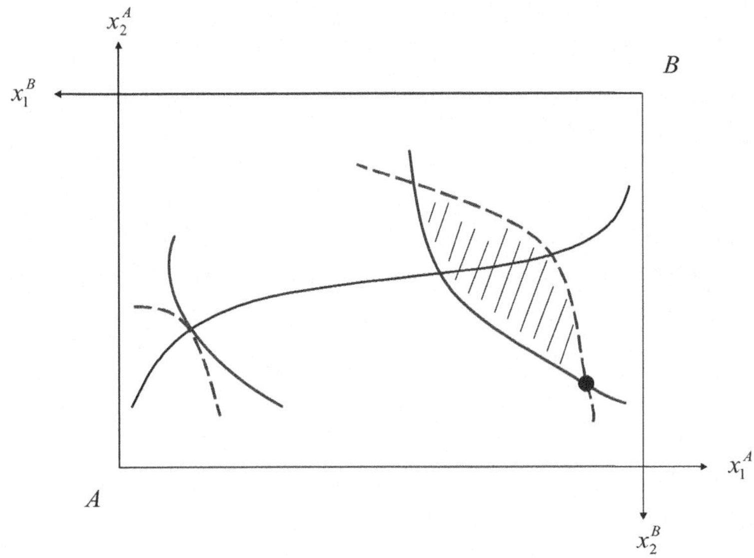

FIGURE XIX.7. Exchange lens and core

A coalition S can block an allocation if it can present an allocation that improves the lot of its members and that can be afforded by S:

DEFINITION XIX.11 (BLOCKABLE ALLOCATION, CORE). *Let*

$$\mathcal{E} = \left(N, G, \left(\omega^i\right)_{i\in N}, \left(\precsim^i\right)_{i\in N}\right)$$

be an exchange economy. A coalition $S \subseteq N$ is said to block an allocation $(y^i)_{i\in N}$, if an allocation $\left(z^i\right)_{i\in N}$ exists such that

- $z^i \succsim^i y^i$ *for all* $i \in S$, $z^i \succ^i y^i$ *for some* $i \in S$ *and*
- $\sum_{i\in S} z^i \leq \sum_{i\in S} \omega^i$

hold.

An allocation is not blockable if there is no coalition that can block it. The set of all feasible and non-blockable allocations is called the core of an exchange economy.

Within the Edgeworth box, the core can be depicted graphically. We see the endowment point and the associated exchange lens in figure XIX.7. Every household (considered a one-man coalition) blocks any allocation that lies below the indifference curve cutting his endowment point. Therefore, the core is contained inside the exchange lens. Both households together block any allocation that is not Pareto efficient. Thus, the core is the intersection of the exchange lens and the contract curve.

We now turn to a remarkable claim:

THEOREM XIX.4. *Assume an exchange economy $\mathcal{E}$ with local non-satiation and weak monotonicity. Every Walras allocation lies in the core.*

PROOF. Consider a Walras allocation $\left(\widehat{x}^i\right)_{i\in N}$. Lemma VI.4 (p. 141) implies

$$\widehat{p} \overset{(1)}{\geq} 0$$

where $\widehat{p}$ is the equilibrium price vector.

Assume, now, that $\left(\widehat{x}^i\right)_{i\in N}$ does not lie in the core. Then, there exists a coalition $S \subseteq N$ that can block $\left(\widehat{x}^i\right)_{i\in N}$. I.e., there is an allocation $\left(z^i\right)_{i\in N}$ such that

- $z^i \succsim^i \widehat{x}^i$ for all $i \in S$, $z^j \succ^j \widehat{x}^j$ for some $j \in S$ and
- $\sum_{i\in S} z^i \leq \sum_{i\in S} \omega^i$.

The second bullet, together with (1), leads to the implication

$$\widehat{p} \cdot \left(\sum_{i\in S} z^i - \sum_{i\in S} \omega^i\right) \leq 0. \tag{XIX.2}$$

The first point implies

$$\widehat{p} \cdot z^i \overset{(2)}{\geq} \widehat{p} \cdot \widehat{x}^i = \widehat{p} \cdot \omega^i \text{ for all } i \in S \text{ (by local nonsatiation) and}$$
$$\widehat{p} \cdot z^j \overset{(3)}{>} \widehat{p} \cdot \widehat{x}^j = \widehat{p} \cdot \omega^j \text{ for some } j \in S \text{ (otherwise, } \widehat{x}^j \text{ is not an optimum).}$$

Summing over all households from S yields

$$\begin{aligned}\widehat{p} \cdot \sum_{i\in S} z^i &= \sum_{i\in S} \widehat{p} \cdot z^i \text{ (distributivity)}\\ &> \sum_{i\in S} \widehat{p} \cdot \omega^i \text{ (above inequalities (2) and (3))}\\ &= \widehat{p} \cdot \sum_{i\in S} \omega^i \text{ (distributivity).}\end{aligned}$$

This inequality can be rewritten as

$$\widehat{p} \cdot \left(\sum_{i\in S} z^i - \sum_{i\in S} \omega^i\right) > 0$$

and contradicts equation XIX.2. □

We now consider figure XIX.8. Agent A's preferences violate non-satiation. He is indifferent between all the bundles in the shaded area that comprises the highlighted endowment point and the price line. The equilibrium point E is the point of tangency between that price line and agent B's indifference curve. This point is not Pareto efficient. Agent A could forego some units of both goods without harming himself.

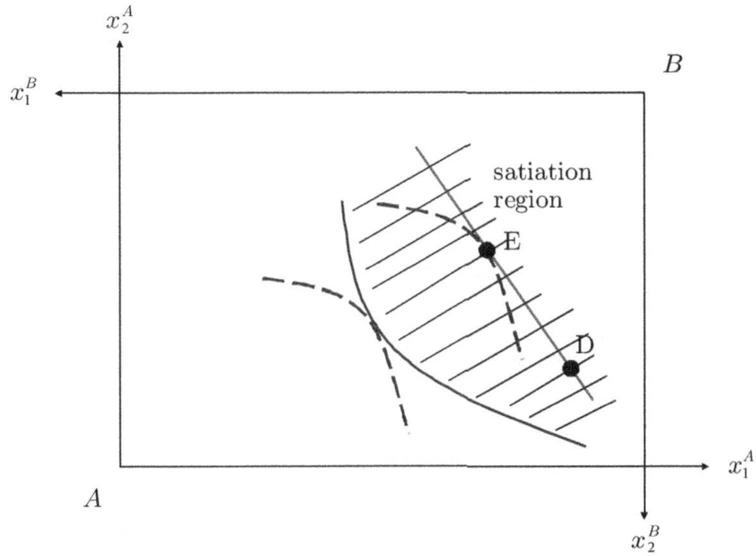

FIGURE XIX.8. A non-efficient equilibrium

4.3. The second welfare theorem. The second welfare theorem turns the first welfare theorem upside down:

- The first welfare theorem says: Walras allocations are Pareto efficient.
- The second welfare theorem claims: Pareto-efficient allocations can be achieved as Walras allocations.

THEOREM XIX.5. *Assume an exchange economy $\mathcal{E}$ with convex and continuous preferences for all consumers and local non-satiation for at least one household. Let $\left(\widehat{x}^i\right)_{i\in N}$ be any Pareto-efficient allocation. Then, there exists a price vector $\widehat{p}$ and an endowment $\left(\omega^i\right)_{i\in N}$ such that $\left(\widehat{x}^i\right)_{i\in N}$ is a Walras allocation for $\widehat{p}$.*

The reader is invited to consider figure XIX.9 and the Pareto opimum highlighted. If point E is given as endowment point, the associated Walras allocation is indeed the Pareto optimum. If, however, the original endowment is D instead of E, we can redistribute endowments by transfering some units of good 1 from B to A.

Figure XIX.10 illustrates why we assume convexity in the above theorem. Agent B does not have convex preferences. At the prices given by the price line, he does not demand his part of the Pareto optimum but some point C.

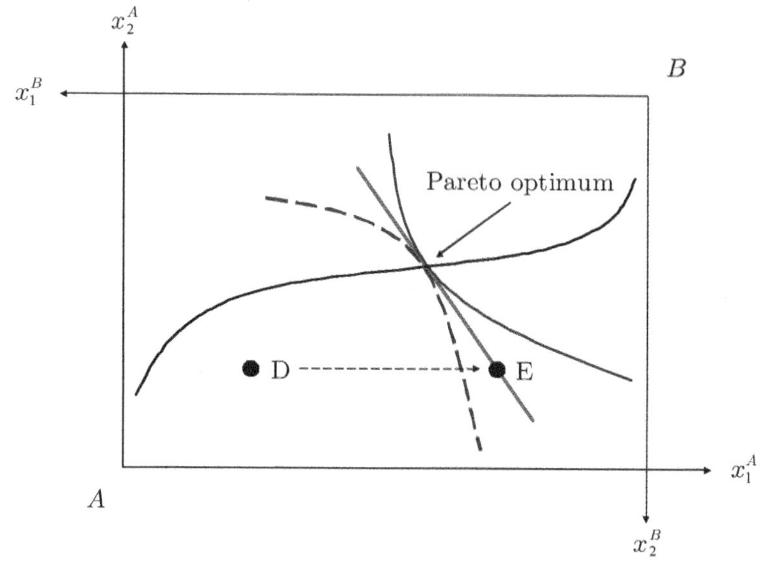

FIGURE XIX.9. The second welfare theorem

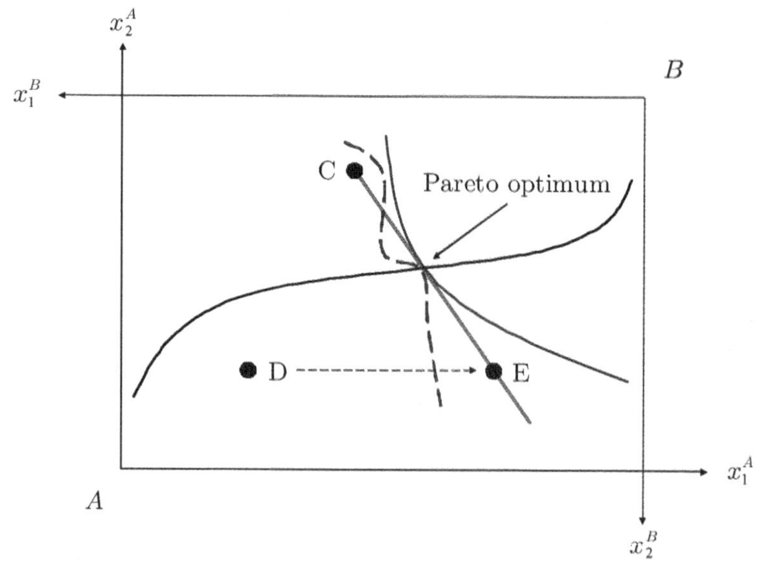

FIGURE XIX.10. Convexity is a necessary condition for the second welfare theorem.

5. Topics and literature

The main topics in this chapter are

- exchange economy

- excess demand
- market clearing
- Walras equilibrium
- first welfare theorem
- second welfare theorem
- Walras' law
- free goods
- Brouwer's fixed-point theorem
- positive theory
- normative theory
- product mix

While the nice textbook by Hildenbrandt & Kirman (1988) deals with an exchange economy, exclusively, Debreu's (1959) famous treatment of General Equilibrium Theory also covers production.

6. Solutions

Exercise XIX.1

The individual is always free to consume ω. If he wants to consume another bundle, the prices are relevant.

Exercise XIX.2

Markets clear for p_1^h, but not for p_1^l. At price p_1^l, individuals A and B want to consume more of good 1 than they possess together. Just note that D^A is to the right of D^B. At price p_1^l, there is excess supply of good 2 :

$$x_2^A + x_2^B < \omega_2^A + \omega_2^B.$$

Exercise XIX.3

From

$$\begin{aligned} z_1(p) + z_2(p) + z_3(p) &= \frac{8}{p} - 4 + \frac{4}{p} - 2 + \frac{12}{p} - 2 \\ &= \frac{24}{p} - 8 \end{aligned}$$

we obtain the market clearing price

$$p^* = 3.$$

For individual 3, we have $z_3(3) = \frac{12}{3} - 2 = 2$. He is a buyer.

Exercise XIX.4

Abba will find the household optimum by

$$px_1^A + x_2^A \stackrel{!}{=} 18p \text{ and } MRS = \frac{MU_1}{MU_2} = \frac{\frac{1}{2\sqrt{x_1^A}}}{1} \stackrel{!}{=} \frac{p}{1}.$$

Solving the second equation for x_1^A, we obtain

$$x_1^A = \frac{1}{4p^2}.$$

Substituting into the first equation yields

$$p\frac{1}{4p^2} + x_2^A = 18p \text{ and hence}$$

$$x_2^A = 18p - \frac{1}{4p}.$$

Bertha's optimal bundle is

$$x_1^B = \frac{1}{4p^2}, x_2^B = 10 - \frac{1}{4p}.$$

Equation $\omega_1^A + \omega_1^B = x_1^A + x_1^B$ can be written as

$$18 = x_1^A + x_1^B = \frac{2}{4p^2}$$

and we find

$$p = \frac{1}{6}.$$

The same price is obtained from

$$10 = x_2^A + x_2^B = 18p - \frac{1}{4p} + 10 - \frac{1}{4p}.$$

Exercise XIX.5

We can rewrite

$$\sum_{i=1}^n x^i\left(\widehat{p}, \omega^i\right) \leq \sum_{i=1}^n \omega^i$$

as

$$\sum_{i=1}^n \left(x_1^i\left(\widehat{p}, \omega^i\right), ..., x_\ell^i\left(\widehat{p}, \omega^i\right)\right) \leq \sum_{i=1}^n \left(\omega_1^i, ..., \omega_\ell^i\right)$$

which just means

$$\sum_{i=1}^n x_g^i\left(\widehat{p}, \omega^i\right) \leq \sum_{i=1}^n \omega_g^i \text{ for all } g = 1, ..., \ell.$$

Exercise XIX.6

a) There are two cases for which M is not compact: If M is not bounded, for example $M = \mathbb{R}$, the continuous function $f(x) = x + 1$ maps M onto M

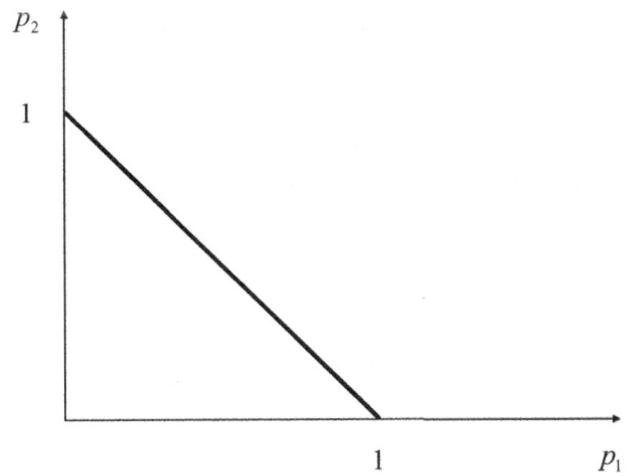

FIGURE XIX.11. The 1-dimensional unit simplex

but does not have a fixed point. For M being open, for example $M = (0, 1)$, the continuous function f given by $f(x) = \frac{1+x}{2}$ has no fixed point.
b) With $M = \left[0, \frac{1}{3}\right] \cup \left[\frac{2}{3}, 1\right]$, the continuous function f given by $f(x) = \begin{cases} 1, \text{ if } 0 \leq x \leq \frac{1}{3} \\ 0, \text{ if } \frac{2}{3} \leq x \leq 1 \end{cases}$ has no fixed point.
c) Let $M = [0, 1]$. Then, the discontinuous function f given by $f = \begin{cases} 1, \text{ if } 0 \leq x \leq \frac{2}{3} \\ 0, \text{ if } \frac{2}{3} < x \leq 1 \end{cases}$ has no fixed point.

Exercise XIX.7

S^1 is the one-dimensional segment as shown in figure XIX.11.

Exercise XIX.8

We obtain

$$MRS^A = \frac{x_2^A}{x_1^A}, MRS^B = \frac{1}{2}, MRT = \left| 2\frac{1}{2\sqrt{10 - x_1}}(-1) \right| = \frac{1}{\sqrt{10 - x_1}}.$$

Hence Pareto optimality requires

$$\frac{x_2^A}{x_1^A} \overset{(2)}{=} \frac{1}{2} \overset{(1)}{=} \frac{1}{\sqrt{10 - x_1}}$$

while Bertha's utility level of 8 is expressed by

$$x_1^B + 2x_2^B \overset{(3)}{=} 8.$$

Solving (1) yields $x_1 = 6$. By the production possibility frontier, we obtain $x_2 = 4$. Therefore, we have four equations in four unknowns:

$$\begin{aligned} x_1^A + x_1^B &= 6, x_2^A + x_2^B = 4, \\ x_1^A &= 2x_2^A \text{ by (2), and} \\ x_1^B &= 8 - 2x_2^B \text{ by (3).} \end{aligned}$$

They are solved by

$$\begin{aligned} x_1^A &= 3, x_2^A = \frac{3}{2} \\ x_1^B &= 3, x_2^B = \frac{5}{2}. \end{aligned}$$

7. Further exercises without solutions

PROBLEM XIX.1. There are two farmers Tim and Bob who harvest and trade corn (c) and wheat (w). Their endowments are $\omega^T = \left(\omega_c^T, \omega_w^T\right) = (10, 10)$ and $\omega^B = \left(\omega_c^B, \omega_w^B\right) = (30, 0)$. Tim's preferences are represented by the utility function $U_T(w, c) = \sqrt[3]{wc^2}$. Bob's utility is a strictly increasing function of wheat. Assume that aggregate excess demand for corn is given by

$$z_c(p_c, p_w) = \frac{-70p_c + 20p_w}{3p_c}$$

(1) Show $z_c(p_c, p_w) = z_c(kp_c, kp_w)$ for all $k > 0$!
(2) Determine the aggregate excess demand function for wheat! Hint: Why can you apply Walras' law (p. 498)?
(3) Determine the price ratio $\frac{p_c}{p_w}$ such that the corn market clears. Applying lemma XIX.2 (p. 499), which prices clear the wheat market?
(4) What is Tim's marginal rate of substitution $MRS = \left|\frac{dw}{dc}\right|$ between corn and wheat in equilibrium?
(5) Is Bob a net supplier of corn?

PROBLEM XIX.2. Assume two states of the world $g = 1, 2$, that occur with probabilities p and $1 - p$, respectively. Consider two players $i = A, B$ with vNM preferences. Draw an exchange Edgeworth box where x_g^i denotes the payoff (money) enjoyed by player i if state of the world g occurs. Assume that agents like high payoffs in every state that occurs with a probability greater than zero. Agent i's endowment ω_g^i is his payoff in the case where the two agents do not interact.

(1) Imagine a bet between the two agents on the realization of the state of the world. For example, player A puts a small amount of his money on state 1. How are bets and allocations linked?
(2) Assume Agent B to be risk neutral and A to be risk averse. What do the indifference curves look like?
(3) Consulting theorem XIX.4 (p. 516), can you confirm the following statement: In equilibrium, agent B provides full insurance to agent A.

CHAPTER XX

Beyond general equilibrium theory

1. Nobel price for Friedrich August von Hayek

After introducing General Equilibrium Theory, we present some applications and shed critical light on GET. We begin by a presentation of envy freeness in the next section. We then present the jungle economy, a Pareto efficient but nasty alternative to GET. We also report on the views of Austrian and other economists as they relate to markets and to GET.

Here, we begin with the Nobel price awarded to von Hayek. Friedrich August von Hayek was born in Vienna in 1899 and died in Freiburg in 1992. He studied law, economics, and psychology. In the beginning of his career, he worked on business cycles. Later on, he became very interested in the possibility of socialist planning and in the functioning of market systems. von Hayek worked in Vienna, at the London School of Economics, and at the universities in Chicago, Freiburg, and Salzburg.

The Sveriges Riksbank Prize in Economic Sciences in Memory of Alfred Nobel in 1974 was awarded to Gunnar Myrdal from Sweden and Friedrich von Hayek "for their pioneering work in the theory of money and economic fluctuations and for their penetrating analysis of the interdependence of economic, social and institutional phenomena". According to the press release by the Royal Swedish Academy of Sciences,

> von Hayek's analysis of the functional efficiency of different economic systems is one of his most significant contributions to economic research in the broader sense. From the mid-thirties he embarked on penetrating studies of the problems of centralized planning. As in all areas where von Hayek has carried out research, he gave a profound historical exposé of the history of doctrines and opinions in this field. He presented new ideas with regard to basic difficulties in "socialistic calculating", and investigated the possibilities of achieving effective results by decentralized "market socialism" in various forms. His guiding principle when comparing various systems is to study how efficiently all the knowledge and all

H. Wiese, *Advanced Microeconomics*,
https://doi.org/10.1007/978-3-658-34959-2_20

the information dispersed among individuals and enterprises is utilized. His conclusion is that only by far-reaching decentralization in a market system with competition and free price-fixing is it possible to make full use of knowledge and information.

2. Envy freeness

The idea of envy freeness is simple. An allocation is envy-free if nobody strictly prefers the bundle of any other person. Thus, envy freeness means that nobody would like to exchange bundles with any other person. Arnsperger (1994) presents a survey.

DEFINITION XX.1 (ENVY-FREE ALLOCATION). *Assume an exchange economy*

$$\mathcal{E} = \left(N, G, \left(\omega^i\right)_{i\in N}, \left(\precsim^i\right)_{i\in N}\right).$$

A feasible allocation $\left(x^i\right)_{i\in N}$ *is called envy-free if*

$$x^i \nsucc^i x^j \text{ for all } i, j \in N$$

holds.

EXERCISE XX.1. *Can you find an envy-free allocation without any preference information?*

Pareto efficiency and envy freeness are independent concepts – neither implies the other.

DEFINITION XX.2 (FAIR ALLOCATION). *Assume an exchange economy*

$$\mathcal{E} = \left(N, G, \left(\omega^i\right)_{i\in N}, \left(\precsim^i\right)_{i\in N}\right).$$

A feasible allocation $\left(x^i\right)_{i\in N}$ *is called fair if it is both envy-free and Pareto-optimal.*

EXERCISE XX.2. *Reconsider the exchange economy of exercise XIV.7 on p. 387 where the efficient allocations are given by* $x_A^1 = x_A^2$. *Which allocations are envy-free? Which are fair? Sketch your answers.*

THEOREM XX.1. *A Walras allocation is fair if all the endowments have the same value.*

PROOF. Assume that the Walras equilibrium with prices $\widehat{p}$ and Walras allocation $\left(\widehat{x}^i\right)_{i=1,\dots,n} = \left(x^i\left(\widehat{p}, \omega^i\right)\right)_{i=1,\dots,n}$ violates envy freeness although

$\widehat{p} \cdot \omega^i = \widehat{p} \cdot \omega^j$ holds for all agents $i, j \in 1, ..., n$. This means that there is an agent i who envies some other agent $j \neq i$:

$$x^i \left(\widehat{p}, \widehat{p} \cdot \omega^i\right) \prec^i x^j \left(\widehat{p}, \widehat{p} \cdot \omega^j\right).$$

Then, agent i cannot afford bundle $x^j \left(\widehat{p}, \widehat{p} \cdot \omega^j\right)$,

$$\widehat{p} \cdot x^j \left(\widehat{p}, \widehat{p} \cdot \omega^j\right) > \widehat{p} \cdot \omega^i,$$

and, by $\widehat{p} \cdot \omega^i = \widehat{p} \cdot \omega^j$, neither can agent j. This is the contradiction we were looking for. □

3. The jungle economy

3.1. Introduction. Typically, we are used to the idea of consuming goods that we buy or exchange. Of course, there are other methods to determine allocations. For example, in some societies the strongest members are able to obtain the lion's share. Indeed, this is the idea behind the jungle economy modelled by Piccione & Rubinstein (2007). We will present their model in this section. Thus, while the previous and following sections are devoted to "consuming what you buy or exchange", we now focus on "consuming what you can steal or grab".

3.2. Definition of a jungle economy.

DEFINITION XX.3 (JUNGLE ECONOMY). *A jungle economy is a tuple*

$$\mathcal{J} = \left(N, S, G, \omega, \left(\precsim^i\right)_{i \in N}, \left(X^i\right)_{i \in N}\right)$$

consisting of

- *the set of agents* $N = \{1, 2, ..., n\}$,
- *a power relation* S *on* N,
- *the finite set of goods* $G = \{1, ..., \ell\}$,
- *a total endowment* $\omega \in \mathbb{R}^\ell_+$,

and for every agent $i \in N$

- *a preference relation* $\precsim^i$ *and*
- *a consumption set* $X^i \subseteq \mathbb{R}^\ell_+$.

In a jungle economy, the n agents from N consume bundles consisting of ℓ goods from G. Consumption is determined by S, ω, $\left(\precsim^i\right)_{i \in N}$, and $\left(X^i\right)_{i \in N}$. First, consumption is influenced by a power relation S. Indeed, the basic principle of a jungle economy is that strong agents take from weak ones whatever they like (preferences) and whatever they can physically get hold of (consumption set). Formally, S is a binary relation such that for any agents i and j from N,

$$iSj$$

means that i is stronger than j. Without loss of generality, we assume

$$1S2S...Sn.$$

Strength is a transitive relation, i.e., iSj and jSk imply iSk for all $i, j, k \in N$.

Second, consumption depends on the total endowment for the whole economy. G and ω define an Edgeworth box of dimension $\ell = |G|$ and size $\omega = (\omega_1, ..., \omega_\ell)$. However, in the jungle economy there are no individual endowments for agents.

Third, preferences also determine the consumption bundles (see chapter IV). Fourth, and lastly, the consumption possibilities of agent i are restricted by the consumption set X^i which we assume to be bounded. The idea is that there are physical limits to consumption. For example, while an agent would like to consume some units of a good (which is a question of his preferences), he may not be able to do so because the respective good is beyond his reach.

Consumption sets imply the need to build on the feasibility concept, defined on p. 389. A feasible allocation in a jungle economy respects both the overall endowment and the consumption sets. In exchange economies we did not consider consumption sets although they are found in the literature. The point is that in a jungle economy without consumption sets, the strongest agent consumes everything, ω (unless we have satiation).

DEFINITION XX.4. *Let* $\mathcal{J} = \left(N, S, G, \omega, \left(\precsim^i\right)_{i \in N}, \left(X^i\right)_{i \in N}\right)$ *be a jungle economy. An allocation* $\left(x^i\right)_{i \in N}$ *in* $\mathcal{J}$ *is called feasible if*

- $\sum_{i=1}^{n} x^i \leq \omega$ *and*
- $x^i \in X^i$, $i \in N$,

hold. $x^0 := \omega - \sum_{i=1}^{n} x^i \geq 0$ *is called the leftover.*

3.3. The jungle equilibrium. Now that we have defined the main principles of a jungle economy, we will go on to do positive (existence and uniqueness of equilibrium) and normative analysis (welfare theorems).

More often than not, positive analysis is equilibrium analysis. For the jungle economy we will ask the question: Can we find an allocation such that no agent can benefit from taking what belongs to a weaker individual?

DEFINITION XX.5. *Let* $\mathcal{J} = \left(N, S, G, \omega, \left(\precsim^i\right)_{i \in N}, \left(X^i\right)_{i \in N}\right)$ *be a jungle economy. A feasible allocation* $\left(\widehat{x}^i\right)_{i \in N}$ *is called a jungle equilibrium if there is no agent* $i \in N$ *and no bundle* $y^i \in X^i$ *such that* $y^i \succ \widehat{x}^i$ *and* $y^i \leq \widehat{x}^i + \hat{x}^0 + \sum_{\substack{j \in N, \\ iSj}} \widehat{x}^j$*, where* $\hat{x}^0 := \omega - \sum_{j \in N} \widehat{x}^j$ *is the leftover.*

According to this recursive definition, a strong agent can grab the leftover and any (parts of) bundles from weaker agents as long as the preferred bundle y^i belongs to his consumption set. $\widehat{x}^i + \hat{x}^0 + \sum_{\substack{j \in N, \\ iSj}} \widehat{x}^j$ can also be expressed by

$$\begin{aligned} & \widehat{x}^i + \hat{x}^0 + \sum_{\substack{j \in N, \\ iSj}} \widehat{x}^j \\ = \; & \widehat{x}^i + \left(\omega - \sum_{j \in N} \widehat{x}^j \right) + \sum_{j=i+1}^{n} \widehat{x}^j \\ = \; & \omega - \sum_{j=1}^{i-1} \widehat{x}^j. \end{aligned}$$

Thus player i is free to consume any bundle left over by those agents stronger than him.

PROPOSITION XX.1 (EXISTENCE AND UNIQUENESS OF JUNGLE EQUILIBRIA). *Let* $\mathcal{J} = \left(N, S, G, \omega, \left(\precsim^i\right)_{i \in N}, \left(X^i\right)_{i \in N}\right)$ *be a jungle economy. Assume without loss of generality* $1S2S...Sn$*. If*

- *every* $X^i, i \in N$*, is bounded, closed, and convex, and*
- *every preference relation* $\precsim^i, i \in N$*, is continuous, strictly convex, and strictly monotonous,*

a unique jungle equilibrium exists. It is given by a feasible allocation $\left(\widehat{x}^i\right)_{i \in N}$ *such that*

$$\widehat{x}^1 \text{ is the best bundle from } \left\{x^1 \in X^1 : x^1 \leq \omega\right\}$$

and

$$\widehat{x}^i \text{ is the best bundle from } \left\{ x^i \in X^i : x^i \leq \omega - \sum_{j=1}^{i-1} \widehat{x}^j \right\}, i = 2, ..., n.$$

Thus, player 1 who is the strongest chooses first. Player 2 then makes his pick from the remainder. Note that player 1 may leave some goods to the other players because of the boundedness of his consumption set and because of satiated preferences. At some stage i, $\left\{x^i \in X^i : x^i \leq \omega - \sum_{j=1}^{i-1} \widehat{x}^j\right\}$ might contain nothing but $0 \in \mathbb{R}^\ell_+$. Then, all the remaining players $i, i+1, ..., n$ have nothing left to consume.

Following Piccione & Rubinstein (2007, p. 888), we will consider a jungle economy with discrete goods, for example houses. The consumption sets imply that every agent can consume (live in) one house only. All the houses are within reach for all agents. Preferences on the consumption sets are strict. The power relation S on N is given by $1S2S...Sn$.

EXERCISE XX.3. *Let* $\mathcal{J} = \left(N, S, G, \omega, \left(\prec^i\right)_{i\in N}, \left(X^i\right)_{i\in N}\right)$ *be the jungle economy described above. Can you define* G, ω, *and* $\left(X^i\right)_{i\in N}$? *Find the jungle equilibrium.*

The proposition can be easily proven in two steps. First, we have to confirm that $\left(\widehat{x}^i\right)_{i\in N}$ is well defined, i.e., that "the best bundle" exists for all players. Second, we will show that $\left(\widehat{x}^i\right)_{i\in N}$ is indeed an equilibrium.

For the first step of the proof we will define "the best bundle" for each player. Note that the properties of the preferences guarantee the existence of continuous utility functions representing those preferences. Player 1's choice set can be written as

$$\left\{x^1 \in X^1 : x^1 \leq \omega\right\} = X^1 \cap \left\{x \in \mathbb{R}_+^\ell : x \leq \omega\right\}$$

Since both X^1 and $\left\{x \in \mathbb{R}_+^\ell : x \leq \omega\right\}$ are bounded and closed, so is their intersection, $\left\{x^1 \in X^1 : x^1 \leq \omega\right\}$. Player 2 can choose from the (bounded and closed) set $\left\{x^2 \in X^2 : x^2 \leq \omega - x^1\right\}$. By an induction argument, the choice sets of the other players are also bounded and closed. All choice sets contain $(0, ..., 0) \in \mathbb{R}_+^\ell$ and are not empty. Therefore, the agents' optimization problems consist in choosing from a bounded and closed set, or, differently put, in maximizing (continuous!) utility on a suchlike set. You might remember from chapter VI (theorem VI.1, p. 152) that any continuous function $f : X \to \mathbb{R}$ on a nonempty, closed, and bounded set $X \subseteq \mathbb{R}^n$ has a maximum on X.

Note that the proposition does not exclude the possibility of several best bundles. However, it can be shown that there is only one best bundle. Since both X^1 and $\left\{x \in \mathbb{R}_+^\ell : x \leq \omega\right\}$ are convex, so is their intersection, $\left\{x^1 \in X^1 : x^1 \leq \omega\right\}$. Similarly, the choice sets of the other players are also convex. By the strict convexity of the preferences, we know that there is only one best bundle (see theorem VI.2, p. 153).

We now deal with the second step of our proof to show that $\left(\widehat{x}^i\right)_{i\in N}$ is an equilibrium. Suppose that it is not an equilibrium. Then, there is an agent $i \in N$ and a bundle $y^i \in X^i$ such that $y^i \succ \widehat{x}^i$ and $y^i \leq \widehat{x}^i + \hat{x}^0 + \sum_{\substack{j\in N,\\ iSj}} \hat{x}^j$ with $\hat{x}^0 = \omega - \sum_{j\in N} \widehat{x}^j$. Such a bundle cannot exist. This is the desired contradiction, which completes the proof of proposition XX.1. In light of the proposition, we are now justified to talk of "the" jungle equilibrium.

3.4. Welfare economics of the jungle equilibrium. Although the jungle equilibrium $\left(\widehat{x}^i\right)_{i\in N}$ might be considered unfair, it is Pareto efficient. Let us assume that a feasible allocation $\left(y^i\right)_{i\in N}$ is a Pareto improvement over $\left(\widehat{x}^i\right)_{i\in N}$. Then, agent 1 (the strongest) will weakly prefer y^1 to $\widehat{x}^1$. Since $\widehat{x}^1$ is the best bundle from the feasible set $\left\{x^1 \in X^1 : x^1 \leq \omega\right\}$, y^1 cannot be better than $\widehat{x}^1$. Indeed, by strict convexity y^1 has to equal to $\widehat{x}^1$. Similarly, we can show that y^i is equal to $\widehat{x}^i$ for all $i \in N$. This results in a contradiction to $\left(y^i\right)_{i\in N}$ being a Pareto improvement over $\left(\widehat{x}^i\right)_{i\in N}$.

PROPOSITION XX.2 (PARETO EFFICIENCY OF THE JUNGLE EQUILIBRIUM). *Let* $\mathcal{J} = \left(N, S, G, \omega, \left(\precsim^i\right)_{i\in N}, \left(X^i\right)_{i\in N}\right)$ *be a jungle economy. Assume that the properties specified in proposition XX.1 hold. Then,* $\left(\widehat{x}^i\right)_{i\in N}$ *is a Pareto-efficient allocation.*

This proposition is the first welfare theorem of the jungle economy. Piccione & Rubinstein (2007) discuss the possibility of a second welfare theorem of the jungle economy.

4. Applications

Under the heading of applications, we report

- how socialist planning was thought to benefit from GET (the economic theory of socialism),
- in which way the ordoliberal school (and some of nowadays competition theory and policy) proposes to apply GET to regulation and competition policy, and
- how empirical analyses and simulation build on GET models to track the complex consequences of policy changes.

4.1. The economic theory of socialism. Some economists have used General Equilibrium Theory to establish the concept of a socialist economy. The Polish economist Oskar Lange (1904-1965) has been a leading representative of this school of thought. A pivotal and well-known piece of his work is Lange (1936).

Lange's socialist economy consists of state-owned firms directed by state officials. These officials have to base their decision on the prices announced by a central planning board. They are asked to react to these prices as price takers, minimizing costs and maximizing profits just as firms do in microeconomic textbooks. According to Lange's theory the central planning board has the same function as the Walras' auctioneer. By a process of trial and error, the central planning board tries to get as close as possible to the equilibrium price vector. Lange (1936, pp. 66) claims:

> ... the accounting prices in a social economy can be determined by the same process of trial and error by which prices on a competitive market are determined. ... The only "equations" which would have to be "solved" would be those of the consumers and the managers of production plants. These are exactly the same "equations" which are solved by the present economy system and the persons who do the "solving" are the same also.

Of course, Oskar Lange's work precedes the microeconomic treatment of the problems of asymmetric information and the present literature on incentives. But even in Lange's time, Friedrich August von Hayek (1937, 1945) and others pointed out that there is no conceivable way that a central planning board could obtain the information (partly contradictory), held by millions of consumers and producers. We will shortly return to the marvel felt by von Hayek when considering the price system in action.

Many people will feel thankful that the General Equilibrium Theory could not be put into practice as proposed by Oskar Lange.

4.2. The regime of competition of the Freiburg School of "Ordoliberalism". Walter Eucken (1891– 1950) was a leading proponent of the Freiburg School of *Ordoliberalism.* Eucken's (1990, p. 254) regime of competition (*Wettbewerbsordnung*) is based on perfect competition. Every economic policy measure is to be judged by the model of perfect competition. This is the basic principle (*Grundprinzip*), the first of a set of principles called constitutive principles (*konstituierende Prinzipien*) (pp. 254 – 291). Other principles belonging to this set are

- monetary stability (Primat der Währungspolitik),
- open markets (Offene Märkte),
- private property (Privateigentum),
- freedom of contract (Vertragsfreiheit),
- accountability (Haftung), and
- economic policy consistency (*Konstanz der Wirtschaftspolitik*).

Apart from the "konstituierende Prinzipien" Eucken's regime of competition is build on the so-called regulating principles (*regulierende Prinzipien*). The author suggests that an anti-monopoly bureaucracy (*Monopolamt*) should deal with monopoly problems:

- Monopolies have to be dissolved or, should dissolution be impossible, to be regulated (see Eucken 1990, p. 294).
- The institutions (firms, unions) that wield power should be forced to act as if perfect competition held. For example, in order to emulate the law of one price, price discrimination is to be outlawed (see Eucken 1990, p. 294). Also, regulation should aim for marginal-cost pricing (see Eucken 1990, p. 297). However, since marginal costs are difficult to ascertain, Eucken (1990, p. 297) suggests that the intersection of average cost and demand be used instead. This is Ramsey pricing, explained in chapter XXI (pp. 556).
- The prices fixed by the Monopolamt are meant to incite firms to reduce costs whenever possible. Eucken (1990, p. 297) observes that a monopoly's production capacities are often outdated and advises the Monopolamt to revise prices from time to time.

For a still readable critique of Eucken's approach, turn to Lenel (1975).

4.3. Computable General Equilibrium Theory. Computable GET sets out to build a dynamic multi-market model where the specific functions and values derive from real-world data. For any given set of parameters (taxes set by government, environmental regulation, climate change), a path of equilibrium prices and quantities is found by empirical analyses and simulations. The prices and quantities are given in numerical form (concrete numbers). Therefore, it is not always easy to tell why a specific policy change had the observed consequences.

5. Austrian critical perspectives and Schumpeter

The Austrian School of Economics criticizes the way competition is presented in the models of perfect competition (and industrial organization). In this section, rather than providing a thorough discussion of the Austrian position, we will focus on some contributions to competition theory by Ludwig von Mises, Friedrich August von Hayek, and Israel Kirzner. In particular, these economists discuss

- equilibrium analysis (they concentrate on the equilibrating forces rather than on the equilibrium itself),
- knowledge assumptions (they stress the importance of dispersed knowledge, imagination, and surprise), and
- the role of the entrepreneur in market processes (the Austrian entrepreneurs do not "mechanically" maximize profits but discover profit opportunities and act as arbitrageurs).

Joseph Schumpeter is not considered a member of the Austrian school; his process of creative destruction is mentioned in the last subsection.

5.1. Ludwig von Mises: the market process. Ludwig Heinrich von Mises (1881 – 1973) was one of the leading Austrian economists. In his major work "Human action", von Mises (1996, pp. 328) writes:

> The driving force of the market process is provided neither by the consumers nor by the owners of the means of production – land, capital goods, and labor – but by the promoting and speculating entrepreneurs. These are people intent upon profiting by taking advantage of differences in prices. Quicker of apprehension and farther-sighted than other men, they look around for sources of profit. They buy where and when they deem prices too low, and they sell where and when they deem prices too high. They approach the owners of the factors of production, and their competition sends the prices of these factors up to the limit corresponding to their anticipation of the future prices of the products. They approach the consumers, and their competition forces prices of consumer goods down to the point at which the whole supply can be sold. Profit-seeking speculation is the driving force of the market as it is the driving force of production.
>
> On the market agitation never stops. The imaginary construction of an evenly rotating economy has no counterpart in reality. There can never emerge a state of affairs in which the sum of the prices of the complementary factors of production, due allowance being made for time preference, equals the prices of the products and no further changes are to be expected. There are always profits to be earned by somebody. The speculators are always enticed by the expectation of profit.

It is not astonishing that von Mises (1996, pp. 356) is not impressed by equilibrium analysis: "The mathematical description of various states of equilibrium is mere play. The problem is the analysis of the market process."

5.2. Friedrich August von Hayek: the price system as a machinery for registering change. Friedrich August von Hayek is concerned with the question of who knows what, and how people obtain information in order to make good decisions. Since society needs to adapt to constant changes, von Hayek (1945, pp. 524) insists on decentral decisions

> because only thus can we ensure that the knowledge of the particular circumstances of time and place will be promptly used. But the "man on the spot" cannot decide solely on the basis of his limited but intimate knowledge of the facts of his immediate surroundings. There still remains the problem of communicating to him such further information as he needs to fit his decisions into the whole pattern of changes of the larger economic system.

According to von Hayek (1945, p. 526), in such circumstances, it is the prices that coordinate actions of people:

> Assume that somewhere in the world a new opportunity for the use of some raw material, say tin, has arisen, or that one of the sources of supply of tin has been eliminated. It does not matter for our purpose – and it is very significant that it does not matter – which of these two causes has made tin more scare. All that the users of tin need to know is that some of the tin they used to consume is now more profitably employed elsewhere, and that in consequence they must economize tin.

For von Hayek (1945, p. 527), the price system is "a kind of machinery for registering change". He goes on to say:

> The marvel is that in a case like that of a scarcity of one raw material, without an order being issued, without more than perhaps a handful of people knowing the cause, tens of thousands of people whose identity could not be ascertained by months of investigation, are made to use the material or its products more sparingly, i.e., they move in the right direction.

One might summarize the important 1945 paper, by saying that Hayek characterizes the price system as a machinery for registering change in a world of dispersed knowledge of particular circumstances.

5.3. Friedrich August von Hayek: competition as discovery procedure. Friedrich August von Hayek is also famous for his 1968 lecture "Der Wettbewerb als Entdeckungsverfahren" at the "Institute for the World Economy" in Kiel. Hayek (2002, p. 9-13) writes:

> ... wherever we make use of competition, this can only be justified by our not knowing the essential circumstances that determine the behavior of the competitors. In sporting events,

> examinations, the awarding of government contracts, or the bestowal of prizes for poems, not to mention science, it would be patently absurd to sponsor a contest if we knew in advance who the winner would be.
>
> ...
>
> [M]arket theory often prevents access to a true understanding of competition by proceeding from the assumption of a "given" quantity of scarce goods. Which goods are scarce, however, or which things are goods, or how scarce or valuable they are, is precisely one of the conditions that competition should discover: in each case it is the preliminary outcomes of the market process that inform individuals where it is worthwhile to search. Utilizing the widely diffused knowledge in a society with an advanced division of labor cannot be based on the condition that individuals know all the concrete uses that can be made of the objects in their environment. Their attention will be directed by the prices the market offers for various goods and services.

5.4. Israel Kirzner: entrepreneurial discovery. Building on von Hayek's and von Mises' ideas, Israel Kirzner's entrepreneurial-discovery theory deals with three interrelated concepts,

- the entrepreneur,
- discovery, and
- rivalrous competition.

Turning to the entrepreneur, Kirzner (1997, p. 70) writes:

> Whereas each neoclassical decision maker operates in a world of *given* price and output data, the Austrian entrepreneur operates to *change* price/output data. In this way ... the entrepreneurial role drives the ever-changing process of the market. Where shortages have existed, we understand the resulting price increases as driven by entrepreneurs recognizing, in the face of the uncertainty of the real world, the profit opportunities available through the expansion of supply through production, or through arbitrage. Except in the never-attained state of complete equilibrium, each market is characterized by opportunities for pure entrepreneurial profit. These opportunities are created by earlier entrepreneurial errors which have resulted in shortages, surplus,

> misallocated resources. The daring, alert entrepreneur discovers these earlier errors, buys where prices are "too low" and sells where prices are "too high". In this way low prices are nudged higher, high prices are nudged lower; price discrepancies are narrowed in the equilibrative direction.

The Kirzner entrepreneur is the one who makes competition a discovery process in Hayek's sense. Kirzner (1997, pp. 71) is careful to distinguish between discovery and search:

> Systematic search can be undertaken for a piece of missing information, but only because the searcher is aware of the nature of what he does not know, and is aware with greater or lesser certainty of the way to find out the missing information. In the economics of search literature, therefore, search is correctly treated as any other deliberate process of production. But it is in the nature of an overlooked profit opportunity that it has been *utterly* overlooked, i.e., that one is not aware at all that one has missed the grasping of any profit. From the neoclassical perspective, therefore, a missed opportunity might seem (except as a result of sheer, fortuitous good luck) to be destined for permanent obscurity.
>
> It is here that the Austrian perspective offers a new insight, into the nature of *surprise* and *discovery*. When one becomes aware of what one had previously overlooked, one has not produced knowledge in any deliberate sense. What has occurred is that one has discovered one's previous (utterly unknown) ignorance. What distinguishes *discovery* (relevant to hitherto unknown profit opportunities) from *successful search* (relevant to the deliberate production of information which one knew one had lacked) is that the former (unlike the latter) involves *surprise* which accompanies the realization that one had overlooked something in fact readily available. ("It was under my very nose!") This feature of *discovery* characterizes the entrepreneurial process of the equilibrating market.

Turning to rivalrous competition, Kirzner (1997, pp. 73) emphasizes

> the dynamically competitive character of such a process. The process is made possible by the freedom of entrepreneurs to enter markets in which they see opportunities for profit. In

> being alert to such opportunities and in grasping them, entrepreneurs are competing with other entrepreneurs. This competition is not the competitive state achieved in neoclassical equilibrium models, in which all market participants are buying or selling identical commodities, at uniform prices. It is, instead, the rivalrous process we encounter in the everyday business world, in which each entrepreneur seeks to outdo his rivals in offering goods to consumers (recognizing that, because those rivals have not been offering the best possible deals to consumers, profit can be made by offering consumers better deals).
>
> It is from this perspective that Austrians stress (i) the discovery potential in rivalrous competition, and (ii) the entrepreneurial character of rivalrous competition.

5.5. Joseph Schumpeter: creative destruction. In 1942, Joseph A. Schumpeter published a book with the title "Capitalism, Socialism and Democracy". Schumpeter argues that socialism rather than capitalism would survive in the long run. The second part of this book (Can Capitalism Survive) contains a chapter on "The Process of Creative Destruction". This process is essential for the capitalist system. Schumpeter (1976, pp. 82) writes:

> Capitalism ... is by nature a form or method of economic change and not only never is but never can be stationary. And this evolutionary character of the capitalist process is not merely due to the fact that economic life goes on in a social and natural environment which changes and by its change alters the data of economic action; this fact is important and these changes (wars, revolutions and so on) often condition industrial change, but they are not its prime movers. Nor is this evolutionary character due to a quasi-automatic increase in population and capital or to the vagaries of monetary systems of which exactly the same thing holds true. The fundamental impulse that sets and keeps the capitalist engine in motion comes from the new consumers' goods, the new methods of production or transportation, the new markets, the new forms of industrial organization that capitalist enterprise creates.

Then, Schumpeter (1976, p. 83) goes on to describe the process of Creative Destruction:

> [T]he history of the productive apparatus of a typical farm, from the beginnings of the rationalization of crop rotation, plowing and fattening to the mechanized thing of today - linking up with elevators and railroads - is a history of revolutions. So is the history of the productive apparatus of the iron and steel industry from the charcoal furnace to our own type of furnace, or the history of the apparatus of power production from the overshot water wheel to the modern power plant, or the history of transportation from the mailcoach to the airplane. The opening up of new markets, foreign or domestic, and the organizational development from the craft shop and factory to such concerns as U.S. Steel illustrate the same process of industrial mutation - if I may use that biological term - that incessantly revolutionizes the economic structure from within, incessantly destroying the old one, incessantly creating a new one. This process of Creative Destruction is the essential fact about capitalism. It is what capitalism consists in and what every capitalist concern has got to live in.

Finally, Schumpeter (1976, pp. 84) criticizes the common view on competition:

> Economists are at long last emerging from the stage in which price competition was all they saw. As soon as quality competition and sales effort are admitted into the sacred precincts of theory, the price variable is ousted from its dominant position. However, it is still competition within a rigid pattern of invariant conditions, methods of production and forms of industrial organization in particular, that practically monopolizes attention. But in capitalist reality as distinguished from its textbook picture, it is not that kind of competition which counts but the competition from the new commodity, the new technology, the new source of supply, the new type of organization ... competition which commands a decisive cost or quality advantage and which strikes not at the margins of the profits and the outputs of the existing firms but at their foundations and their very lives. This kind of competition is as much more effective than the other as a bombardement is in comparision with forcing a door, and so much more important that it becomes a matter

> of comparative indifference whether competition in the ordinary sense functions more or less promptly: the powerful lever that in the long run expands output and brings down prices is in any case made of other stuff.

Schumpeter (1976, p. 85) was well aware of the importance of potential competition: “It disciplines before it attacks.”

Despite the similarities between Schumpeter's and Kirzner's approach, Kirzner (1973, p. 127) notes: “For Schumpeter the entrepreneur is the disruptive, disequilibrating force that dislodges the market from the somnolence of equilibrium; for us the entrepreneur is the equilibrating force whose activity responds to the existing tensions and provides those corrections for which the unexploited opportunities have been crying out.” Concerning Schumpeter's critique of his fellow economists ("price competition was all they saw"), Kirzner (1973, p. 129) writes that “for Schumpeter price competition exemplifies the *non*entrepreneurial, pedestrian kind of competition (which he wishes to relegate to the background), whereas the dynamic, entrepreneurial type of competition (which for Schumpeter is the essence of the capitalist process) is exemplified by the new commodity and new technology. For us, the *process* of price competition is as entrepreneurial and dynamic as that represented by the new commodity, new technique, or new type of organization.”

6. A critical review of GET

6.1. General equilibrium theory. General equilibrium analysis is an important part of economic theory. Using GET we can analyze any number of markets simultaneously. In accomodating contingent markets GET can also deal with risk (given probabilities). GET has some important shortcomings, some of them related to the Austrian worldview.

- Although many economists call GET a decentralized allocation mechanism, in another sense it is very centralized because it involves the Walrasian auctioneer.
- GET deals with many markets (specified by attributes, place, time, contingencies) all of which are assumed to have many sellers and buyers. However, in the real world, there are not many buyers or sellers of white pianos to be delivered in Leipzig in September 2042 should it rain the day before.
- Increasing returns to scale are a common real-world phenomenon, but contradict convex production sets. Therefore, it is unlikely

that many small producers will enter the market. This makes price-taking behavior a strong assumption.

- GET checks whether equilibrium prices exist but does not explain how prices are formed. Thus, arbitrageurs or real estate agents (as in "real" markets) have no role to play. While excess demand may well lead to price increases, price takers cannot perform this function. Of course, this is part of the Austrian critique of perfect competition.
- Following up on the previous point, it is an open question how the auctioneer suggests price vectors, based on the information of excess demands. Even plausible tâtonnement processes cannot always converge towards Walras equilibria. Thus, exogenous shocks will not necessarily lead to a new equilibrium.
- The quality of goods is no problem tackled by GET. Indeed, when people are prepared to pay a certain price for a good, they can be sure to obtain the quality agreed upon. Thus, all the problems with which principal-agent theory deals are assumed away.
- Contracts in GET are complete and simple: Goods are exchanged for other goods or against money. In contrast, contracting in everyday life is seldom done on the basis of complete contracts. Bowles (2004, p. 10) claims that norms and power often replace contracts. "An employment contract does not specify any particular level of effort, but the employee's work ethic or fear of job termination or peer pressure from workmates may accomplish what contractual enforcement cannot."

  From the point of view of social exchange theory, Walrasian exchange is but a very small part of social exchange. Social exchange often takes place in long time intervals and it is not always clear to the participants who owes what to whom. Social exchange relations exist in markets, between neighbors, colleagues, or politicians.
- GET is based on utility maximization and is hence susceptible to criticism levied against "rationality" (intransitivity, loss aversion, inconsistency in temporal discounting, overvaluation of low probability events).

6.2. The two welfare theorems. The strong point of the economic system, envisioned by GET, is the impersonal nature of economic transactions. Agents just have to observe the price vector and do not need to do complicated deals with other (possibly several) agents. Bowles (2004,

p. 208) calls this a "utopian capitalism" in spite of the negligence of distributive justice. Indeed, GET depicts a utopian state of affairs in many respects: no theft, no quality problems, no market concentration.

Let us turn to the first welfare theorem.

- Pareto optimality and extreme inequality of consumption or income can go hand in hand.
- Our comment on the first welfare theorem was: Free markets are wonderful. The reader will remember that the first welfare theorem holds in the jungle, too. This may make us think twice whether Pareto efficiency is a reason to jubilate.
- Maybe, free markets are wonderful for reasons not directly connected to the welfare theorems. According to von Hayek (1945, p. 527), the price system is "a kind of machinery for registering change" (see pp. 534 above). Hayek marvels about the ability of the prices to transport the information that prompt millions of people to move into the right direction. Of course, GET has something to say about these reactions. Indeed, comparative statics will let us know what "happens" if some raw material has uses not thought about "before". In GET, people do not "move in the right direction", but jump to the new equilibrium bundle.

The second welfare theorem shows that Pareto efficient allocations can be achieved by suitable redistributions of endowments.

- The second welfare theorem allows to separate efficiency from distribution arguments. Redistribution is done via transfers, the market cares for efficiency. But: if the government knows which efficient allocations to aim at, what does it need a market for?
- If redistributive measures are pending, the agents can be expected to carry out preemptive measures.

7. Topics

The main topics in this chapter are

- envy freeness
- jungle economy
- Walter Eucken
- Ludwig von Mises
- Friedrich August von Hayek
- Joseph Schumpeter
- Israel Kirzner
- Oscar Lange: The economic theory of socialism
- computable GET

8. Solutions

Exercise XX.1

If every agent has the same bundle, envy freeness is ensured. For example,

$$x^i = \frac{\omega}{n}, i = 1, ..., n,$$

is an envy-free allocation.

Exercise XX.2

The envy-free allocations have to fulfill both

$$u^A\left(x_1^A, x_2^A\right) \geq u^A\left(x_1^B, x_2^B\right) \text{ (for individual } A)$$

and

$$u^B\left(x_1^B, x_2^B\right) \geq u^B\left(x_1^A, x_2^A\right) \text{ (for individual } B)$$

Since the utility functions are identical, we find

$$u^A\left(x_1^A, x_2^A\right) \geq u^A\left(x_1^B, x_2^B\right) = u^B\left(x_1^B, x_2^B\right) \geq u^B\left(x_1^A, x_2^A\right) = u^A\left(x_1^A, x_2^A\right)$$

and hence

$$u^A\left(x_1^A, x_2^A\right) = u^A\left(x_1^B, x_2^B\right).$$

This implies

$$x_1^A x_2^A = \left(100 - x_1^A\right)\left(100 - x_2^A\right),$$

whence we have

$$x_2^A = 100 - x_1^A.$$

Figure XX.1 sketches the envy-free allocations and the efficient ones. There is only one fair allocation: $\left(x_1^A, x_2^A\right) = \left(x_1^B, x_2^B\right) = (50, 50)$.

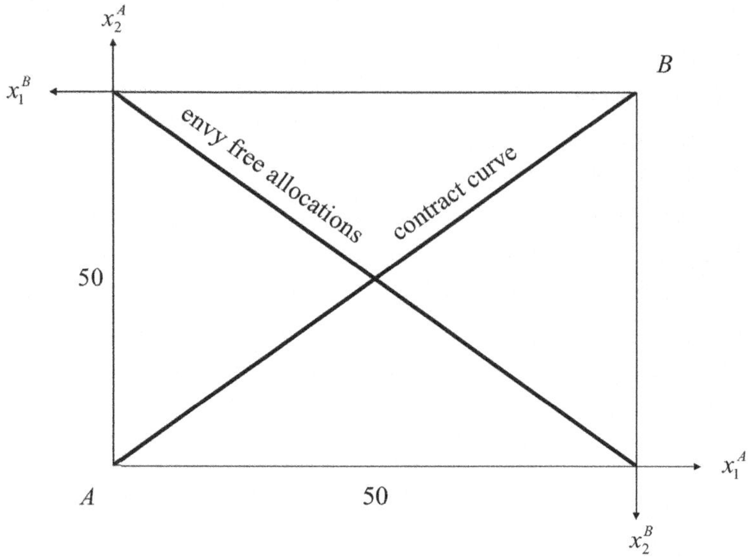

FIGURE XX.1. Envy free and efficient allocations

Exercise XX.3

We have $G = \{1, ..., \ell\}$ (as usual), the total endowment $\omega = (1, ..., 1) \in \mathbb{R}_+^\ell$, and, for every agent $i \in N$, the consumption set

$$X^i = \{(1, 0, ..., 0), ..., (0, ..., 0, 1, 0), (0, ..., 0, 1)\} \subseteq \mathbb{R}_+^\ell.$$

Since preferences are strict, the strongest agent will pick his most preferred house, then the second strongest agent will have his choice of house. In case of $n \leq \ell$, every agent will obtain a house. Otherwise ($n > \ell$), only the strongest ℓ agents will obtain a house, while the remaining $n - \ell$ agents do not find one.

9. Further exercises without solutions

PROBLEM XX.1. Are the Nash outcomes of a first-price auction envy-free? Are the Nash outcomes of a second-price auction envy-free?

CHAPTER XXI

Introduction to competition policy and regulation

Competition policy and regulation deal with many different problems and models. In this chapter, we briefly discuss the concept of a market, the structure-conduct-performance paradigm, Ramsey prices, and similar topics.

1. Themes

What can the government do to make the market mechanism work better for the consumers? Since we structure the problem this way, it is natural to ask: Where and why does the market not work to the consumers' satisfaction? Basically, consumers are interested in (low) prices and (high) quality. Therefore, relevant questions focus on the following areas:

- How are market prices affected by the number and size of the firms?
- Should firms be allowed to merge?
- How does the number of firms on the market affect innovation?
- How do potential competitors discipline the actual competitors?
- Which liability rules will increase product safety and what are the costs?
- Should the government mandate that the underlying structure for network industries (rails, electricity, water) be operated by firms different from those that actually offer the services?
- How are prices for public utilities to be set? Should cross subsidies be allowed?

Competition theory and regulation is a wide field, indeed. However, there are some aspects outside its domain. For example, income and tax policy and social welfare are generally not treated under this heading. Also, competition theory and regulation starts from the premise that markets are, in general, capable of solving our economic problems. Therefore, the theory does not deal with socialist planning or outright nationalization.

H. Wiese, *Advanced Microeconomics*,
https://doi.org/10.1007/978-3-658-34959-2_21

2. Markets

2.1. The relevant market. Competition theory and policy are concerned with markets. However, it is very unclear what constitutes a market. The German anti-cartel law (Gesetz gegen Wettbewerbsbeschränkungen, GWB) uses the phrase of "relevant market". How do we find out what is the relevant market?

2.1.1. *Cross price elasticity of demand.* First of all, the relevant market should contain all products that are close substitutes. We ascertain substitutes using the cross price elasticity of demand:

$$\varepsilon_{x_g,p_k} = \frac{\frac{\partial x_g}{x_g}}{\frac{\partial p_k}{p_k}} = \frac{\partial x_g}{\partial p_k}\frac{p_k}{x_g}$$

In case of $\varepsilon_{x_g,p_k} > 0$, goods g and k are called substitutes (see, however, the discussion in chapter VII, pp. 189). If the price of good k is lowered, demand for good g goes down. If the cross elasticity is above a certain threshold, good k is said to belong to the market of g.

2.1.2. *Supply-side substitutes.* Second, the relevant market might be extended so as to include supply-side substitutes. For example, a firm, producing tables from wood, may consider making wooden toys if the price of these toys increases. In this sense, immediate entrants (if not their products) may also be reckoned to be part of the toy market.

2.1.3. *SSNIP-Test.* SSNIP stands for "small but significant non-transitory increase in prices" and has been introduced by the US department of justice. For instance, consider the question whether butter and margarine belong to the same market. If, hypothetically, all the producers of butter merged, would it be in the interest of the newly-formed butter monopolist to increase the price of butter by about 5 to 10%? If the answer is "yes", margarine is not a sufficiently strong substitute for butter. If the answer is no, margarine is a strong substitute for butter. Then, butter and margarine belong to one market. Of course, with the SSNIP test we can look at the hypothetical merger of butter and margarine producers to examine whether honey, also, belongs to the butter/margarine market.

2.1.4. *Price correlation test.* Stigler & Sherwin (1985) propose the price-correlation test. The basic idea is as follows: If two goods belong to the same market, their prices should follow a similar time path.

2.2. Measures of concentration. Once we have defined the relevant market, we can consider its concentration. Measures of concentration often refer to market shares. If firm i's output is x_i, its market share is given by

$$s_i := \frac{x_i}{X},$$

where X denotes the sum of outputs of all n firms in the industry, i.e., $X = \sum_{i=1}^{n} x_i$.

A very simple measure of concentration is the rate of concentration C_k. It adds up the market shares of the k largest firms. Assuming $s_1 \geq s_2 \geq ...$ (which is just a matter of renaming), the k-rate of concentration C_k is given by

$$C_k = \sum_{i=1}^{k} s_i.$$

EXERCISE XXI.1. *Determine the C_2 rate of concentration for the following examples:*

(1) *Two firms with equal market shares.*
(2) *Three firms with market shares of $s_1 = 0.8$, $s_2 = 0.1$ and $s_3 = 0.1$.*
(3) *Three firms with market shares of $s_1 = 0.6$, $s_2 = 0.2$ and $s_3 = 0.2$.*

For n equally large firms, we obtain

$$C_k = \frac{k}{n}, k \leq n.$$

The more firms there are, the lower the rate of concentration. In general, measures of concentration yield 1 for the monopoly case and 0 for perfect competition. The rates of concentration fulfill this desideratum. The monopoly case leads to $k = n = 1$ and $\frac{k}{n} = 1$, while perfect competition means $n \to \infty$ and $\lim_{n \to \infty} \frac{k}{n} = 0$. One may not like the fact that the merger of two firms will not change C_k if the merged firm still does not belong to the k largest firms. This peculiarity is avoided by the Herfindahl index, to which we now turn.

The Herfindahl index H is another prominent concentration index. Rightfully, it should be named the Hirschman (1964) index; it is also often called the Herfindahl-Hirschman index. It is calculated by squaring the market shares of all firms in an industry and summing them up:

$$H = \sum_{i=1}^{n} \left(\frac{x_i}{X}\right)^2 = \sum_{i=1}^{n} s_i^2.$$

First, we check the monopoly and perfect competition cases. The monopoly case yields $H = 1^2 = 1$. If we have n equally large firms in the industry, we find

$$H = \sum_{i=1}^{n} \left(\frac{1}{n}\right)^2 = n \cdot \frac{1}{n^2} = \frac{1}{n},$$

a very nice result indeed. Perfect competition yields $H = 0$ by $\lim_{n\to\infty} \frac{1}{n} = 0$.

EXERCISE XXI.2. *Determine the Herfindahl index for the following examples:*

(1) *Two firms with equal market shares.*
(2) *Three firms with market shares of $s_1 = 0.8$, $s_2 = 0.1$ and $s_3 = 0.1$.*
(3) *Three firms with market shares of $s_1 = 0.6$, $s_2 = 0.2$ and $s_3 = 0.2$.*

EXERCISE XXI.3. *Can we be sure that the Herfindahl index increases when two firms merge?*

The Herfindahl index obviously takes account not only of the number of firms but also of the disparity of market shares. Indeed, the Herfindahl index is a function of the number of firms in the market, n, and the variation coefficient, V:

$$H = \frac{1 + V^2}{n}.$$

The variation coefficient is defined by

$$V = \frac{\text{standard deviation}}{\text{mean}} = \frac{\sqrt{\frac{1}{n}\sum_{i=1}^{n}\left(x_i - \frac{X}{n}\right)^2}}{\frac{X}{n}}.$$

EXERCISE XXI.4. *Can you show $H = \frac{1+V^2}{n}$? Hint: Express V^2 in terms of n and H!*

3. Models

3.1. Introduction. Regulation and competition policy aims to provide methods to make the market mechanism work. Obviously, we need to predict market outcomes (positive theory) and to judge them (normative theory). In this book, we consider the following models relevant for regulation and competition issues:

- perfect competition and the first welfare theorem (chapter XIX),
- Cournot monopoly model (chapter XI),
- oligopoly models such as the Bertrand, Cournot, and Stackelberg oligopoly models (chapter XI, also this chapter, pp. 551),
- structure-conduct-performance paradigm (this chapter, pp. 551),
- natural monopoly and Ramsey pricing (this chapter, pp. 556), and
- rate-of-return restriction in the private sector (this chapter, pp. 556).

3.2. The structure-conduct-performance paradigm. The structure-conduct-performance paradigm provides an effective approach to analyze industrial performance. Established by Mason (1939) and Bain (1956), this model uses three main categories:

- Structure
  - How many firms are there in the market/industry, how concentrated is the market/industry?
  - Do potential competitors face entry barriers?
  - Are the products differentiated?
- Conduct
  - Are prices close to marginal costs?
  - Are the products of high quality? Are they differentiated?
  - What distributional channels do the firms use? Do they offer their products in major cities only?
  - How much do firms spend on advertising?
  - How much do firms spend on research and development?

  Basically, conduct refers to the four p known from marketing: price, product, place, promotion.
- Performance
  - Do firms make profits?
  - Are the products safe?
  - Are prices close to marginal costs?
  - Do the firms successfully innovate (process or product innovation)?

The structure-conduct-performance paradigm is based on a simple idea. Structure determines conduct and conduct determines performance. For example, a monopoly (structure: one seller only) chooses the profit-maximizing price (conduct: marginal cost equals marginal revenue) so that the monopolist is well-off while consumers suffer (performance: monopoly profits, deadweight loss).

Of course, this is an incomplete picture. Profits may be used to finance research and development (performance influences conduct). Successful innovation may alter the industry structure through lower costs or different products (performance influcences structure). A third example: prices, advertising, or product differentiation (conduct) may be used to deter entry (structure).

3.3. Cournot, concentration, and monopoly power. There is a simple answer to the question why colluding might be frowned upon from the standpoint of economic welfare. Collusion means that several firms

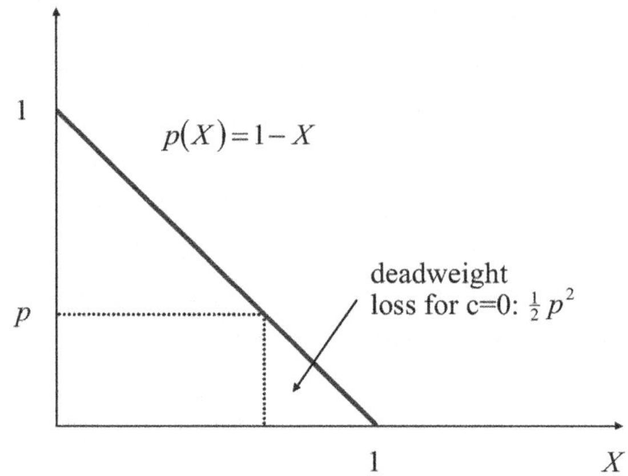

FIGURE XXI.1. The inverse linear demand curve

act as one firm. This is bad because according to the Cournot model the deadweight loss is higher for fewer firms. First, we show this by way of a very simple Cournot model with identical and constant marginal costs (subsection 3.3.1). In that model, all firms produce the same output and concentration is just a negative function of the number of firms. In general, not only the number of firms, but also their sizes, matter for concentration and market outcomes. Therefore, we have looked at different concentration measures in subsection 2.2 before linking one of these measures, the Herfindahl index, to the Cournot model in the upcoming subsection 3.3.2.

3.3.1. *Cournot and welfare.* The Cournot duopoly considers two firms producing outputs x_1 and x_2, respectively. The market output is $X = x_1 + x_2$. As in chapter XI, we assume a linear inverse demand function given by

$$p(X) = a - bX$$

and depicted in figure XXI.1 (we return to the deadweight loss in a minute).

Moreover, marginal costs are constant, at c_1 and c_2, respectively. The two firms' profit functions are

$$\begin{aligned} \Pi_1(x_1, x_2) &= (a - b(x_1 + x_2))\, x_1 - c_1 x_1 \text{ and} \\ \Pi_2(x_1, x_2) &= (a - b(x_1 + x_2))\, x_2 - c_2 x_2. \end{aligned}$$

In order to simplify the profit functions even further, we assume $a = 1$, $b = 1$, and $c_1 = c_2 = 0$ and obtain

$$\begin{aligned} \Pi_1(x_1, x_2) &= (1 - (x_1 + x_2))\, x_1 \text{ and} \\ \Pi_2(x_1, x_2) &= (1 - (x_1 + x_2))\, x_2. \end{aligned}$$

In case of identical unit costs below the prohibitive price, both firms produce a positive quantity in equilibrium. We find the reaction functions x_1^R and x_2^R, given by

$$\begin{aligned} x_1^R(x_2) &= \underset{x_1}{\operatorname{argmax}}\, \Pi_1(x_1, x_2) = \frac{1 - x_2}{2} \text{ and} \\ x_2^R(x_1) &= \underset{x_2}{\operatorname{argmax}}\, \Pi_2(x_1, x_2) = \frac{1 - x_1}{2}. \end{aligned}$$

The Cournot-Nash equilibrium is then the strategy vector $\left(x_1^C, x_2^C\right) = \left(\frac{1}{3}, \frac{1}{3}\right)$. Also, we find

$$\begin{aligned} X^C &= x_1^C + x_2^C = \frac{2}{3}, \\ p^C &= \frac{1}{3}, \\ \Pi_1^C &= \Pi_2^C = \frac{1}{3} \cdot \frac{1}{3} = \frac{1}{9}. \end{aligned}$$

EXERCISE XXI.5. *Calculate the reaction function of firm 1 in case of three firms. Use the above linear inverse demand function and identical and constant unit costs $c = 0$. Assuming a symmetric equilibrium, apply $x_1 = x_2 = x_3$ to that reaction function in order to arrive at the symmetric Nash equilibrium.*

Figure XXI.1 above reveals that a price p (above constant marginal costs of zero!) leads to the deadweight loss of $\frac{1}{2}p^2$. We can now show that the number of firms in the Cournot model is negatively related to the deadweight loss.

EXERCISE XXI.6. *Calculate the monopoly output and compare the deadweight loss occurring for 1, 2, and 3 firms.*

3.3.2. *Cournot, Lerner, and Herfindahl.* The aversion to collusion is due to the increasing deadweight loss. Alternatively, we can look at an important measure of monopoly power, the so-called Lerner degree of monopoly or Lerner measure. For an individual firm, the Lerner measure is equal to the price-cost margin

$$\frac{p - MC}{p}.$$

It equals zero for perfect competition and answers the question by how much a firm can increase its price above marginal cost. For the whole industry with n firms, the Lerner measure is the weighted average of the individual

Lerner measures where the market shares are used as weights:

$$\sum_{i=1}^{n} s_i \frac{p - MC_i}{p}$$

We will show that there is a close relation between the industry Lerner measure and the Herfindahl index.

Consider one specific firm i in Cournot competition. Firm i's revenue is given by

$$R(x_i) = p(X)\, x_i.$$

In the Cournot model, firms simultaneously choose their outputs so that $dx_j/dx_i = 0$ for $i \neq j$ and hence $dX/dx_i = 1$. Therefore, firm i's marginal revenue is

$$\frac{dR}{dx_i} = p + x_i \frac{dp}{dX}\frac{dX}{dx_i} = p + x_i \frac{dp}{dX}.$$

It can be expressed as a function of the firm's market share $s_i = \frac{x_i}{X}$ and the price elasticity of demand

$$\varepsilon_{X,p} = \frac{\frac{dX}{X}}{\frac{dp}{p}} = \frac{dX}{dp}\frac{p}{X}.$$

Since $\varepsilon_{X,p}$ is negative for negatively sloped demand functions (inverse or not), we often write $-|\varepsilon_{X,p}|$ instead of $\varepsilon_{X,p}$. Now, by way of some simple manipulations we get

$$\begin{aligned}
MR_i(x_i) &= p + x_i \frac{dp}{dX} \\
&= p\left(1 + \frac{x_i}{p}\frac{dp}{dX}\right) \text{ (factor out } p\text{)} \\
&= p\left(1 + \frac{x_i}{X}\frac{X}{p}\frac{dp}{dX}\right) \text{ (multiply by } X/X\text{)} \\
&= p\left(1 - s_i \frac{1}{|\varepsilon_{X,p}|}\right) \text{ (apply definitions).}
\end{aligned}$$

The profit maximization in a Cournot equilibrium requires the equalization of marginal revenue and marginal cost. Hence, the Lerner index for firm i in equilibrium is equal to

$$\frac{p - MC_i}{p} = \frac{p - p\left(1 - \frac{s_i}{|\varepsilon_{X,p}|}\right)}{p} = \frac{s_i}{|\varepsilon_{X,p}|}, \tag{XXI.1}$$

while the industry Lerner index equals

$$\sum_{i=1}^{n} s_i \frac{p - MC_i}{p} = \sum_{i=1}^{n} s_i \frac{s_i}{|\varepsilon_{X,p}|} = \frac{1}{|\varepsilon_{X,p}|}\sum_{i=1}^{n} s_i^2 = \frac{H}{|\varepsilon_{X,p}|}. \tag{XXI.2}$$

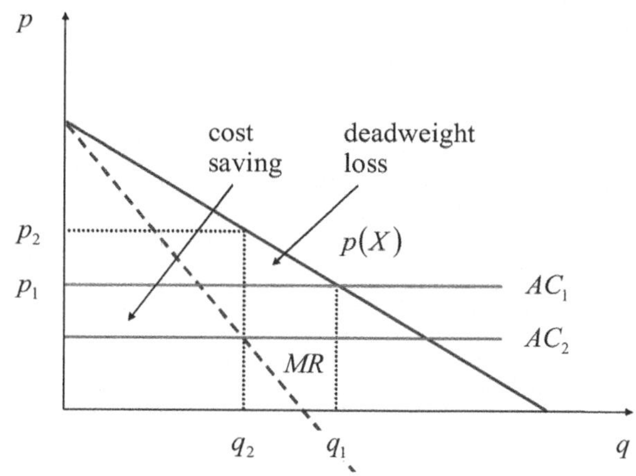

FIGURE XXI.2. Deadweight loss versus efficiency gain

Thus, the industry Lerner degree of monopoly power is the higher,

- the less elastic market demand and
- the more concentrated the market.

3.3.3. *Williamson's naive tradeoff model.* In our model there is no difference between collusion and merger of firms. Although competition policy typically opposes collusion, it has no clearcut view on mergers. The reason is that mergers may yield economies of scale. Figure XXI.2, taken from Williamson (1968, pp. 21), shows the ensuing tradeoff. Suppose that two firms (in Bertrand competiton) have average (and marginal) costs AC_1. Now, they merge which results in a cost saving of $AC_1 - AC_2$ per quantity produced. Now costs are lower and the merged firm uses the $MR = MC$ rule. On the one hand, there is a deadweight loss with respect to the old costs AC_1. On the other hand, the firms produce cheaper because costs are lower. The net effect is unclear.

Williamson (1968, pp. 26) also comments on the danger that a specific merger could start a merger trend which may be undesirable. Other economists of the Chicago school of economics tend to stress the efficiency gains, for example Bork & Bowman, Jr. (1965, p. 594): "The difficulty with stopping a trend toward a more concentrated condition at a very early stage is that the existence of the trend is prima facie evidence that greater concentration is socially desirable. The trend indicates that there are emerging efficiencies or economies of scale – whether due to engineering and production developments or to new control and management techniques ... "

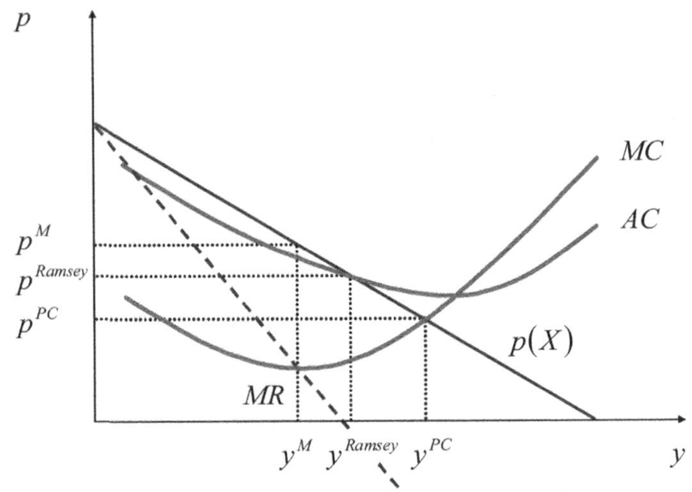

FIGURE XXI.3. A conflict between non-negative profits and welfare maximization

3.4. Natural monopoly and Ramsey prices. Some industries can be said to be natural monopolies. This means that we have a cost structure such that one firm is best suited to serve the market at minimal cost. Consider figure XXI.3 where you see three prices:

- the monopoly price p^M,
- the marginal-cost, or perfect-competition, price p^{PC} and, finally,
- the Ramsey price p^{Ramsey}.

While the marginal-cost price p^{PC} is welfare-maximizing, it may well lead to negative profits as can be seen from the figure. If negative profits are to be avoided for some reason, one may consider the following optimization problem: Choose the welfare-maximizing prices compatible with nonnegative profits. These prices are called Ramsey prices. Indeed, price p^{Ramsey} does the trick.

One may consider applying Ramsey pricing to public utilities (such as electricity or water supply). However, as a practical rule, Ramsey pricing has its drawbacks. First, the informational requirements (cost functions, elasticities!) are prohibitive. Second, in the model the cost functions are given and not subject to the behaviour of the public utilities. Thus, the incentives of public utilities do no play any role. Third, Ramsey pricing does not take account of distributional issues.

3.5. Restricting the rate of return in the private sector. Picking up on the problem described in the previous section, the government might be concerned about excessive profits in the private sector. It may try to impose maximal cash-flow returns or maximal returns on profits. The famous Averch-Johnson model shows that such a profit constraint may lead to an inefficient use of the factors of production.

We use the following definitions:

$$\begin{aligned}\text{cash-flow return} &= \frac{\text{revenue} - \text{labor costs}}{\text{capital}},\\ \text{return on profit} &= \frac{\text{profit}}{\text{capital}}.\end{aligned}$$

Let K and L be the factors of capital and labor, respectively, and let i and w be the factor prices. By $R(K,L)$, we denote the revenue obtainable from the input of K units of capital and L units of labor. Then, the firm's profit is given by

$$\Pi(K,L) = R(K,L) - wL - iK.$$

If the government imposes a maximal cash-flow return of s, capital and labor are to be chosen in accordance with

$$\frac{R(K,L) - wL}{K} \leq s.$$

By

$$\begin{aligned}&\frac{R(K,L) - wL}{K} \leq s\\ \Leftrightarrow\quad & R(K,L) - wL \leq sK\\ \Leftrightarrow\quad & R(K,L) - wL - iK \leq (s-i)K\\ \Leftrightarrow\quad & \frac{\Pi(K,L)}{K} \leq s - i\end{aligned}$$

a cash-flow return of s corresponds to a profit return of $s - i$. In this sense, a cash-flow return constraint is equivalent to a profit return constraint.

According to the last inequality, the maximal profit attainable is

$$\pi(K,L) = (s-i)K.$$

Now, since the use of more capital allows the firm a higher profit, it has an incentive to substitute labor by capital. This is the result found by Averch & Johnson (1962) and hinted at in figure XXI.4. In that figure, profit is depicted as a function of capital K where labor $L(K)$ is chosen optimally for the respective level of capital.

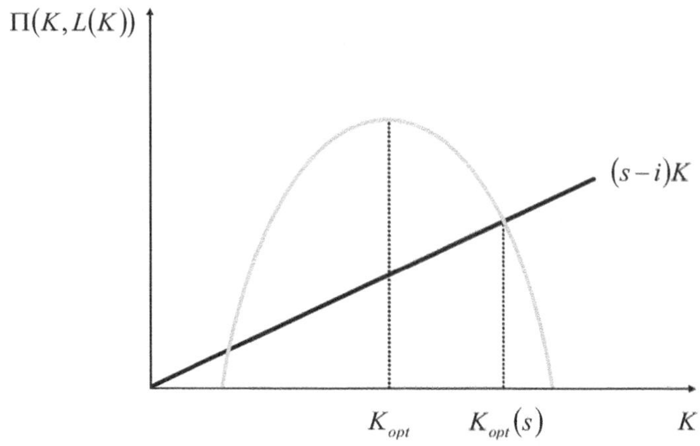

FIGURE XXI.4. Overuse of capital

4. Overall concepts of competition (policy)

4.1. Classical liberalism. Adam Smith favors open markets and the abolishment of monopoly privileges, granted by the government. He does not advocate a specific anti-cartel or anti-merger policy.

However, Smith identifies sectors where competition would not work and where the government has to provide these goods: streets, bridges, canals, ports, postal services, water.

4.2. Perfect competition and GET. General equilibrium theory and perfect competition (chapter XIX) provide important perspectives on competition. According to this theory, many small firms choose quantities by equalizing marginal cost and price.

4.3. Freiburg school of "ordoliberalism". Walter Eucken (1990, pp. 255-299) not only favors liberal principles such as open markets, private property, and freedom of contract, but also advocates a "Monopolamt". In chapter XX (pp. 532) we show how Eucken wanted to apply the concept of perfect competition to competition policy. The German "Act against Restraints of Competition" (Gesetz gegen Wettbewerbsbeschränkungen, GWB) is partly inspired by Eucken's ideas.

4.4. Chicago school of antitrust policy. The Chicago School of Antitrust Policy argues that cartels, mergers, and other business practices, that many economists like Eucken criticize, are beneficial to consumers (see also p. 555). From this point of view the main problem is that monopoly power is bestowed by the government. The Chicago School representatives argue that the model of perfect competition has no prominent status. This is also true for the Harvard School.

According to William Landes (see Kitch 1983, p. 193), Ronald Coase, a famous member of the Chicago school, said "he had gotten tired of antitrust because when the prices went up the judges said it was monopoly, when the prices went down, they said it was predatory pricing, and when they stayed the same, they said it was tacit collusion."

4.5. Harvard school of workable competition. The Harvard school of workable competition has an empirical focus. Starting from the structure-conduct-performance paradigm (pp. 551), the Harvard scholars try to identify structural elements that lead to a good or bad performance.

4.6. Contestable markets. While the model of perfect competition deals with actual competition, contestable-markets theory focuses on potential competition in an extreme manner.

4.7. Kantzenbach's model of optimal competition intensity. Erhard Kantzenbach (1966) argues that competition intensity depends on how fast firms react to advances of other competitors (lower prices, higher quality etc.). He suggests that in case of many firms on the market, advances of individual competitors are not important to the other firms. From this point of view (potential competition intensity), a small number of firms is more conducive to competition intensity than a larger one. However, since cartelization is more likely with a small number of firms, the so-called effective competition intensity is maximized for 4 to 6 competitors. An early critique of Kantzenbach's approach is Hoppmann (1966).

4.8. The Austrian school. The Austrian School of Economics does not subscribe to the way competition is portrayed in models of perfect competition (and industrial organization). Instead,

- Friedrich Hayek insists on the importance of ongoing change in an economy and on competition as a discovery process (what would the use of competition be if all the relevant facts were given?) while
- Israel Kirzner stresses the importance of entrepreneurial discovery.

In chapter XX (pp. 533) we present central quotes from this strand of thought.

4.9. Joseph Schumpeter. Joseph Schumpeter describes economic and technical change as a process of creative destruction (see pp. 538).

4.10. Freedom and competition. Hoppmann (1966) stresses freedom of competition ("Wettbewerbsfreiheit"). While competition may further economic welfare, this fact should not be the main argument in favor of the freedom of economic actors to engage in competition.

5. Topics

The main topics in this chapter are

- relevant market
- cross price elasticity of demand
- supply-side substitutes
- SSNIP-Test
- price correlation test
- measures of concentration
- rate of concentration
- Herfindahl index
- perfect competition and the first welfare theorem
- Lerner index of monopoly power
- Williamson's naive tradeoff model
- oligopoly models (due to Bertrand, Cournot and Stackelberg)
- structure-conduct-performance paradigm
- natural monopoly and Ramsey pricing
- rate-of-return restriction in the private sector (Averch-Johnson theorem)
- classical liberalism
- Freiburg school of "ordoliberalism"
- Chicago school of antitrust policy
- Harvard school of workable competition
- contestable markets
- Kantzenbach's model of optimal competition intensity
- Austrian school
- competition laws

6. Solutions

Exercise XXI.1

Did you find

(1) $\frac{1}{2} + \frac{1}{2} = 1$,
(2) $0.8 + 0.1 = 0.9$, and
(3) $0.6 + 0.2 = 0.8$?

Exercise XXI.2

You have found

(1) $H = 0.5$,
(2) $H = 0.66$, and
(3) $H = 0.44$.

Exercise XXI.3

If the two firms 1 and 2 merge, the Herfindahl index increases by

$$(s_1 + s_2)^2 - \left[s_1^2 + s_2^2\right] = 2s_1 s_2 \geq 0$$

with strict inequality iff $s_1 > 0$ and $s_2 > 0$ hold.

Exercise XXI.4

First we find

$$\begin{aligned}
V^2 &= \frac{\frac{1}{n}\sum_{i=1}^n \left(x_i - \frac{X}{n}\right)^2}{\frac{X^2}{n^2}} = \frac{\frac{1}{n}\sum_{i=1}^n \left(x_i^2 + \frac{X^2}{n^2} - 2x_i \frac{X}{n}\right)}{\frac{X^2}{n^2}} \\
&= \frac{n^2}{n}\sum_{i=1}^n \left(\frac{x_i^2 + \frac{X^2}{n^2} - 2x_i\frac{X}{n}}{X^2}\right) = n\sum_{i=1}^n \left(\left(\frac{x_i}{X}\right)^2 + \frac{1}{n^2} - 2\frac{x_i}{X}\frac{1}{n}\right) \\
&= n\sum_{i=1}^n \left(\frac{x_i}{X}\right)^2 + n\sum_{i=1}^n \frac{1}{n^2} + n\sum_{i=1}^n \left(-2\frac{x_i}{X}\frac{1}{n}\right) \\
&= nH + n\left(n\frac{1}{n^2}\right) - 2\sum_{i=1}^n \frac{x_i}{X} \\
&= nH + 1 - 2 = nH - 1
\end{aligned}$$

and then the required formula.

Exercise XXI.5

Firm 1's profit function is given by

$$\Pi_1(x_1, x_2, x_3) = (1 - (x_1 + x_2 + x_3))\, x_1.$$

Forming the derivative with respect to x_1 and solving for x_1 yields

$$x_1^R(x_2, x_3) = \underset{x_1}{\operatorname{argmax}}\, \Pi_1(x_1, x_2, x_3) = \frac{1 - (x_2 + x_3)}{2}.$$

Assuming a symmetric equilibrium, we obtain $x_1 = \frac{1-(x_1+x_1)}{2}$ and hence $\left(x_1^C, x_2^C, x_3^C\right) = \left(\frac{1}{4}, \frac{1}{4}, \frac{1}{4}\right)$. Also, we find

$$\begin{aligned}
X^C &= x_1^C + x_2^C + x_3^C = \frac{3}{4}, \\
p^C &= \frac{1}{4}, \\
\Pi_1^C &= \Pi_2^C = \Pi_3^C = \frac{1}{4} \cdot \frac{1}{4} = \frac{1}{16}.
\end{aligned}$$

Exercise XXI.6

The monopoly profit function is given by $\Pi(X) = (1 - X)\, X$ so that $X^C = \frac{1}{2}$ maximizes the monopoly's profit. The following table yields the desired comparison:

number of firms	X^C	price	welfare loss
1	$\frac{1}{2}$	$\frac{1}{2}$	$\frac{1}{8}$
2	$\frac{2}{3}$	$\frac{1}{3}$	$\frac{1}{18}$
3	$\frac{3}{4}$	$\frac{1}{4}$	$\frac{1}{32}$

7. Further exercises without solutions

PROBLEM XXI.1. A labor union campaigns for a rule that forces companies to have a profit share of at least σ spend on the worker saleries. Can we draw a similar conclusion as in the Averch-Johnson model?

Part G

Contracts and principal-agent theories

Contract theory is the topic of this last part. In most models, there is a (badly informed) principal and a (well-informed) agent. If the informational asymmetry concerns the type of the agent, we have an adverse-selection model (chapter XXII). The archetypical example is a worker whose ability is unknown to the prospective employer. The second class of asymmetric-information models considers actions undertaken by an agent. For example, the agent may work diligently or be a lazy sod. Since the principal cannot observe (or verify) the action, we talk about hidden-action models (chapter XXIII).

CHAPTER XXII

Adverse selection

In chapter XVIII, we consider a principal who defines a mechanism for agents whose types he does not know. We pursue this theme in this chapter. The models presented here differ from the mechanism design model in (at least) two respects:

- Apart from the principal, we often consider one player (often called the agent), only.
- We explicitly introduce the participation constraint for the agent. Thus, the agent can say "no" to all the proposals put forth by the principal.

These two characteristics hold for adverse-selection models as well as for hidden-action models (that we consider in the next chapter). This chapter describes the problem of adverse selection and presents two different ways to mitigate it. In screening models, the principal moves first and offers different contracts so that, ideally, different types reveal themselves. In signaling models, the agent moves first and undertakes an effort to convince the principal that he is of a good type. This chapter borrows from Varian (1992, chapter 25) and Rasmusen (2001, chapter 9).

1. Introduction and an example

In 2001, the Sveriges Riksbank Prize in Economic Sciences in Memory of Alfred Nobel was awarded to the US economists George A. Akerlof (University of California at Berkeley), A. Michael Spence (Stanford University), and Joseph E. Stiglitz (Columbia University) "for their analyses of markets with asymmetric information." According to the press release by the Royal Swedish Academy of Sciences,

> [m]any markets are characterized by asymmetric information: actors on one side of the market have much better information than those on the other. Borrowers know more than lenders about their repayment prospects, managers and boards know more than shareholders about the firm's profitability, and prospective clients know more than insurance

H. Wiese, *Advanced Microeconomics*,
https://doi.org/10.1007/978-3-658-34959-2_22

> companies about their accident risk. During the 1970s, this year's Laureates laid the foundation for a general theory of markets with asymmetric information. Applications have been abundant, ranging from traditional agricultural markets to modern financial markets. The Laureates' contributions form the core of modern information economics.
>
> George Akerlof demonstrated how a market where sellers have more information than buyers about product quality can contract into an adverse selection of low-quality products. He also pointed out that informational problems are commonplace and important. Akerlof's pioneering contribution thus showed how asymmetric information of borrowers and lenders may explain skyrocketing borrowing rates on local Third World markets; but it also dealt with the difficulties for the elderly to find individual medical insurance and with labour-market discrimination of minorities.
>
> Michael Spence identified an important form of adjustment by individual market participants, where the better informed take costly actions in an attempt to improve on their market outcome by credibly transmitting information to the poorly informed. Spence showed when such signaling will actually work. While his own research emphasized education as a productivity signal in job markets, subsequent research has suggested many other applications, e.g., how firms may use dividends to signal their profitability to agents in the stock market.
>
> Joseph Stiglitz clarified the opposite type of market adjustment, where poorly informed agents extract information from the better informed, such as the screening performed by insurance companies dividing customers into risk classes by offering a menu of contracts where higher deductibles can be exchanged for significantly lower premiums. In a number of contributions about different markets, Stiglitz has shown that asymmetric information can provide the key to understanding many observed market phenomena, including unemployment and credit rationing.

Following the work done by the Nobel laureates and others, we consider informational asymmetries that are already present before the players decide

Screening

Principal chooses menu of contracts.	Nature chooses the agent's type.	Agent chooses signal.	Agent decides which contract (if any) to accept.

Signaling

Nature chooses the agent's type.	Agent chooses signal.	Principal observes the signal, but not the type and chooses contract.	Agent decides whether to accept or not.

FIGURE XXII.1. Screening and signaling

whether or not to accept the contract or which contract to accept. The so-called adverse-selection models deal with these problems. For example,

- the ability of a worker is known to the worker (agent) but not to the firm (principal) who considers to hire the worker,
- the car driver (agent) is better informed than the insurance company (principal) about the driver's accident-proneness, and, finally,
- the owner of a used car for sale (agent) may have a very good idea about the quality of that car while the potential buyer (principal) does not.

The problem of adverse selection is this: for a given wage, a given insurance premium, or a given price for a used car, the badly qualified workers, the high-risk insurees, and the owners of bad cars are more eager to enter into a contract than the opposite types of agents. The qualified workers have alternative employment possibilities, the low-risk insurees do not need the insurance as badly, and the good cars are of use to their owners. At first sight, the informational asymmetry is a problem for the badly informed party, the principal. However, the principal's problem immediately turns into a problem for the agent. It is the agent who needs to convince the principal that he is of a "good type".

Two different classes of models exist (see figure XXII.1):

- In screening models, the principal moves first. He offers a menu of contracts to the agent. Some contracts are more attractive to good types, others more attractive to bad types. In this manner, the principal tries to attenuate his informational deficit.

- In signaling models, the agent moves first. Depending on his type, signals he might send are more or less costly. On observing these signals, the principal offers a contract.

Screening and signaling are important activities in labor markets. Workers send a signal by obtaining university or other degrees. Firms screen by offering different kinds of bonus systems.

2. A polypsonistic labor market

2.1. The market model. The model of adverse selection is originally due to Akerlof (1970) who considers the market for used cars. Our example refers to a labor-market model taken from Mas-Colell et al. (1995, p. 437-443).

We begin with a short informal description of the labor market. Labor is the only input. We have many identical firms with a binary (yes or no) unit demand for labor. Thus, the firms want to employ one worker or none. In a sense, there are many more firms than potential workers. Therefore, some firms may end up without a worker. The workers are not homogeneous but differ in their productivities:

- The marginal productivity t of any worker is constant. As a handy shortcut, we address the worker with productivity t as worker t. There is a continuum of workers on the type interval $T := \left[\underline{t}, \overline{t}\right]$ with $0 \leq \underline{t} < \overline{t} < \infty$.
- Worker t's marginal productivity is the same in every firm that hires the worker.
- $r(t)$ is the opportunity cost of employment for worker t. $r(t)$ can be interpreted as worker t's productivity when working on his own.

We now proceed to the formal definition:

Definition XXII.1 (polypsonistic labor market). *A polypsonistic labor market is a tuple* $\Gamma = \left(P, T, \tau, r, A, (u_t)_{t \in T}, u_P\right)$ *where*

- P *is the principal (or the set of identical principals),*
- $T = \left[\underline{t}, \overline{t}\right]$ *is the set of agents respectively types,*
- τ *is a probability (or density) distribution on* T,
- $r : T \to \mathbb{R}$ *is the reservation-wage function,*
- $A = \{\mathbf{y}, \mathbf{n}\}$ *is the action set for each agent with actions* $\mathbf{y}$ *("yes" or "accept") and* $\mathbf{n}$ *("no" or "decline"),*
- $u_t : \mathbb{R} \times A \to \mathbb{R}$ *is agent* t*'s payoff function defined by*

$$u_t(w, a) = \begin{cases} w, & a = \mathbf{y} \\ r(t), & a = \mathbf{n} \end{cases}$$

and

- $u_P : \mathbb{R}\times 2^T \to \mathbb{R}$ *is the principal's payoff function defined by*

$$u_P(w, W) = E[t : t \in W] - w$$

In this definition, w stands for wage rate and $W \subseteq T$ for the set of workers.

$$E[t : t \in W] := \int_{t\in W} t \frac{\tau(t)}{\int_{\bar{t}\in W} \tau(\bar{t})\, d\bar{t}} dt$$

is the average productivity of all the types who are employed. Take note of the conditional density. We let $E[t : t \in \emptyset] := E[t : t \in T]$.

DEFINITION XXII.2 (AGENT STRATEGY). *Let* Γ *be a polypsonistic labor market. An agent strategy is a function* $s : T \to A$. *The function* $s^w : T \to A$, *given by*

$$s^w(t) = \begin{cases} \mathbf{y}, & w \geq r(t) \\ \mathbf{n}, & w < r(t) \end{cases}$$

is called the optimal agent strategy at wage rate w.

2.2. Observable productivity. We first consider the situation where the productivity level is readily observable. By perfect competition between the principals, every worker is paid his marginal product:

$$w^*(t) = t, t \in T$$

Using the optimal agent strategy defined in the previous subsection, we obtain

$$s^{w^*(t)=t}(t) = \begin{cases} \mathbf{y}, & t \geq r(t) \\ \mathbf{n}, & t < r(t) \end{cases}$$

so that workers obeying $r(t) \leq t$ get employed. These are the workers who produce less on their own than in the firm. This is a Pareto-efficient outcome. Note, also, that the firms do not make any profit.

2.3. Unobservable productivities. If the productivities are not observable, all the workers obtain the same wage. At wage w, those workers $t \in T$ get employed whose opportunity cost is lower than the wage:

DEFINITION XXII.3 (LABOR SUPPLY). *Let* Γ *be a polypsonistic labor market and* s^w *the optimal agent strategy. Then,*

$$T(w) := \{t \in T : s^w(t) = \mathbf{y}\}$$

is called the labor supply.

Of course, we have $T(w) = \{t \in T : w \geq r(t)\}$. Note that the labor supply definition implies an optimal response by the agents to the wage set by the principal. This is one ingredient into our equilibrium definition. The other is the zero-profit condition:

DEFINITION XXII.4 (EQUILIBRIUM). *Let Γ be a polypsonistic labor market under nonobservability. A wage $\widehat{w}$ and the worker set $\hat{W}$ form an equilibrium if we have*

- $t \in \hat{W} \Leftrightarrow s^{\widehat{w}}(t) = \mathbf{y}$ *and*
- $\widehat{w} = E\left[t : t \in \hat{W}\right]$.

Thus, our definition of a polypsonistic labor market in equilibrium has two requirements. At wage rate $\widehat{w}$, the agents who want to be employed are employed and the other agents are not. This condition can also be written as $\hat{W} = T(\widehat{w})$. Every polypsonist's expected payoff is zero. This condition makes sure that no firm has an incentive to enter or leave the market.

By

$$\begin{aligned} \widehat{w} &= E\left[t : t \in \hat{W}\right] \\ &= E\left[t : s^{\widehat{w}}(t) = \mathbf{y}\right] \\ &= E\left[t : t \in T(\widehat{w})\right] \\ &= E\left[t : \widehat{w} \geq r(t)\right] \end{aligned}$$

the two conditions can be summarized in just one equation:

$$\widehat{w} = E\left[t : \widehat{w} \geq r(t)\right]. \qquad \text{(XXII.1)}$$

Thus, the equilibrium is characterized by a wage rate $\widehat{w}$ such that the workers, who want to be employed at that wage rate, have an average productivity of $\widehat{w}$. This fixed-point equation plays an important role in the following subsections.

2.4. Inefficient equilibria. Adverse selection may induce the inefficiency that (i) all workers should be employed, but (ii) are not in equilibrium. We consider the following class of examples:

- We have $r(t) \leq t$ for all $t \in T$, i.e., all agents should be employed.
- r is a monotonicly increasing function, i.e., productive workers have better outside options.

The equilibrium condition XXII.1

$$\widehat{w} = E\left[t : \widehat{w} \geq r(t)\right]$$

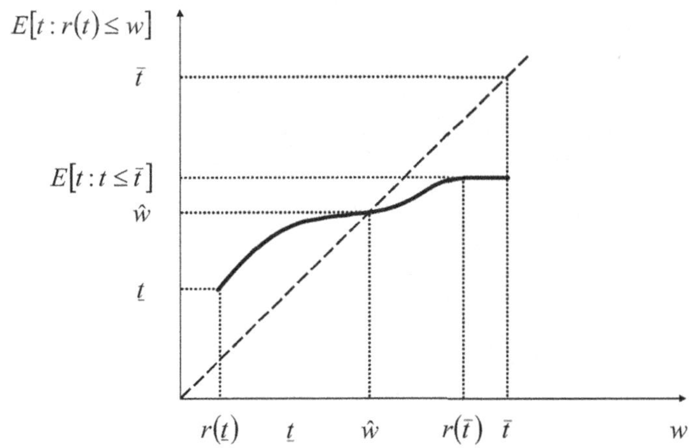

FIGURE XXII.2. Adverse selection: first case

can now be visualized as the intersection point of the average-productivity curve $E[t : r(t) \leq w]$ and the 45-degree line. See figure XXII.2, which is taken from Mas-Colell et al. (1995, Figure 13.B.1, p. 441). You see the wage at the abscissa. If the wage rate is at $r(\underline{t})$, the agent with the lowest productivity is employed and the average productivity is $\underline{t} \geq r(\underline{t})$. By the above assumptions, an increasing wage rate implies that increasingly able workers accept the contract and that average productivity increases (remember marginal something larger than average something on p. 223). Of course, the average productivity of all employees, $E[t : t \leq \bar{t}]$, is below the productivity of the most productive workers $\bar{t}$. Figure XXII.2 represents the case of $r(\bar{t}) > E[t : t \leq \bar{t}]$. If the firm(s) wanted to ensure the employment of the most productive worker, also, the wage rate would have to rise up to $r(\bar{t})$, so that the wage would be above the average productivity.

Figure XXII.3 depicts the case of $r(\bar{t}) < E[t : t \leq \bar{t}]$. Here, there are not many badly qualified workers so that the curve of average productivity is rather steep. Despite asymmetric information, we might obtain an efficient outcome. Although the wage $r(\bar{t})$ is sufficient to employ all the workers, the average productivity is above this wage rate. Entry of new firms drives up the workers' wage. Some of the firms may not be able to secure a worker for themselves. However, this is not a problem because the firms do not make any positive profit in equilibrium.

Consider, finally, figure XXII.4 where the market collapses wholesale. A numerical example may help to understand the underlying mechanism:

- We have $r(t) = \alpha t$, where $\frac{1}{2} < \alpha < 1$.

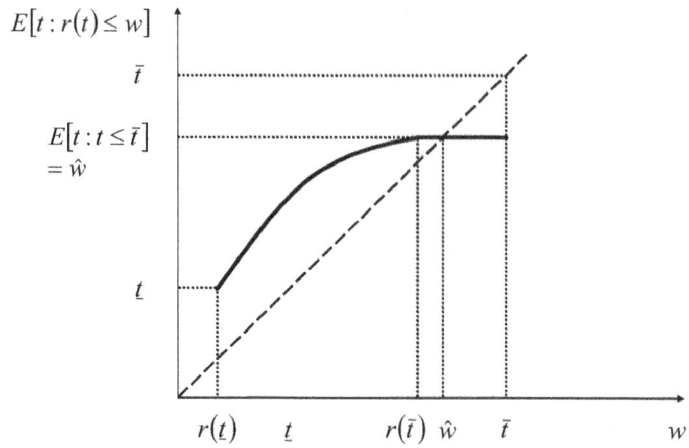

FIGURE XXII.3. Adverse selection: second case

- t is distributed equally on $[0, 2]$.

In this case, we have

- $r(\underline{t}) = r(0) = 0 = \underline{t}$,
- $r(t) = \alpha t < t$ for all $t > 0$ (i.e., all the agents should go to work)
- $E[t : r(t) \leq w] = \frac{1}{2}\frac{w}{\alpha} < w$.
- Why average productivity $\frac{1}{2}\frac{w}{\alpha}$?
  - The most productive worker has reservation price $w = r(t) = \alpha t$ and
  - thus the productivity $\frac{w}{\alpha}$.
- By $E[t : r(t) \leq w] < w$ firms would expect to suffer losses.

3. A polypsonistic labor market with education

3.1. Introduction. Asymmetric information can lead to inefficient outcomes as we have seen in the previous section. Signaling is an action undertaken by the informed party. However, just saying "I am a good type" does not help. The reason is that bad types may have an incentive to claim the same if the principal treats good types better than bad types. Useful signals need to sort good and bad types. For this to happen, the incentives to associate with one group or another need to depend on the types. We use a simple two-type model to shed light on screening and signaling. The underlying seminal paper is Spence (1973).

3.2. The market model. In rough terms, our model looks like this. As in the previous model, labor is the only input and the many identical

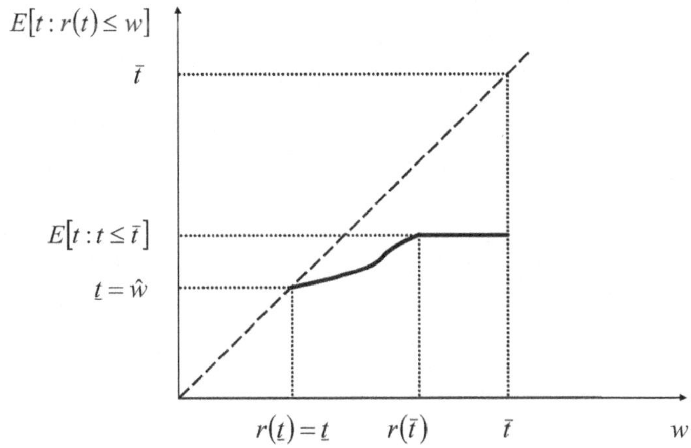

FIGURE XXII.4. Adverse selection: third case

firms have a binary unit demand for labor. We now assume that workers need to work, but are free to choose the best wage offered to them.

The signaling/screening aspect refers to the educational effort the prospective workers exert. Workers come in two types, "good" and "bad". We assume that good workers suffer less from schooling. As we are focussing on the informational function of education, we assume that education has no effect on productivity.

Schooling and the cost of schooling are modeled in the following manner:

- We have two groups of workers, the good ones with high productivity $t_h > 0$ and the bad ones with low productivity $t_l > 0$ where $t_h > t_l$ holds.
- The workers' time in schools and universities is denoted by $a \geq 0$ (*education*). The workers have a disutility from schooling of $c(t)\,a$ where $c_h := c(t_h)$ refers to high-productivity workers and $c_l := c(t_l)$ to those with low productivity. We assume $c_l > c_h > 0$.

We now proceed to the formal definition:

DEFINITION XXII.5 (POLYPSONISTIC LABOR MARKET WITH EDUCATION). *A tuple $\Gamma = \left(P, T, \tau, A, (u_t)_{t \in T}, u_P\right)$ is a polypsonistic labor market with education where*

- *P is the principal (or the set of identical principals),*
- *$T = \{t_h, t_l\}$ is the set of types (agents),*
- *τ is a probability distribution on T with $\tau_h := \tau(t_h)$ and $\tau_l := \tau(t_l) = 1 - \tau_h$ denoting the respective portions of workers,*

- $A = \mathbb{R}_+$ *is the action set for each agent with actions* $a \in A$ *denoting the number of school years,*
- $u_t : \mathbb{R} \times A \to \mathbb{R}$ *is the payoff function for agent* t, $t \in \{t_h, t_l\}$, *defined by*

$$u_t(w, a) = w - c(t)\, a$$

and

- $u_P : \{t_h, t_l\} \times \mathbb{R} \to \mathbb{R}$ *is the principal's nonprobabilistic payoff function defined by*

$$u_P(t, w) = t - w$$

where, again, w *stands for wage rate.*

This is the general model of a polypsonistic labor market with education.

3.3. Observable productivity. If the productivity is readily observable, the equilibirum wages are

$$\begin{aligned} w^*(t_h) &= t_h > \tau_h t_h + \tau_l t_l \text{ and} \\ w^*(t_l) &= t_l < \tau_h t_h + \tau_l t_l \end{aligned}$$

and every worker gets his marginal productivity. In that case, there is no need to suffer the burden of education and we have a Pareto-efficient outcome.

3.4. Unobservable productivity (pooling equilibrium). Assume, now, that the productivities are not observable. We have two types of equilibria. Equilibria are called separating if different types are treated differently, and pooling, if all the types are treated the same.

We cannot obtain a separating equilibrium if the principals do not pay different wages to agents of different education. By the zero-profit condition, the wage rate offered to all agents is

$$w^* = \tau_h t_h + \tau_l t_l$$

In that case, all workers choose

$$a^* = 0.$$

(w^*, a^*) is indeed an equilibrium irrespective of the sequence of events:

- either the principals offer the uniform wage rate first and then the workers decide on a,
- or the workers first decide on a and then the principals offer a uniform wage rate.

Thus, wage differentiation is necessary for a separating equilibrium. Note that the good workers obtain less than their average productivity and the bad workers obtain more. Therefore, the good workers have an incentive to be screened or to signal their good quality.

4. A polypsonistic labor market with education and screening

4.1. Sequence and strategies. According to figure XXII.1, screening is a four-stage model:

(1) The principals choose wage contracts which specify the wage as a function of schooling. While any function $w : A \to \mathbb{R}$ is open to the principals a priori, we restrict attention to binary wage contracts where a minimum schooling of some a^+ leads to a high wage and a schooling below a^+ to a low wage.
(2) Nature chooses each worker's type, t_h or t_l.
(3) The workers decide about their education effort $a \in A$ after learning their type and after learning the wage functions.
(4) Finally, the workers decide which firm to choose. We do not explicitly model this last stage. We simply take it for granted that the workers choose a firm with a maximal wage rate, given the years of schooling.

As an extensive-form game, this game is one of perfect information. The workers learn their types and the principals do not move after nature's decision. The sequence of events described above entails the strategies open to the agents and the principals.

DEFINITION XXII.6 (SCREENING STRATEGIES). *Let Γ be a polypsonistic labor market with education and screening. A principal strategy is a tuple $s_P = (a^+, w_h, w_l)$ from $A \times \mathbb{R} \times \mathbb{R}$ that defines a wage function $w : A \to \mathbb{R}$ given by*

$$w(a) = \begin{cases} w_h, & a \geq a^+ \\ w_l, & a < a^+ \end{cases}$$

An agent strategy is a function $s : T \times A \times \mathbb{R} \times \mathbb{R} \to A$, where

$$s\left(\underbrace{t}_{type}, \underbrace{a^+}_{\substack{minimal\\schooling}}, \underbrace{w_h}_{high\ wage}, \underbrace{w_l}_{low\ wage}\right) \in A$$

is the educational effort chosen by the agent.

4.2. Separating equilibria. We now deal with the conditions for a separating equilibrium (where different types are treated differently):

DEFINITION XXII.7 (SCREENING EQUILIBRIA). *Let Γ be a polypsonistic labor market with education and screening. The strategy combination $(\hat{s}_P, \hat{s}) = ((\hat{a}^+, \hat{w}_h, \hat{w}_l), \hat{s})$ (with the principal strategy $\hat{s}_P$ and the agent strategy $\hat{s}$) is a separating equilibrium*

- *if the principal differentiates wages and maximizes his profits which are zero, i.e., if $\hat{s}_P = (\hat{a}^+, \hat{w}_h, \hat{w}_l)$ obeys $\hat{w}_h = t_h$ and $\hat{w}_l = t_l$,*
- *if the different types act differently:*

$$\hat{s}\left(t, \hat{a}^+, \hat{w}_h, \hat{w}_l\right) \begin{cases} \geq a^+, & t = t_h \\ < a^+ & t = t_l \end{cases}$$

  and
- *if the agents maximize their payoff, i.e., if*

$$\hat{s}\left(t, a^+, w_h, w_l\right) = \arg\max_{a \in A} \begin{cases} w_h - c(t)\, a, & a \geq a^+ \\ w_l - c(t)\, a, & a < a^+ \end{cases}$$

  holds for all $(t, a^+, w_h, w_l) \in T \times A \times \mathbb{R} \times \mathbb{R}$.

Thus, in a separating screening equilibrium,

- the principals pay productivity wages and earn zero payoff,
- agent t_h chooses an education level equal to or higher than a^+, while agent t_l goes to school for less than a^+ years, and
- every agent chooses his number of school years in a payoff-maximizing fashion.

If $(\hat{s}_P, \hat{s})$ is a separating equilibrium where the good types suffer education and the bad types do not, the agent's payoff maximization implies

$$\hat{s}\left(t, \hat{a}^+, \hat{w}_h, \hat{w}_l\right) = \begin{cases} a^+, & t = t_h \\ 0, & t = t_l \end{cases} \tag{XXII.2}$$

which is immediate by $c_l > 0$ and $c_h > 0$. We now need to make sure that the agents do indeed separate themselves in this manner. Consider the highly-productive agent t_h. He prefers to follow equation XXII.2 if

$$\underbrace{w_h - c_h a^+}_{\substack{\text{The productive} \\ \text{agent's payoff} \\ \text{in case of} \\ \text{extensive schooling}}} \quad \geq \quad \underbrace{w_l - c_h \cdot 0}_{\substack{\text{The productive} \\ \text{agent's payoff} \\ \text{in case of} \\ \text{no schooling}}}$$

or, solving for a^+,

$$a^+ \leq \frac{w_h - w_l}{c_h}$$

holds. This condition makes sure that the salary difference $w_h - w_l$ outweights the productive type's cost of education $c_h a^+$.

The unproductive agent t_l is happy not to choose any education in case of

$$\underbrace{w_h - c_l a^+}_{\substack{\text{The unproductive}\\ \text{agent's payoff}\\ \text{in case of}\\ \text{extensive schooling}}} \leq \underbrace{w_l - c_l \cdot 0}_{\substack{\text{The unproductive}\\ \text{agent's payoff}\\ \text{in case of}\\ \text{no schooling}}}$$

or, solving again for a^+,

$$a^+ \geq \frac{w_h - w_l}{c_l}.$$

EXERCISE XXII.1. *What is the intuitive idea behind the above inequality?*

Given the fact of separation, the zero-profit condition implies $\hat{w}_h = t_h$ and $\hat{w}_l = t_l$. So far, we have identified necessary conditions for a separating equilibrium. These are also sufficient:

LEMMA XXII.1. *Let Γ be a polypsonistic labor market with education and screening. The separating equilibria $(\hat{s}_P, \hat{s}) = ((\hat{a}^+, \hat{w}_h, \hat{w}_l), \hat{s})$ are given by*

- *$\hat{s}_P = (\hat{a}^+, t_h, t_l)$ where a^+ fulfills $\frac{t_h - t_l}{c_l} \leq a^+ \leq \frac{t_h - t_l}{c_h}$ and*
- *$\hat{s}$ defined by eq. XXII.2.*

By $c_l > c_h$, the a^+-interval is nonempty. Of course, the productive types prefer a level of education at the lower end of this interval. Since education has no productive function in this model, any a^+ above $\frac{w_h - w_l}{c_l}$ is Pareto inefficient. If a principal fixes a higher a^+ than other principals, the former will not attract any worker.

It is not difficult to see that the lemma is true. Given the principals' (minimal) level of a^+ within the given range, the agents cannot do any better than following equation XXII.2. The productive workers choose $a = a^+$ and the unproductive ones choose $a = 0$.

Note also that the equilibria specified in lemma XXII.1 are subgame perfect. The agents choices are optimal given any principal strategy, not just an equilibrium principal strategy.

EXERCISE XXII.2. *For $\varepsilon > 0$, consider the principals' wage function*

$$w^*(a) = \begin{cases} t_h, & a \geq \frac{t_h - t_l}{c_l} - \varepsilon \\ t_l, & a < \frac{t_h - t_l}{c_l} - \varepsilon \end{cases}$$

Why can it not be part of any separating equilibrium?

In a somewhat similar fashion, a polypsonistic labor market with education and signaling can also be constructed. In a suchlike model, the workers, not the principals, are the first movers. Without going into any formal details, the productive workers can choose the minimum level of education necessary to have unproductive agents shy away from education. Therefore, the problem of inefficiency observed in the screening model, is absent with signaling.

5. Revisiting the revelation principle

The informed agents have to bear the cost of education in the screening and signaling models. Why is it not sufficient for the agents to just tell their productivity?

Consider a message game where the message consists of telling one's type. Assume that wage depends on the type message such that a high wage is earned by the productive workers and a low wage for the unproductive ones. Then, every agent has an incentive to announce the high type. We know from the revelation principle that we can construct a truthful message game that is as good as the game considered here. The construction is simple. All the agents truthfully announce their type and the mechanism transforms this truthful message into the message "high productivity". Obviously, a separating equilibrium cannot be had in a message game.

The reason why screening and signaling works is that different types have different incentives to undergo the cost of schooling. "Actions speak louder than words."

6. Topics and literature

The main topics in this chapter are

- adverse selection
- hidden characteristics
- labor market
- education
- screening
- signaling
- separating equilibrium
- pooling equilibrium

The German textbook (focussing on organization and management) by Kräkel (2007) is to be recommended.

7. Solutions

Exercise XXII.1

By $c_l a^+ \geq w_h - w_l$, the unproductive type's cost of a^+ years of schooling is higher than the additionl wage obtainable by this effort.

Exercise XXII.2

Given this wage function, the low-productivity workers would not choose $a = 0$ but rather $a = \frac{t_h - t_l}{c_l} - \varepsilon$. This can be seen from

$$\begin{aligned} & t_h - \left(\frac{t_h - t_l}{c_l} - \varepsilon\right) c_l \\ = \; & t_h - (t_h - t_l) + \varepsilon c_l \\ = \; & t_l + \varepsilon c_l \\ > \; & t_l + 0 \cdot c_l. \end{aligned}$$

The productive types choose $\frac{t_h - t_l}{c_l} - \varepsilon$ a fortiori. Thus, we cannot have a separating equilibrium with this education threshold level.

CHAPTER XXIII

Hidden action

1. Introduction

Pursuing asymmetric information, we now turn to hidden action. In this chapter, the agent is to perform some task for the principal. Then, the following problem arises: the insuree may be careless about the insured object once he has obtained the insurance. Another example: workers (or managers) do not exert the high effort that the manager (or the owners) expect. Thus, the asymmetry of information (has the worker exerted sufficient effort) occurs after the agent has been employed. This constellation is called a principal-agent situation or principal-agent problem. This chapter borrows from Varian (1992, chapter 25) and Rasmusen (2001, chapter 7).

Indeed, the problem arises because the output is assumed to be a function of both the agent's effort and of chance. Since the effort is not observable, the payment to the agent (as specified in the contract) is a function of the output, but not of effort. Consider figure XXIII.1.

We use the hidden-action model to introduce the participation constraint that has played no serious role in chapter XVIII on mechanism design or chapter XXII on adverse selection.

Normally, the principal-agent problem is described as the principal's maximization problem subject to two conditions. The first is the agent's participation constraint. He enters into the contract only if he expects a payoff higher than his reservation utility. Once employed, the agent chooses among several actions. The action the principal would like to induce has to

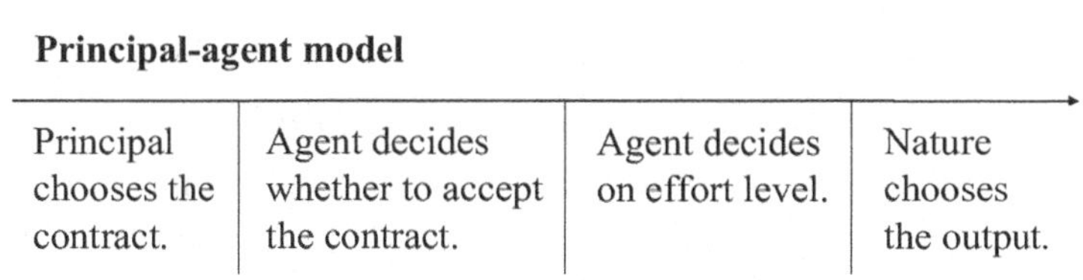

FIGURE XXIII.1. Hidden action

H. Wiese, *Advanced Microeconomics*,
https://doi.org/10.1007/978-3-658-34959-2_23

be a best action given the contract and the probability distribution dictated by nature. This second side condition is called incentive compatibility.

Principal-agent models typically assume that the principal is risk neutral and the agent risk averse. Focusing on the distribution of risk, it is therefore optimal that the agent does not bear any risk but obtains a fixed payment. However, in order to incite the agent not to be lazy, it may be necessary to have the agent bear some risk.

We present the basic model in sections 2 and 3. If the principal can observe the agent's action or deduce the action from the output, an efficient output is possible; this is shown in section 4. We turn to the asymmetric-information case in section 5. A special case, with only two outputs, is treated in section 6.

2. The principal-agent model

We present a principal-agent model with one principal and one agent. The employed agent (after agreeing to the contract proposed by the principal) chooses an effort level $e \in E$ and incurs cost of effort $c(e)$. His non-probabilistic payoff is $w - c(e)$ where w is the wage rate. The agent is free to reject the contract in which case he obtains a reservation utility denoted by $\overline{u}$. The principal's nonprobabilistic payoff is $x - w$ where $x \in X$ is the output or net profit accruing to the principal, i.e., his profit net of the wage payable to the agent. In case of uncertainty (which we assume), the effort e generates a probability distribution ξ_e on the set X of the principal's net profits. Thus, ξ_e is an element of Ξ, the set of probability distributions on X. Thus, we have this definition:

DEFINITION XXIII.1 (PRINCIPAL-AGENT PROBLEM). *A tuple* $\Gamma = (\{P, A\}, E, X, (\xi_e)_{e \in E}, c, \overline{u})$ *is called a principal-agent problem where*

- *P is the principal,*
- *A is the agent,*
- *$E = \mathbb{R}_+$ is the agent's action set (his effort level),*
- *$c : E \to \mathbb{R}$ is the agent's cost-of-effort function,*
- *X is the output set or the set of net profits,*
- *ξ_e is the probability distribution on X generated by effort level e,*
- *the principal's nonprobabilistic payoff is given by*

$$x - w$$

with $x \in X$ and wage rate $w \in \mathbb{R}$,

- *the agent's nonprobabilistic payoff is given by*

$$w - c(e)$$

and

- *the agent's reservation utility is* $\overline{u} \in \mathbb{R}$.

3. Sequence, strategies, and solution strategy

According to figure XXIII.1, the principal-agent problem is modeled as a four-stage game:

(1) The principal chooses a wage function which specifies the wage as a function of the output. This wage function is also called a contract.
(2) The agent decides whether to accept the contract.
(3) The agent decides on his effort level.
(4) Nature chooses the output and thus the payoffs for both principal and agent.

This sequence defines a game of perfect information (!) leading to the following strategies for the two players:

DEFINITION XXIII.2 (STRATEGIES). *Let* Γ *be a principal-agent problem. The principal's strategy is a wage function* $s_P = w : X \to \mathbb{R}$. *Let* S_P *be the set of those wage functions. The agent's strategy is a function* $s_A : S_P \to \{\mathbf{y}, \mathbf{n}\} \times E$, *where* $\mathbf{y}$ *means "yes" or "accept" and* $\mathbf{n}$ *stands for "no" or "decline". This is the agent's participation decision. The set of those strategies is denoted by* S_A. s_A *is sometimes written as* $\left(s_A^{\{\mathbf{y},\,\mathbf{n}\}}, s_A^E\right)$ *with* $s_A^{\{\mathbf{y},\,\mathbf{n}\}}(s_P) \in \{\mathbf{y}, \mathbf{n}\}$ *and* $s_A^E(s_P) \in E$.

The principal can foresee the agent's reaction to any wage function he offers. In particular, any wage function has the agent accept or reject the contract and, in case of acceptance, choose a preferred effort level. Differently put: we look for a subgame-perfect equilibrium. Our solution strategy to the principal-agent problem focuses on the effort level of an agent who accepts a contract. Imagine that the principal aims for an effort level $b \in E$ (a best effort level). Then, the principal maximizes his payoff under two conditions:

- the agent needs to prefer accepting the contract and exerting effort level b to not accepting the contract (this is the participation constraint) and
- the agent needs to prefer effort level b to any other effort level $e \in E$ (the incentive constraint).

Thus, the two constraints have to hold for any wage function and ensure subgame perfection. Of course, the effort level b might not be optimal from the principal's point of view. After reiterating the above maximization problem for all the effort levels $e \in E$ open to the agent, the principal can compare the respective payoffs and choose the maximal one. Since any wage function leads to a specific effort level, this solution strategy works.

4. Observable effort

Observability of effort may mean two different things. Either the principal can directly observe the agent's effort. Or the principal observes the output and can deduce the effort unequivocally. Then, the principal can propose a payment scheme with domain E or X, just as he pleases. We assume domain X, in accordance with the above definition of the principal's strategy. Assume that the principal wants the agent to choose effort level $b \in E$. Then, his maximization problem is

$$\max_w \left(x\left(b\right) - w\left(x\left(b\right)\right)\right)$$

subject to the side conditions

$$\begin{array}{lr} w\left(x\left(b\right)\right) - c\left(b\right) \geq \overline{u}, & \text{participation constraint} \\ w\left(x\left(b\right)\right) - c\left(b\right) \geq w\left(x\left(e\right)\right) - c\left(e\right) \text{ for all } e \in E, & \text{incentive constraint} \end{array}$$

There is no need to give more to the agent than the reservation utility.

$$w\left(x\left(b\right)\right) = \overline{u} + c\left(b\right) \tag{XXIII.1}$$

is the minimal wage that fulfills the participation constraint. We consider the incentive constraint shortly.

The optimal effort chosen by the principal (!) is

$$e^* = \arg\max_e \left(x\left(e\right) - \left(\overline{u} + c\left(e\right)\right)\right)$$

where e^* is obtainable (in good-natured problems) from

$$\underbrace{\frac{dx}{de}}_{\text{marginal output}} \overset{!}{=} \underbrace{\frac{dc}{de}}_{\text{marginal cost}}.$$

It is easy to achieve the incentive constraint. The principal needs to pay a very low wage for outputs other than $x\left(e^*\right)$ or efforts other than e^*. For example, a boiling-in-oil contract guarantees the incentive constraint:

$$w\left(x\right) = \begin{cases} \overline{u} + c\left(e^*\right), & x = x\left(e^*\right) \\ -\infty & x \neq x\left(e^*\right) \end{cases}$$

EXERCISE XXIII.1. *Instead of* $-\infty$, *what is the maximal wage fulfilling the incentive constraint for* e^**?*

The payoffs are $x(e^*) - \overline{u} - c(e^*)$ for the principal and $\overline{u}$ for the agent. Thus, the higher the agent's reservation utility, the lower the principal's payoff. The sum of the payoffs is $x(e^*) - c(e^*)$ and hence the payoff that the principal could achieve if he were his own agent.

Selling the production possibility to the agent is a second road to the optimal solution. In that case, the agent obtains the residual income, i.e., output minus cost of effort minus the price of the business. Differently put, the agent is the "residual claimant".

EXERCISE XXIII.2. *If the principal can make a take-it-or-leave-it offer, which price will he charge?*

5. Unobservable effort

5.1. The model. Given a principal-agent problem $\Gamma = (\{P, A\}, E, X, (\xi_e)_{e\in E}, c, \overline{u})$, we assume that the principal knows the probability distribution ξ_e generated by effort level $e \in E$. In general, this knowledge plus the specific output is not sufficient to reconstruct the effort level itself. Since (and if) the principal cannot observe the effort, he bases his wage payments w on the output.

DEFINITION XXIII.3 (PRINCIPAL-AGENT MODEL). *Let* $\Gamma = (\{P, A\}, E, X, (\xi_e)_{e\in E}, u_P, c, u_A, u, \overline{u})$ *be a principal-agent problem according to XXIII.1 with the strategies defined in XXIII.2. The principal-agent model with n outputs is given by*

- *the output set* $X = \{x_1, ..., x_n\}$,
- *the principal's utility function* u_P *is given by* $u_P(s_P, s_A) =$
$$\begin{cases} \sum_{x\in X} \xi_{s_A^E(s_P)}(x)\,(x - w(x)), & s_A^{\{\mathbf{y},\,\mathbf{n}\}}(s_P) = \mathbf{y} \\ 0, & \text{otherwise} \end{cases}$$
- *the agent's utility function* u_A *is given by* $u_A(s_P, s_A) =$
$$\begin{cases} \sum_{x\in X} \xi_{s_A^E(s_P)}(x)\,u(w(x)) - c\left(s_A^E(s_P)\right), & s_A^{\{\mathbf{y},\,\mathbf{n}\}}(s_P) = \mathbf{y} \\ \overline{u}, & \text{otherwise} \end{cases}$$
where $u : \mathbb{R} \to \mathbb{R}$ *(not* u_A*) is a vNM utility function obeying* $u' > 0$ *and* $u'' < 0$.

Note that the agent's expected utility function u_A is somewhat special in that the cost of effort can be separated from the utility with respect to the wage earnings. The agent is assumed to be risk averse.

We now try to solve the principal-agent model. The two side conditions for action $b \in E$ are

$$\sum_{x \in X} \xi_b(x) u(w(x)) - c(b) \geq \overline{u}, \qquad \text{participation constraint}$$

$$\begin{aligned} &\sum_{x \in X} \xi_b(x) u(w(x)) - c(b) \\ \geq &\sum_{x \in X} \xi_e(x) u(w(x)) - c(e) \text{ for all } e \in E, \end{aligned} \qquad \text{incentive constraint}$$

5.2. Applying the Lagrangean method to the participation constraint. In a first step, we assume that the incentive constraint poses no problem. Let $w_i := w(x_i)$ for all $i = 1, ..., n$. Then, the principal's maximization problem is

$$\max_{w_1, ..., w_n} \sum_{i=1}^{n} \xi_b(x_i)(x_i - w_i)$$

subject to the participation constraint

$$\sum_{i=1}^{n} \xi_b(x_i) u(w_i) - c(b) \geq \overline{u}.$$

The principal maximizes his payoff by fulfilling the participation constraint as an equality. You know that such problems may be attacked by the Lagrangean method (see chapter VI). The Lagrangean of this problem is

$$\begin{aligned} & L(w_1, w_2, ..., w_n, \lambda) \\ = & \sum_{i=1}^{n} \xi_b(x_i)(x_i - w_i) + \lambda \left(\sum_{i=1}^{n} \xi_b(x_i) u(w_i) - c(b) - \overline{u} \right). \end{aligned}$$

The Lagrange multiplier $\lambda > 0$ indicates the additional payoff accruing to the principal if the participation constraint is relaxed. Reducing the reservation utility by one unit increases the principal's payoff by

$$\lambda = -\frac{du_P}{d\overline{u}}$$

which, as you know from pp. 156, is not quite, but basically correct.

The partial derivatives with respect to w_i $(i = 1, ..., n)$ yield

$$\frac{\partial L}{\partial w_i} = \underbrace{-\xi_b(x_i)}_{\substack{\text{wage payments increase} \\ \text{with probability } \xi_b(x_i)}} + \lambda \underbrace{\xi_b(x_i) u'(w_i)}_{\substack{\text{participation constraint} \\ \text{is relaxed}}} \overset{!}{=} 0.$$

An increase of w_i (i.e., in case of output x_i) by one unit reduces the expected profit by $\xi_b(x_i)$ because the wage payments are increased by one unit with probability $\xi_b(x_i)$. This is the bad news. The good news is that a wage increase eases the participation constraint by $\xi_b(x_i)\, u'(w_i)$; multiply by λ to obtain the profit increase.

In case of $\xi_b(x_i) > 0$, we find $u'(w_i) \overset{!}{=} \frac{1}{\lambda}$. Thus, the wages are the same for all outputs that matter. By $u'' < 0$, $u'(w_i) \overset{!}{=} \frac{1}{\lambda}$ means that the wage is independent of the output. Hence, the risk-averse agent is not exposed to any risk at all.

5.3. Applying the Kuhn-Tucker method to the incentive constraint. A constant wage is not optimal if the incentive constraint is binding. The principal's optimization problem leads to the Lagrangean

$$\begin{aligned}
& L(w_1, w_2, ..., w_n, \lambda, \mu) \\
= & \sum_{i=1}^{n} \xi_b(x_i)(x_i - w_i) \\
& +\lambda \left(\sum_{i=1}^{n} \xi_b(x_i)\, u(w_i) - c(b) - \overline{u} \right) \text{ (participation constraint)} \\
& +\mu_{e'} \left(\sum_{i=1}^{n} \xi_b(x_i)\, u(w_i) - c(b) - \left(\sum_{i=1}^{n} \xi_{e'}(x_i)\, u(w_i) - c(e') \right) \right) \\
& +\mu_{e''} \left(\sum_{i=1}^{n} \xi_b(x_i)\, u(w_i) - c(b) - \left(\sum_{i=1}^{n} \xi_{e''}(x_i)\, u(w_i) - c(e'') \right) \right) \\
& +\text{... (all the other incentive constraints)}
\end{aligned}$$

The Lagrange multipliers $\mu_{e'} > 0$, $\mu_{e''} > 0$ reflect the principal's marginal payoff for relaxing the incentive constraint with respect to effort e', e'' etc. We cannot, in general, be sure that all the incentive constraints are binding. Kuhn-Tucker optimization theory says that the product

$$\mu_e \left(\sum_{i=1}^{n} \xi_b(x_i)\, u(w_i) - c(b) - \left(\sum_{i=1}^{n} \xi_e(x_i)\, u(w_i) - c(e) \right) \right)$$

has to equal zero for every effort level $e \in E$. We differentiate the Lagrange function with respect to x_i to obtain

$$\frac{\partial L}{\partial w_i} = \underbrace{-\xi_b(x_i)}_{\substack{\text{wage payments increase} \\ \text{with probability } \xi_b(x_i)}} + \lambda \underbrace{\xi_b(x_i)\, u'(w_i)}_{\substack{\text{participation constraint} \\ \text{is relaxed}}}$$

$$+\mu_{e'}\underbrace{\overbrace{(\xi_b(x_i)-\xi_{e'}(x_i))}^{\text{assumption: positive}}u'(w_i)}_{\substack{\text{incentive constraint}\\ \text{is relaxed}}}$$

$$+\mu_{e''}\underbrace{\overbrace{(\xi_b(x_i)-\xi_{e''}(x_i))}^{\text{assumption: negative}}u'(w_i)}_{\substack{\text{incentive constraint}\\ \text{is exacerbated}}}+\ldots\overset{!}{=}0$$

Assume, now, $\mu_{e'}>0$ and $\mu_{e''}>0$. Then, the restrictions are binding. If we have $\xi_b(x_i)-\xi_{e'}(x_i)=0$, an increase in w_i does not ease the incentive constraint. However, $\xi_b(x_i)-\xi_{e'}(x_i)>0$ implies that a wage increase for output x_i tends to make effort level b more attrative than effort level e' because $(\xi_b(x_i)-\xi_{e'}(x_i))\,u'(w_i)$ is greater than zero, too.

Consider the assumption $\xi_b(x_i)-\xi_{e''}(x_i)<0$. In that case, an increase in w_i makes it more difficult to fulfill the incentive constraint for e'' because e'' leads to x_i with a higher probability than b.

Assume the special case of only two effort levels b and e. Then, the above maximization condition implies

$$-\xi_b(x_i)+\lambda\xi_b(x_i)\,u'(w_i)+\mu_e(\xi_b(x_i)-\xi_e(x_i))\,u'(w_i)\overset{!}{=}0$$

and finally

$$u'(w_i)\overset{!}{=}\frac{1}{\lambda+\mu_e\frac{\xi_b(x_i)-\xi_e(x_i)}{\xi_b(x_i)}}.$$

Thus, in case of $\mu_e>0$ (incentive constraint binding), we get this intuitive result: If action b brings about output x_i with a greater probability than action e, wage w_i should be relatively high in order to give the agent an incentive to choose b rather than e. Formally, the inequality implies that $u'(w_i)$ is smaller for $\mu_e>0$ than for $\mu_e=0$. Just sketch a concave vNM utility function so that you see why a small u' implies a large w_i.

6. Special case: two outputs

6.1. The model. We now turn to the special case of two output levels and two actions, only. The actions are denoted by e and b and the output levels by x_1 and x_2. We assume

- Output x_2 is higher than output $x_1: x_1<x_2$,
- b makes x_2 more likely than $e: \xi_b(x_2)>\xi_e(x_2)$, and
- b is the principal's preferred action.

EXERCISE XXIII.3. *Do* $x_1 < x_2$ *and* $\xi_b(x_2) > \xi_e(x_2)$ *imply that the principal aims for b rather than e?*

Can you confirm the following proposition?

PROPOSITION XXIII.1. *If f is a strictly monotonically increasing function, f is injective. If, furthermore, f is concave,* f^{-1} *is convex.*

In the above sections, we have the principal fix wages that depend on the output, $w = w(x)$ and we work with vNM utility $u(w)$. In this section, we assume that the principal fixes the vNM utility levels rather than the wages. That is, $w(u)$ is the wage level necessary in order to give vNM utility u to the agent. Thus, w as a function of u is the inverse of u as a function of w. The inverse function $w = u^{-1}$ exists because u is a strictly monotonically increasing function. Note that the vNM utility is understood as a gross utility where the cost of effort still need to be subtracted. If u is concave, $w = u^{-1}$ is convex.

If, now, the principal aims at effort level b, his maximal payoff is

$$\pi(b) = \max_{u_1,u_2} \xi_b(x_1)[x_1 - w(u_1)] + \xi_b(x_2)[x_2 - w(u_2)]$$

subject to the two side conditions

$$\begin{array}{ll} \xi_b(x_1)u_1 + \xi_b(x_2)u_2 - c(b) \geq \overline{u}, & \text{part. c.} \\ \xi_b(x_1)u_1 + \xi_b(x_2)u_2 - c(b) \geq \xi_e(x_1)u_1 + \xi_e(x_2)u_2 - c(e), & \text{inc. c.} \end{array}$$

The reason for all this trouble is that the graphical description turns out to be easier. We can reformulate the side conditions:

$$\begin{array}{ll} u_2 \geq \frac{\overline{u}+c(b)}{\xi_b(x_2)} - \frac{\xi_b(x_1)}{\xi_b(x_2)}u_1, & \text{participation constraint} \\ u_2 \geq u_1 + \frac{c(b)-c(e)}{\xi_b(x_2)-\xi_e(x_2)}, & \text{incentive constraint} \end{array}$$

EXERCISE XXIII.4. *Please, confirm the reformulation of the incentive constraint. Hint: you can employ the assumption* $\xi_b(x_2) > \xi_e(x_2)$ *and* $\xi_e(x_1) + \xi_e(x_2) = 1$ *for all* $e \in E$.

6.2. The indifference curves. Figure XXIII.2 depicts indifference curves for effort level b and effort level e. They are negatively sloped. If the agent obtains more in one case, his wage and/or utility in the other case can be reduced along an indifference curve. Assuming a constant expected utility $\widetilde{u}$, the indifference curve for effort level e is given by

$$\widetilde{u} = \xi_e(x_1)u_1 + \xi_e(x_2)u_2 - c(e)$$

or, after solving for u_2,

$$u_2 = \frac{\widetilde{u} + c(e)}{\xi_e(x_2)} - \frac{\xi_e(x_1)}{\xi_e(x_2)}u_1. \qquad \text{(XXIII.2)}$$

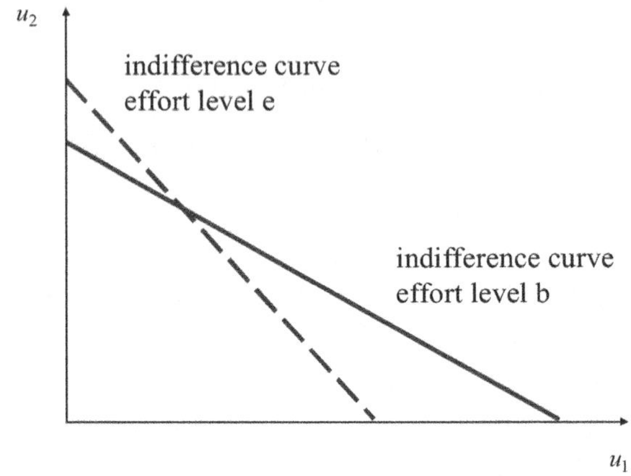

FIGURE XXIII.2. The indifference curves for action e are steeper than those for action b.

Analogously, we obtain the indifference curve for effort level b

$$u_2 = \frac{\widetilde{u} + c(b)}{\xi_b(x_2)} - \frac{\xi_b(x_1)}{\xi_b(x_2)} u_1. \tag{XXIII.3}$$

The slope of these indifference curves is readily interpreted. If the agent foregoes one unit of u_1 (in case of output x_1), he needs $\frac{\xi_e(x_1)}{\xi_e(x_2)}$ or $\frac{\xi_b(x_1)}{\xi_b(x_2)}$ additional units of u_2 in order remain on the given indifference curve. By $\xi_b(x_2) > \xi_e(x_2)$ the indifference curves for b are flatter than those for e. The reason is that b produces u_2 with a higher probability than e. See figure XXIII.2 (adapted from Varian (1992, figure 25.1, p. 450)) for an illustration.

EXERCISE XXIII.5. *What is the relationship between the indifference curves and the participation constraint for effort level b?*

Figure XXIII.3 (adapted from Varian (1992, figure 25.1, p. 450)) depicts the participation-constraint line for effort level b and the incentive-constraint line (for choosing b over e). The incentive line has slope 1 and lies above the 45°- line if $c(b) - c(e) > 0$ holds and below if $c(b) - c(e) < 0$ is true. The positive slope ensures that the utility difference $u_2 - u_1$ does not fall below $\frac{c(b)-c(e)}{\xi_b(x_2)-\xi_e(x_2)}$.

In order to fulfill both constraints, the utility levels u_1 and u_2 have to be chosen inside the highlighted area of figure XXIII.3.

6.3. The principal's iso-profit lines. We find the principal's optimal utility combination by adding the iso-profit lines to figure XXIII.3.

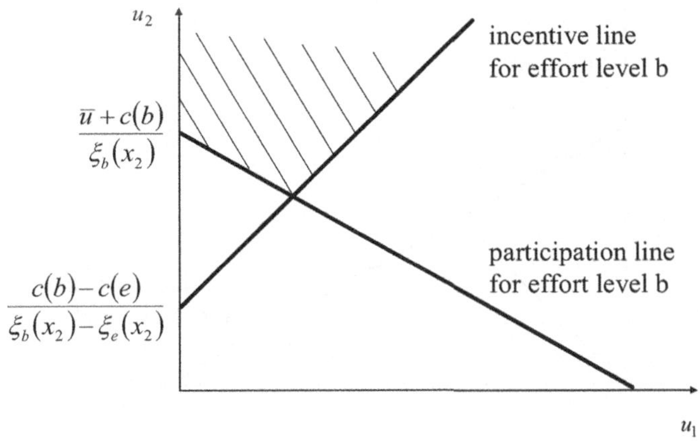

FIGURE XXIII.3. Both constraints are fulfilled in the the highlighted area.

Considering the principal's profit

$$\pi(u_1, u_2) = \xi_b(x_1)[x_1 - w(u_1)] + \xi_b(x_2)[x_2 - w(u_2)],$$

the slope of the iso-profit lines is given by

$$\frac{du_2}{du_1} = -\frac{\frac{\partial\pi}{\partial u_1}}{\frac{\partial\pi}{\partial u_2}} = -\frac{\xi_b(x_1)\, w'(u_1)}{\xi_b(x_2)\, w'(u_2)}, \qquad \text{(XXIII.4)}$$

as you know from preference theory.

Inspecting equation XXIII.4, we observe:

(1) The iso-profit lines are negatively sloped because $w'(u_1)$ and $w'(u_2)$ are positive.
(2) The nearer the iso-profit lines are to the origin, the lower the wages and the higher the profit.
(3) The iso-profit lines are bent away from the origin: An increase in u_1 leads to
   - an increase in $w'(u_1)$ (by the convexity of w),
   - a decrease in u_2 (by the negative slope of the iso-profit line) and hence
   - a decrease in $w'(u_2)$ (again by the convexity of w)

   so that the absolute value of the slope increases with u_1.
(4) In the special case of $u_1 = u_2$ (along the 45°- line), the iso-profit line's slope is $-\frac{\xi_b(x_1)}{\xi_b(x_2)}$. Compare with equation XXIII.3 to confirm that, along the 45°- line, the iso-profit line's slope equals the slopes

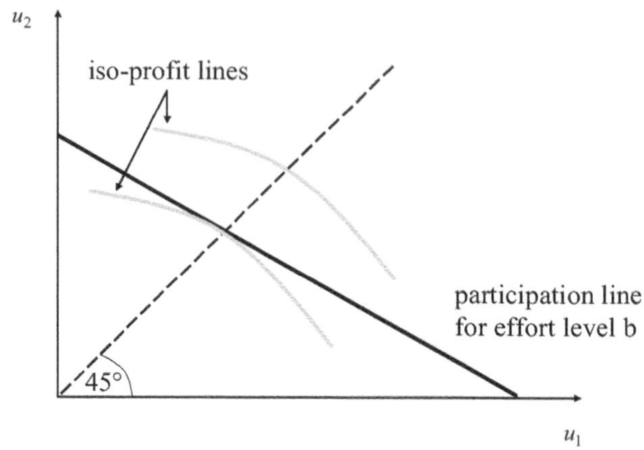

FIGURE XXIII.4. An isoproft line touches the participation line at $u_1 = u_2$

of the agent's indifference curves, both with respect to a given action b.

6.4. Solving the principal-agent problem. Assume that the principal aims for effort level b. Figure XXIII.4 shows how an iso-profit line touches the agent's participation constraint. If incentive compatibility is not an issue, the same wage should be chosen for both outputs. Then, the agent obtains the wage necessary to make him participate.

Comparing figures XXIII.4 and XXIII.3 makes clear that a utility combination along the 45-degree line may not always fulfill the incentive constaint. We need to consider two cases:

- If $u_1 + \frac{c(b)-c(e)}{\xi_b(x_2)-\xi_e(x_2)} \leq u_1$ holds, the incentive line is below the 45°-line and we have $c_b \leq c_e$. Then, it is easy to fulfill the incentive constraint because the agent's cost for b are lower than for e. In this case, the incentive constraint does not prevent Pareto efficiency. Figure XXIII.5 (adapted from Varian (1992, figure 25.2.B, p. 451)) depicts this first-best situation.
- In contrast, figure XXIII.6 (adapted from Varian (1992, figure 25.2.A, p. 451)) depicts the situation with $u_1 + \frac{c(b)-c(e)}{\xi_b(x_2)-\xi_e(x_2)} > u_1$, i.e., $c_b > c_e$. Here, the incentive line lies above the 45-degree line and the optimal risk sharing at $u_1 = u_2$ is not possible. For reasons of incentive compatibility, the agent has to bear a part of the risk. Sufficiently high wage differences give him the incentive to choose the costly effort b. This is the second-best solution. Given

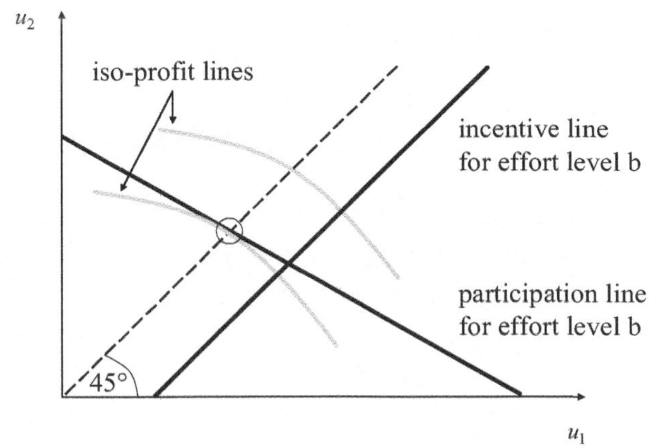

FIGURE XXIII.5. The first-best solution is obtained despite asymmetric information

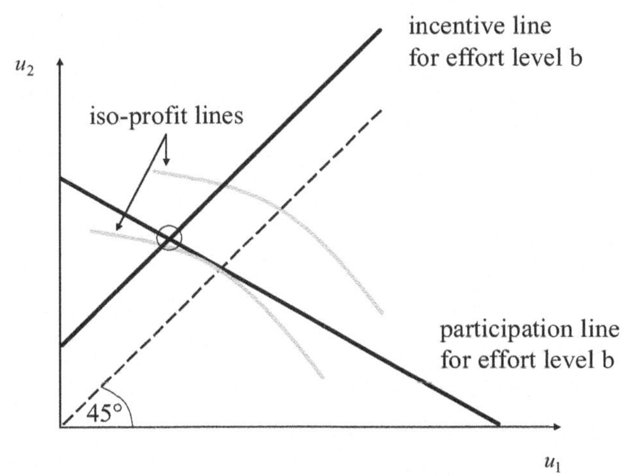

FIGURE XXIII.6. The incentive-compatibility constraint is binding

the problem caused by moral hazard, the utility combination indicated in the figure is the best solution obtainable.

EXERCISE XXIII.6. *Can you show that the first-best solution is realizable if the agent is risk neutral? Hint: Examine the principal's iso-profit lines.*

EXERCISE XXIII.7. *Consider the following example (taken from Milgrom & Roberts 1992, pp. 200-203):*

- *We have two outputs* 10 *and* 30.

- *The agent has two effort levels,* 1 *and* 2. *Effort level* 2 *makes output 30 more likely than effort level* 1 :

Effort level	*Output* $x = 10$	*Output* $x = 30$
$e = 1$	$\xi_1(10) = 2/3$	$\xi_1(30) = 1/3$
$e = 2$	$\xi_2(10) = 1/3$	$\xi_2(30) = 2/3$

- *The agent is risk averse with vNM utility function* $u(w, e) = \sqrt{w} - (e - 1)$. *Note that the vNM utility includes the cost of effort. The reservation utility is* $\overline{u} = 1$.
- *The principal has the profit function* π *given by* $\pi(w, x) = x - w$.
- *In case of unobservable effort, the principal's wage function is given by* $w(10) \equiv w_l$, $w(30) \equiv w_h$.

Solve the principal-agent problem by going through these questions:

(1) *Observable effort:*
  - *If the principal aims for* $e = 1$, *what is his optimal wage function?*
  - *If the principal aims for* $e = 2$, *what is his optimal wage function?*
  - *Should the principal aim for effort level* 1 *or* 2?

(2) *Unobservable effort,* $e = 2$
  - *Write down the participation constraint in terms of* $\sqrt{w_l}$ *and* $\sqrt{w_h}$.
  - *Write down the incentive constraint in terms of* $\sqrt{w_l}$ *and* $\sqrt{w_h}$.
  - *Depict the two constraints by putting* $\sqrt{w_l}$ *on the abscissa and* $\sqrt{w_h}$ *on the ordinate.*
  - *Determine* w_l *and* w_h!

(3) *Unobservable effort,* $e = 1$
Is the principal's profit higher for $e = 1$ *than for* $e = 2$?

(4) *What is the optimal contract for these probabilities:*

Effort level	*Output* $x = 10$	*Output* $x = 30$
$e = 1$	$\xi_1(10) = 2/3$	$\xi_1(30) = 1/3$
$e = 2$	$\xi_2(10) = 0$	$\xi_2(30) = 1$

7. More complex principal-agent structures

So far, we have looked at two-tier principal-agent structures. Tirole (1986) points to three-tier structures. An agent has to perform a task for the principal who uses the services of an intermediate supervisor. Consider these examples:

	principal	supervisor	agent
production unit	manager	foreman	worker
regulation	government	regulating authority	firm
PhD procedure	faculty council	professor	PhD stud.
professorship	ministry of educ.	dean/rector	professor

Perhaps for reasons of time, competence, or cost efficiency,

- the principal cannot directly supervise the agent (otherwise, the second layer becomes superfluous) and
- the supervisor cannot take on the role of the principal (otherwise, the principal could sell his firm to the supervisor).

Surely, the principal cannot always be sure that the supervisor acts in his interests:

- Sometimes, the agent's achievements reflect on the supervisor. This may explain why some doctoral theses gain too much praise.
- The supervisor and the agent collude against the principal. For example, the supervisor (secretly) announces the controlling activity he is to perform shortly. Then, the bad information unagreeable to both parties cannot be generated.
- In case of corruption, secret side payments play a role.

8. Topics and literature

The main topics in this chapter are

- hidden actions
- participation constraint
- incentive constraint
- effort

The German textbook (focussing on organization and management) by Kräkel (2007) is to be recommended.

9. Solutions

Exercise XXIII.1

If the agent chooses e^*, his payoff is $w(x(e^*)) - c(e^*)$, while any other effort e yields the payoff $w(e) - c(e)$. In order to guarantee $w(x(e^*)) - c(e^*) \geq w(e) - c(e)$ for all $e \in E$, we need

$$\begin{aligned} w(e) &\leq w(x(e^*)) - c(e^*) + c(e) \\ &= \overline{u} + c(e^*) - c(e^*) + c(e) \text{ (see eq. XXIII.1, p. 586)} \\ &= \overline{u} + c(e) \text{ for all } e \in E \\ &= \overline{u} + \min_{e \in E} c(e) \end{aligned}$$

Exercise XXIII.2

Given the price p for the business, the agent who exerts effort $e \in E$, obtains

$$x(e) - c(e) - p$$

The optimal effort level e^* is again given by $\left|\frac{dx}{de}\right|_{e^*} \overset{!}{=} \left|\frac{dc}{de}\right|_{e^*}$. The agent will not buy the business unless we have

$$\begin{aligned} & x(e^*) - c(e^*) - p \geq \overline{u} \text{ or} \\ \iff & p \leq x(e^*) - c(e^*) - \overline{u}. \end{aligned}$$

Thus, the principal can ask for the price $x(e^*) - c(e^*) - \overline{u}$. His payoff is the same as in the main text.

Exercise XXIII.3

No. The agent's cost for b may be considerably higher than the agent's cost for e. Since the principal needs to observe the participation constraint, b may make a very high wage necessary.

Exercise XXIII.4

From

$$\xi_b(x_1)\,u_1 + \xi_b(x_2)\,u_2 - c(b) \geq \xi_e(x_1)\,u_1 + \xi_e(x_2)\,u_2 - c(e)$$

we obtain

$$(\xi_b(x_2) - \xi_e(x_2))\,u_2 \geq (\xi_e(x_1) - \xi_b(x_1))\,u_1 + c(b) - c(e)$$

and by $\xi_b(x_2) - \xi_e(x_2) = \xi_e(x_1) - \xi_b(x_1)$ the desired inequality

$$u_2 \geq u_1 + \frac{c(b) - c(e)}{\xi_b(x_2) - \xi_e(x_2)}.$$

Exercise XXIII.5

The participation constraint for b is given by $u_2 \geq \frac{\overline{u}+c(b)}{\xi_b(x_2)} - \frac{\xi_b(x_1)}{\xi_b(x_2)}u_1$. These are all those utility combinations (u_1, u_2) that lie on or above the indifference curve with constant utility $\widetilde{u} = \overline{u}$.

Exercise XXIII.6

In case of risk neutrality, we have a linear vNM utility function u (with argument w). The inverse function w (with argument u) is also linear. Thus, we obtain $w'(u_1) = w'(u_2)$ and the slope $-\frac{\xi_b(x_1)}{\xi_b(x_2)}$ of the principal's isoprofit lines. One of these isoprofit lines lies on the participation line. Since risk-bearing by the agent is no problem in case of risk neutrality, any point on the participation line that also fulfills the incentive constraint will do.

Exercise XXIII.7

(1) Observable effort:

- If the principal aims for $e = 1$, he needs to take care of the participation constraint, only:

$$\sqrt{w} - (e-1) \geq \overline{u}.$$

The wage rate $w = 1$ fulfilling this constaint automatically takes care of the incentive problem.

- In case of observable effort, it is easy to force $e = 2$. The wage rate of $w_{e=2} = 4$ guarantees the participation constraint $\sqrt{w_{e=2}} - (2-1) \geq 1$. The incentive constraint is $\sqrt{w_{e=2}} - (2-1) \geq \sqrt{w_{e=1}} - (1-1)$ which can be rewritten as

$$\begin{aligned}\sqrt{w_{e=1}} &\leq \sqrt{w_{e=2}} - 1\\ &= \sqrt{4} - 1\\ &= 1.\end{aligned}$$

Thus, the wage function

$$w = \begin{cases} 4, & e = 2 \\ 1, & e = 1 \end{cases}$$

is optimal.

- $e = 1$ and $w = 1$ implies the expected profit

$$\begin{aligned} \pi(e = 1) &= \frac{2}{3} \cdot 10 + \frac{1}{3} \cdot 30 - 1 \\ &= \frac{47}{3} \end{aligned}$$

while $e = 2$ and $w = 4$ leads to

$$\begin{aligned} \pi(e = 2) &= \frac{1}{3} \cdot 10 + \frac{2}{3} \cdot 30 - 4 \\ &= \frac{58}{3} \\ &> \frac{47}{3}. \end{aligned}$$

The principal should aim for $e = 2$.

(2) Unobservable effort, $e = 2$

In case of unobservability, the wage needs to be a function of output, not effort. w_l is the wage for the low output 10 and w_h is the wage for the high output 30.

- The agent's participation constraint for the high effort 2 is

$$\begin{aligned} & \frac{1}{3} u(w_l, 2) + \frac{2}{3} u(w_h, 2) \\ = & \frac{1}{3} \left(\sqrt{w_l} - 1 \right) + \frac{2}{3} \left(\sqrt{w_h} - 1 \right) \\ = & \frac{1}{3} \sqrt{w_l} + \frac{2}{3} \sqrt{w_h} - 1 \\ \geq & 1, \end{aligned}$$

or

$$\sqrt{w_h} \geq 3 - \frac{1}{2} \sqrt{w_l}.$$

- The incentive constraint for effort 2 rather than 1 is

$$\begin{aligned} & \frac{1}{3} \sqrt{w_l} + \frac{2}{3} \sqrt{w_h} - 1 \\ = & \frac{1}{3} u(w_l, 2) + \frac{2}{3} u(w_h, 2) \\ \geq & \frac{2}{3} u(w_l, 1) + \frac{1}{3} u(w_h, 1) \\ = & \frac{2}{3} \sqrt{w_l} + \frac{1}{3} \sqrt{w_h}, \end{aligned}$$

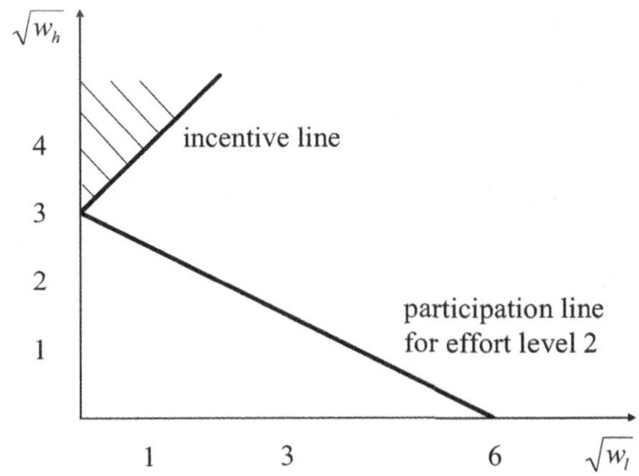

FIGURE XXIII.7. Participation and incentive constraints in our simple example

which can also be written as

$$\sqrt{w_h} \geq 3 + \sqrt{w_l}.$$

- Participation and incentive constraints are depicted in figure XXIII.7.

- From figure XXIII.7, we learn that the principal should not pay a positive wage to the agent in case of $x = 10$. We have $\sqrt{w_h} = 3$ and $\sqrt{w_l} = 0$ or the wage function

$$w = \begin{cases} 9, & x = 30 \\ 0, & x = 10 \end{cases}.$$

The principal's profit is

$$\begin{aligned} \pi(e = 2) &= \frac{1}{3} \cdot (10 - 0) + \frac{2}{3} \cdot (30 - 9) \\ &= \frac{52}{3}. \end{aligned}$$

(3) Unobservable effort, $e = 1$
Very similar to the case of observable effort, if the effort level 1 is aimed for, the incentive constraint is no problem. We know that $w = 1$ fulfills the participation constraint and leads to the profit $\frac{47}{3}$. By $\frac{52}{3} > \frac{47}{3}$ the principal should go for $e = 2$. Note $\frac{58}{3} > \frac{52}{3}$, i.e., observability leads to a higher profit. After all, $e = 2$ is a second-best solution, only.

(4) The new probabilities reduce the principal's uncertainty. The high effort precludes the low output. Here, a boiling-in-oil contract is optimal:

$$w = \begin{cases} 4, & x = 30 \\ 0, & x = 10 \end{cases}$$

fulfills the participation constraint because the agent has the (expected) payoff $\sqrt{4} - (2 - 1) = 1 = \bar{u}$. Effort level $e = 1$ leads to the expected utility $\frac{2}{3}\sqrt{0} + \frac{1}{3}\sqrt{4} = \frac{2}{3} < 1$.

Index

H. Wiese, *Advanced Microeconomics*,
https://doi.org/10.1007/978-3-658-34959-2

Bibliography

Akerlof, G. A. (1970). The market for lemons: Quality uncertainty and the market mechanism, *Quarterly Journal of Economics* **84**: 488–500.

Arnsperger, C. (1994). Envy-freeness and distributive justice, *Journal of Economic Surveys* **8**: 155–186.

Aumann, R. J. (1974). Subjectivity and correlation in randomized strategies, *Journal of Mathematical Economics* **1**: 67–96.

Aumann, R. J. (1977). The St. Petersburg paradox: A discussion of some recent comments, *Journal of Economic Theory* **14**: 443–445.

Aumann, R. J. & Myerson, R. B. (1988). Endogenous formation of links between players and of coalitions: An application of the Shapley value, *in* A. E. Roth (ed.), *The Shapley Value*, Cambridge University Press, Cambridge et al., pp. 175–191.

Averch, H. & Johnson, L. L. (1962). Behavior of the firm under regulatory constraint, *American Economic Review* **52**: 1052–1069.

Bain, J. (1956). *Barriers to New Competition*, Harvard University Press, Cambridge (MA).

Baron, D. P. (1989). Design of regulatory mechanisms and institutions, *in* R. Schmalensee & R. D. Willig (eds), *Handbook of Industrial Organization*, North-Holland, Amsterdam et al., chapter 24, pp. 1347–1447.

Basu, K. (1994). The travelers's dilemma: Paradoxes of rationality in game theory, *American Economic Review, Papers and Proceedings* **84**: 391–395.

Bester, H. (2017). *Theorie der Industrieökonomik*, 7 edn, Springer Gabler, Berlin et al.

Bork, R. H. & Bowman, Jr., W. S. (1965). The crisis in antitrust, *Columbia Law Review* **65**: 363–376.

Bowles, S. (2004). *Microeconomics: Behavior, Institutions, and Evolution*, Princeton University Press, Princeton and Oxford.

Brams, S. J. (1983). *Superior Beings*, Springer, New York et al.

Brander, J. A. (1981). Intra-industry trade in identical commodities, *Journal of International Economics* **11**: 1–14.

Brander, J. A. & Spencer, B. J. (1983). International R&D rivalry and industrial strategy, *Review of Economic Studies* **50**: 707–722.

Brander, J. A. & Spencer, B. J. (1981). Tariffs and the extraction of foreign monopoly rents under potential entry, *Canadian Journal of Economics* pp. 371–389.

Brander, J. & Krugman, P. (1983). A 'reciprocal dumping' model of international trade, *Journal of International Economics* **15**: 313–321.

Chandakant, K. & Fowler, J. (1975). *Choice of Friends, Tales from the Hitopadesha*, Amar Chitra Katha Pvt.

H. Wiese, *Advanced Microeconomics*,
https://doi.org/10.1007/978-3-658-34959-2

Chiang, A. C. & Wainwright, K. (2005). *Fundamental Methods of Mathematical Economics*, Mcgraw-Hill Higher Education, New York et al.

Coase, R. H. (1960). The problem of social cost, *The Journal of Law and Economics* **3**: 1–44.

de la Fuente, A. (2000). *Mathematical Methods and Models for Economists*, Cambridge University Press, Cambridge.

Debreu, G. (1959). *Theory of Value. An Axiomatic Analysis of Economic Equilibrium*, Wiley, New York.

Eaton, J. & Grossman, G. M. (1986). Optimal trade and industrial policy under oligopoly, *Quarterly Journal of Economics* pp. 383–406.

Edgeworth, F. Y. (1881). *Mathematical Psychics*, Paul Kegan, London.

Eucken, W. (1990). *Grundsätze der Wirtschaftspolitik*, 6 edn, J.C.B. Mohr (Paul Siebeck), Tübingen. edited by Edith Eucken and K. Paul Hensel.

Fitzgerald, J. L. (2004). *The Mahabharata, Vol. 7*, University of Chicago Press.

Fudenberg, D. & Tirole, J. (1991). *Game Theory*, MIT Press, Cambridge (MA)/London.

Gibbons, R. (1992). *A Primer in Game Theory*, Harvester Wheatsheaf, New York et al.

Gillies, D. (1959). Solutions to general nonzero sum games, *Annals of Mathematics* **40**: 47–87.

Gravelle, H. & Rees, R. (1992). *Microeconomics*, 2. edn, Longman, London/New York.

Gravelle, H. & Rees, R. (2004). *Microeconomics*, 3. edn, Prentice Hall, London/New York.

Harsanyi, J. C. (1973). Games with randomly disturbed payoffs: A new rationale for mixed-strategy equilibrium points, *International Journal of Game Theory* **2**: 1–23.

Harsanyi, J. C. & Selten, R. (1988). *A General Theory of Equilibrium Selection in Games*, The MIT Press, Cambridge (MA)/London.

Hayek, F. A. v. (2002). Competition as a discovery procedure, *The Quarterly Journal of Austrian Economics* **5**: 9–23. translated from German (Wettbewerb als Entdeckungsverfahren) by Marcellus S. Snow.

Helpman, E. & Krugman, P. R. (1989). *Trade Policy and Market Structure*, The MIT Press, Cambridge (MA)/London.

Hildenbrandt, W. & Kirman, A. P. (1988). *Equilibrium Analysis: Variations on Themes by Edgeworth and Walras*, North-Holland, Amsterdam et al.

Hirschman, A. O. (1964). The paternity of an index, *American Economic Review* **56**: 761–762.

Hoppmann, E. (1966). Das Konzept der optimalen Wettbewerbsintensität - Rivalität oder Freiheit des Wettbewerbs: Zum Problem eines wettbewerbspolitisch adäquaten Ansatzes der Wettbewerbstheorie, *Jahrbücher für Nationalökonomie und Statistik* **179**: 286–323.

Kale, M. R. & Kale, V. (1967). *Hitopadesha of Narayana*, 6 edn, Motilal Banarsidass Publishers.

Kantzenbach, E. (1966). *Die Funktionsfähigkeit des Wettbewerbs*, Vandenhoeck u. Ruprecht, Göttingen.

Kirzner, I. M. (1973). *Competition and Entrepreneurship*, University of Chicago Press, Chicago/London.

Kirzner, I. M. (1997). Entrepreneurial discovery and the competitive market process: An austrian approach, *Journal of Economic Literature* **35**: 60–85.

Kitch, E. (1983). The fire of truth: A remembrance of law and economics at Chicago, 1932-1970, *Journal of Law and Economics* **26**: 163–234.

Kräkel, M. (2007). *Organisation und Management*, 3 edn, Mohr Siebeck, Tübingen.

Kreps, D. M. & Scheinkman, J. A. (1983). Quantity precommitment and Bertrand competition yield Cournot outcomes, *Bell Journal of Economics* **14**: 326–337.

Lange, O. (1936). On the economic theory of socialism, *Review of Economic Studies* **4**: 53–71.

Leach, J. (2004). *A Course in Public Economics*, Cambridge University Press, Cambridge et al.

Lenel, H.-O. (1975). Vollständiger und freier Wettbewerb als Leitbilder für die Wettbewerbspolitik gegenüber mächtigen Unternehmen, *in* H. Sauermann & E.-J. Mestmäcker (eds), *Wirtschaftsordnung und Staatsverfassung. Festschrift für Franz Böhm zum 80. Geburtstag*, Mohr Siebeck, Tübingen, pp. 317– 340.

Mas-Colell, A., Whinston, M. D. & Green, J. R. (1995). *Microeconomic Theory*, Oxford University Press, New York/Oxford.

Mason, E. S. (1939). Price and production policies of large-scale enterprise, *American Economic Review* **29**: 61–74.

Menger, K. (1967). The role of uncertainty in economics, *in* M. Shubik (ed.), *Essays in Mathematical Economics in Honor of Oskar Morgenstern*, Princeton University Press, Princeton, pp. 211–231. first appearance in Zeitschrift für Nationalökonomie 1934, translated from German.

Milgrom, P. & Roberts, J. (1992). *Economics, Organization and Management*, Prentice-Hall International, Englewood Cliffs.

Morris, P. (1994). *Introduction to Game Theory*, Springer-Verlag, New York et al.

Moulin, H. (1995). *Cooperative Microeconomics: A Game-Theoretic Introduction*, Prentice Hall, London et al.

Myerson, R. B. (1991). *Game Theory: Analysis of Conflict*, Harvard University Press, Cambridge (MA)/London.

Myerson, R. & Satterthwaite, M. A. (1983). Efficient mechanisms for bilateral trading, *Journal of Economic Theory* **29**: 265–281.

Nash, J. F. (1950). Equilibrium points in n-person games, *Proceedings of the National Academy of Sciences of the USA* **36**: 48–49.

Nash, J. F. (1951). Non-cooperative games, *Annals of Mathematics* **54**: 286–295.

Nozick, R. (1969). Newcomb's problem and two principles of choice, *in* N. Rescher, D. Davidson & C. G. Hempel (eds), *Essays in Honor of Carl G. Hempel*, Reidel, Dordrecht, pp. 114–146.

Olivelle, P. (2006). *The five discourses on worldly wisdom (by Vishnusharman)*, New York University Press and JJC Foundation.

Pfähler, W. & Wiese, H. (2008). *Unternehmensstrategien im Wettbewerb: Eine spieltheoretische Analyse*, 3 edn, Springer, Berlin et al.

Piccione, M. & Rubinstein, A. (2007). Equilibrium in the jungle, *Economic Journal* .

Rasmusen, E. (2001). *Games and Information*, 3 edn, Blackwell, Cambridge (MA)/Oxford.

Rubinstein, A. (1982). Perfect equilibrium in a bargaining model, *Econometrica* **50**(1): 97–109.

Schumpeter, J. A. (1976). *Capitalism, Socialism and Democracy*, 5 edn, Routledge, London/New York.

Shapley, L. S. (1953). A value for n-person games, *in* H. Kuhn & A. Tucker (eds), *Contributions to the Theory of Games*, Vol. II, Princeton University Press, Princeton, pp. 307–317.

Shubik, M. (1981). Game theory models and methods in political economy, *in* K. J. Arrow & M. D. Intriligator (eds), *Handbook of Mathematical Economics*, Vol. III, Elsevier, Amsterdam et al., chapter 7, pp. 285–330.

Spence, M. A. (1973). Job market signalling, *Quarterly Journal of Economics* **87**: 355–374.

Stigler, G. J. & Sherwin, R. A. (1985). The extent of the market, *Journal of Law and Economics* **28**: 555–585.

Tirole, J. (1986). Hierarchies and bureaucracies: On the role of collusion in organizations, *Journal of Law, Economics, and Organization* **2**: 181–214.

Varian, H. R. (1992). *Microeconomic Analysis*, 3 edn, W. W. Norton & Company, New York/London.

Varian, H. R. (2010). *Intermediate Microeconomics*, 8 edn, W. W. Norton & Company, New York/London.

von Hayek, F. A. (1937). Economics and knowledge, *Economica* **4**: 33–54.

von Hayek, F. A. (1945). The use of knowledge in society, *American Economic Review* **35**: 519–530.

von Mises, L. (1996). *Human Action*, 4 edn, Fox & Wilkes, San Francisco. 1. edition 1949.

Wegehenkel, L. (1980). *Coase-Theorem und Marktsystem*, J.C.B. Mohr (Paul Siebeck), Tübingen.

Wiese, H. (2005). *Kooperative Spieltheorie*, R. Oldenbourg, München/Wien.

Wiese, H. (2012). Backward induction in Indian animal tales, *International Journal of Hindu Studies* **16**: 93–103.

Williamson, O. E. (1968). Economies as an antitrust defense: The welfare tradeoffs, *American Economic Review* **58**: 18–36.

Wilson, R. (1971). Computing equilibria of n-person games, *SIAM Journal on Applied Mathematics* **21**: 80–87.

Young, H. P. (1994). Cost allocation, *in* R. J. Aumann & S. Hart (eds), *Handbook of Game Theory with Economic Applications*, Vol. II, Elsevier, Amsterdam et al., chapter 34, pp. 1193–1235.

The manufacturer's authorised representative in the EU is Springer Nature Customer Service Centre GmbH, Europaplatz 3, 69115 Heidelberg, Germany. If you have any concerns regarding our products, please contact ProductSafety@springernature.com

Printed and bound by CPI Group (UK) Ltd, Croydon, CR0 4YY
30/09/2026
02225195-0020